Particle Size Analysis

Particle Size Analysis

Edited by
N.G. Stanley-Wood
Department of Chemical Engineering, University of Bradford

R.W. Lines
Coulter Electronics Ltd., Luton

ROYAL
SOCIETY OF
CHEMISTRY

The Proceedings of the 25th Anniversary Conference organised by the Particle Characterisation Group of the Analytical Division of The Royal Society of Chemistry. 17th–19th September 1991, University of Technology, Loughborough.

Special Publication No. 102

ISBN 0-85186-487-2

A catalogue record for this book is available from the British Library

Published by The Royal Society of Chemistry,
Thomas Graham House, Science Park, Cambridge CB4 4WF

Preface

Although this Seventh PSA Conference, held under the auspices of the Royal Society of Chemistry, was also our 25th (Silver) Anniversary Conference, one of the first symposia in Britain which provided an intellectual interest in powders and solid particles was held in London in 1947. It was held under the auspices of the Society of Chemical Industries and the Institution of Chemical Engineers. At the opening of the 1947 Symposium it was stated that, as far as the President of the Institution was aware, no previous attempts had been made to survey the difficult subject of particle size analysis. From that symposium, the efforts of a few active powder technologists led to the formation of a Particle Size Analysis Group, which was eventually to become the Particle Characterisation Group of the Analytical Division of the Royal Society of Chemistry. The Group's first international meeting was held 1966, at Loughborough University of Technology, and covered 3 days.

The interest in the measurement of size, size distribution, shape, shape distribution and surface area of powders and all forms of particulate matter has increased enormously since those days, along with the technologies of measurement. The disciplines of Particle Characterisation and Powder technology, which initially emanated from the UK and continental Europe, have now become global in their ramifications and acceptability. This can be illustrated by the range of nationalities of the contributing authors and the countries which they represented, i.e. Australia, Belgium, Canada, Germany (both the former East and West), Italy, Japan, Norway, Netherlands, Poland, Portugal, United Kingdom, United States of America, and the (then) USSR, with delegates also attending from Finland, Sweden, and even Guernsey!

Throughout most of those   countries there   are now active scientific and engineering Particle Technology groups, which not only hold national meetings on a regular basis but also co-sponsor our PSA Conferences, as well as other international conferences. The co-sponsorship of this 25th Anniversary Conference, came from the GAMS group [France], GVC VDI [Germany], the Belgian Particle Technology Club, the Finnish Fine Particle Society and the Japanese Society of Powder Technology, as well as the Particle Technology Subject Group of the Institute of Chemical Engineers [UK].

This 25th PSA Conference gave authors not only the opportunity to review the development of particle characterisation over the past 25 years, but also provided the Keynote lecturers with the stimulus to speculate on the future. These Keynote lectures, highlighting the changes and advances in various fields of

Particle Characterisation, were presented by widely known experts in their specialised fields [Scarlett; Sing; Jimbo; Azzopardi; Weiner; Rood; Kaye; Lines; Leschonski; Allen] before each scientific session.

In the first few Conferences [1966 (Loughborough), 1970 (Bradford) and 1977 (Salford)], many scientific contributions were on the subjects of sieving, sedimentation and microscopy. In this 25th Anniversary Conference [the seventh PSA], there was only one paper on sieving [Meloy and Williams] but, although the sieving technique was discussed, the paper was mainly directed towards the measurement of the shape of particles rather than the measurement of their sizes.

The concept of shape measurement has always been of intellectual interest and academic novelty, and this Conference produced not only a Keynote contribution on the methodologies of shape measurements [Kaye] but, possibly for the first time, two presentations on the preparation and production of differently shaped particles [Marshall and Mitchell; Gowland and Wilshire].

When the first PSA conference was held 1966, microscopy was being used, together with a scanning flying spot technique to take the tedium out of particle size measurements. During the next PSA conferences [1970, 1977] there was a steady decline in the scientific papers presented using this technique for counting and sizing irregularly shaped particles until, at the Salford Conference [1977], the technique of fractal analysis of particles was introduced by Kaye as a method which could be applied to describe numerically the shape of irregular particles, and microscopy then again became a viable characterisation technique. At PSA 91 there were 8 papers on such shape analysis [Marshall and Mitchell; Gowland and Wilshire; Kaye; Morris; Bottlinger; Dunnett, Goodbody and Stanisstreet; Whiteman and Ridgway; Meloy and Williams], one of which was a Keynote lecture.

The sedimentation methods of particle size analysis, both gravitational and centrifugal, have, over the years, generated many contributions and are still extensively used in many traditional industries. PSA 91 saw however only three such contributions [Allen (2); Bernhardt], one which was a Keynote lecture reviewing the historical developments of this fundamental method. The Andreasen sedimentation technique and sieving continue as essential methods used for the characterisation of many certified reference materials.

As in many previous PSA conferences, the metrology of particle characterisation by the electrical sensing zone method was again prominent [Berge, Feder and Jøssang]. The emphasis these days is however on the mass calibration of the instrument, sometimes also using certified or standard reference materials

[Figueiredo, Rasteiro and Ferreira; Merkus, Jansma, Scarlett and Figueiredo], a topic which was noted as being one of the major contributions to the Conference.

Some of the most advanced techniques which dominated PSA 91 were the application of either light diffraction [Azzopardi; Atkinson et al.], light scattering [Zege and Kokhanovsky; Sommer, Harrison and Montague], light scattering with Doppler [Rudoff, Sankar and Bachalo; Manasse, Jiang, Wreidt and Bauckhage; Naqwi and Durst], light obstruction [Umhauer; Akers, Rushton, Sinclair and Stenhouse], or photon correlation spectroscopy [Finsy, De Groen, Deriewmaeker, Geladé and Joosten; Van der Meeren, Vanderdeelan and Baert], techniques which have grown considerably in industrial usage since the earlier PSA conferences.

Also during each of the more recent Conferences [1970 - 1991] there has been a scientific session on on-line, in-stream, particle size analysis. PSA 91 showed a marked increase in the number of papers which used computer interfaced laser diffraction, laser light scattering instruments in often an experimental format, or a critical appraisal of commercially available techniques. Such an increase in contributions, together with new interpretations on data obtained from process streams and systems [Bernard, Andries and Scarlett; Svarovsky and Svarovsky; Heidenreich; Stinz and Riebel], illustrate the progress being made in achieving this long sought-after ideal of in situ, in-process, particle characterisation.

The extent that particle characterisation techniques can be used in other technological fields was illustrated by the inclusion at this Conference of two papers which have taken the principles of solid particle characterisation and applied them successfully to biological systems [Groves; Dunnett et al.] and to the pharmaceutical industry [Atkinson and White; Atkinson, Greenway, Holland, Merrifield and Scott; Caramella, Ferrari, Bonferoni and Bertoni].

Although the symbol PSA stands for Particle Size Analysis, there has always been included in each PSA Conference a scientific session on surface area estimation. At PSA 91, two main thrusts were to be found; one being on the measurement on porosity of powders in terms of meso- and micro-porosity [Sing; Buczek; Cowan and Wenman] and the other, because of the global need to have traceability and mandatory powder product specifications, on the characterisation and comparability of reference materials [Stanley-Wood, Osborne and Till] and the need for Standards [Lines]. The particle size distributions reported by several types of measuring instruments for the same materials were presented [Jimbo, Tsubaki and Yamamoto].

For the first time in 25 years the PSA Organising Committee proposed that, within a PSA conference, a poster session should be held. This innovative plan for the PSA 91 conference must be rated as an immense success especially for young and new experimenters in powder technology [for example, from the Delft School] who were given the opportunity to put forward ideas which were in the process of being developed and to benefit from exposure to a scientific community looking for future and innovative methodologies and metrologies.

With Information Transfer being essential to the wellbeing of any academic, technological, or commercial organisation or country, the communication of ideas, information and technological knowledge which may be on the perimeter of particle characterisation or at the interface of powder technology and bulk powder handling, has now become more rapid and more international. The onus on members within the Particle Characterisation Group to arrange and organise every four years a worthwhile and prodigious international conference demands a great many hours of commitment, a heavy work-load from a volunteer committee, and particularly from the local organiser. This, having been recognised, has recently generated an opinion and a desire that recommends and necessitates a change from such a U.K.-based event to a more European or global conference. To this end, the Particle Characterisation Group Committee has considered that future conferences should be encompassed within an event which can highlight Particle Characterisation as part of the overall areas and fields of Powder Technology.

To date, two World Congresses of Powder Technology have been held, one in Germany [1986] and one in Japan [1990]. In 1994, the Third World Congress is to be held in the United States of America, possibly on the West Coast, whilst plans are being prepared for a possible 4th World Congress, in 1998, within Europe.

The Particle Characterisation Group Committee believes that the progress of particle characterisation and the interests of those particle technologists currently supporting the Group can best be served if future PSA conferences can be incorporated within such a World Congress. This would give workers in the field of fine particle characterisation an opportunity to exchange concepts, knowledge and experience with other workers involved on the bulk or macro powder handling scale, and so bring together the whole gambit of particle and powder technologies.

Whatever the future of international PSA Conferences, the Committee will continue to provide several one- or two-day technical meetings each year, on a whole range of topics, at locations throughout the country for the benefit of members.

Editors:-

Nayland Stanley-Wood and Roy Lines.

PREVIOUS PSA CONFERENCE PUBLICATIONS:

PARTICLE SIZE ANALYSIS, 14-16 September 1966, Loughborough University of Technology; Society for Analytical Chemistry, London, 1967, 368 pp.

PARTICLE SIZE ANALYSIS 1970, 9-11 September 1970, University of Bradford; eds. M.J. Groves and J.L. Wyatt-Sargent, Society for Analytical Chemistry, London, 1972, 430 pp.

PARTICLE SIZE ANALYSIS, 12-15 September 1977, University of Salford; ed. M.J. Groves, Heyden & Son Ltd., London, 1978, 492 pp.   ISBN 0 85501 158 0.

PARTICLE SIZE ANALYSIS 1981, 21-24 September 1981, Loughborough University of Technology; eds. N. Stanley-Wood and T. Allen, John Wiley & Sons, Chichester, 1982, 461 pp. ISBN 0 471 26221 8.

PARTICLE SIZE ANALYSIS 1985, 16-19 September 1985, University of Bradford; ed. P.J. Lloyd, John Wiley & Sons, Chichester, 1987, 669 pp.   ISBN 0 471 90832 0.

PARTICLE SIZE ANALYSIS 1988, 19-20 April 1988, University of Surrey, Guildford; ed. P.J. Lloyd, John Wiley & Sons, Chichester, 1988, 361 pp.   ISBN 0 471 91997 7.

Contents

Posters

25 Years of Particle Size Conferences

B. Scarlett

DEPARTMENT OF CHEMICAL ENGINEERING, TECHNICAL UNIVERSITY, P.O. BOX 5045, DELFT, 2600 GA, THE NETHERLANDS

1   1966 AND ALL THAT

It is almost precisely twenty five years since the first of these conferences, organised then by the Particle Size Sub-Committee of the Society of Analytical Chemistry which has now grown into the Particle Size Characterisation Group of the Analytical Division of the Royal Society of Chemistry. The sub-committee had recently completed a review of 74 different methods of particle size analysis and these efforts were the inspiration both for the first conference and for the formation of the permanent subject group. This is the seventh conference in the series and it is entirely appropriate that it should be held in the same place, at Loughborough University. The proceedings of each of the conferences is permanently documented by a bound book of conference proceedings. Several authors feature in all the seven conferences.

The closing address at the first conference was given by Professor Harold Heywood [1], an address which was based on many years of experience and which could be given, almost without modification, at the end of this conference and still be entirely appropriate. Thus, some truths do not change and although they must be learnt anew by each succeeding generation, they remain essentially the same.
The developments which he foresaw have largely occurred over the intervening twenty five years but it may be that now those basic messages have even more practical importance and that the rate of development grows exponentially. In this paper, my intention is to re-state some of the message of Heywood and to illustrate it in the circumstances of today.

2   WHY OH WHY ?

The first quotation I would like to take concerns

why particle size measurement is important, why it
commands the attention of professional groups, companies, journals and conferences specifically devoted to
its development. Heywood wrote:-

**"However, it must be realised that particle size analy-
sis is not an objective in itself but is a means to an
end, the end being the correlation of powder properties
with some process of manufacture, usage or preparation."**

A particle size measurement does not have a meaning unless the objective of the measurement is also
specified. Thus, the techniques which should be used
depend entirely on the accuracy which is required and
the circumstances of place and time in which the measurements must be made. There is no such thing as the
"best" particle size technique unless the circumstances
are also specified. A particle analysis laboratory
cannot operate on the same basis as some normal analytical laboratories, samples come in and numbers go out.
In a particle analysis, the question must always first
be addressed. "What do we need to know?" Another way of
saying this is that, eventually, particle analysis is
an engineering tool, not a basic science. Engineers
must use all the known laws of science to solve the
particular problem with which they are confronted in
the most elegant manner.

In this respect, it may be that the circumstances
are more difficult than they were twenty five years
ago. At that time, the number of instruments was more
restricted than now as was certainly the number of
companies who supplied those instruments. Current instrument manufacturers must try to provide a package of
hardware and software which will handle a large number
of powders and applications. The commercially available
instruments are thus, inevitably, a compromise between
what is desirable and what is possible, the same instrument being sometimes too sophisticated and sometimes inadequate. To be a manufacturer of particle measuring instruments is not an easy life and it can only
get worse because the tendency will be for the applications to become more diverse and sophisticated. Why
should the applications become more diverse? That comes
partly from the development of new technologies, of
course, but it also comes from a gradually increasing
sophistication in the use of the measurements. (Fig.
1).

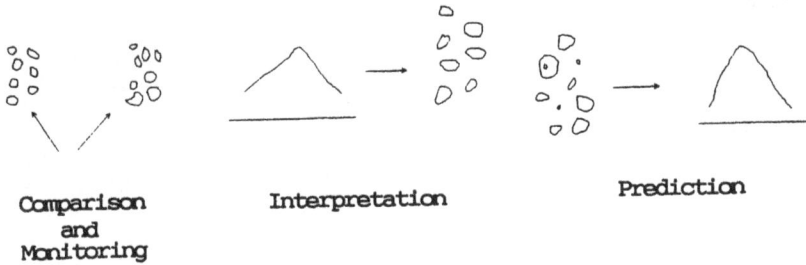

Figure 1   Particle characterisaction - why?

At the simplest level we use particle size measurements
to monitor their concentration or to control the reproducibility of a product. Thus, we compare what we find
with what we expect and if the two do not coincide we
reject the product. The science of powder technology,
however, is concerned to use the microscopic properties
of the system, for example the particle size distribution, to interpret the bulk behaviour of the powder. If
it is to be used in dilute circumstances, then the bulk
behaviour can be derived by integrating the behaviour
of the individual particles but usually this is not so
and the relationship between the microscopic and macroscopic properties must take account of the particle
interactions. By observing the difference in particle
size distribution of samples which exhibit a different
bulk behaviour, we begin to make a "correlation" between the two which, whether empirical or theoretical,
quantitative or qualitative, involves interpretation of
the mechanisms involved. Somewhere between these two
purposes usually lies the purpose of a particle size
measurement. There is, however, a far more ambitious
level at which powder technology must eventually operate and, as yet, rarely does. That is to design the
particles and the particle mixture to produce required
properties, to use the relationships between microscopic and macroscopic properties in a predictive manner.
It is the more rigorous use of particle size measurements which introduces the real diversity and which
requires the measurements to be carefully matched to
the problem. The increased diversity does not alter the
basic needs which Heywood described.
He wrote:-

**"Finally, we should examine the problems awaiting solu-
tion in the study of particle sizing and applications
to industrial production".**

He then went on to classify the problems into three
groups. The first was:-

Standards through Standards

"First there is the development of some standard method
of analysis which gives an absolute measure of particle
size distribution. This may already exist but, if so,
the validity needs establishing beyond question. Time
of operation is not a vital factor for such a procedure, as it would mainly be used for fundamental research".

Of course, the fundamental method has always existed and is indeed terribly tedious. In a review of
the 1981 conference, Leschonski [2] wrote:-

"It would, therefore, be most desirable to describe
size more often, based on the volume of the particle,
i.e. by its volume diameter".

Thus, if it were possible to measure the volume of
each individual particle, for example by weighing and
if the density is known then the result can be presented either as a distribution by number, or a distribution by mass, of the equivalent volume diameter. Thus,
in order to carry out an "absolute" method of particle
size measurement, all that is required is an accurate
balance. Of course, the method is impractical and, as
the particles become smaller, impossible but it is the
ideal against which other methods can be assessed and
can be calibrated. Over the past twenty five years, the
view has increasingly prevailed that we should regard
the equivalent volume diameter of the particle as the
basic size and that other equivalents are dependent
upon both the size and shape of the particle. To accept
this view requires only two simple steps. First to
accept that a measurement of the size of a particle
assigns only one scalar number to a particle and is,
therefore, bound to contain a limited amount of information about the particle. The 'size' of the particle
means, basically, how much material is contained in the
particle, that is its volume. The second step is to
accept that other equivalent spheres appertain to particular conditions. For example, the equivalent settling diameter depends on the Reynolds number and thus
on the fluid in which it is settling. The equivalent
light scatter diameter depends on the scattering parameters and thus the wavelength of the incident light
and the refractive index of the surrounding medium. Any
equivalent diameter relates exactly only to the precise
circumstances in which it is determined and the translation of it to different circumstances requires some
assumptions to be made.

In our modern circumstances, the need for an absolute method of measurement can be interpreted as the
need for calibration and reference materials which are
certified on the basis of traceable measurements, measurements which can eventually be related back to the
standard kilogram and standard metre. In creating the
coarser BCR fractions Leschonski's suggestion, based in
turn upon the suggestion of Andreason [3], has been
adopted and the samples are certified in terms of the
equivalent volume diameter, traceable to the standard
kilogram. Some of the samples of standard spherical
particles which are available can also claim to be
certified on the basis of measurements directly traceable, in that case to the standard metre. The current
efforts of organisations such as BCR and IFPRI to create more reference and calibration materials must be encouraged and intensified. At the same time, standards
must be written, and adopted by ISO, which describe
clearly how those materials should be used to calibrate
and control the commonly used methods of particle size
measurement. In this way, the standard method of analysis will not only exist, it will also become available.
When materials which have been certified by a traceable
method are presented to another instrument, we are in
fact comparing that instrument with Heywood's absolute
method. If we simply compare, the materials are being
used as a reference material whereas if the results are
used to adjust the output of the instrument then we are
calibrating. In general, distributions of particles are
more useful as reference materials and monosized fractions to calibrate but I believe that the procedures
which we must develop will require both and will present them, mixed in different proportions, in order to
assess three parameters:-

> reproducibility
> accuracy
> sensitivity.

The existing BCR materials are already being used to
assess reproducibility and accuracy, for example as
reported in this conference by Allen and Davis [4].
Before we can adequately assess all three factors and,
if necessary calibrate, we need also the spherical
reference materials which are currently being produced
as well as much narrower fractions of spherical particles. It is also necessary to elaborate the use of
those materials and the development and description of
these various uses of the materials is just as important as the certification of the materials. The ISO
standards are the obvious medium for this second task.
Because the range of materials which is available at
present is quite restricted, the way in which we use
them is similarly restricted.

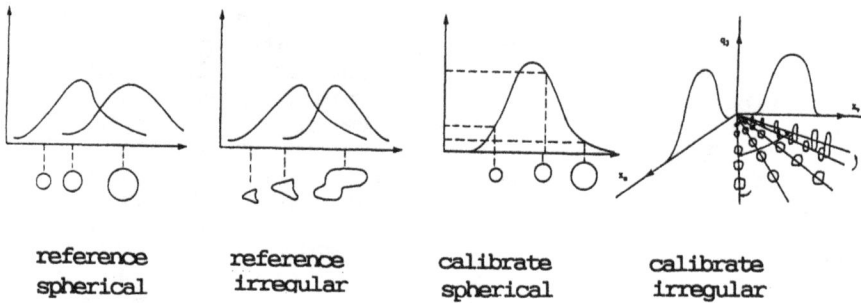

reference
spherical

reference
irregular

calibrate
spherical

calibrate
irregular

Figure 2   Particle standards

When used to assess reproducibility, there is not a great difference between spherical and irregular reference materials. However, in assessing accuracy there is some difference. In using spherical particles we are comparing what the instrument produces with what it is expected to produce. This is because the concept of the instrument is almost certainly based on the idea that the answer will be in terms of some equivalent spherical diameter. On the other hand, if a sample of irregular particles is used as a reference material and the recorded results diverge from the certified results, the explanation can often be advanced that a different equivalent sphere has been measured. We should regard a big divergence as an indication that the instrument must be calibrated to measure particle size, that is equivalent volume diameter. In one way then, spherical materials present a fairer reference material for an instrument in that it can be expected to produce the correct result, irregular particles probe its intrinsic accuracy. When it comes to using the materials to calibrate, on the other hand, the relative role of spherical and irregular particles is reversed. The spherical particles calibrate the size axis accurately only to measure spherical particles. If accuracy is required for other materials, then the calibration must be achieved with that same material. In order to calibrate a method for particle size, monosized fractions must be created. For spherical particles this is relatively easy but to create samples of irregular particles, all of which have the same equivalent volume diameter, is difficult and requires further development of the methodology. Furthermore, a calibration must concern not only the size axis but also the distribution axis. This can also be achieved by the use of monosized fractions, mixed in different proportions.

Thus, we see that not only has the absolute method always existed but also that, with imagination, we can compare and calibrate our more practical methods to it. However, we must also now extend our first tenet, that the particle size measurement is not an end in itself but must be related to the usage. Reproducibility, accuracy and sensitivity are also not absolute objectives and must also be related to the usage.

Beware the Software

The next need which Heywood listed was:-

"Secondly, there is the development of methods of analysis for routine procedures, for which repetitive accuracy is most important, though absolute accuracy should not be ignored".

The past twenty five years have seen an explosion in the diversity of the equipment available and a complete step change in the design and convenience of those instruments. I think that it is true to say that the instruments of today do not really use any physical principles or laws which were not used twenty five years ago. Rather the development has come about by the availability of modern components, lasers, optic fibres, photodiodes, and piezo crystals for example. Above all, however, engineering has been revolutionised by the development of the computer and this generality applies also to particle size measurement. The ability to carry out complex calculations and to record and manipulate large amounts of data in a short time has completely changed the concept of what is possible in a particle size measurement and this is a process which is still in its infancy. It is this development which has also compounded the problem. Previously the basic problem was that techniques which are based on different physical principles give different answers with the same sample. Now, instruments which are ostensibly the same give different results, both in substance and form, due to the disparate software packages which process and manipulate the basic data and eventually produce it in an attractive and convincing form.

If we have now accepted the basic tenet that both the measurement and its quality must match the problem, and we have, and if also the problems are becoming more numerous and sophisticated it follows that the diversity and sophistication of the measurement techniques will continue to increase. Preversely, this will not be in the hardware of the instruments which may well become, increasingly, standard arrays of photodiodes, optic fibre bundles, piezo transducers and conductance electrodes. Rather the difference will be in the software. The basic configuration of the instruments cannot change much.

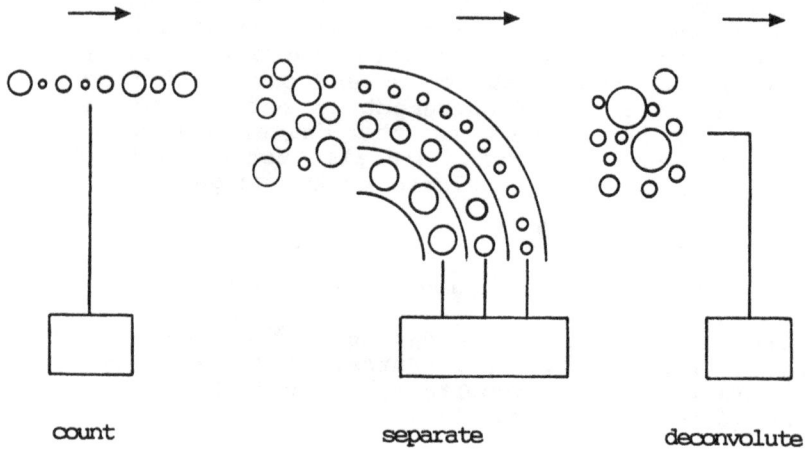

Figure 3   Particle characterisation - how?

On the one hand we can try to present the particles to the detector individually and count and characterise them or, on the other hand, we can maintain the particles as a complete mixture and then must address the problem of deconvoluting the signal. In general the first method produces more accuracy if a number distribution is required, the second can probably handle a larger sample. A combination of the two can always produce a better result than either individually. The intermediate state involves making some sort of partial separation, either within the measuring region or a separation in time, as in the classic sedimentation techniques. These basic possibilities will not change but what has and will continue to change is the number of signals which can be presented to the sample and the number of detectors with which it can be surrounded. If all these signals are used to estimate, check, interogate in a rigorous manner then the quality of our results can only improve. If they are so manipulated, smoothed, adjusted that the final result resembles what we wish we had rather than the reality, then we would be better to stick to the sieve and the microscope. Which of these situations appertains depends upon our software package. If I must identify today one problem which is significantly different than in Heywood's day, it is that the software packages we use are still rather closed and it is not obvious precisely what they do to the data. The development which must occur is that an instrument must be able to utilise a whole library of software which is added to and borrowed from by a host of people as in any library. The present situation is that the library often has one book, written by the librarian.

In-situ knowledge is Power

Heywood's third problem read as follows:-

"Thirdly, there will be a need for some sort of sizing analysis procedure that can form part of a fully automated production unit. This will involve automatic sampling, size analysis and feedback of the information to the process control".

The early efforts to achieve on-line measurement were largely devoted to trying to adapt laboratory methods of particle size measurement to an on-line use, usually confined to on-line monitoring.

Today, the picture is far broader and the emphasis is shifting. This is the age of information technology and it is that technology which is leading the pace of the others, including our own. Consider the various situations in which we measure particle size distribution.

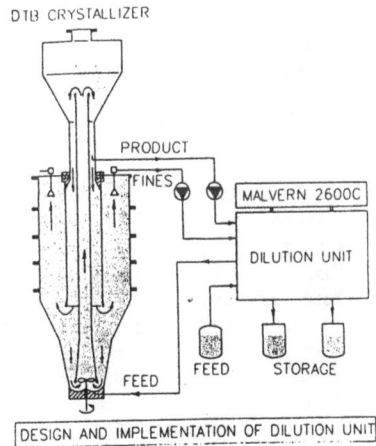

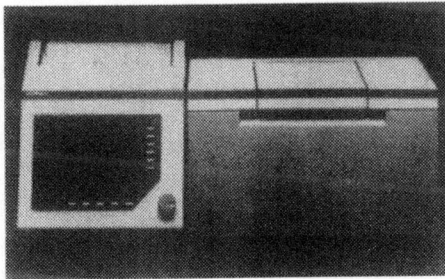

Off-line

On-line

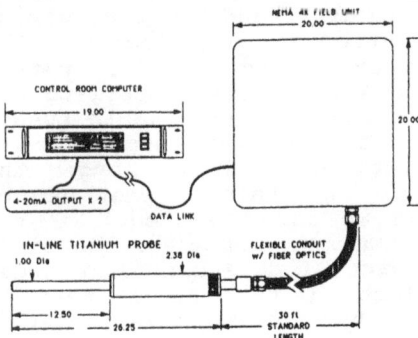

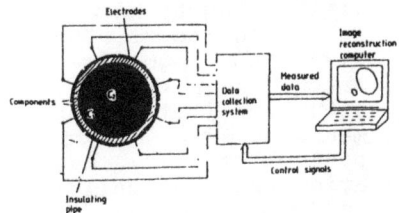

In-line

In-Situ

Figure 4 Particle characterisation - where?

In the off-line situation, particle size measurements can be used to control the quality of a product or to retrospectively examine a sample from a process. The inaccuracy is always with the sampling and the difficulty in knowing whether what is analysed is what originally existed. The step to on-line analysis is really only a step of automation and robotics, sampling and analysing closer by both in time and space what could be achieved by a laboratory analysis. The in-line measurement, on the other hand, minimises the sampling step and tries to insert a probe into the stream or the apparatus and to make a localised measurement. The next step is the in-situ measurement in which the properties of the particles in a local region can be deduced by interpreting the signals of sensors remote from that region. Twenty five years ago valiant efforts were being made to step from the off-line to the on-line situation and publications were describing methods of on-line sieving, sedimentation, permeametry and others. Today, many plants operate with on-line monitoring and several in-line instruments are commercially available. In the research papers, valiant efforts are being made move to the in-situ situation, first with laser-doppler methods and more recently with all the developments in tomography. Twenty five years from now I assume that the in-situ situation will be common place. Full automation will also transend passive monitoring. What does this imply for us? In terms of the measuring technique, there is no real difference between the off-line and the in-situ situation. There is no difference in principle whether a probe is inserted into a process stream or into a beaker in the laboratory. A ring of sensors can equally well surround a process pipe or a beaker in the laboratory. The difference in these situations lies in the time scale within which the information is available. As we progress through the off-line, on-line, in-line, in situ situations so we progress through a possible time scale of hours, minutes, seconds and milliseconds. This improvement means, for example, that we will progress from knowing the average particle size distribution within a vessel to knowing how it varies from region to region in the vessel and how that variation is changing over a period of milliseconds. The amount of information which can be generated is potentially staggering and will enable process modelling and automation to be advanced to a level unthinkable twenty five years ago. Never has it been more necessary to match the system to the problem. It is no longer the situation that we must think that this is what can be measured, how realistic can we make the model or the control strategy. Rather now are we coming to the situation where an unlimited amount of data can be recorded and the question is how much of that do we need to create the model or implement the control strategy.

5   THE EQUIVALENT CRYSTAL SPHERE

In 1981 the opening address to the conference was given by Professor Brian Kaye [5]. He wrote:-

"Being a prophet is a dangerous occupation so I will limit myself to comments on methodology in a restricted number of areas. First, let me predict that 'CAT scanning' will soon be part of the fineparticle technologies repertoire".

He referred to the tomography systems which are now, ten years later, so topical. I shall, as always, listen to the advice of Professor Kaye and make only two comments about the next twenty five years. The reason to do so is not to try to force the future, rather to recommend that we should not too closely restrict our thinking.

In the first instance, we must realise that particles are becoming more complex. Powder technology is concerned not only with characterising natural or simple materials, but in the modern industries with making designed particles each of which may be a complex, encapsulated mixture.

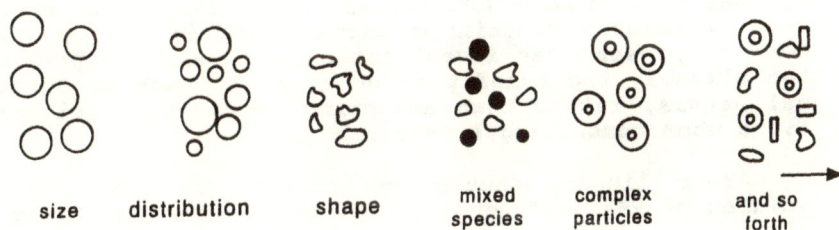

| size | distribution | shape | mixed species | complex particles | and so forth |

Figure   5   Particle Characterisation - What?

Models of uniform spheres can tell us much but cannot simulate very complex behaviour. By the same token defined behaviour often cannot be achieved by simple mixtures of sizes and shapes but must be possible if we are able to create mixtures of complex particles. Thus, for some applications, the measurements must be much more complex than a simple particle size distribution. It is some ten years ago that the organisers of this conference changed their name from the Particle Size Analysis Group to the Particle Characterisation Group. The characteristics of a complex mixture may well require both physical and chemical data to be combined to represent the total structure of the mixture.

The second point is that this, in turn, places strains on the manner in which we present the data. The traditional manner of representing a particle size distribution as a graph is pictorial, convenient and easy. In reality, the graph represents a probability of finding a particular particle. In powder technology we are dealing basically with populations of particles and the subject is based on statistics, the probability that we shall find a particular element of material in a particular place and at a particular time. As we increasingly use computers for simulation, modelling and control the data must be presented to the computer in the form which suits it best, which is as numbers not as pictures. We shall have the power with the instruments to generate vast quantities of data. The real value and interpretation of the data lies in its statistical nature.

A third and easy prediction. The conferences of the next twenty five years will be at least as interesting and productive as the past twenty five years. The theme of my paper is that we must, and shall, adapt all the modern technology as it develops to make better and better measurements and applications. However, some basic facts do not change. One final quotation from the 1966 address of Harold Heywood:-

"I feel that present authors are neglecting some important researches containing matter that is very pertinent to present day investigations. I cannot emphasise too strongly the benefit of going right back to original papers, rather than accepting second-hand reiterations when conducting a research".

Your library should now contain at least seven volumes of PSA conferences, volumes of original papers.

6 REFERENCES

1. H. Heywood. Proc. 1st Particle Size Anal.Cont. Sept. 1966. p. 355-359 (Heffer)

2. K.L. Leschonski. Proc 4th Particle Size Anal.Conf. Sept. 1981. p. 437-445 (Wiley Heyden).

3. A.H.M. Andreason. VOI Forschungsh. 399 Ausg.B.10 1939. pl.25.

4. T. Allen & R. Davis. Proc. 4th Particle Size Anal. Conf. Sept. 1991.

5. B.H. Kaye, Proc. 4th Particle Size Anal.Conf. Sept. 1981. p. 3-17 (Wiley Heyden).

Adsorption Methods for Surface Area Determination

K. S. W. Sing

CENTRE DE THERMODYNAMIQUE ET DE MICROCALORIMÉTRIE DU C.N.R.S., 26, RUE DU 141ÈME
R.I.A., 13003 MARSEILLE, FRANCE

1. INTRODUCTION

Adsorption occurs when a clean solid surface is brought into contact with a gas or liquid. The forces involved are physical or chemical and hence give rise to physical adsorption (physisorption) or chemisorption. Physisorption is always spontaneous, but its magnitude (i.e. the increase in concentration or density of one or more components in the interfacial layer) is dependent on the nature of the system, the surface area and porosity of the solid adsorbent and the conditions of measurement (pressure or concentration and temperature). In principle, it is therefore possible to employ physisorption measurements for the determination of the total surface area of a porous solid provided that the other variables can be controlled or their influence taken into account. On the other hand, chemisorption is highly specific and under favourable conditions provides a means of determining the surface area of an active component (e.g. the metal area of a supported catalyst). This aspect is not discussed in the present review.

Although certain adsorption techniques have been in use for over fifty years for determining the surface area of powders and porous materials, the interpretation of the adsorption data is still under discussion. Most procedures depend on the evaluation of the monolayer capacity, n_m (i.e. the amount of gas or solute required to cover the surface with a complete single layer of adsorbed molecules). To obtain the surface area, A, from n_m it is necessary to adopt a standardised value for the effective molecular area, a_m, of the adsorbate in the completed monolayer. Many attempts have been made to check the validity of this approach and to establish the conditions under which a particular adsorption method may be used with confidence. Unfortunately, it is often difficult to find a truly independent method which is capable of providing an evaluation of the surface area with the degree of accuracy required. Furthermore, it is necessary to keep in mind that the concept of surface area becomes questionable in the case of highly porous materials.

In the context of adsorption, it has been found useful to classify porous solids in terms of their pore widths[1]. Thus, a *microporous* solid has pores of width < 2 nm ; a *mesoporous* solid has pore widths in the range 2 - 50 nm ; a *macroporous* solid has pores of width > 50 nm.

2. PHYSISORPTION OF GASES

2.1. Methodology

The amount of gas adsorbed is determined with the aid of either (a) a volumetric method, which involves measurement of the amount of gas removed from the gas phase, or (b) a gravimetric method, which involves the direct determination of the increase in mass of the adsorbent. In practice, either method may be operated in a static or dynamic mode, but it is important to achieve at least quasi-equilibrium conditions. A carrier gas technique is useful for some measurements provided that the adsorption of the carrier gas is negligible.

The adsorption isotherm is usually constructed point-by-point by the admission to the adsorbent (maintained at constant temperature) of successives charges of gas. Each point on the isotherm is recorded when the residual gas pressure has become constant and equilibrium has been attained. If a quasi-equilibrium technique is employed[2], it is essential to confirm that the results are not affected by change in flow rate.

Prior to the determination of an adsorption isotherm all previously physisorbed material must be removed from the adsorbent surface. This is achieved by either outgassing or flushing the adsorbent with an inert gas at elevated temperature. To obtain reproducible isotherms it is necessary to control the outgassing conditions to within limits which depend on the nature of the system. For this purpose it is often useful to undertake a preliminary temperature programmed gravimetric study[3].

The low-temperature adsorption of nitrogen is the generally recommended technique in all the standard methods proposed in recent years for the determination of surface area by means of gas adsorption[1,4]. For a number of reasons nitrogen isotherms at the temperature of the normal boiling point of nitrogen (*i.e.* at ca 77 K) are usually determined volumetrically and routine measurements are now often made with the aid of fully automated equipment. If the surface area determination is to be made on a sample which is already known to be non-porous, macroporous or mesoporous, it is sufficient to record the isotherm data to a maximum relative pressure, p/p^0, of ~ 0.3. However, it is important to note that if the technique is to be fully exploited for the characterization of pore structure, the equilibrium p/p^0 should be taken as close to saturation as possible, before the desorption curve is recorded[5].

2.2. Classification of nitrogen isotherms

Inspection of the overall shape of a nitrogen isotherm can provide a first useful indication of the nature of the adsorbent porosity and hence of the most appropriate procedure to be adopted for the analysis of the isotherm data. The hypothetical isotherms in Fig. 1 are representative of nitrogen isotherms obtained with a wide variety of adsorbents[6]. The general features of these isotherms are consistent with Types I, II and IV, in the classification already proposed by the International Union of Pure and Applied Chemistry[1].

Isotherms of Type Ia or Ib are obtained with microporous adsorbents having very small external areas (*i.e.* the area *outside* the micropores). The initial steep region is associated with the filling of pores of molecular dimensions (width < ~ 0.7 nm). The high adsorption affinity, which is a special feature of the Type Ia isotherm, is due mainly to an enhanced energy of adsorption associated with the strong adsorbent-adsorbate interactions in the very narrow pores. In the case of the Type Ib isotherm, the more gradual approach to the plateau is the result of the filling of wider micropores (of width ~ 0.7 - 2 nm) by a secondary process involving cooperative adsorbate-adsorbate interactions.

   The reversible Type IIa isotherm is the normal form of nitrogen
isotherm given by a non-porous or macroporous adsorbent and is indicative
of unrestricted monolayer-multilayer adsorption. The adsorption branch of a
Type IIb isotherm appears to have the same characteristic Type II shape as
a normal monolayer-multilayer isotherm, but the multilayer section of the
desorption branch is quite different -giving rise to a form of adsorption
hysteresis. Isotherms of this type are generally given by aggregates of
platy particles or solids containing slit-shaped mesopores.

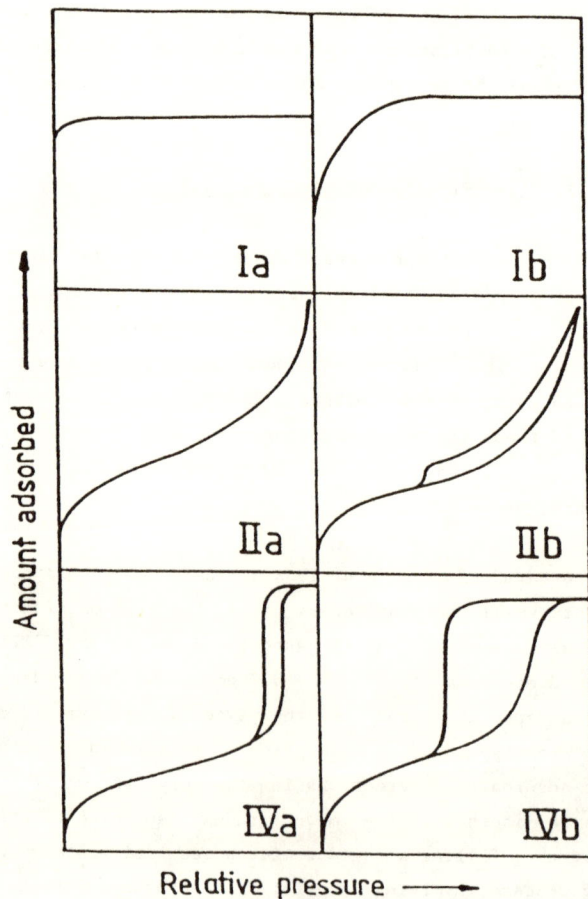

Figure 1.   Types of nitrogen adsorption isotherms

The Type IVa isotherm is given by a well-defined mesoporous solid having a fairly narrow distribution of pore size. In the monolayer and initial multilayer region (*i.e.* up to $p/p^0 \sim 0.4$) this isotherm follows the same path as the corresponding Type IIa isotherm determined on the equivalent area of a non-porous surface of similar structure. In this case capillary condensation is responsible for the upward deviation of the isotherm in the multilayer region. Such isotherms are obtained only with those mesoporous solids having uniform pore structures.

The Type IVb isotherm is representative of the behaviour of many inorganic oxide gels which possess an assortment of interconnected pores of different size and shape. The steep part of the desorption branch was originally thought to indicate a narrow distribution of pore size, but a hysteresis loop of this shape is now known to be associated with a highly complex pore structure.

2.3. The BET method

In spite of its limitations, the Brunauer-Emmett-Teller (BET) gas adsorption method[7] has become the most widely used procedure for the determination of the surface area of ultra-fine powders and porous materials.

The BET equation is normally applied in its linear form

$$\frac{p}{n \, (p^0 - p)} = \frac{1}{n_m C} + \frac{(C - 1)}{n_m C} \frac{p}{p^0} \tag{1}$$

where n is the amount adsorbed at p/p^0, n_m is the monolayer capacity and C is an empirical constant. It follows that there should be a linear relation between $p/n(p^0 - p)$ and p/p^0 (*i.e.* the BET plot). In practice, the range of linearity is always restricted to a limited part of the isotherm[4] -rarely outside the p/p^0 range of 0.05 - 0.30.

In the early work of Brunauer and Emmett[8], Point B -the beginning of the almost linear middle range of a Type II isotherm- was identified as the stage of monolayer completion. If the value of C is sufficiently high ($C \geqslant$ ca 80) it is possible to assess the uptake at Point B by visual inspection of the isotherm and in that case it is generally found that good agreement is obtained with the value of n_m derived from the BET plot. On the other hand, if C is low (say, C < ca 20), Point B cannot be identified as a single point on the isotherm and the theoretical significance of the derived value of n_m is uncertain[4].

The second stage in the application of the BET method is the calculation of the surface area, A(BET), (*i.e.* the "BET area") from n_m. As mentioned earlier, this requires a knowledge of the molecular cross-sectional area, a_m -or more strictly the average area occupied by the adsorbate molecule in the complete monolayer. Thus

$$A (BET) = n_m \cdot L \cdot a_m \qquad (2)$$

where L is the Avogadro constant.

It is usually assumed that the nitrogen monolayer at 77 K is in a close-packed liquid state, giving $a_m(N_2) = 0.162$ nm^2. With a wide range of non-microporous adsorbents the use of this value gives BET areas which appear to be valid since they agree moderately well with values of surface area determined by other methods (to within ~ 20%). However, it should be appreciated that the existence of a strictly constant value of $a_m(N_2)$ is unlikely since the packing of molecules in the completed monolayer is dependent on the adsorbent-adsorbate interactions and the structure of the adsorbent surface. The experimental evidence available indicates that, in the case of nitrogen adsorption at 77 K, this may vary by up to ~ 30%.

The BET analysis does not take account of micropore filling which is associated with Type I isotherms.. There is now general agreement that the BET method cannot be used to obtain a reliable assessment of the internal surface area of an adsorbent exhibiting molecular sieve properties[5,6] (*i.e.* one giving a Type Ia nitrogen isotherm). In this case the BET-area provides only a "fingerprint" of the adsorbent activity which may be useful for comparative purposes, *e.g.* in a patent specification. Authoritative opinions differ, however, on the applicability of the BET analysis to Type Ib isotherms. Recent work has indicated that the method can be used to assess the surface area of the wider micropores (of width ~ 1-2 nm) provided that there are very few narrow micropores which would distort the isotherm in the monolayer range[9].

The measurement of the amount of nitrogen adsorbed becomes increasingly more difficult as the surface area of the adsorbent is decreased to below ~ 5 $m^2 g^{-1}$. Because of its low p^0 at liquid nitrogen temperature, krypton has been used to minimise the dead space correction. Although krypton adsorption has been found to be a useful technique for routine or comparative measurements, it cannot be relied on to provide a highly accurate evaluation of A. Since the Kr monolayer capacity is often difficult to assess, the effective value of a_m appears to vary from one surface to another[4]. A similar difficulty is encountered with argon at 77 K.

2.4. Empirical procedures for isotherm analysis

No current theory is capable of providing a satisfactory mathematical description of experimental adsorption data over the complete range of p/p^0. This failure is primarily due to the complexity of the surface structure and intermolecular forces and the added complications associated with surface roughness and porosity. Such considerations would lead us to believe that each adsorbent would exhibit a unique isotherm for a given gas at a certain temperature. On the other hand, in the absence of pore-filling or surface roughness effects we might expect that a group of chemically similar solids (*e.g.* oxides) would exhibit closely related physisorption behaviour. A number of experimental studies[4,10] have confirmed that this is especially evident with respect to nitrogen adsorption at 77 K.

The most obvious way of comparing the shape of a group of isotherms is to plot them in a reduced form, *i.e.* either as the amount adsorbed per unit area or as a dimensionless quantity, n/n_x, where n_x is the amount adsorbed at a pre-selected p/p^0. The latter approach has been found preferable since it does not depend on any assumptions concerning the determination of surface area[11]. Indeed, it has been found that reduced nitrogen isotherms on a wide range of non-porous solids show remarkably close correspondence in their *multilayer* range[10]. This behaviour is not exhibited to the same extent by other gases such as Ar and Kr. The reason for the "universal" character of the N_2 multilayer is still not entirely clear, but it appears to be associated with the particular nature of the asdorbate-adsorbate interactions. Thus, it seems that the initial influence of the surface structure in controlling the adsorbent-adsorbate interactions is overcome at an early stage of multilayer development.

Lippens and de Boer[12] devised a simple and convenient way of comparing the shape of one isotherm with that of another. However, their t-method suffers from the disadvantage that it is dependent on the use of the BET method to obtain the standard multilayer thickness (t-curve). An improvement was developed in the form of the α_s-method[11], which provides a more rigorous approach and allows a more refined analysis of the isotherm shape[5,9,13]. The amount of gas adsorbed is plotted against α_s, the reduced standard adsorption at the corresponding p/p^0 (with $\alpha_s - 1$ at $p/p^0 - 0.4$). The standard data are determined experimentally on a series of carefully selected reference materials, which give unrestricted monolayer-multilayer adsorption. If the appropriate standard data are selected for the particular gas-solid system, it is therefore possible to identify the influence of porosity (micropore filling or capillary condensation) on the path of the isotherm. In this manner it is possible to check the validity of the BET-area and in favourable cases to evaluate the total surface area, the internal area and the external area[5,11].

3. ADSORPTION AT THE LIQUID/SOLID INTERFACE

3.1. Adsorption from solution

Adsorption at the liquid/solid interface is responsible for the decrease in solute concentration which occurs when a powder or porous solid is immersed in a suitable solution (*e.g.* of a dyestuff or surface active agent). A rigorous thermodynamic treatment of the experimental data is not straightforward because solvent is also adsorbed at the interface (to an unknown extent) and because the boundary of the adsorbed layer is difficult to locate exactly[14]. Fortunately, a simple interpretation is acceptable for many purposes provided that the solution is dilute and the adsorption of solute much greater than that of the solvent.

Adsorption from solution measurements are obviously facilitated if an instrumental technique (*e.g.* a spectrophotometer or refractometer) can be used to follow the change in solute concentration. However, it is not always easy to achieve high accuracy and reproducibility, and it is therefore necessary to take account of a number of practical considerations and precautions[14] The traditional experimental method is to add a known mass of the solid to a measured amount of solution of known composition. The container is then sealed and subjected to some form of agitation (shaking or preferably tumbling), whilst being held at constant temperature. Samples of the supernatant liquid are withdrawn and analysed. In such an experiment it is not easy to establish whether equilibrium has been established and in fact this may take from one hour to a few days! Problems may also arise in the separation of the adsorbent and the supernatant liquid. The procedure is laborious and cumbersome, because a fresh sample is required for the determination of each point on the isotherm.

A more elegant approach is to circulate the solution through a bed of adsorbent and monitor the concentration continuously (*e.g.* by passage through a UV cell). This technique has a number of important advantages including the ease of thermostatting the adsorption cell, monitoring the attainment of equilibrium and in the adjustment of concentration. Whichever method is used care must be taken in controlling the pH, ionic strength etc and in checking the purity of the solvent. In particular, a small quantity of water in an organic solvent is likely to have a large effect on the solute adsorption.

Different solution-solid systems have been found to give a variety of isotherm shapes. The classification introduced by Giles and his co-workers[15] still provides a useful basis for the comparison of different systems. In the context of surface area determination we need refer to only a few of these characteristic shapes and to note the superficial similarity between certain solution-solid and gas-solid systems.

Some solute isotherms (the L-type) are concave to the concentration axis over a wide range of concentration. In certain cases it is possible to apply an empirical equation of Langmuir form

$$\frac{n}{n_L} = \frac{b\ C_A}{1 + b\ C_A} \qquad\qquad (3)$$

where n is the amount of solute adsorbed at the equilibrium concentration C_A, n_L is the amount adsorbed at the plateau and b is an empirical constant. Although L-type isotherms exhibit the characteristic "Langmuir" shape, the range of adsorption over which Eq. (3) is obeyed is usually quite limited.

It is at once evident that there is a remarkable degree of similarity between the shapes of L-type solute isotherms and Type I physisorption isotherms[1]. However, this similarity is misleading since the adsorption mechanisms involved are likely to be quite different. We have seen already that Type I *physisorption* isotherms for gas-solid systems are normally associated with micropore filling. In contrast, the plateau of an L-type solute isotherm usually corresponds to monolayer completion. In this respect, solute adsorption appears to correspond more closely to the classical Langmuir mechanism. If this is indeed the case it would seem to be possible to calculate the surface area from n_L by the application of a simple equation of the same form as Eq. (2).

Unfortunately, there are a number of reasons why this apparently straightforward approach does not provide a reliable method for the determination of surface area : -

(a) Unless the plateau is well-defined, the use of Eq. (3) in a range of low concentration to obtain n_L is unlikely to yield the true value of the monolayer capacity of solute ;

(b) Most surfaces are heterogeneous and interact specifically with solute molecules and therefore the surface coverage is not uniform ;

(c) The orientation of the adsorbate molecules may vary considerably from one surface to another and in the case of large molecules the adsorbate-adsorbate interaction often determines the structure of the adsorbed phase which may extend beyond the monolayer or involve the adsorption of micelles ;

(d) Incorporation of solvent in the monolayer may result in an appreciable variation in the effective molecular area, a_m.

The measurements of de Boer and his co-workers[16] were amongst the first to clearly demonstrate the effect of the solvent on the adsorption of a long-chain acid (lauric acid by activated alumina). The competition was pronounced with benzene and especially diethyl ether whereas pentane did not interfere to any significant extent. Thus, the isotherm of lauric acid in pentane solution showed a constant amount adsorbed (*i.e.* plateau) over the complete range of recorded concentration. This type of behaviour (Type H in the Giles classification) is characteristic of very high adsorption affinity and is generally due to relatively strong adsorbent-adsorbate interactions -as with gas adsorption. However, the solute monolayer is unlikely to be close-packed and therefore a_m is dependent on the surface chemistry. In some cases, there is also evidence of cooperative adsorption[17].

For many years it was established practice to use dyestuffs in aqueous solution for surface area determination. Methylene blue was often employed, but it has been shown that the apparent molecular area is not constant from one type of surface to another and that it cannot be predicted[18]. Giles recommended[19] the use of p-nitrophenol as the solute. This has the advantage that it is fairly small and is soluble in both water and organic solvents (*e.g.* n-hexane or xylene), but it is evident that its application is dependent on calibration by means of a set of reference materials.

Detailed investigations by Fuerstenau and his co-workers[20] of the adsorption of cationic surfactants in the context of mineral flotation have helped to improve our understanding of certain mechanisms of solute adsorption. Thus, at low concentration the positively charged surfactant ions adsorb on negatively charged surface sites through electrostatic attraction, but as the concentration is increased the adsorbate-adsorbate interaction becomes increasingly important and the hydrocarbon chains begin to associate with the formation of two-dimensional aggregates (termed "hemimicelles"). The role of surface heterogeneity in obscuring the two-dimensional phase transformations at the solution/solid interface was studied by Cases[21]. It was found that the steps at very low surface coverage associated with adsorbate-adsorbate interactions on a uniform surface were hardly detectable when the surface became heterogeneous as a result of grinding. As in the case of gas adsorption, it appears that at low surface coverage smooth isotherms are generally associated with energetic heterogeneity in the adsorbent-adsorbate interactions.

Very few attempts have been made to investigate the influence of porosity on the course of adsorption from solution. Recent work in this laboratory[22] has shown that in favourable cases it is possible to extend the application of the α_s-method for the analysis of solute isotherms. The analysis gave a clear indication of the effect of primary micropore filling in distorting the isotherm shape when applied to certain L-type isotherms (*e.g.* of iodine and salicylic acid adsorption by activated charcoals), but difficulties were encountered when the non-porous reference material gave H-shaped or S-shaped isotherms (*i.e.* either very high or very low affinity).

3.2. Heat of immersion

The heat of immersion (or heat of wetting) is defined as the amount of heat evolved when a known mass of outgassed solid is completely immersed (but not dissolved) in a given liquid. When determined under conditions of constant pressure and temperature it may be regarded as the enthalpy of immersion, $\Delta_{imm}H$, which is in turn directly related to the integral enthalpy of adsorption for the particular vapour-solid system.

In the absence of complicating factors, such as micropore filling effects, the magnitude of $\Delta_{imm}H$ is dependent on the nature of the liquid-solid interactions and the extent of the available surface. Thus

$$\Delta_{imm}H = A \cdot \Delta \hat{h} \tag{4}$$

where $\Delta \hat{h}$ is now the areal enthalpy of immersion for the given liquid-solid system and A is the area of the liquid/solid interface.

Immersion calorimetry has been used for many years as a convenient method for the semi-routine characterisation of activated carbons and some oxide adsorbents. It would appear to offer the possibility of determining the surface area by a single measurement -provided that $\Delta \hat{h}$ is known for the liquid-solid system. The technique appears to have the advantage that it is less time-consuming than, say, gas adsorption. If the surface area is sufficiently large (*i.e.* A > 10 m^2), the amount of heat released is not difficult to register with the aid of a modern calorimeter.

This apparent simplicity of technique is, however, somewhat deceptive. The recorded heats will be of little scientific value unless the measurements are made under carefully controlled conditions : to achieve thermodynamic validity, the changes of state must be clearly defined[23]. Furthermore, the required levels of accuracy and reproducibility cannot be attained unless certain precautions are taken in the handling of samples and the operation of the calorimeter[24].

To ensure complete wetting, the solid sample must be adequately evacuated before it is exposed to the given liquid. The usual procedure is to insert a known amount into a glass bulb which is then attached to a vacuum system. The outgassing conditions (rate of heating, temperature and residual pressure) are controlled in the same manner as for physisorption measurements, but the sample bulb is then sealed off from the vacuum system. After it has been placed under the liquid in the calorimeter cell and time allowed for the temperature to equilibrate, a small brittle tip of the bulb is broken to allow the liquid to reach the solid surface.

Various commercial calorimeters are now available for routine heat of immersion measurements. For research it is preferable to use a calorimetric technique which is consistent with thermodynamic requirements[23]. We recommend the employment of a Tian-Calvet type of microcalorimeter[24], which by means of two thermopiles composed of a large number of thermocouple junctions allows the heat flux to be measured accurately at practically constant temperature ($\Delta T < 10^{-3}$K). Whichever technique is used, the experiments must be devised in a manner which will allow the evaluation of a number of corrective terms due to partial evaporation of the liquid, bulb breaking, stirring and effect of atmospheric pressure. In practice, this does not present difficulties because the detailed procedures and calculations are described in the literature[24].

It was originally assumed that the areal enthalpy of immersion in a given liquid was almost independent of the nature of the solid surface and therefore that the determination of the surface area was quite straightforward. Although this approach is now known to be oversimplified, it might still appear to be possible to employ immersion calorimetry for surface area determination if a set of standard values of $\Delta \hat{h}$ could be made available. It turns out, however, that even this procedure must be used with caution because it does not allow for the complexities of the adsorbent structure and molecular interactions.

Recent systematic studies[25] have shown that the measured values of $\Delta_{i\,mm}H$ may be affected to an unknown extent by : -

(a) Changes in the surface chemistry ;

(b) Enhancement of the adsorption energy by micropore filling ;

(c) Activated entry of molecules through narrow pore entrances or between aggregated particles ;

(d) Restricted entry into pores through molecular sieving.

With some adsorbents, such as activated carbons[13,26], several of these effects may operate together and then it becomes very difficult (if not impossible) to calculate the surface area from the heat of immersion data. Furthermore, in the light of these findings we must conclude that many of the early estimates of $\Delta \hat{h}$ are of doubtful validity since they were based on measurements made with inadequately characterised materials.

A remarkable attempt was made by Harkins and Jura[27] nearly fifty years ago to overcome some of these difficulties. Their idea was to cover a non-porous adsorbent with a multilayer thick enough to have a liquid-like surface. The pre-coated adsorbent should therefore exhibit a surface enthalpy identical to that of the liquid and immersion in the same liquid should liberate an amount of energy equivalent to the removal of this surface. In principle, therefore, it would seem a fairly simple matter to calculate the surface area from the heat of immersion of the coated solid. The measurements Harkins and Jura appeared to indicate that 5-7 molecular layers were required to overcome the influence of the solid surface and reduce the surface energy to that of the liquid. Since this layer thickness would correspond to a very high relative pressure, it would be difficult to avoid some capillary condensation -even with non-porous powdered materials- and therefore the method appeared to have very limited applicability.

A more recent investigation[28] has revealed that this problem can be overcome by using water as the liquid since two molecular layers are sufficient to effectively "screen" the underlying surface of many adsorbents. These results have led to a modification of the original Harkins-Jura "absolute" method for surface area determination and they make it possible to apply the technique to mesoporous solids (by avoiding the complication of capillary condensation). Obviously, the approach cannot be used in isolation to study micropore filling, activated entry or molecular sieving, but it becomes a powerful tool when combined with gas adsorption.

3.3. Liquid flow calorimetry

Flow calorimetry is another technique which has been developed in recent years for the study of interactions at the liquid/solid interface and has been used also as an indirect means of determining surface areas. Some flow calorimeters are of relatively simple design whereas others are more sophisticated : it is not surprising to find that the quality of the thermal data is dependent on the type of equipment and the *modus operandi*.

The flow microcalorimeter designed by Groszek[29] is an excellent example of the former type which makes use of a single temperature sensor (a thermistor) to register the flow of heat between the adsorbent and a surrounding metal block. The device is remarkably sensitive to change in temperature, but is is not designed to monitor the *total* heat flow resulting from solvent-solid and solute-solid interactions at the liquid/solid interface. Although this equipment cannot be regarded as a precision microcalorimeter, it is easy to operate and is capable of giving very useful semi-quantitative information. The sample is normally flushed with pure solvent (*e.g.* n-heptane) and the injection of the solute (*e.g.* butanol) then produces a thermal effect, the magnitude of which depends on the experimental conditions (liquid flow rate and heat capacity sample mass and grain size etc.). Under controlled conditions, the thermal effect is proportional to the surface area of the sample -provided, of course, that the surface density of active sites does not change from one sample to another. The application of this technique has been extended by Groszek to provide an effective means of comparing the nature of the surface sites on partially graphitized[30] and microporous[31] carbons.

The more refined flow microcalorimeters[32] -especially those based on the Tian-Calvet principle- can be used to determine accurate values of the enthalpy of displacement of the solvent (*e.g.* heptane) by the solute (*e.g.* an alcohol, acid or long-chain alkane). Satisfactory results of this kind can be obtained with samples of specific area $\geqslant 10$ m^{2}g^{-1}. For surface area determination the limitation is no longer experimental, but is now governed by the interpretation of the experimental data (as in the case of the heat of immersion). Thus, two samples of, say, kaolin with the same BET-nitrogen areas will usually give different enthalpies of displacement (for a given solvent-solute system). This situation arises because the recorded changes of enthalpy are dependent on the energetics of adsorption and displacement and not simply on the magnitude of the surface area.

It is evident that liquid flow calorimetry should be regarded as a complementary -rather than an alternative- approach to the BET method. Indeed, there is considerable scope for further work on the characterization of the adsorption site energy distribution of a wide range of adsorbents and advanced materials (*e.g.* catalysts, ceramics and molecular sieves).

4. GENERAL DISCUSSION AND CONCLUSIONS

It is evident that the area of a solid surface becomes more difficult to define and evaluate as its magnitude is increased above a certain level. One reason for this difficulty is that materials with large specific surfaces (above ~ 500 m^{2}g^{-1}) are almost invariably microporous. Of course, some adsorbents of much lower area are also microporous, but generally surface areas in the region of 200 - 300 m^{2}g^{-1} are associated with mesopore structures.

In recent years the application of fractal analysis[33] to surface and colloid science has reinforced the need for caution in the interpretation of adsorption data. From a fractal standpoint, the estimated surface area is unlikely to have a single value but instead be dependent on the size of the adsorbate molecule (*i.e.* on the a_m value). Furthermore, it is the *fractal dimension* which is the characteristic "roughness parameter" of the adsorbent and this is estimated from the plot of monolayer capacity against the adsorbate molecular area.

At first sight, the fractal approach appears to be more sophisticated than the traditional one of adjusting the values of a_m -usually to produce agreement with $a_m(N_2) = 0.162$ nm^2. On further consideration, however, it becomes apparent that an oversimplified application of fractal analysis may tend to obscure rather than clarify the interpretation of the adsorption data. In practice, there are two complicating factors : (a) the calculated values of n_m are not always reliable (especially those obtained by the BET method) and (b) the mechanism of pore filling is dependent on the ratio of pore width/molecular diameter and therefore may not remain constant for a given porous solid. Until further progress is made in our understanding of the mechanisms of physisorption, the interpretation of the fractal dimension must be a matter for discussion.

Although there is now general agreement that the BET-nitrogen method is the preferred procedure for the determination of the surface area of ultra-fine powders and mesoporous solids, there remains the difficult question of whether nitrogen adsorption can be used to assess the area within the *wider* group of micropores. On the basis of our present understanding of the mechanisms of surface coverage· and pore filling, we tentatively propose that the method can be used to determine the *internal* surface area provided that the effective pore width is at least ~ 1 nm. This proposal follows from the hypothesis that secondary (cooperative) micropore filling involves an *initial* stage of monolayer adsorption and in our view the evaluation is best made by the α_s -method.

In principle, there is no upper limit of pore size which would restrict the use of nitrogen adsorption, but for experimental reasons it becomes increasingly difficult to undertake physisorption measurements if the specific surface is smaller than ~ 5 m^{2}g^{-1}. This presents a difficult problem in the study of coarse powders and many macroporous materials of technological importance.

No other readily available method can offer the same scope, reliability and accuracy as gas adsorption. Adsorption from solution measurements are relatively easy to carry out, but often difficult to interpret ; they cannot be recommended for general use, but are essential for some applications involving treatment of liquid media. Although enthalpies of immersion are more difficult to determine, they can provide useful information provided that isotherm data are also available. Liquid flow calorimetry is becoming popular for the characterization of powders and porous solids, but the technique requires refinement if it is to be used to obtain thermodynamic data.

Finally, we must refer to work in progress on the selection and characterization of reference adsorbents[11,13]. Graphitized carbon blacks and ultra-fine silicas are amongst the few non-porous solids of moderately high surface area ($\sim$ 50 - 200 $m^2 g^{-1}$) which have been systematically studied in a number of laboratories. The standard data obtained so far have been found to be of particular value for the analysis of adsorption isotherms and heat data determined on a wide range of porous carbons and silicas. Efforts are being made also to select other materials having stable and well-defined mesopore or micropore structures[34]. With the aid of these reference pore structures it is hoped that it will be possible to make further progress in the application of adsorption techniques for surface area determination.

REFERENCES

1. K.S.W. Sing, D.H. Everett, R.A.W. Haul, L. Moscou, R.A. Pierotti, J. Rouquerol and T. Siemieniewska, *Pure Appl. Chem.* 1985, 57, 603

2. J. Rouquerol, F. Rouquerol, Y. Grillet and R.J. Ward, in "Characterization of Porous Solids" (Eds. K. Unger, J. Rouquerol, K.S.W. Sing and H. Kral), Elsevier, Amsterdam, 1988, p.67

3. J. Rouquerol, *Thermochimica Acta* 1989, 144, 209

4. S.J. Gregg and K.S.W. Sing, "Adsorption, Surface Area and Porosity", 2nd edn., Academic Press, London, 1982

5. K.S.W. Sing, *Colloids and Surfaces* 1989, 38, 113

6. K.S.W. Sing, in "Fundamentals of Adsorption" (Eds. A.B. Mersmann and S.E. Scholl), Engineering Foundation, New York, 1991, p.69

7. S. Brunauer, P.H. Emmett and E. Teller, *J. Amer. Chem. Soc.* 1938, 60, 309

8. P.H. Emmett and S. Brunauer, *J. Amer. Chem. Soc.* 1937, 59, 1553

9. P.J.M. Carrott, F.C. Drummond, M.B. Kenny, R.A. Roberts and K.S.W. Sing, *Colloids and Surfaces* 1989, 37, 1

10. P.J.M. Carrott and K.S.W. Sing, *Pure Appl. Chem.* 1989, 61, 1835

11. K.S.W. Sing, in "Surface Area Determination" (Eds. D.H. Everett and R.H. Ottewill), Butterworths, London, 1970, p.25

12. B.C. Lippens and J.H. de Boer, *J. Catalysis* 1965, 4, 319

13. J. Fernandez-Colinas, R. Denoyel, Y. Grillet, F. Rouquerol and J. Rouquerol, *Langmuir* 1989, 5, 1205

14. D.H. Everett, *Pure Appl. Chem.* 1986, 58, 967

15. C.H. Giles, D. Smith and A. Huiston, *J. Colloid Interface Sci.* 1974, 47, 755

16. J.H. de Boer, G.M.M. Houben, B.C. Lippens, W.H. Meijs and W.K.A. Walgrave, *J. Catalysis* 1962, 1, 1

17. C.H. Giles, A.P. D'Silva and I.A. Easton, *J. Colloid Interface Sci.* 1974, 47, 766

18. S.J. Gregg and K.S.W. Sing, "Adsorption Surface Area and Porosity", 1st edn., Academic Press, 1967, p.293

19. C.H. Giles and A.P. D'Silva, *Trans. Faraday Soc.* 1969, 65, 1943

20. D.W. Fuerstenau and R. Herrera-Urbina, in "Cationic Surfactants" (Eds. D.N. Rubingh and P.M. Holland), Marcel Dekker, New York, 1991, p.407

21. J.M. Cases, Industrie Minérale-Minéralurgie, 1979; Octobre, 1

22. J. Fernandez-Colinas, R. Denoyel and J. Rouquerol in "Characterization of Porous Solids II" (Eds. F. Rodriguez-Reinoso, J. Rouquerol, K.S.W. Sing and K.K. Unger), Elsevier, Amsterdam, 1991, p.399

23. C. Létoquart, F. Rouquerol and J. Rouquerol, *J. Chim. Phys.* 1973, 70, 559

24. J. Rouquerol, *Thermochimica Acta* 1985, 96, 377

25. J. Fernandez-Colinas, R. Denoyel, Y. Grillet, J. Vandermeersch, J.L. Reymonet, F. Rouquerol and J. Rouquerol, "Fundamentals of Adsorption" (Eds. A.B. Mersmann and S.E. Scholl), Engineering Foundation, New York, 1991, p.261

26. D. Atkinson, A.I. McLeod, K.S.W. Sing and A. Capon, *Carbon* 1982, 20, 339

27. W.D. Harkins and G. Jura, *J. Amer. Chem. Soc.* 1944, 66, 1362

28. S. Partyka, F. Rouquerol and J. Rouquerol, *J. Colloid Interface Sci.* 1979, 68, 21

29. A.J. Groszek, *Proc. Roy. Soc.* 1970, A314, 473

30. A.J. Groszek, *Carbon* 1987, 25, 717

31. A.J. Groszek, *Carbon* 1989, 27, 33

32. R. Denoyel, F. Rouquerol and J. Rouquerol, *J. Colloid Interface Sci.* 1990, 136, 375

33. D. Avnir and D. Farin in "The Fractal Approach to Heterogeneous Chemistry", (Ed. D. Avnir), John Wiley, 1989, p.271

34. D.H. Everett, M. Haynes, N. Pernicone, J. Rouquerol, N. Stanley-Wood, K.K. Unger, unpublished work

Porous Structure Throughout Active Carbon Particles

B. Buczek

INSTITUTE OF ENERGOCHEMISTRY OF COAL AND PHYSICOCHEMISTRY OF SORBENTS,
UNIVERSITY OF MINING AND METALLURGY, 30-059 CRACOW, POLAND

1   INTRODUCTION

Most active carbons have been commercially produced by
steam or carbon dioxide reaction with carbonaceous materials. The activating gas in the process is thought to
penetrate the particle as a result of diffusion accompanied by chemical reactions. The endothermic character of
reactions may lead to a temperature decrease in the reaction zone while heat flows slowly to the deepest layers
of the particle. Because of these factors, the burn-off
of the carbonaceous substance, may be a function of the
position within the particle.
     Fundamental work on the gas reactions of carbon can
be found in Walker et al's paper[1]. Some aspects of non-
-uniform porous structure within active carbons are also
presented in papers[2,3]. Nevertheless, no extensive research showing the changes in burn-off and porous structure within a particle of active carbon as a result of
the steam activation processes, is described anywhere
in the literature.

2   EXPERIMENTAL

Materials

     Investigations were carried out on selected commer-
cially manufactured active carbons. The granular active
carbons were obtained using the following processes:

A - from hard coal by carbonization and steam acti-
      vation at 1173-1227K,
R - same method of preparation as carbon A, except
      in this case from peat at 1227K,
H - wood carbon, activated first with zinc chlori-
      de and then by steam at 873K.

During the activation process the carbon particles
were kept in the batch reactors for the same lenghts of
time, i.e. they were not subjected to any residence time
distribution. In addition the active carbon particles are
characterised by a very narrow size range (1.0÷1.5,

1.5÷2.0 and 4.0÷4.5 mm respectively). This means we can
assume that the overall burn-off for all particles is a
constant quantity.

Sampling from active carbon particles

A successive method was used to remove the layers
from the carbon particles. The method is based on intensive abrasion in a spouted bed[4]. A diagram of the experiment used is shown in Figure 1.

The active carbon particles (A,R and H) were subjected separately to abrasion for a chosen length of time.
The powdered samples were taken from positions more and
more remote from the external surface of the particles.
The remaining granular core samples were also obtained
such that external layers of different thicknesses were
removed. In preliminary investigations, optimum parameters for processing the particles in the spouted bed were
determined. This enabled abrasion of the material from
the particle surface to occur without crushing them. The
shape and dimensions of the active carbon particles change, as a result of working them in a spouted bed.

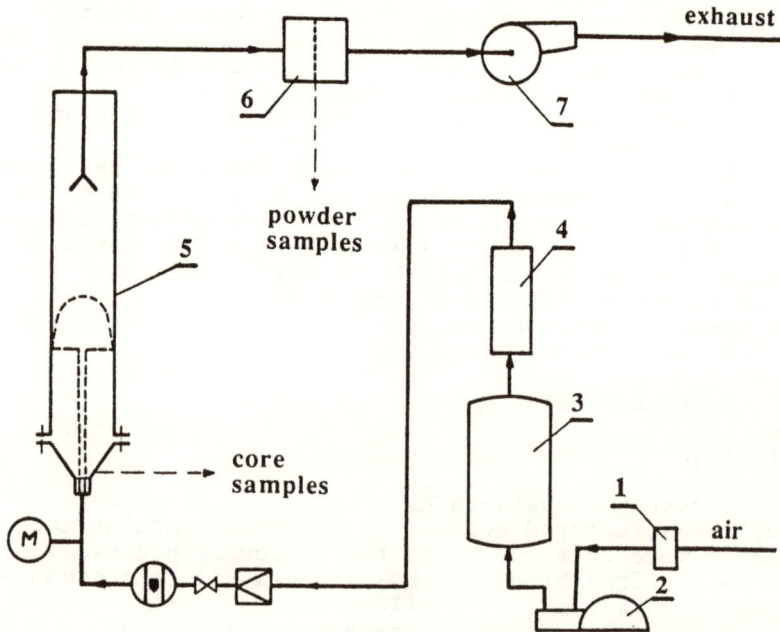

Figure 1.   Experimental equipment used for abrading the
             active carbon particles in spouted bed 1, pre-
             liminary filter;2, compressor;3, pressure ves-
             sel;4, adsorption filter;5, spouted bed column;
             6, bag filter;7, fan.

Table 1 gives details of experimental material obtained by abrading the initial active carbon particles, together with abrasion range for cores and powders. Thus the aim of the experiments was first to obtain particle samples from the three active carbons and then to analyse their properties.

Table 1. Experimental material obtained through abrasion of the active carbons particles

Carbon	Abrasion range, wt %	Form of sample	Symbol
A	0	Initial particles	●
A20	0 - 20.85		◗
A40	0 - 40.39	Cores	◑
A60	0 - 60.08		◑
A80	0 - 82.59		⊙
AP1	0 - 12.17		○
AP2	12.17 - 20.85		◒
AP3	20.85 - 31.30		⊘
AP4	31.30 - 40.39	Powders	⊕
AP5	40.39 - 50.57		⊖
AP6	50.57 - 60.08		⊖
AP7	60.08 - 82.59		⊗
R	0	Initial particles	■
R20	0 - 20.55		◧
R40	0 - 42.30		◩
R60	0 - 63.42	Cores	▮
R80	0 - 83.70		⊡
RP1	0 - 11.50		⊔
RP2	11.50 - 20.55		◪
RP3	20.55 - 32.25		◿
RP4	32.25 - 42.30	Powders	⊞
RP5	42.30 - 51.38		⊞
RP6	51.38 - 63.42		⊟
RP7	63.42 - 83.70		⊠
H	0	Initial particles	▲
H40	0 - 41.11	Cores	▲
H80	0 - 80.03		◬
HP1	0 - 41.11	Powders	△
HP2	41.11 - 80.03		◺

3 RESULTS AND DISCUSSION

Burn-off versus position inside the particles

Many experiments have shown that during the activation process, when the overall burn-off is increasing, the ash content in active carbon increases whereas its apparent density of active carbon decreases. Providing this behaviour actually occurs within the particle locally, it is possible to determine the properties of samples obtained from the A,R and H carbon particles. The results are shown in Figure 2. The properties of the samples are assigned to a specific radial position within the carbon particle, which is expressed in terms of the ratio of the radius (r) from which the sample was taken, to the initial active carbon particle radius (r_0). Both the ash content (A^a) and apparent density (d_p) of the samples indicate the burn-off decreases towards the centre of the A and R active carbon particles.

Based on the results of laboratory studies on steam activation of the carbonizate feed to produce carbon A burn-off of the active carbon samples was determined by interpolation. Burn-off was shown to range between 0.45-0.8 when the apparent density was used to calculate it. A similar result was obtained when the known content of ash was considered in the calculations.

For the R carbon samples, burn-off is difficult to determine accurately because of the lack of experimental data. Nevertheless, by way of analogy, the range of burn-off variations within these particles can be estimated to be twice as narrow as A carbon particles.

In the case of active carbon H particles, burn-off is uniform.

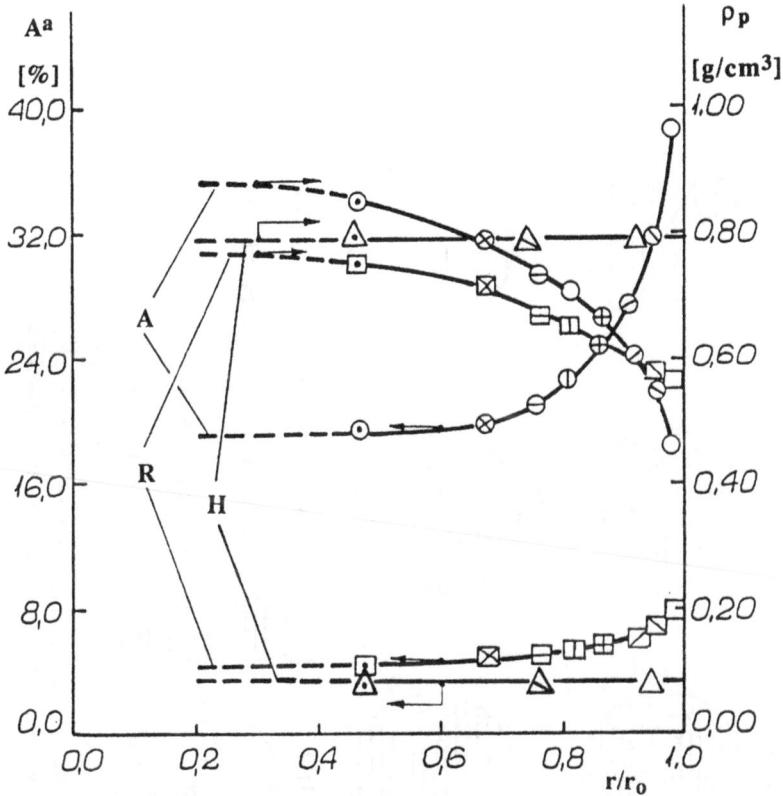

Figure 2. Ash content (A^a) and apparent density (ρ_p) variations within A and R active carbon particles.

Porous structure of carbon particles with external layers of different thicknesses removed

Active carbons and the core samples were investigated by densimetry, mercury porosimetry and adsorption. This made it possible to obtain a synthetic image of their texture within the micro- and macropores. Pore volume, accessible to helium was determined from measurements of true and apparent densities. Mercury porosimetry was used to analyse pores between 100-7500 nm. Benzene adsorption isotherm and the Kelvin equation were used to determine pore radii 1.5-100 nm.

A graphic illustration of the results using pie chart diagrams is shown in Figure 3. These charts illustrate each pore type in the total pore volume present in the carbon particles.

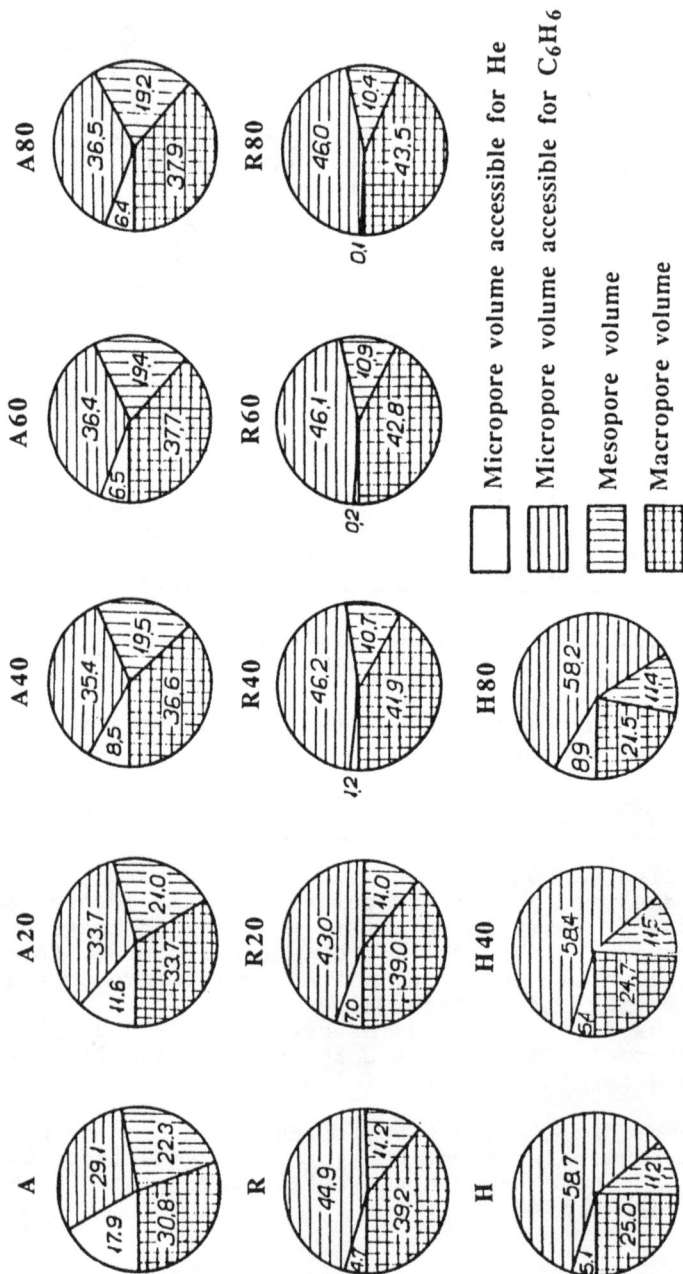

Figure 3. Pore types as percentage of the total pore volume for carbon particles with external layers of varying thickness removed

Micropore volume accessible for He

Micropore volume accessible for C_6H_6

Mesopore volume

Macropore volume

Evaluating the data for the carbon series A to A80 and R80 show that as advancement is made further into the particle:

(i) there is a decrease in the micropore percentage being the sum of micropore volume accessible to benzene and helium;

(ii) the fraction of benzene-accessible micropores in the core samples with 40, 60 and 80 percent of material removed is similar, whereas in the other series of carbons it is somewhat lower;

(iii) in both A and R carbon series, there is a fall in mesopore volume and an opposite change in the macropore percentage.

Such porous structure can be understood by considering; the effect of mass transfer resistances to steam//carbon reaction, the different volume increases of each pore type which depends on the actual reactivity of the carbon, and the reaction position within the particle.

For the carbon series H to H80 only slight changes of texture were observable.

4 CONCLUSIONS

It has been shown that there exists a correlation between the properties of samples obtained via an abrasion technique, to the position within the active carbon particles. Anisotropy of porous structure is due to burn-off of the carbonaceous substance which reduces radially from the outer surface of the particle to its inner core. The drop of burn-off, in turn, is the result of mass transfer resistance caused by the high temperature activation process.

It is possible to separate experimentally from active carbon particles, fragments which have physical and chemical properties dependent on the position within the particle. Removing outer layers which have undergone a high burn-off, leads to a better quantity of active carbon particles with increased adsorption capacity and mechanical strength.[5]

REFERENCES

1. P.L.Walker Jr., F.Rusinko Jr. and L.G.Austin, Advances in Catalysis and Related Subjects 1959, 11, 133
2. H.Jankowska, A.Światkowski, Z.Witkiewicz and A.Jarmoluk, Biul. Wojsk. Akad. Tech. 1977, 26, 103
3. A.Korta, Koks, Smola, Gaz, 1982, 27,78
4. B.Buczek, Powder Technology, 1983, 35, 113
5. B.Buczek and L.Czepirski, Adsorption Science & Technology, 1987, 4, 217.

British Standards for Particle, Surface and Pore Characterisation

R. W. Lines

CHAIRMAN OF BSI COMMITTEE GME/29/4, C/O COULTER ELECTRONICS LTD.
NORTHWELL DRIVE, LUTON, BEDS. LU3 3RH, UK

1 ABSTRACT

British Standards committee GME/29/4 is concerned with "Particle sizing methods other than sieving". Its work covers not only particle size distribution analysis but also methods of estimation of surface area and pore size distribution, as well as a glossary of terms relevant to those subjects.

The current British Standards in the BS 3406 series (particle size analysis) and BS 4359 series (surface area) are reviewed, and a progress report is given on the status of several new and revised Standards which are in preparation.

2 INTRODUCTION

British Standards committee GME/29/4 (formerly CPE/10/4) is charged with "Sizing by Methods other than Sieving". Test sieves and test sieving are covered by committees GME/29/1 and GME/29/2, under the chairmanship of Professor J.W. Mullin, and two recent Standards from those Committees have been

BS 410 : 1986 : Specification for test sieves.

BS 1796 : **Part 1** : 1989 (Identical with ISO 2591-1: 1988) : Test sieving, Part 1, Methods using test sieves of woven wire cloth and perforated metal plate.

Others are in the course of preparation.

GME/29/4 has considered Standards in the series :

BS 3406 : Methods for the Determination of Particle Size Distribution.

BS 4359 : Methods for the Determination of Specific Surface Area of Powders.

BS 2955 : Glossary of Terms,

and has commenced a new series concerned with porosity and pore size distribution. Former chairmen have been Dr. T. Allen and Professor B. Scarlett.

3 BS 3406 : METHODS FOR THE DETERMINATION OF PARTICLE SIZE DISTRIBUTION.

Table 1 shows the BS 3406 series.

Table 1   Methods for the Determination of Particle Size Distribution

BS 3406 : Part 1 : 1986 : Guide to powder sampling.

BS 3406 : Part 2 : 1984 : Recommendations for gravitational liquid sedimentation methods for powders and suspensions.

BS 3406 : Part 3 : 1963 (reconfirmed 1983) : Air elutriation methods.

BS 3406 : Part 4 : 1963 (reconfirmed 1985) : Optical microscope method.

BS 3406 : Part 5 : 1983 : Recommendations for electrical sensing zone method (the Coulter principle).

BS 3406 : Part 6 : 1985 : Recommendations for centrifugal liquid sedimentation methods for powders and suspensions.

BS 3406 : Part 7 : 1988 : Recommendations for single particle light interaction methods.

BS 3406 : Part 1

This is a practical guide to taking representative subdivisions of the sample delivered to the laboratory, but it also includes guidelines on procedures for obtaining the gross and laboratory samples. It recommends and describes scoop sampling, chute riffling, rotary sampling and cone and quartering.

Methods for preparing small (<0.5cm^3) test portions are included, for instance by suspension sampling, as well as comments on the preliminary examination of the sample and precautions to be taken. Clauses on the principles of sampling, sampling strategy, and the selection of sampling methods have been added, and comparative sampling efficiency data are presented.

BS 3406 : Part 2

This Part is concerned with the gravitational liquid sedimentation methods. The fixed position pipette incremental (Andreasen) method is treated as the prime standard technique since it gives the mass distribution directly, however the flexibility of the variable position pipette may be preferred for some routine applications. Sedimentation balance designs are included, as are those designs which monitor the mass change during sedimentation by means of x-rays or incandescent light. The use of x-rays gives directly the change of mass of material (for materials which absorb x-rays; i.e. those with atomic number above 13), whereas with light a complex correction for the extinction coefficient is required, or a value of unity arbitrarily accepted in order to allow the reporting of comparative results only.

The Standard gives a list of suggested dispersion and suspension fluids. It is recommended that gravitational sedimentation is not used below 1µm; for the low- and sub-micrometre size range centrifugal techniques should be used (see BS 3406 : Part 6).

BS 3406 : Part 3

This part covers air elutriation methods, such as the Gonell, the Roller (modified ASTM), and the miniature elutriation methods. As these are rarely, if ever, used for particle size distribution determination nowadays, it is likely that this Standard will be withdrawn.

BS 3406 : Part 4

This is still currently the 1963 Standard, the "Optical microscope method". Work on updating it re-commenced in 1987, and it is undergoing a major revision; not that it was originally in any way flawed, but the use of microscopes fitted with draw tubes (as specified) is no longer common, and automatic image analysis has become very important. Full attention is being given to the use of automatic and semi-automatic image analysis methods as well as to the means of generating the image, whether by light microscopy or by scanning or transmission electron microscopy. The Draft is expected to be submitted for Public Comment during 1991.

BS 3406 : Part 5

Introduced in 1983, this Part gives recommendations for the electrical sensing zone method (the Coulter principle). It is recommended that the primary calibration technique is that of "mass integration", where a known volume of particles under test is used to calibrate the volumetric size response directly. This allows the method to be self-calibrating and to approach being absolute. This calibration method accounts for any potential errors in instrument response due to particle shape, porosity or conductivity. A secondary method of calibration, by using latex spheres which have been measured by another technique, is given as an acceptable alternative. Whichever method is used, it should be stated on the report form.

This Standard supplies a simple response theory in order to give a basic understanding but it is recognised that more complex theory exists, as results can be produced by some designs of instrument more linearly than simple theory predicts. The Standard also gives a suggested list of electrolyte solutions for some two hundred of the more commonly found powdered materials.

BS 3406 : Part 6

This Part covers centrifugal sedimentation techniques, suitable for the size range of some 5 or 10μm down to about 0.05μm. Potential errors due to concentration effects are noted, and the user is recommended not to exceed a concentration of 0.2% by volume, and also to verify the results by repeating the analysis at half of that concentration. The methods recommended and detailed are the x-ray centrifuge, the photo-centrifuge, the fixed pipette centrifuge and the decantation centrifuge.

The same list of dispersing and suspending liquids is given as are in BS 3406 : Part 2.

BS 3406 : Part 7

Part 7 describes recommended methods for the use of single particle light interaction methods, such as those which are widely used for particle contamination monitoring, in liquids and in gases. This Standard recommends calibration by the use of latex particles; the now suspect AC Fine Test Dust method was deliberately not included. This decision was in line with other Standards, such as BS 5540: Part 6 "Evaluating Particulate Contamination of Hydraulic Fluids : Method of Calibrating Liquid Automatic Particle-Count Instruments" (now near to final draft).

BS 3406 : Future Developments

A few years ago, a proposal was made for a new Standard on Recommended Methods for the multiple particle light interaction (laser diffraction) technique. It was felt at that time, not only by GME/29/4 but also by the parent committee GME/29, that the state of the art in diffraction methods was not well enough advanced to allow standardisation, and so to ensure that similar results could be obtained from different marques and models of these analysers. This is partly because different manufacturers use different algorithms to solve the complex inversion matrix. Standardisation in this area is certainly required; the question revolves around how best to achieve it. The matter is being kept under review.

More recently, the same comments were applied to photon correlation spectroscopy, and again these problems will have to be tackled if industry requires such a Standard.

4 BS 4359 :   DETERMINATION OF THE SPECIFIC SURFACE AREA OF POWDERS.

Table 2 shows the BS 4359 series.

Table 2   Determination of the Specific Surface Area of Powders

BS 4359 : Part 1 : 1984 : Recommendation for gas adsorption (BET) methods.

BS 4359 : Part 2 : 1982 (reconfirmed 1987) : Recommended air permeability methods.

BS 4359 : [Part 3 : 1970 : Calculation from the particle size distribution - WITHDRAWN].

BS 4359 : Part 1

Part 1 gives recommendations for the gas adsorption (BET) methods. It concentrated rather more in two areas, the redesign of the burette of a basic apparatus, and the updating of nomenclature. It includes the continuous flow method.

The BET equation is normally used in the form

$$\frac{x}{n^{s}(1-x)} = \frac{1}{n_m^{s} C} + \frac{(C-1)x}{n_m^{s} C}$$

where x is the relative pressure, n^{s} is the number of moles of gas adsorbed, n_m^{s} is the relative number of moles of gas to form a complete mono-layer, and C is a constant.

It is normal to quote the range of applicability of the equation in interpreting experimental data as lying between x=0.05 and x=0.35 but the region of fit should be found from the linear part of the plot of

$$\frac{x}{n^{s}(1-x)} \quad \text{against x}$$

It is therefore desirable to have the maximum number of equally spaced points lying inside this region of fit if the region is to be determined reliably, and consistent values are to be obtained between various determinations. The burette recommended in this specification was designed to achieve this objective and was the result of a computer simulation by Dr. M.J. Jaycock of Loughborough University.

BS 4359 : Part 2

The recommended air permeability methods, in contrast

to the gas adsorption (BET) method, use the effective
(rather than the true) solid density for their calculations,
thereby ignoring porosity. Consequently, the specific
surface obtained from permeability measurements is properly called the "effective permeability volume specific
surface", denoted S_v. This gives a measure of the fineness
of a powder, which is what is sometimes required.

A technical change of the basic equation used to relate permeability data to specific area was made in this
revision. The original Standard was based on the Kozeny-
Carman equation and referred the user to other publications
for procedures to follow in the transitional zone between
wholly viscous flow and flow due to molecular diffusion.
(The extra component of flow in this transitional zone is
known as slip flow). The revision is based on one of the
equations that takes slip flow into account, and the Carman
-Arnell equation was chosen. For coarse powders, where
the effect of slip flow is small, or for some industrial
purposes where the value due to viscous flow alone is sufficient for quality control, that part of the Carman-Arnell
equation giving the contribution of viscous flow is identical to the Kozeny-Carman equation.

The methods in Part 2 are applicable to powders with
permeability volume specific surfaces in the range $2x10^4$
to $5x10^7 m^2/m^3$, ($0.02-50m^2/cm^3$).

BS 4359 : Part 3

This 1970 Standard described methods for the calculation of surface area from a particle size distribution,
but it has recently been withdrawn. Although there was
nothing mathematically wrong with the methods described in
this Standard, it was felt that surface area must be measured and not simply estimated from the particle size distribution, so the methods could no longer be supported.

5 OTHER BRITISH STANDARDS IN PREPARATION

These are listed in Table 3.

BS 2955

The glossary of terms relating to powders is being
completely updated. Since the original 1958 issue, a
large number of new terms have found their way into particle technology - those concerned with automatic image analysis for example. It is expected that this revision will
be available as a Draft for Public Comment during 1991.

BS XXXX Pore Size Distribution

In 1988, GME/29/4 received authority to commence work
on a new series of Standards to be concerned with the estimation of pore size distribution.

Table 3   Other British Standards in Preparation

BS 2955 : 1958 : Glossary of Terms Relating to Powders -
                 UNDER REVISION.

               : Proposed new title : Glossary of Terms
                 Relating to Particle Technology.

BS XXXX : Recommended Methods for the Evaluation of Poros-
          ity and Pore Size Distribution -
          NEW STANDARDS, provisional titles :

        : Part 1 : Mercury porosimetry.

        : Part 2 : Gas adsorption.

        : Part 3 : Challenge test.

        : Part 4 : Liquid expulsion.

Part 1, the mercury porosimetry technique, is at the
Draft for Public Comment stage. This technique involves
filling the pores with mercury under pressure, the pressure
applied being a function of the pore entrance diameter
and the volume of mercury intruded being taken as representing the volume of the pore of that diameter. The
method is suitable for many materials with pores in the
approximate diameter range of 0.003-400µm, and especially
in the range 0.1-100µm.

Work is well advanced on **Part 2,** the gas adsorption
technique. Using this method the pores are characterised
by adsorbing a gas, such as nitrogen, at low temperature,
such as at liquid nitrogen temperature. The method is used
for the approximate pore diameter range of 0.0004-0.05µm,
and is an extension of the surface area estimation technique (BS 4359:Part 1). The Standard is drafted for the
use of nitrogen gas only, and describes the calculation of
mesopore size distribution (2-50nm) as well as the determination of micropore volume (<2nm).

Some preliminary work has commenced on **Part 3,** the
estimation of pore size by particle challenge testing.
That method involves the use of particles of known size,
e.g. latex beads, bacterial cells and/or macromolecules,
in order to quantify the effective pore size of a filter
material. It may be, however, that a recently formed CEN
committee (CEN/TC 272 'Metrology of microfiltration or
ultrafiltration membranes (Test Methods and Terminology)')
may render further B.S.I. work on Parts 3 and 4 unnecessary.

6   ACKNOWLEDGEMENTS

The author acknowledges the assistance of chairmen of
various working parties and Mr. R. Spiers, Secretary to
GME/29/4, in providing assistance in the preparation of
this report. Mr. Lines represents the Royal Society of

Chemistry on this BSI committee.

7   SOURCE OF BRITISH STANDARDS

British Standards are available from BSI Sales Department, Linford Wood, Milton Keynes, MK14 6LE.

The Comparison of Porosity Measurements Using Different Instruments and Techniques

N. G. Stanley-Wood[1], N. Osborne[1] and M. Till[2]

[1] DEPARTMENT OF CHEMICAL ENGINEERING, UNIVERSITY OF BRADFORD,
WEST YORKSHIRE BD7 1DP, UK
[2] MINISTRY OF DEFENCE, AWE(A), ALDERMASTON, READING RG7 4RP, UK

1.0 INTRODUCTION

The strength, toughness and elastic modulus together with other technological properties of porous solids are influenced by the pore volume-fraction within a porous matrix (1). A pore size distribution rather than the magnitude of the volume-fraction may be, however, a more relevant porous parameter to describe the wide range of porous products, such as catalysts, sinters, filters and compacted materials containing a porous network. Agreement between the various techniques as well as the different instruments currently available is not always possible and the evidence for agreement tends to be conflicting.

Comparison of surface areas and pore size distributions obtained from nitrogen adsorption, using the classical Kelvin model and various other cylindrical and non-cylindrical pore models (2,3) with the surface areas and pore size distributions obtained from mercury intrusion (2,4) have been reported for a wide range of materials in the pore size range 0.002 μm to 0.1 μm diameter (5-11). Some workers (4), using silica in the BET surface area range of 50-400 m^2g^{-1}, have found a linear relationship between the nitrogen adsorption surface areas with the surface areas obtained from mercury intrusion. The ratio of mercury penetration to nitrogen BET surface area was unity only at surface area values of circa 200 m^2g^{-1}. The ratio increased for lower surface area and decreased for higher surface area silicas. In a later paper by the same workers (6) the mercury intrusion surface area measured by the Rootare-Penzlow equation (4) for a series of silicas – which had pore diameters of circa 15, 30 and 60 nanometre, each one being accompanied with three pore volumes – was higher in every case than the BET surface areas. The more macroporous samples showed a closer mercury/nitrogen BET surface area agreement, but never a value of unity.

Tomanova, Zbuzek, Jerakek and Schneider (7) compared mercury and nitrogen adsorption pore size distributions, calculated from the Broekhoff and de Boer equations (14), obtained from a series of controlled pore glasses (CPG). There was always an overestimation of the average pore size irrespective of the cylindrical model used or the branch of the low temperature isotherm (adsorption or desorption) chosen. Nor did the isotherm comport to a slit-shaped pore model. All the adsorption average pore sizes were less than the manufacturers quoted CPG sizes.

Eighty percent of the cases investigated by de Wit and Scholten (8,9,10) the nitrogen capillary condensation pore sizes calculated from the Broekhoff and de Boer model were greater than those obtained by mercury intrusion. This was similar to the results of Ihm and Ruckenstein (11). It was reasoned that the smaller modal pore sizes observed with mercury intrusion were due to the penetration of mercury into a comprehensive network of pores which was influenced by the narrowest entry point into a pore volume.

A more realistic representation of porous media now acknowledged is one in which the void structure of a solid is regarded as a series of various sized inter-connected channels and not a series of non-intersecting cylindrical pores (12). In this three dimensional Pore-Throat model the progress of mercury intrusion is believed to be controlled by volumeless cylindrical throats to a network of spherical pores, whilst the size of the voids or pores control the extrusion of mercury. Since each pore is accessible through more than one throat the volume of the pore is measured by the largest entry throat size. This presupposes that larger throat sizes dominate the sequential porosimetry process with subsequent pore filling from progressively smaller throats. There is a secondary phenomena which also occurs sequentially, one in which a pore volume can not be reached or filled until mercury has been pressure intruded via a small throat. Hence the volume of such pores will be allocated to the class size of the most restricted entry. It is reasoned therefore that a shift in the mercury distribution of pore sizes will occur towards the small pore size dimensions. Any non-converging pore size network, such as a convergent-divergent channel into a pore volume, may not however produce such a sequential fill.

Nitrogen adsorption is not, however, a sequential phenomena since pores of all sizes are accessible to nitrogen vapour from the onset of adsorption. Capillary condensation occurs from small to large pores and thus has a different mechanism of pore filling to mercury intrusion. Although the final pore distributions obtained from either adsorption or mercury porosimetry may be similar in certain size ranges and networks, the process of pore filling is not directly comparable.

In current mercury porosimetry instrumentation the process of intrusion into a porous network can be achieved by two distinct methods. One relies upon an equilibrium pressure step, which waits for a series of voids/pores to be filled, the time allocated for this "pseudo-equilibrium" step being of the order of seconds. Since the nature of porous materials is not uniform the time necessary for pressure equilibrium to occur varies from pressure step to pressure step. Care must therefore be taken to ensure pressure equilibration has occurred prior to a subsequent pressure increase. A second and alternate method for mercury intrusion is one in which pores/voids are filled by the steady and continuous application of pressure to cause mercury to flow into a porous network.

Comparisons of surface areas and pore size distributions have been obtained between capillary condensation and equilibrium pressure step mercury intrusion techniques, but no direct comparison has been made between different instrumentation and various pore size characterisation techniques. The objective of this paper is therefore to measure the surface areas and pore size distributions of two commercially available porous solids using two different mercury penetration instruments. The results obtained from both instruments are then compared with those obtained from low temperature nitrogen adsorption and capillary condensation together with pore size distributions obtained by image analysis.

2.0 EXPERIMENTAL

(a) Mercury Porosimeters

Stepwise Pressurisation. A Micromeritics Pore Sizer 9310 was used in conjunction with Micromeritics software and a PC XT computer over the pressure range 1.6 psia to circa 30,000 psia. The pressure values were defined from a selectable Pressure Table within the program. The default values for mercury contact angle and mercury surface tension were 130° and 485 mNm^{-1} respectively. The stepwise pressurisation equilibrium times can be selected from zero to 30+ seconds, the default time being 10 seconds.

There is also the facility, in the dynamic mode, to apply a steady pressure increase at rates of up to a maximum of 100 psia per minute at zero equilibrium time.

Continuous Pressurisation. A Quantachrome Autoscan 33 was operated via Quantachrome software and a PC XT computer over the pressure range 20 psia to circa 33,000 psia. The rate of high pressurisation could be varied between 550 psia per minute to 3300 psia per minute. The default values for mercury contact angle and mercury surface tension were 140° and 480 mNm^{-1}. These, however, could be reset to the defaults of the stepwise pressurisation instrument (130° and 485 mNm^{-1} respectively).

The range of pore diameter that can be measured for both instrument at sub- and supra-atmospheric pressures were from circa 100 micrometres to circa 3.0 nanometres.

(b) Nitrogen Adsorption

The low temperature nitrogen adsorption isotherms were obtained at 77K gravimetrically from a C.I. electronic, Mark 2, microbalance which was housed in a pressure system with an ultimate vacuum of less than 10^{-5} mmHg. The pressure transducers were either Bell and Howell (0-1000 mb) or Transamerica Instruments (0-35 mb). An IBM XT compatable microcomputer was interfaced with the pressure transducers and controlled the microbalance which enabled continuous monitoring of the vacuum system and reproducible measurement of the adsorptive onto the chosen materials.

The size range of pore diameter from nitrogen capillary condensation, the Kelvin (radius) equation and micro-, meso- and macro-pore analysis was in the range 0.2 to 100 nanometres.

(c)   Image Analysis

Electron microscope photographs of the fractured, gold coated, porous materials were obtained from a Hitachi 520 electron- microscope at Imperial Chemicals Industries, Runcorn. Image analysis of these photographs was achieved with the use of an AMS Optomax V Image Analyser which could identify the pore and particle number and the dimensions of area, perimeter, spherical diameter and breadth amongst other particle characters.

(d)   Reference Materials

The materials chosen for this investigation were a molecular sieve and a commercially accepted reference material but not a reference standard.   These materials were both chemically and physically stable and required no special pre-preparation other than evacuation preceding measurement by mercury porosimetry or nitrogen adsorption.   Random one gram riffled samples were taken from the source materials which were subsequently sub-divided, by riffling, to a test weight from which surface areas and pore size distributions were measured by mercury penetration, nitrogen adsorption and image analysis.

Pore Volume Reference Material 820.   This was a pelleted silica-alumina catalyst, supplied by Micromeritics Inc., which at 60,000 psia had the following characteristics

Total Intrusion Volume (V)	0.4997 ± 0.0518 cm^{3}g^{-1}
Total Pore Area (A)	221.5 ± 31.3 m^{2}g^{-1}
Average Pore Diameter (4V/A)	9.0 nm ± 0.8 nm

Molecular Sieve 5A.   This was a $^1/_{16}$ inch extruded pellet supplied by BDH, Poole prepared from aluminium calcium silicate powder (particle size $0.5 - 50$ μm) which contained pores of 0.5 nanometre diameter and used commercially for its' properties of molecular sieving.

3.   RESULTS

In the characterisation of porous material the interest remains in the measurement of both surface area and the distribution of pores within the total pore volume.   The arbitiary classification of pore dimension adopted by IUPAC (15) is based on the pore width:

(a)   Micropores have widths of up to 2 nm
(b)   Mesopores have widths in the range 2-50 nm
(c)   Macropores have widths greater than 50 nm (0.05 μm).

While the technique of nitrogen adsorption at 77K is now generally accepted as the preferred technique for the routine determination of surface area and mesopore size distribution, mercury intrusion remains the technique for macropores although interpretation of the results is complicated by the network effect of sequential intrusions.

(a)   Surface Area

Nitrogen adsorption isotherms were obtained after degassing

the chosen materials at 298K for 16 hours at a vacuum pressure of less than 10^{-3} mmHg (0.13 kPa). The surface area was calculated from the linear region of the BET equation.

Mercury penetration curves were generated from the pressure-volume measurements using either a Stepwise Penetration (Micromeritics) or a Continuous Penetration (Quantachrome) technique. Corrections were made, within each of the computer programs, for hydrostatic pressure in the larger pore size range and for compressibility for the smaller pore sizes. The surface area (S_M) was calculated from the Rootare-Poenzlow equation as expressed by the computer program of Rootare and Spencer (16). Table I shows the surface area values obtained from nitrogen adsorption (S_{BET}), pore size summation (S_c) and mercury peentration (S_M). Materials 5A and 820 showed no agreement between the techniques of nitrogen adsorption and mercury penetration. Some agreement was found between the stepwise and continuous intrusion instruments. Summation of the nitrogen pore size surface area for 5A material did however show agreement with that obtained from mercury although 820 material did not. This agreement was dependent, however, upon the ultimate pore size chosen within the summation.

Table 1   Surface Areas and Pore Size Diameters

Instrument	MERCURY PENETRATION				NITROGEN ADSORPTION				
	Surface Area	Intrusion Volume	Average* Pore dia.	Median Pore dia. (volume)	B E T Surface Area	Condensed Volume	Average∮ Pore Diameter	Median Pore Diameter (Volume)	Cumulative Surface Area
	S_M	V_M	d_M^A	d_M^M	S_{BET}	V_{N_2}	$d_{N_2}^A$	$d_{N_2}^M$	S_c
	[m²g⁻¹]	[cm³g⁻¹]	[μm]	[μm]	[m²g⁻¹]	[cm³g⁻¹]	[μm]	[μm]	[m²g⁻¹]
Pore Sizer 9130	5A 13.10	.2769	.0845	.3435					
Auto Scan 33	5A 13.53	.2020	.0845	.3080	9.22	.0159	.00690	.00367	13.2
Pore Sizer 9130	820 58.62	.3306	.0226	.0631					
Auto Scan 33	820 59.41	.2465	.0166	.0218	168.3	.2252	.00535	.00142	164.0

IMAGE ANALYSIS				
Mean Spherical Diameter	Median Spherical Diameter	Mean Pore Diameter	Median Area Diameter	* Average = $4V_M/S_M$
$\bar{d}_s$	d_s^M	$\bar{a}_A$	d_A^M	∮ Average = $4V_{N_2}/S_{BET}$
[μm]	[μm]	[μm]	[μm]	
5A .592	.392	.781	.089	
820 .433	.240	.569	.034	

The emphasis of this investigation is, however, upon the comparison of stepwise with dynamic intrusion pore size measurement rather than surface area evaluation.

(b)   Pore Size Distribution

   <u>Image Analysis</u>. Pore size distribution can generally be represented by plotting a derivative of the internal pore volume with respect to the pore dimension. In the case of adsorption this is usually a radius whilst with mercury intrusion it is generally a pore diameter (d_M). All comparisons shown in Figures 1 to 6 are based on pore diameter.

   The volume of a pore within a particle seen by a two dimensional image can be represented in various ways. For irregularly shaped pores one of the more acceptable ways would be to assume unit thickness and express the volume in terms of area. Comparison of the nitrogen adsorption and mercury intrusion data and calculation of pore sizes is based, however, on cylindrical models.   The pore area, obtained from the electron microscope photographs, was taken as a means of describing the pore size distribution in materials 5A and 820 (Figures 1-6).

   <u>Nitrogen Capillary Condensation</u>. Application of the Kelvin equation generally assumes circular pores but in reality for non-circular shapes the Kelvin equation evaluates a volume-surface capillary ratio.   In view of the uncertainty of the Kelvin equation in terms of the variation observed between the adsorbed and the bulk physical properties of nitrogen, together with the inconsistency of t-curves and BET coefficients excessive refinement of the pore size distribution model has little warranty (13,17) and thus the modelless treatment was chosen (18).

   The computational model used assumes using the adsorption branch, that all pores with a radius larger than a specific value, r_k, determined from the Kelvin equation and relative pressure of the isotherm, will contain an adsorbed layer of thickness t. Pores smaller than r_k will be filled with capilllary condensed liquid. The pore size distribution (dN2) for 5A and 820 materials are presented in Figures 1 - 6.   Calculation of pore sizes from the adsorption branch of the isotherm is not usually continued below a pore diameter of 1.0 nm because of the uncertain validity of the Kelvin equation.

   4.0   DISCUSSION

(a)   Comparison of Techniques in the Two Reference Materials

   From Figures 1 - 3, which show the comparison of pore diameter distribution obtained from the image analysis of molecular sieve material 5A with the distributions measured by nitrogen capillary condensation and mercury intrusion, it can be seen, that over the larger pore diameter size range, the modal pore size valves from image analysis and mercury intrusion are similar (d_I = 0.5 and d_M = 0.3 μm respectively).

   Since it is known that molecular sieve 5A is prepared from primary particles of 0.5 to 50 μm it maybe concluded that the wide

5A Molecular Sieve

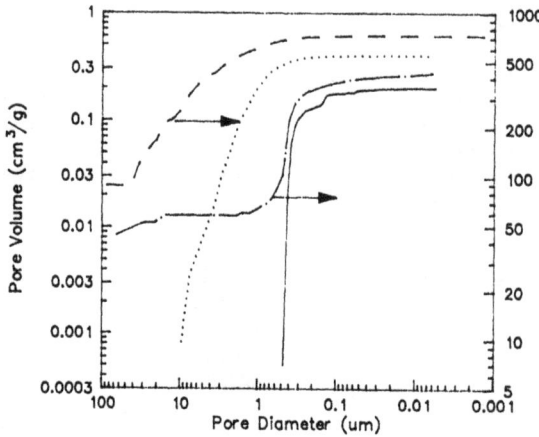

Figure 1

Total Intrusion Volume and Cumulative Number versus Pore Diameter

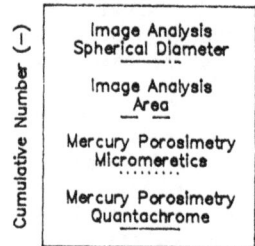

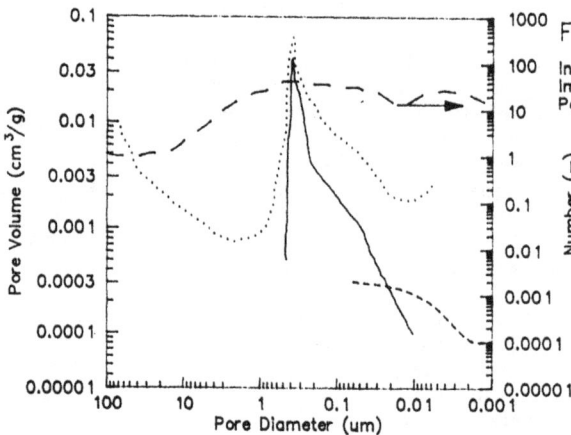

Image Analysis Spherical Diameter	
Image Analysis Area	
Mercury Porosimetry Micromeretics	
Mercury Porosimetry Quantachrome	

Figure 2

Incremental Pore Volume and Image Analysis Number versus Pore Diameter

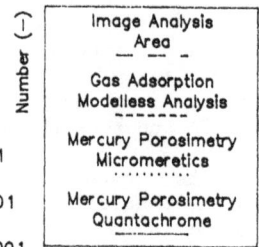

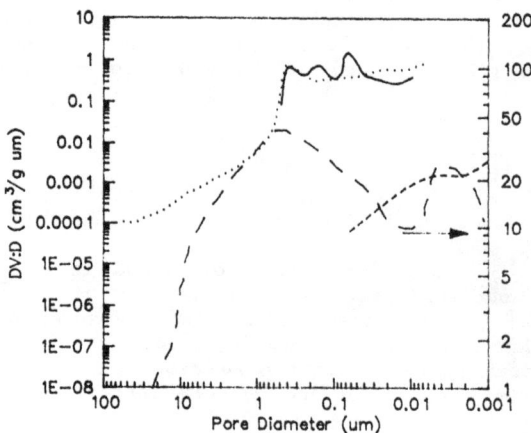

Image Analysis Area	
Gas Adsorption Modelless Analysis	
Mercury Porosimetry Micromeretics	
Mercury Porosimetry Quantachrome	

Figure 3

Differential Mean Size Distribution and Image Analysis Number versus Pore Diameter

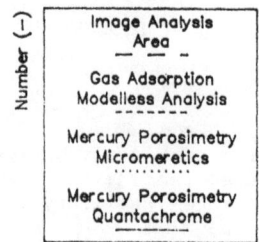

Image Analysis Area	
Gas Adsorption Modelless Analysis	
Mercury Porosimetry Micromeretics	
Mercury Porosimetry Quantachrome	

Reference Material 820

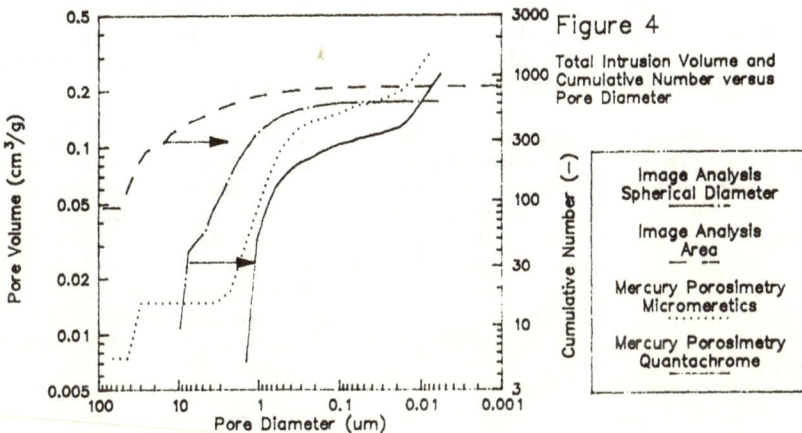

Figure 4

Total Intrusion Volume and Cumulative Number versus Pore Diameter

Image Analysis Spherical Diameter

Image Analysis Area

Mercury Porosimetry Micromeretics

Mercury Porosimetry Quantachrome

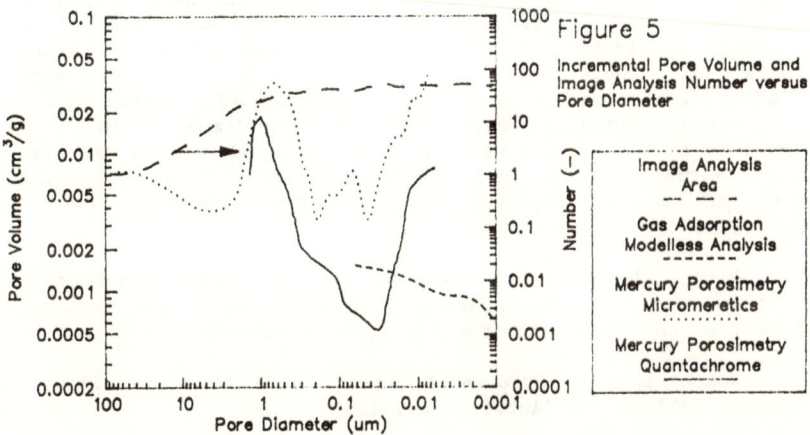

Figure 5

Incremental Pore Volume and Image Analysis Number versus Pore Diameter

Image Analysis Area

Gas Adsorption Modelless Analysis

Mercury Porosimetry Micromeretics

Mercury Porosimetry Quantachrome

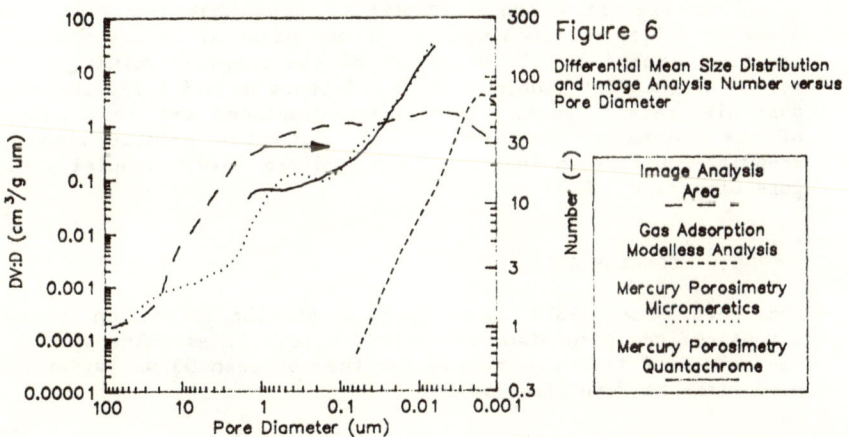

Figure 6

Differential Mean Size Distribution and Image Analysis Number versus Pore Diameter

Image Analysis Area

Gas Adsorption Modelless Analysis

Mercury Porosimetry Micromeretics

Mercury Porosimetry Quantachrome

distribution of approximately 1 μm sizes seen optically and the narrow distribution measured by intrusion are from the intergranulate spaces within the extruded material.

As the range of pore diameters decreases from 0.1 to 0.01 μm (10 nm) all three techniques show a decrease in the number or volume of pores. The nitrogen capillary condensation distribution (d_{N2}) predicting smaller pore diameters than mercury.

Below a pore size of 10 nm, in the intragranulate pore size region, the nitrogen pore diameter distribution trend is similar to that obtained by image analysis but the observed mercury sizes terminate at circa 10 nanometres. Direct comparison of the adsorption sizes below 1.0 nm (0.001 μm) with image analysis must be judged carefully because of the limitations imposed by the validity of the Kelvin equation. Both nitrogen adsorption and image analysis distributions show an increase and presence in the 0.5 nm range (Figure 3) which is the nominal size of 5A molecular sieve material.

In the case of reference material 820 mercury intrusion and image analysis of the intergranulate space size range (0.5 - 0.01 μm) was not in agreement with that observed from nitrogen adsorption (Figure 6). The modal pore size valves for nitrogen being just above 0.001 μm (1 nm) while with the mercury distribution no peck (modal) value was detected. Agreement in the increase in number and pore volume of pore diameters in the size range 50 to 8 nm occurs with both nitrogen and mercury techniques. The eventual decrease in pore volume at pore diameters less than 0.01 μm is only observed with nitrogen adsorption and image analysis.

With both reference materials the spread of the mercury size distributions for both inter- and intra-granualte pore diameter is less than that seen with the nitrogen and mercury techniques.

(b) Comparison of Stepwise-Equilibrium (Pore Sizer 9310) and Dynamic Equilibrium (Autoscan 33) Instruments

From Figures 2 and 3 it can be seen that the modal pore diameter valves obtained with reference material 5A are virtually identical except for the magnitude of the intruded volume and the spread of the distribution. With reference material 820 the modal pore diameters (Figures 5 and 6) are displaced and the magnitude of the volume of pore space once again was greater with the stepwise-equilibrium instrument (Pore Sizer 9310) especially with pore diameters < 0.01 μm.

5.0 CONCLUSIONS

The Pore Sizer 9310 gave self consistent intrusion volumes throughout the pore diameter range 3.0 to 0.01 μm. The spread of the values of intrusion volume for the Autoscan 33 was wider than seen with the Pore Sizer 9310.

Although the instruments gave different intrusion volumes the pore diameter modal valves for both chosen reference materials were similar.

Reference material 820 gave a greater degree of agreement with the 3 techniques of nitrogen capillary condensation, mercury intrusion and image analysis than molecular sieve material 5A. The molecular sieve material seemed to be restricted to intragranulate pore sizes less than 1 nanometre which is at the limit of validity for mercury intrusion and interpretation of capillary condensation by the Kelvin Equation.

REFERENCES

1. Birchall, J.D., Howard, A.J. and Kendall, K., *Nature*, 1981, 289, 388.

2. Stanley-Wood, N.G., "Enlargement and Compaction of Particulate Solids", Ed. N.G. Stanley-Wood, Butterworth & Co. Ltd., 1983, Chapter 2, p.43-119.

3. Broekhoff, J.C.P., "Physical and Chemical Aspects of Adsorbents and Catalysts", Ed. B.C. Linsen, Academic Press 1970, Chapter 1, p.1-59.

4. Rootare, H.H. and Prenzlow, C.F., *J. Phys. Chem.*, 1967, 71 2734.

5. Milburn, D.R., Adkins, B.D. and Davis, B.H., "Fundamentals of Adsorption", 1987, p.401.

6. Milburn, D.R., Adkins, B.D. and Davis, B.H., IUPAC Symposium on Characterisation of Porous Solids (COPS II), Alicante, Spain, May 1990.

7. Tomanova, D., Zbuzek, B., Jarakek, K. and Schneider, P., *Collect. Czechoslovak Chem. Commun.*, 1981, 46, 2060-2067.

8. Scholten, J.J.F., Bars, A.M. and Kiel, A.M., *J. Catalysis*, 1975, 36 pp.23-29.

9. De Wit, L.A. and Scholten, J.J.F., *J. Catalysis*, 1975, 36, 30-35.

10. De Wit, L.A. and Scholten, J.J.F., *J. Catalysis*, 1975, 36, 36-47.

11. Ihm, S.K. and Ruckenstein, E., *J. Colloid Interface Sci.*, 1977, 61, 146.

12. Conner, W.C., Cevallos-Candau, J.F. and Weist, E.L., *Langmuir*, 1986, 2, 151-154.

13. Conner, W.C., Lane, A.M., Ng, K.M. and Goldbeatt, M., *J. Catal.*, 1983, 83, 336.

14. Broekhoff, J.C.P. and de Boer, J.H., J. Catal., 1967, 9, 8,15., *J. Catal.*, 1968, 10, 158,368,377,391.

15. Sing, K.S.W., Everett, D.H., Haul, R.A.W., Moscou, L., Pierotti, R.A. Rouguerol, J. and Siemieniewska, T., *J. Pure Appl. Chem.*, 1985, 57, 603.

16. Rootare, H.M. and Spencer, J., *Powder Technol.*, 1972, 6, 17.

17. Dollimore, D. and Heal, G.R., *J. Colloid Interface Sci.*, 1973, 42, 233.

18. Brunauer, S., Mikhail, R.S. and Bodor, E.E., *J. Colloid Inter. Sci.*, 1967, 24, 451 and 1967, 25, 353.

High Resolution Analysis of Microporous Materials

M. P. Cowan[1] and R. A. Wenman[2]

[1] COULTER ELECTRONICS LTD., NORTHWELL DRIVE, LUTON, BEDS. LU3 3RH, UK
[2] COULTER CORPORATION, HIALEAH, FLORIDA, USA

1 INTRODUCTION

Microporous materials have become of increasing economic importance as new processes and products are developed. Carbons have been used for hundreds of years for the removal of odours and colours from gases and solutions. Methods for their manufacture have been developed so that by controlling their pore structure a range of properties is available. Predominantly, carbons form slit-shaped pores, ideal for the adsorption of flat molecules of carbon dioxide and of aromatic organic substances, hence their common use in vapour extraction.

The zeolite molecular sieves are constructed largely from silicon, aluminium and oxygen. They are highly porous, filled with microscopic channels and chambers. The regular structure enables sieving at predetermined molecular sizes leading to their extensive use for the separation and concentration of gases such as nitrogen and oxygen. This approach has led to the development of equipment for the production of medical oxygen, small enough to be installed in patients' homes. Economically, zeolites are important in the petrochemical industry. It has been reported that the yield of petrol (gasoline) has been improved by up to 20% since their introduction. Development of new zeolites is expected to increase the yield further by cracking larger hydrocarbons than can currently be handled. Other novel applications are constantly being found for these versatile structures. Instrumentation capable of characterising these novel materials has become available in the past few years, aiding their development.

Gas adsorption, where a gas or vapour is brought into contact with a clean solid surface, has been widely used for several decades for the characterisation of surfaces. Properties such as surface area, pore volume and pore size distribution can all be calculated from published models. The starting point for all of these models is the raw data, usually plotted as volume or mass of adsorbate versus the sample pressure or as the ratio of sample pressure to

adsorbate saturation pressure. The resultant plot is
known as the isotherm, which is the only measurement.

The catalysis, adsorption or sieving properties of
any material depend on the internal surface of the
crystal or powder. It is within the extensive channels
or chambers that substrates are brought into close
contact with active sites, or it is the physical size of
channels which blocks movement of the molecules. Pores
have been classified by the International Union of Pure
and Applied Chemistry (1) into three major groups :-

Micropores : less than 2nm (20Å) width.
Mesopores : between 2 and 50nm (20 and 500Å).
Macropores : more than 50nm (500Å).

In the microporous region the pore width is approaching molecular dimensions, for instance the effective
width of the nitrogen molecule is given as 0.364nm. It
is clear that the pore wall adsorption fields are almost
overlapping and adsorbate molecules are in close proximity
to them. This results in enhanced uptake of adsorbate at
very low relative pressures. Horvath and Kawazoe (2)
calculated that, for pores in carbons, pores filled at the
low relative pressure shown in Table 1.

Table 1   Table of Relative Pressure P/Po Against Effective
          Pore Size for Micropores.

P/Po	Effective Pore Size (nm)
1.46×10^{-7}	0.40
6.47×10^{-7}	0.43
2.39×10^{-6}	0.46
1.05×10^{-5}	0.50
1.54×10^{-4}	0.60
2.95×10^{-3}	0.80
2.22×10^{-2}	1.10
3.15×10^{-1}	3.00

The first requirement for any instrument which
measures the microporous region has to be the ability
to measure at low relative pressures. Samples need to
be prepared by evacuating to lower pressures than those
shown in Table 1 to ensure that all of the pores are
empty. The analysis must start at very low relative
pressure, and the sample holder and manifold design
must be of very high integrity to maintain high vacuum.
Sufficient data points must be taken in the low relative
pressure, micropore, region to give adequate characterisation of the isotherm. Previously available commercial
instruments have tended to use large doses of gas,
resulting in just one or two data points in the microporous region.

The COULTER® OMNISORP™ series of analysers are able
to use two of the three main methods for "volumetric"

adsorption measurement. These three main methods may
be classified as STATIC, CONTINUOUS FLOW and DYNAMIC.
Both the static and continuous flow (quasi equilibrated)
techniques may be described as vacuum volumetric methods,
whereas DYNAMIC technique (known by some workers as
continuous flow) does not use vacuum technology,but instead
employs a non-adsorbing carrier gas/adsorptive mixture.
To avoid confusion, we recommend that the term DYNAMIC
be applied to all of the chromatographic type of sorption
methods commonly used for rapid quality control analysis.

The classical static equilibrated method,as
implemented by most manufacturers, doses the adsorptive
onto the sample in individual steps each providing an
equilibrated data point. Data points are accumulated at
ever increasing pressures until the isotherm is complete.
The COULTER® OMNISORP™ instrument is able to perform
static equilibrated measurements with very high
resolution. This is accomplished with a mass flow
controller, which acts as a very sensitive proportional
valve.

The CONTINUOUS FLOW method doses adsorptive onto
the sample at a constant and pre-determined flow rate
which is low enough to enable quasi-equilibration to
occur. The flexibility of this technique is enhanced
through the use of a mass flow controller,as is used
in the OMNISORP™ instruments. Rouquérol et al (3) have
shown that no appreciable difference is found in the
isotherms when compared to static method if the correct
mass flow controller setting is used. They also describe
how this setting can be confirmed. Theoretically
resolution with this method is infinite;in practice it is
normally unnecessary to exceed 1000 data points but the
great advantage of this method is that it is fast. Analysis
times are usually significantly less than with the
static method.

The instrumentation used is not the only controlling
factor on the resolution. Venero and Chion (4) showed
that choice of adsorbate was important when they demonstrated that nitrogen was unable to distinguish subtle
changes in pore sizes for several zeolites. They concluded
that preferential adsorption on the surface was caused
by surface to quadrupole interactions. Argon, which
does not have a quadrupole, is able to show these changes,
often known as transition pressures. Hathaway and
Davis (5) have shown that adsorbent temperatures is
also important. They demonstrate that, with argon
as adsorbate, isotherm resolution is very significantly
improved by holding the adsorbent at the temperature of
liquid argon.

Some prior knowledge of the system under test is an advantage when choosing adsorptive and interpreting the results. The effective diameter of the argon molecule is 0.34nm. Clearly it is possible that with some molecular sieves, argon will be too big to enter the micropores. For example, with zeolite A it is possible to substitute the sodium ions (4A) for potassium ions (3A), and so restrict the pore to an effective diameter in the region of 0.3nm and exclude argon. At this pore size, the only practical adsorbate is water, the effective diameter of which is 0.26nm. Careful choice of adsorbate is necessary otherwise a microporous sample can appear to be non-porous.

2 EXPERIMENTAL

Analysis was carried out using an OMNISORP™ 360 CX, a four port analyser with chemisorption capability. The instrument is capable of both static and continuous flow analysis using a wide range of gases and vapours, without modification. Three types of experiment were performed :-

1) A continuous flow measurement using nitrogen as adsorptive flowing at $0.3cm^3$/m onto a sample of carbon powder held at the temperature of liquid nitrogen. Results are shown in Figures 1, 2 and 3.

2) A static measurement using 1.5 torr doses of argon as adsorptive onto a mixture of zeolites held at the temperature of liquid argon. results are shown in Figures 4 and 5.

3) A static measurement using small doses of water vapour as adsorptive onto zeolite 3A held in an ice bath. The isotherm is shown in Figure 6.

A sample tube containing adsorptive in liquid and vapour phases was maintained in the Dewar flask close to the adsorbent, to give the saturation pressure. For each data point, a measurement was taken of the saturation pressure and recorded.

3 RESULTS AND DISCUSSION

In Figure 1, the carbon powder isotherm is shown composed of approximately 1000 data points. Figure 2 is a linear expansion of the low relative pressures up to P/Po of 0.0005. The plot shows several hundred data points effectively characterising the microporous region. It

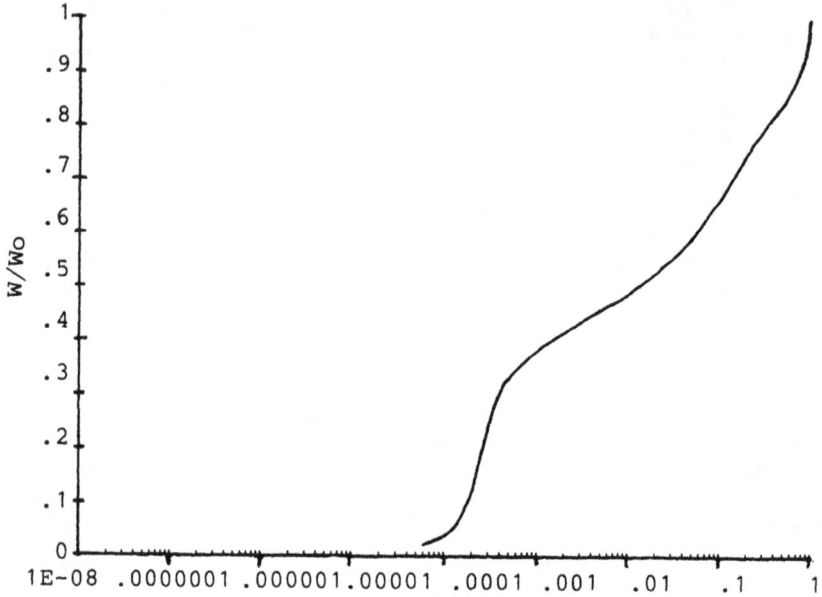

Figure 1. Nitrogen adsorption isotherm for a carbon sample at 77K. The volume of adsorptive is plotted as a fraction of the total adsorbed at saturation, against the log relative pressure.

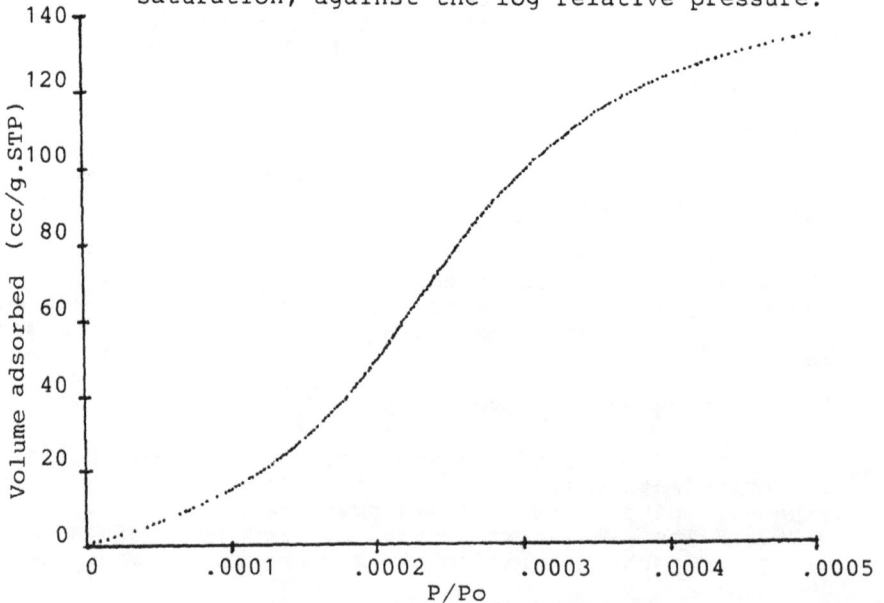

Figure 2. Linear expansion of the low relative pressure region of Figure 1 to show the high resolution of the isotherm in the region of micropore filling. Each dot is a data point.

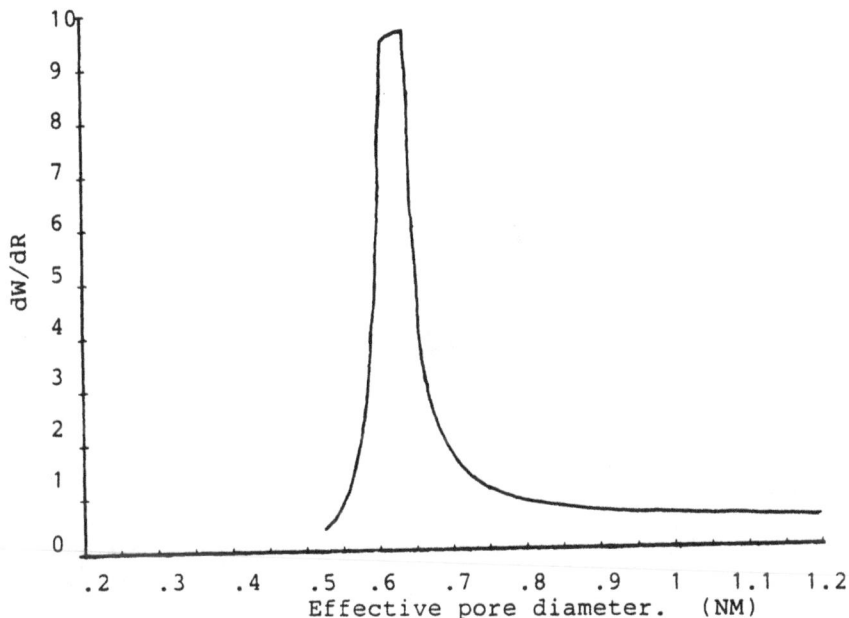

Figure 3. Micropore volume plot for the carbon sample
           using the Horvath and Kawazoe method. The volume
           of gas adsorbed as a fraction of the total volume
           plotted in the microporous region has been plotted
           against the effective pore diameter.

is clear from this plot that the region of pore filling is
between P/Po values of 0.0002 and 0.0003 which, by cross
reference to Table 1, gives a pore diameter of just greater
than 0.6nm. The Horvath and Kawazoe interpretation of the
data reports a peak in the pore size distribution of 0.625nm
(Figure 3).

        For some microporous adsorbents where the pore size is
approaching molecular dimensions, equilibration times can
be extensive and continuous flow methods are not appropriate.
Figure 4 shows the adsorption isotherm for a zeolite
mixture using small doses of argon and true equilibration
using a static method. Again a large number of data points
have been taken at low relative pressures. The expanded
Horvath and Kawazoe plot (Figure 5) separates two pore
structures of similar size from the smaller of the two
components of the mixture (the larger component is not
reported).

        Figure 6 shows a uniquely detailed water isotherm for
zeolite 3A, the potassium substituted zeolite A. It is
not possible to characterise this material with either argon
or nitrogen because both are too large to enter the microporous structure. In the example shown, approximately
fifty data points characterise the microporous region.
However, to our knowledge, satisfactory models have not been
developed to interpret the data.

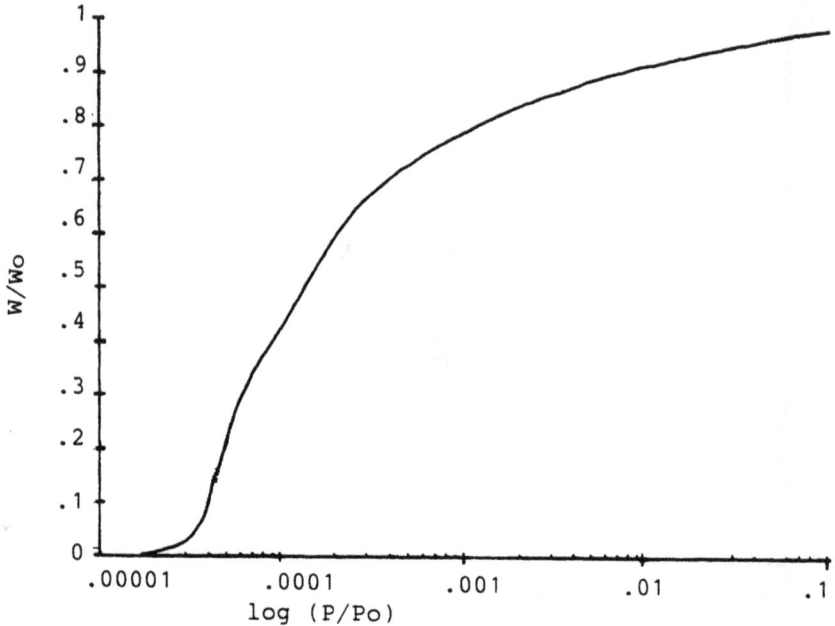

Figure 4. Argon adsorption isotherm for a mixture of
zeolites at 87K. The volume of adsorptive is plotted
as a fraction of the total adsorbed at saturation,
against the log relative pressure.

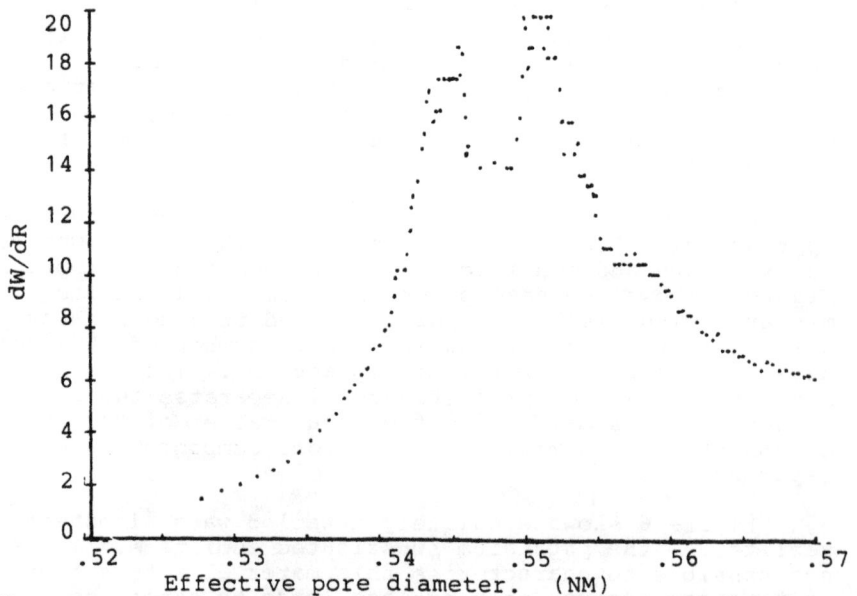

Figure 5. Linear expansion of the Horvath and Kawazoe
plot for the mixture of zeolites. Two features of
the mixture have been separated by the high resolution
analysis.

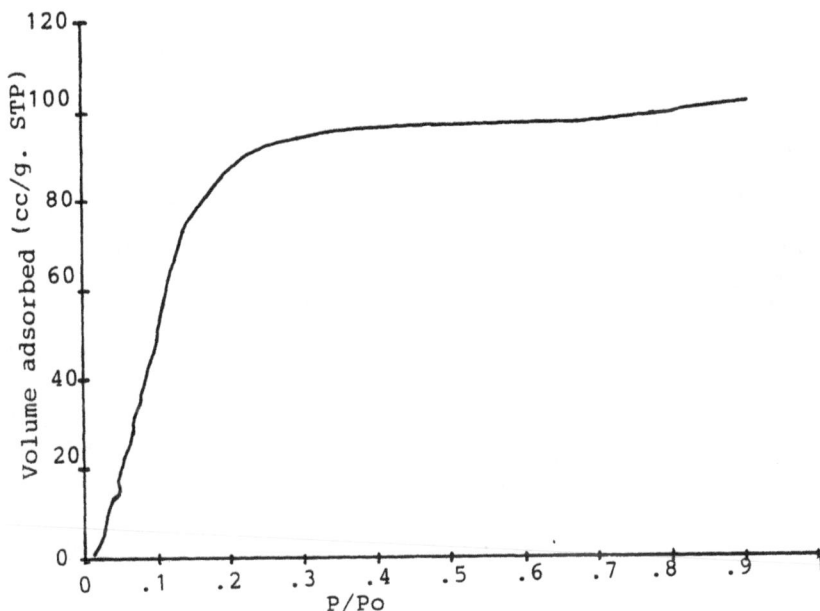

Figure 6. Water vapour adsorption isotherm for zeolite 3A.
The volume adsorbed has been plotted against the
relative pressure.

4   CONCLUSION

In this paper we have proposed definitions for the three
main methods of gas adsorption in an attempt to clarify
the confusion evident from published literature. We have
also shown that, for microporous analysis, careful selection
of analysis conditions is necessary to achieve meaningful
results. Using commercially available instrumentation,
we have demonstrated uniquely detailed isotherms of
zeolite materials using both argon and water as probe
molecules. Commercial instruments that can measure and
record a large number of data points at very low relative
pressure will play an important role in the future
development of gas adsorption characterisation.

REFERENCES

1. K.S.W. Sing, D.H. Everett, R.A.W. Haul, L. Moscou, R.A. Pierotti, J. Rouquérol and T. Siemieniewska, Pure Appl. Chem., 1985, 57, 603.
2. G. Horvath and K. Kawazoe, J. Chemical Engineering Jap., 1983, 16, 470.
3. J. Rouquérol, F. Rouquérol, Y. Grillet and R.J. Ward, 'Characterisation of Porous Solids', Elsevier, The Netherlands, 1988, 67.
4. A.F. Venero and J.N. Chion, Mat. Res. Society Symp. Prod., 1988, 111,
5. P.E. Hathaway and M.E. Davis, Catalysis Letters, 1990, 5, 333.

Comparisons of the Measured Results of Particle Size by Several Kinds of Measuring Instruments (Report of Japanese Working Parties)

G. Jimbo[1], J. Tsubaki[2] and H. Yamamoto[3]

[1] NAGOYA UNIVERSITY, NAGOYA 464-01, JAPAN
[2] JAPAN FINE CERAMICS CENTER, NAGOYA 456, JAPAN
[3] SOKA UNIVERSITY, HACHIOJI 192, JAPAN

1 INTRODUCTION

Recent development of particle size analyzers, both in their functions and in variety of types, is still proceeding remarkably and steadily. New principles and machines are still being introduced into this essential basic technique of powder technology and then the number of types of analyzers has increased largely based on several kinds of different principles.

Accordingly it is not surprising that there have always been problems of poor coincidence among the data obtained by different methods and also by different operators. Therefore since more than two decades ago, there have been many projects to compare the data by different methods and operators for the purpose of finding out the causes of errors and scattering of data in size measurements.[1-4]

Also in Japan, there have been four working groups on size measurement for general powders, namely powders not specified into some industrial fields, since early 1960s. This paper intends to introduce the results obtained by these groups and to indicate the present state of size measurement technique and main points to be considered to improve it.

2 WORKING GROUP ACTIVITIES IN SIZE MEASUREMENT IN JAPAN

The working groups on size measurement in Japan are shown in Table 1, in which three have been organized by the Society of Powder Technology, Japan (SPT), and the last one is now just working. The authors of this paper are all the organizers of each of these three groups.

Table 1 Working Groups on size measurement in Japan

Period	Organization	Organizer	Materials	Number of types of measurement	Number of members	Publication
1961-1963	1st Working Group on Size Measurement (Soc. Powder Techn., Japan)		Abrasive fused alumina, $CaCo_3$, Clay, Fine dust, Glass beads, Silicate sand, TiO_2, ZnO	14	30	ref.(5)
1983-1987	2nd Working Group on Size Measurement (Soc. Powder Techn., Japan)	G. Jimbo (Nagoya Univ.)	Mica, Limestone, Abrasive fused alumina (I, II), $CaCO_3$, $\alpha-Fe_2O_3$	30	32 groups	ref.(6)
1989-1991	JFCC Multiclient Project on Particle Size Analysis (Japan Fine Ceramics Center)	J. Tsubaki (JFCC)	Raw materials of advanced ceramics	19	46(groups and individuals)	data are not open
1989-	Working Group on Sub-micron Size Measurement (Soc. Powder Techn., Japan)	H. Yamamoto (Tokyo Univ. →Soka Univ.)	Mainly about sub-micron size powder	30 (10 principles)	55(groups and individuals)	on working

Here in this paper only the summarized results by the 2nd Working Party of SPT are shown. But a couple of things related to other groups also must be written. The results by these group activities affected the following development of sizers. In the results by the 1st working party, it was clearly shown that the data of air-phase measurements both by sedimentation and classification deviated and scattered largely. After that air-phase measurement almost disappeared except air-jet sieving, until recently dry laser method developed. Also these results affected the makers' activities, and many machines were modified and improved by the suggestions made based on these results. Comparing the results by the 2nd Working Group and JFCC project, we easily found that after the former's activities and even during the latter's, many important improvements, perhaps mainly in software, have been made by the makers, and then the results improved remarkably.

In such a sense, the results shown in the following chapter are already old. The sizes used (Table 2) are also rather old. If we measure the same samples with the same types of instruments, we can obtain better results with good coincidence and good reproducibility. But at the same time it can be said from the results by JFCC project, which are not open, that the summarized conclusions from the results by the 2nd Working Party shown in chapter 4 are valid in principle, though some other factors, which are important in sub-micron size range and also for synthesized ceramics raw materials, must be taken into account.

3   SUMMARIZED RESULTS IN LARGER THAN 1 μm SIZE RANGE MEASUREMENT

Summarized results of the measurements by the 2nd Working Group of the Society of Powder Technology, Japan, are shown in Figures 1 to 8. All results are shown with 50% average size with the width of distribution between cumulative 20% and 80% size. The alphabetical characters denote the principles of measuring instruments used and the numbers I to IV show the type of sizers.

Mica (Figure 1) is selected from its good dispersability and special disk-like shape.

Three results with alumina powders (Figures 2, 3 and 4) show the effect of the pattern of size distribution. First measurement of limestone powder (Kansuistone) showed very poor results (Figure 5), and then repeat measurement was made by standardizing the operating conditions, especially the preparation of dispersant, but the second results did not show sufficient improvement (Figure 6). Ferrite powder measured was smaller than 1 μm, and so this was out of this working

Table 2   Instruments used

Principle		Instrument
Laser diffraction scattering		Microtrac SRA/SPA, Malvern (modified by Nakayama, Gunma Pref. Univ.) CILAS Granulometer 715, SK-Laser Micronsizer PRO-7000 (Seishin)
Electrical sensing zone method		Coulter Counter
Sieving		Microsieve, Air jet sieve Sonic sieve etc.
Sedimen- tation	Pipette	Andreasen pipette
	Sedimentation balance	SA-2 (Shimadzu), Sedimenputer SPT-G (Hosokawa), Self-made (Kousaka, Osaka Pref. Univ.)
	Number counting	Self-made (Kousaka, Osaka Pref. Univ.)
	Specific gravity	Hydrometer, RS-1000 (Shimadzu)
	Unbalance method	Sedimenputer SPT-C (Hosokawa)
	Photo extinction	CAPA-500 (Horiba), CP-50, SA-CP2, SA-CP3 (Shimadzu), Micron Photosizer-SK, SKN-1000, SKA-5000, SKC-2000 (Seishin)
	X-ray attenuation	Sedigraph 5000-OD, 5000D, 5000-01, 5000-02 (Shimadzu)
Image analysis		Luzex 500, Microscope (manual)
Inertia		Cascade impactor

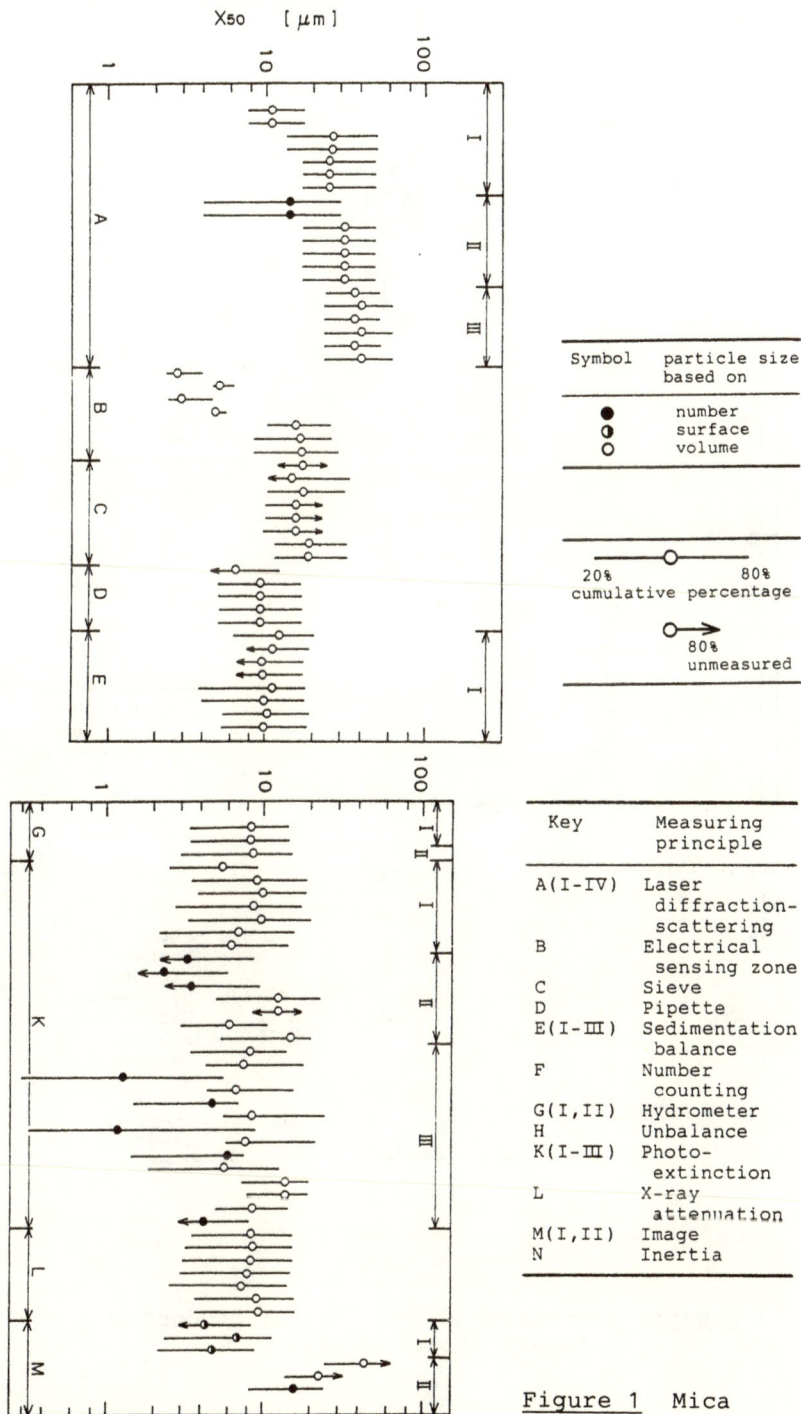

Figure 1 Mica

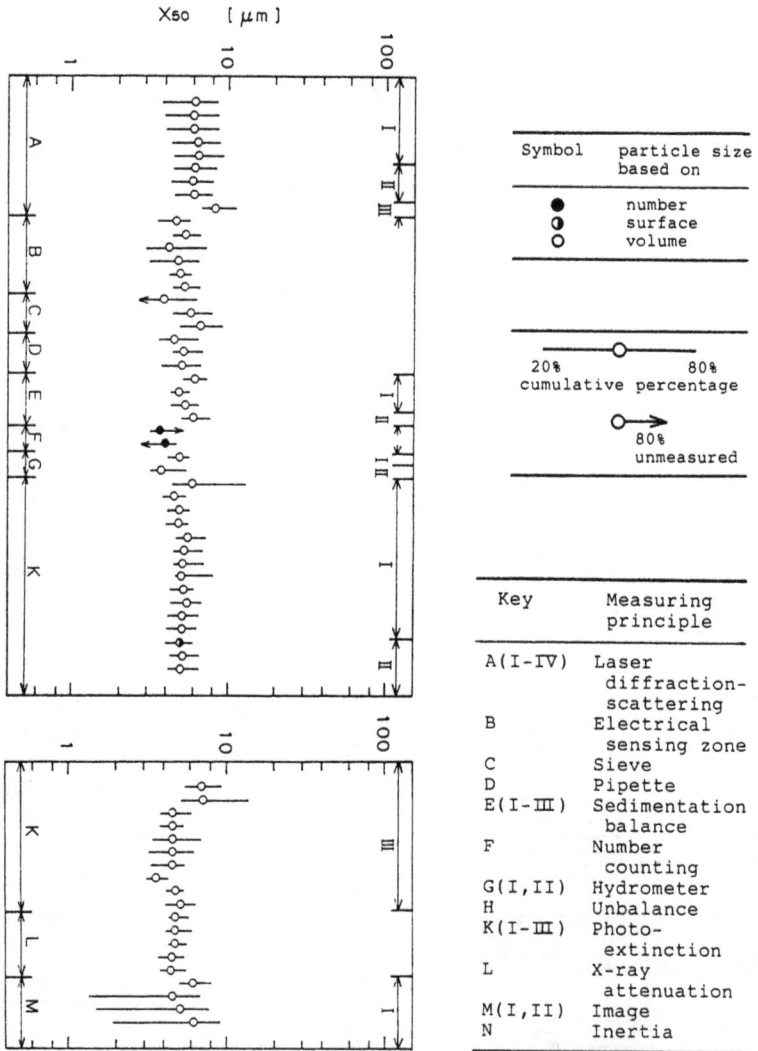

Figure 2   Abrasive fused alumina powder
(sharp distribution)

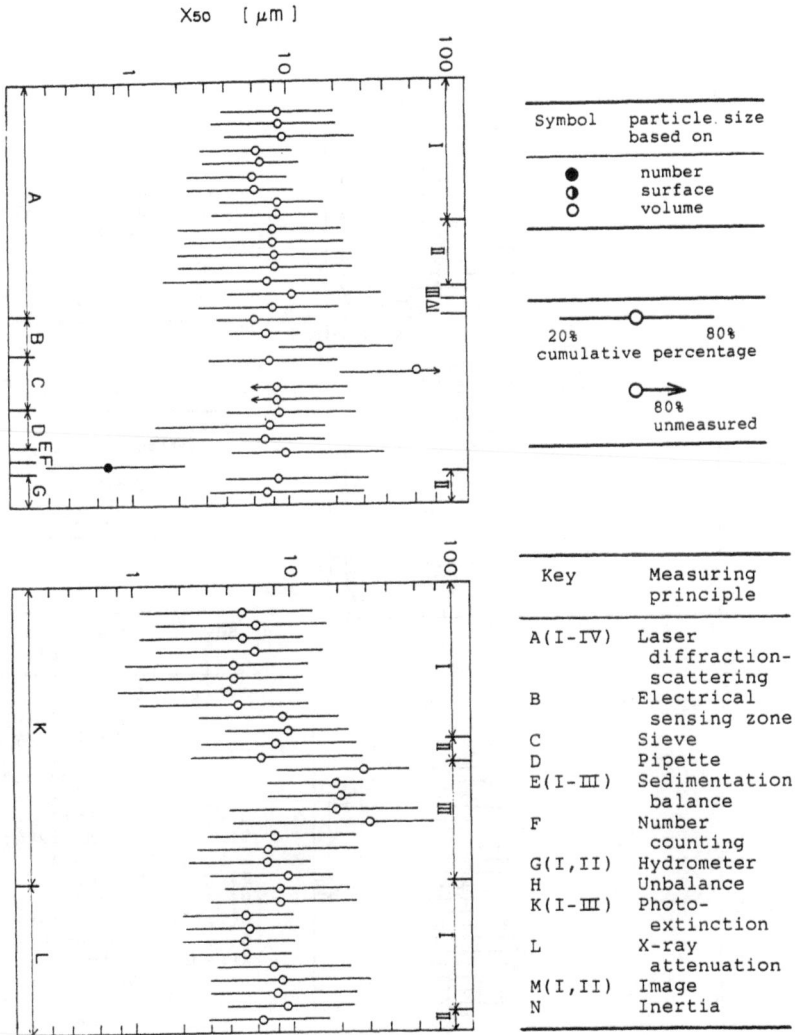

Figure 3   Abrasive fused alumina powder
            (wide distribution)

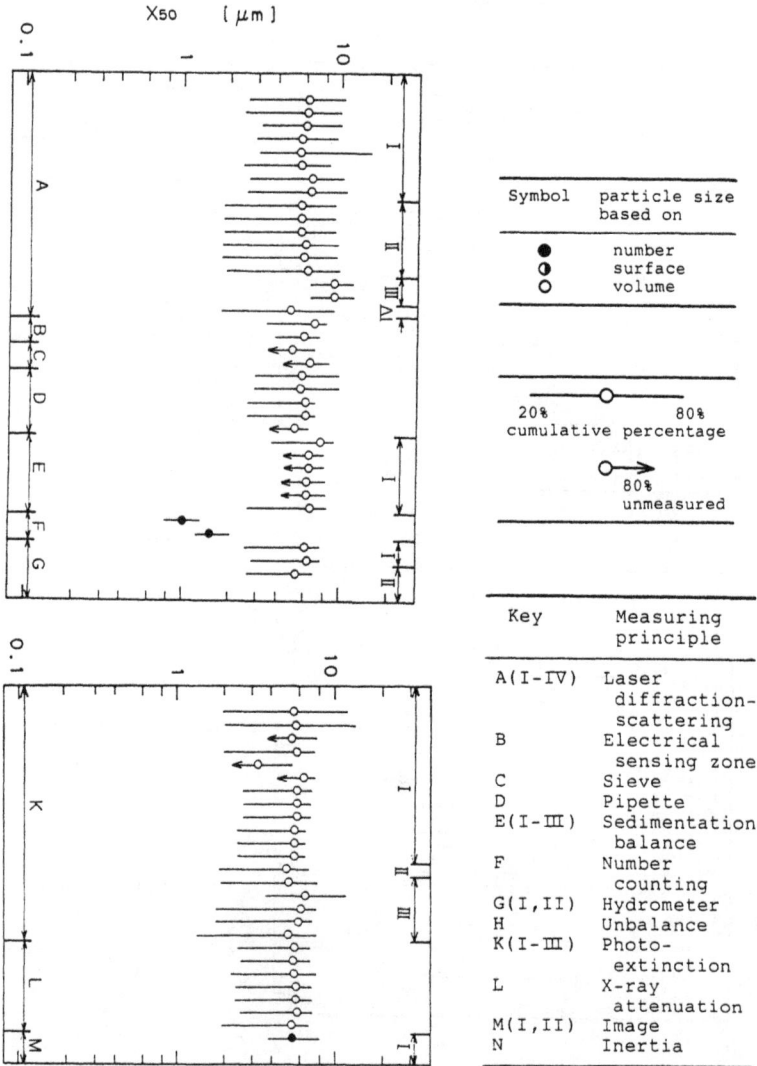

Figure 4   Abrasive fused alumina powder
(bi-modal distribution)

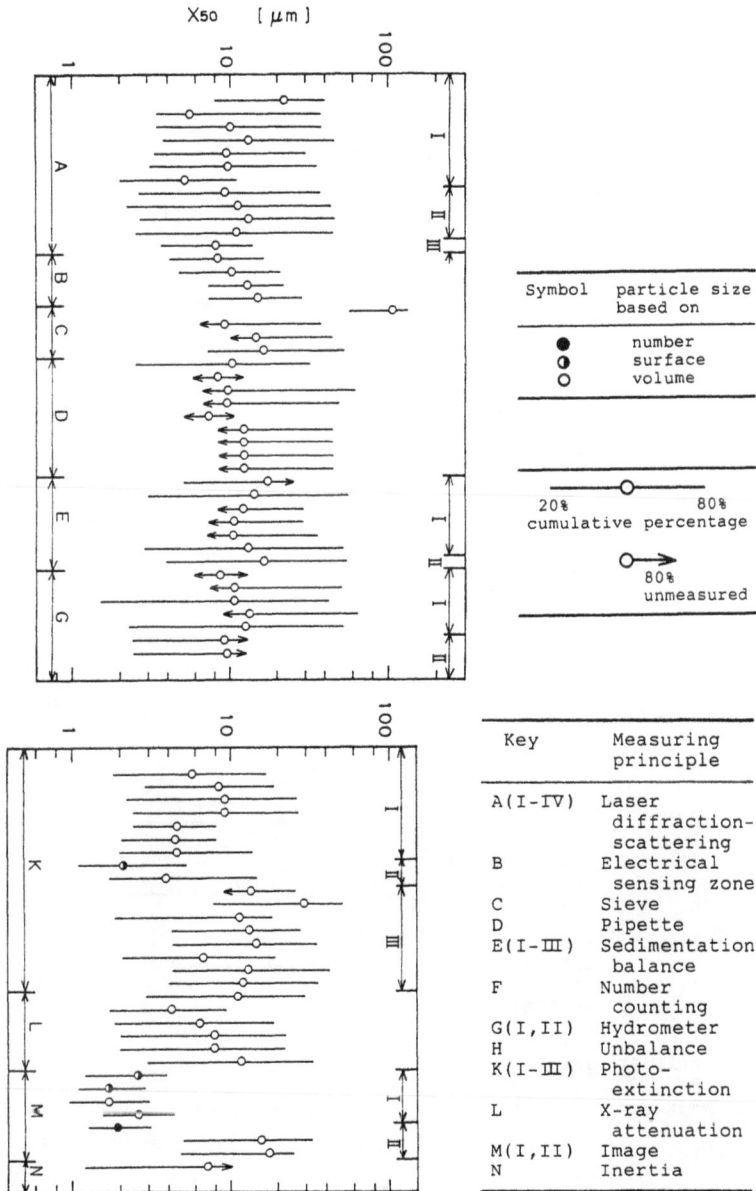

Figure 5 Kansui-stone (Limestone powder)

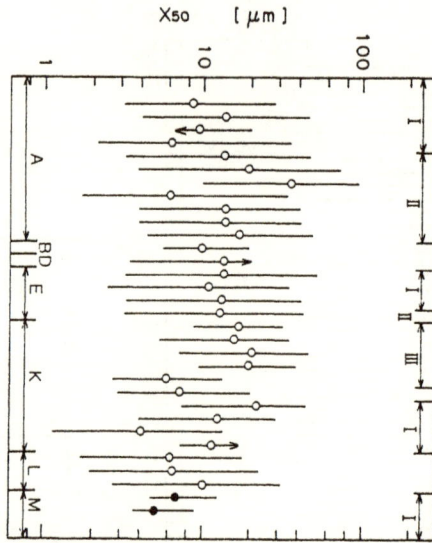

Key	Measuring principle
A(I-IV)	Laser diffraction-scattering
B	Electrical sensing zone
C	Sieve
D	Pipette
E(I-III)	Sedimentation balance
F	Number counting
G(I,II)	Hydrometer
H	Unbalance
K(I-III)	Photo-extinction
L	X-ray attenuation
M(I,II)	Image
N	Inertia

Symbol	particle size based on
●	number
◉	surface
○	volume

Figure 6   Kansui-stone (Limestone powder), repeat

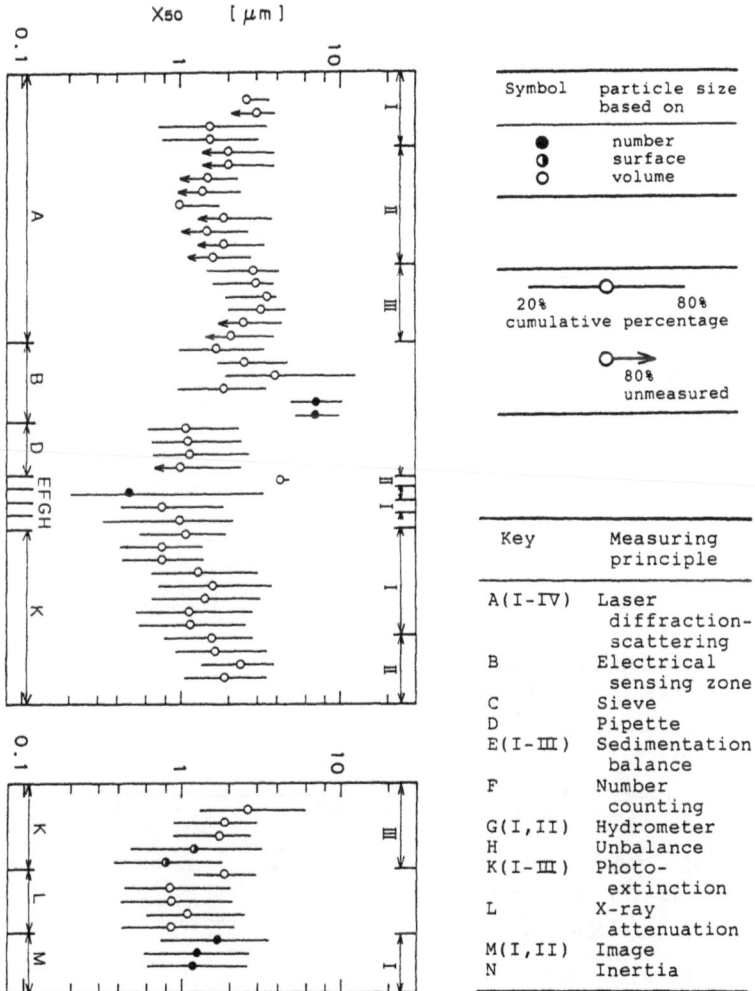

Figure 7   Calcium carbonate powder

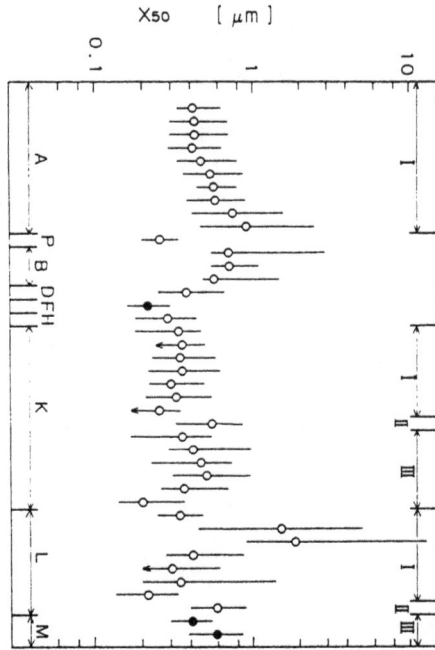

Figure 8   Ferrite powder (α -Fe$_2$O$_3$)

group's activity (Figure 8).

4 CONCLUSIONS

From the results of average size as shown in Figures 1 to 8, and also from size distribution charts, together with some results by other working parties, the following conclusions can be summarized.

(1) When sample powder particles can be dispersed well and their shapes are bulky, the coincidence between different methods and reproducibility are generally good enough. They can be well reliable.

(2) The effect of particle shape is larger than first thought, especially between instruments based on different principles. The difference between laser diffractometry and sedimentation methods is a good example. But the correction factor will be obtained.

(3) Wide size distribution causes poor results, even though dispersion and shape of the particles are good enough. Especially coarser size fraction is very dangerous. Sometimes pre-classification may be necessary.

(4) Dispersion of powder sample into liquid is very important. Ultra-sonic dispersion with chemical dispersant is essential, though the quantitative effect has not yet been found.

(5) The amount of sample is important. In many cases, too small a number of particles to be measured caused poor results, especially when powders are easily segregated. Especially cohesive powder and wide size distribution are dangerous.

(6) Powders with difficult characters, such as cohesiveness, poor dispersability and wide size distribution, almost always cause bad results. Standardizing of measurement conditions will be necessary. Too short measuring time must be avoided.

(7) Poor separation of bi-modal size distribution is seen in some cases. This function will become more important in the future.

REFERENCES

1. ASTM 'Fineness of Cement'(Symposium at San Francisco in 1968), Special Technical Publication, 1970, 473
2. Batel and Wilhelm, Einfuhrung die Korngrossenmesstechnik (The Society for Analytical Chemistry,

London 1968) (Fachausschusses fur Staubtechnik in VDI), Springer-Verlag, 1971

3. B. Koglin, W. Alex u. H. Jodicke, First European Symposium on Particle Size Measurement (Nurnberg, Sept. 1975) Preprints, 1975, 377
4. T. Allen and R. Davies, 4. European Symposium Particle Characterization, Nurnberg, Apr. 1989, Preprint Part 1, 1989, 17
5. Society of Powder Technology ed., 'Ryudo sokutei gijutsu' (Size measurement techniques), Yokando, 1965 and Nikkan Kogyo, 1975
6. Text of Seminar on Particle Size Measurement, Comprehensive Comparison and Investigation of Recently Developed Size Measurement Method, Soc. Powder Techn., Japan, July 1987 in Tokyo

The Preparation and Characterization of Particle Shape Standards

I. A. Marshall and J. P. Mitchell

CHEMICAL PHYSICS DEPARTMENT, SAFETY AND PERFORMANCE DIVISION, AEA THERMAL
REACTOR SERVICES, WINFRITH TECHNOLOGY CENTRE, DORCHESTER, DORSET DT2 8DH, UK

Introduction

Gramme quantities of regular non-spherical particles
of natrojarosite ($NaFe_3(SO_4)_2(OH)_6$) in the range from 2
to 20 μm volume equivalent diameter have been prepared by
controlled crystal growth from ultra-pure aqueous
solutions. The particles precipitate from solution as
single crystals possessing the geometry of a truncated
cube. Particle monocrystallinity and chemical
composition have been determined by x-ray diffraction.
Energy dispersive spectroscopy coupled with scanning
electron microscopy has also been utilised to confirm the
particle chemical composition obtained from x-ray
diffraction. Particle density has been measured by
helium displacement pycnometry, and found to be in good
agreement with the literature value quoted for
natrojarosite. The dimensions of individual particles
have been determined by optical microscopy-image analysis
and used in conjunction with Euclidian geometry to
calculate particle volume equivalent diameters. Particle
aerodynamic diameters have been measured under Stokesian
conditions using a calibrated Timbrell aerosol
spectrometer. The dynamic shape factor of the particles
is constant at 1.19 $\pm$ 0.06 in the size range from 2 to
20 μm volume equivalent diameter.

Preparation

50 cm^3 aliquots of an aqueous solution containing
iron (III) nitrate (0.18M) and sodium sulphate (0.27M)
were mixed and maintained for 1.5 to 18.5 h in glass
culture tubes at 80 $\pm$ 0.5 °C within a temperature-
controlled oil bath. As many as twenty tubes were used
to carry out preparations in parallel, so that the
eventual yield of particles in some preparations exceeded
4 g. Immediately at the end of the specified reaction
time (Table 1), each reaction tube was cooled to room
temperature to quench the crystallisation process. The
precipitated natrojarosite crystals were separated from
the liquor by vacuum filtration onto acetate membrane

<u>Table 1</u> Natrojarosite Particles: Preparative Details

Identification Code	Reaction Time (h)	Microscopy-Image Analysis		Yield of Particles (g)
		CMD_{max} (μm)	Geometric Standard Deviation (σ_g)	
FS3	9.50	10.3	1.11	4.5
FS4	6.50	7.7	1.12	4.1
FS5	4.75	6.0	1.14	3.8
FS6	3.50	4.9	1.09	4.2

Ageing temperature: $80 \pm 0.5°C$

Solution composition: $Fe(NO_3)_3$ 0.18 M
Na_2SO_4 0.27 M

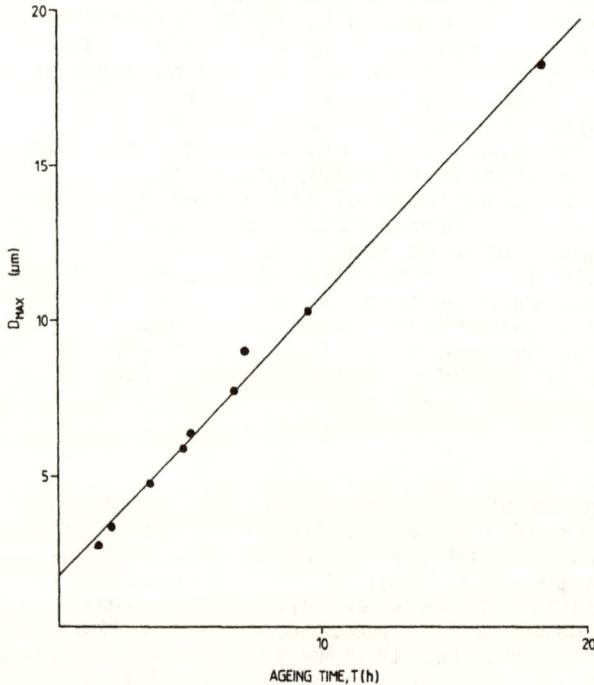

<u>Figure 1</u> Variation of particle size with ageing time

filters (0.45 μm mean pore diameter (Millipore (UK) Ltd, London), before washing with ultra-pure water and drying in a desiccator for at least 24 h prior to use.

Microscopy-Image Analysis

Typical size distribution data were obtained by microscopy-image analysis as listed in Table 1, and represent the mean sizes of more than 300 particles. The value of D_{max} (distance between opposite vertices) varied in direct proportion to the reaction time of the solutions in the oil bath (Figure 1). The particles from each batch preparation were always highly monodisperse (geometric standard deviation of D_{max} (σ_g) < 1.15), and had the same regular truncated cube geometry irrespective of size. The ratio of D_{max} to the distance between opposite edges of the hexagonal projection (D_{min}) is a measure of the regularity of the particles (1.15 for a perfect hexagon); values between 1.13 and 1.15 were obtained for several hundred particles from various batches (Table 2), irrespective of particle size. Representative electron micrographs (Figure 2) illustrate the range of particle sizes that were produced.

Table 2   Profiles of Natrojarosite Particles By Microscopy-Image Analysis

Reaction Time (h)	D_{max} (μm)	D_{min} (μm)	$\dfrac{D_{max}}{D_{min}}$
9.50	10.30	9.12	1.13
6.50	7.70	6.70	1.15
4.75	6.00	5.22	1.15
3.50	4.90	4.30	1.14

X-Ray Diffraction

The chemical composition of a 500 mg sample of the particles was determined by x-ray diffraction (Nicolet-12/V). The diffraction pattern (Figure 3a) is comparable with that for natrojarosite (Figure 3b). Although the x-ray beam was too wide to resolve the diffraction pattern arising from single particles, the resulting peaks were sharp, indicative of a highly crystalline structure. The particle composition was confirmed by energy dispersive spectroscopy (EDS) of a sample examined in a scanning electron microscope (JEOL JSM-35).

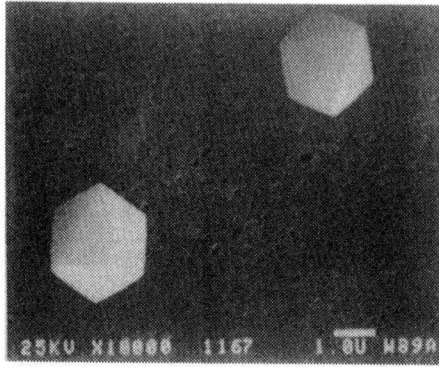

(a) $D_{MAX} = 2 \cdot 7 \mu m$

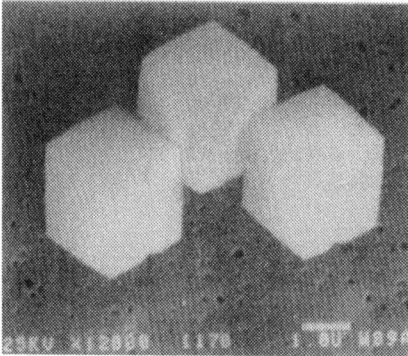

(b) $D_{MAX} = 3 \cdot 3 \mu m$

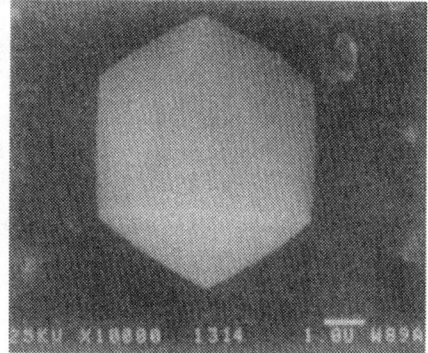

(c) $D_{MAX} = 6 \cdot 3 \mu m$

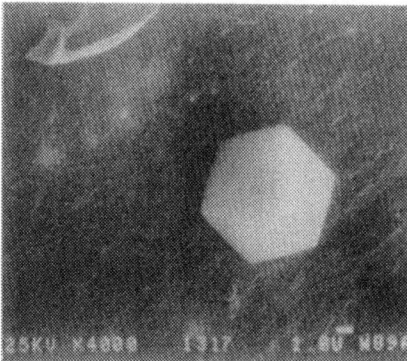

(d) $D_{MAX} = 9 \cdot 0 \mu m$

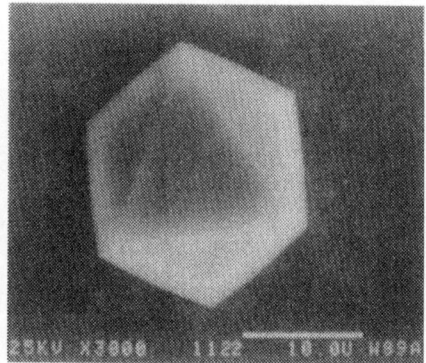

(e) $D_{MAX} = 18 \cdot 3 \mu m$

Figure 2   Natrojarosite particles

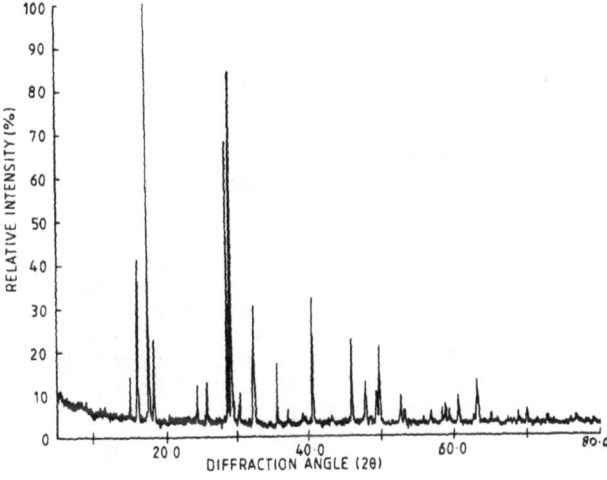

(a) spectrum of material prepared in present study

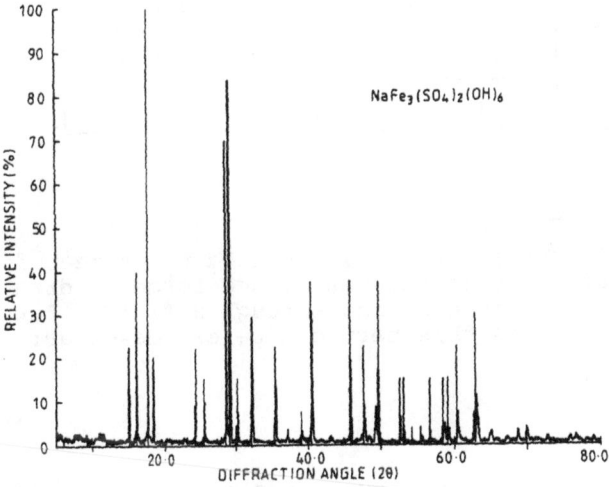

(b) reference spectrum

Figure 3 X-ray diffraction spectra

Helium Pycnometry

Particle density is an important parameter in defining the aerodynamic behaviour of airborne particles[1]. Small quantities of selected batches (ca. 1g of FS3, FS4, FS5 and FS6) were placed in the sample chamber of a helium pycnometer (model 1305 Multivolume Helium Pycnometer, Micromeritics Ltd, Basingstoke, UK), and the particle density determined from the volume of helium displaced, assuming there were no inaccessible voids within the particles. The four measurements (each of which was the mean of 4 to 6 separate determinations (Table 3)) were within the density range of 2.91 to 3.26 x 10^3 kg m^{-3} quoted for natrojarosite[2]. These studies confirm the validity of the x-ray diffraction measurements, and also indicated that the particles were substantially free of internal voids.

Table 3 Material Densities of Natrojarosite Particles

Identification Code	Mass (g)	Volume (cm³)	Density (kg m⁻³ x 10³)
FS3	3.29	1.09	3.01 ± 0.05
FS4	0.47	0.15	3.22 ± 0.05
FS5	0.57	0.19	3.06 ± 0.05
FS6	0.64	0.20	3.14 ± 0.05

Dynamic Shape Factor

The dynamic shape factor (χ) is used to correct for the influence of shape on the resistance force or drag experienced by a particle moving through a fluid. For micron-sized particles this term can be expressed as:

$$\chi = \frac{(D_v)^2 \, \rho_p}{(D_{ae})^2 \, \rho_o} \tag{1}$$

where D_v is particle volume equivalent diameter, ρ_p is particle density (kg m^{-3}), D_{ae} is particle aerodynamic diameter and ρ_o is 10^3 kg m^{-3}. Particle aerodynamic diameter is obtained directly from measurements performed under Stokesian conditions with a calibrated Timbrell spectrometer[3]. Particle volume equivalent diameter can be calculated from microscopy-image analysis measurements of D_{max} as described below.

The natrojarosite crystal habit is rhombohedral, and can be characterised by three dimensions (Figure 4a):

(i) a, untruncated edge length,
(ii) b, truncated edge length,
(iii) c, edge of the equilateral triangular face parallel to the axis of symmetry A-A'.

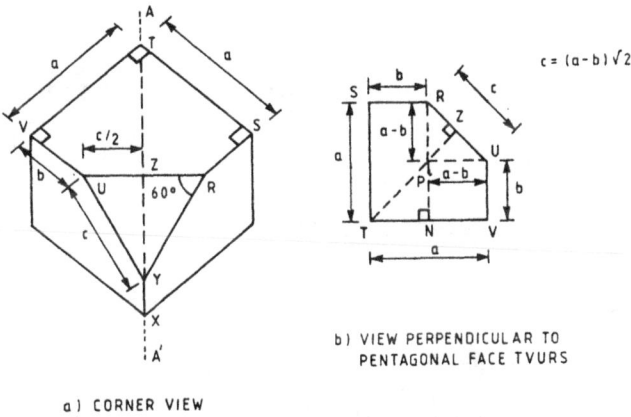

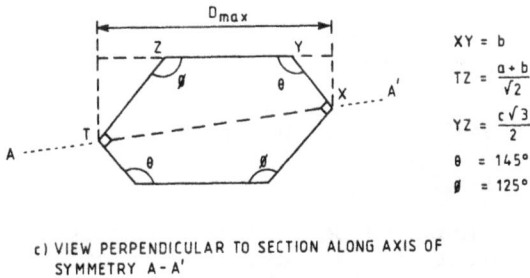

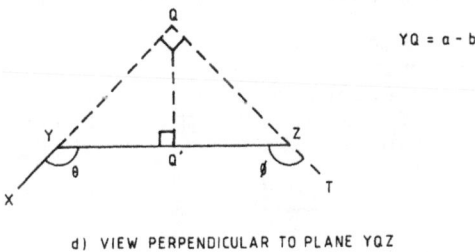

a) CORNER VIEW

$$c = (a-b)\sqrt{2}$$

b) VIEW PERPENDICULAR TO PENTAGONAL FACE TVURS

$$XY = b$$

$$TZ = \frac{a+b}{\sqrt{2}}$$

$$YZ = \frac{c\sqrt{3}}{2}$$

$$\theta = 145°$$

$$\phi = 125°$$

c) VIEW PERPENDICULAR TO SECTION ALONG AXIS OF SYMMETRY A-A'

$$YQ = a-b$$

d) VIEW PERPENDICULAR TO PLANE YQZ

Figure 4 **Single crystal geometry**

The relationship between the parameters "a", "b" and "c" is established by considering one of the pentagonal facets TVURS (Figure 4b). RN and UP are constructed perpendicular to edges SR and UV that are orthogonal, and form the right-angled triangle RPU by intersection with edge UR. Hence

$$c = (a-b)\ \sqrt{2} \qquad\qquad\qquad\qquad (2)$$

by the application of Pythagoras' theorem. D_{max} can be calculated in terms of the parameter "a" by analysis of a section through the particle and the axis of symmetry A-A' (Figure 4c):

$$D_{max} = [TZ\ \cos(180°-\psi)] + [YZ] + [XY\ \cos(180°-\theta)]\ (3)$$

TZ bisects UR and RZT is a right angle (Figure 4b). Applying Pythagoras' theorem to triangles PZR and PNT:

$$TZ = \frac{(a+b)}{\sqrt{2}} \qquad\qquad\qquad\qquad (4)$$

XY is perpendicular to UR (Figure 4a) and:

$$YZ = \frac{(c\ \sqrt{3})}{2} \qquad\qquad\qquad\qquad (5)$$

applying Pythagoras' theorem to triangle UYZ. XY is equal to parameter "b" (Figure 4c), and ψ and θ are 125° 16' and 144° 44' respectively. Equation (3) may therefore be rewritten:

$$D_{max} = \left[\frac{a+b}{\sqrt{6}}\right] + \left[\frac{c\ \sqrt{3}}{2}\right] + \left[\frac{b\ \sqrt{2}}{\sqrt{3}}\right] \qquad (6)$$

Substituting equation (2) into equation (6):

$$D_{max} = 1.63\ a \qquad\qquad\qquad\qquad (7)$$

The ratio D_{max}/c was shown to be 1.38 from microscopic examination of a large number of particles of different sizes. Hence "a", "b" and "c" can be derived from D_{max} using this relationship together with equations 2 and 7.

The volume occupied per particle (V_p) can be calculated in terms of the parameters "a" and "b" by subtracting the volume of the missing tetrahedra from the volume of a cube. The base area of one of the tetrahedra is defined by the triangle URY (Figure 4a), and:

$$\text{area URY} = \frac{c^2\ \sqrt{3}}{4} = \frac{(a-b)^2\ \sqrt{3}}{2} \qquad (8)$$

The perpendicular height of the tetrahedron is obtained by constructing a line QQ' orthogonal to the line YZ that bisects triangle URY (Figure 4d). Q' is located at the centre of gravity of the triangle URY and:

$$YQ' = \frac{2\ YZ}{3} = \frac{c}{\sqrt{3}} = \frac{(a-b)\ \sqrt{2}}{\sqrt{3}} \tag{9}$$

YQ is equal to (a-b). Applying Pythagoras' theorem to triangle YQQ'

$$QQ' = \frac{(a-b)}{\sqrt{3}} \tag{10}$$

and the volume (V_t) of the tetrahedron (QURY) is therefore:

$$V_t = \frac{1}{3} \cdot \frac{c^2\sqrt{3}}{4} \cdot \frac{(a-b)}{\sqrt{4}} = \frac{(a-b)^3}{6} \tag{11}$$

<u>Table 4</u> Dynamic Shape Factors of Natrojarosite Particles

ID Code	CMD_{max} (μm) (from microscopy)	D_v (μm)	D_{ae}* (μm) (mid-point value from Timbrell spectrometer Calibration[+])	Density (kg m^{-3} x 10^3)	χ
FS3	11.3	8.0	13.3 ± 0.4	3.01	1.10
	10.5	7.4	11.9 ± 0.6	3.01	1.17
	9.7	6.9	10.8 ± 0.7	3.01	1.22
	9.1	6.4	9.8 ± 0.7	3.01	1.29
FS4	8.5	6.0	9.0 ± 0.8	3.22	1.44
	7.8	5.5	8.3 ± 0.8	3.22	1.42
	7.0	5.0	7.7 ± 0.8	3.22	1.31
FS5	6.7	4.8	7.2 ± 0.9	3.06	1.32
	6.5	4.6	6.8 ± 0.9	3.06	1.38
FS6	5.9	4.2	6.5 ± 0.9	3.14	1.28
				Mean (χ) Uncertainty	1.29 ± 0.16

* Reference 3

[+] uncertainty limits represent ± 2 standard deviations about D_{ae}

and:

$$V_p = (a^3 - 2V_t) = a^3 - \frac{(a-b)^3}{3}$$ (12)

By definition:

$$D_v = \left(\frac{6\ V_p}{\pi} \right)^{1/3}$$ (13)

D_v was calculated for particles of known D_{ae} that were collected in a calibrated Timbrell spectrometer at various deposition distances ranging from 30 to 75 mm. Individual determinations of χ varied from 1.09 to 1.43 with a weighted mean value ($\bar{\chi}$) of 1.19 ± 0.06 (Table 4). This weighted mean was calculated by placing increased significance on those measurements with the smallest uncertainties. The large uncertainty limits reflect both the limited precision of the Timbrell spectrometer[3] and the precision to which D_{max} could be determined by the image analyser. However, within these limits, particle dynamic shape factor is independent of particle size over the size range from 2 to 20 μm volume equivalent diameter.

Acknowledgement

This work was jointly funded by an extra-mural contract from the OMHL under Contract Number 2210/R42.35 and the Corporate Research Programme of AEA Technology.

References

1.   W. C. Hinds, 'Aerosol Technology', J. Wiley and Sons, New York, 1982.

2.   R. C. Weast, 'CRC Handbook of Chemistry and Physics', 66th Edition, CRC Press Inc., Boca Raton, Florida, 1986.

3.   I. A. Marshall, J. P. Mitchell and W. D. Griffiths, J. Aerosol Sci., 1990, 21(7), 969.

Issues Encountered in the Size Characterization of Biological and Pharmaceutical Particulate Systems

M. J. Groves

INSTITUTE FOR TUBERCULOSIS RESEARCH, UNIVERSITY OF ILLINOIS AT CHICAGO,
115 SOUTH SANGAMON STREET, CHICAGO, ILLINOIS 60607, USA

1 SUMMARY

The special needs of size characterization methods for the biological
and pharmaceutical systems that are now so important in the medical
sciences today are reviewed. Submicrometre and subcolloidal
particulate systems are more frequently encountered that do not have
the sharply defined interfaces familiar to the analyst of 25 years
ago. Philosophically it is pointed out that many particles with
indeterminate interfaces will move under the application of external
forces as if they were inside a 'sphere of influence' and it is, in
fact, the dimensions of this sphere that are measured. Under these
conditions it is valid to measure the particle size characteristics
of 'particles' down to the molecular dimensions, at least, of some of
the larger protein molecules, and some methods will certain reach
down into these size regions without excessive difficulty.

Advances in these areas are being made but in general these are
not attributable to newer methods of measurement or principles of
detection but rather due to a better appreciation and understanding
of the significance of the results obtained by the application of
some older and well established methodologies.

2 INTRODUCTION

Twenty five years after the foundation of the Particle Size Analysis
Group many of the issues facing the founding fathers and mothers of
this unique and valuable networking and discussion organization have
been resolved or are being evaluated from another perspective. Some
issues, however, apparently will not go away. One specific problem
has been the way in which the word 'analysis' is being slowly
displaced by the term 'characterization'. I believe that this is an
implicit admission amongst many scientists involved in this broad-
ranging field that precise measurements have become ambiguous and,
certainly in some cases, blurred in meaning. In a particular system
(pun intended), 'analysis' is hardly an appropriate term when
different measurements, all of them intrinsically correct and valid,
can be obtained by different instrumental principles when applied to
different samples of an otherwise identical material.

Nowhere is this more of a problem than in the pharmaceutical and biological fields that have received such massive attention over the past decade with the introduction of genetically engineered drugs and the applications of biotechnology in general. A related issue that I will have to raise at this point seems to be self-evident but, on closer inspection, is not at all clear, concerns the definition of the term 'particle. If we examine, say, a current textbook in the field (1) what we soon realize is that the author, for perfectly good reasons, is mainly concerned with evaluating inorganic materials such as ores and pigments. While the pharmaceutical and biotechnological products used today may contain some inorganic materials as inert fillers, the components of essential interest to the analyst will usually be organic in nature and often with extremely complex chemical structures. Physico-chemical properties of these materials will be critically influenced by the state of dispersion and, conversely, the methods of measurement of the state of dispersion will depend to a considerable degree by those properties. It is at this point that the term 'characterization' becomes more appropriate but it is important to realize that the term 'particle' requires very careful definition in order to assist the analyst in making his or her choice of instrumental principle, and other scientists to interpret the results of the experiment.

The Meaning of the Word 'Particle'

In a slightly different context I have discussed this term and derived from a consensus of various dictionaries (from both sides of the Atlantic) the following (in slightly paraphrased form)(2):-

A particle is a discrete entity of solid or liquid matter existing in a dispersed state with a diameter at or below approximately 50μm.

For reasons that I will explain later, gas bubbles were specifically excluded although, in the technically broader coverage of the subject represented by this conference, it will be evident that bubbles could also be included in any definition. The definition also specifically includes an upper size limitation which, as most will appreciate, is roughly the smallest size of particle visible to the unaided eye. Some authors feel that a particle is a particle up to approximately 1000μm diameter, after which it becomes a granule or some other suitable collective term. Nevertheless, in the area that I am describing there are rarely any systems with particles larger than 50-100μm so this definition does have some attractions in my own specialized field. There is, however, an implicit understanding in all the textbooks I have examined, as well as in the definition I have tentatively suggested above, that the discrete particles, irrespective of their size, have clearly definable boundaries between the dispersed phase and the surrounding continuum. These sharp boundaries are assumed to exist, whether or not the interface is between any combination of gas, liquid or solid. For many particulate systems this is obviously the situation. Unfortunately, for many of the pharmaceutical and biological systems requiring evaluation today, this distinction begins to disappear, especially as the particles become smaller.

As particles approach the colloidal size range-arbitrarily defined as being between 100nm and 10μm diameter- surfaces tend to become more diffuse as the underlying material becomes hydrated or charged. In addition, other materials such as surface active agents or charged ionic species tend to either adhere to the surface or become closely associated with it, and these associations may, in turn, influence the actual behavior of the dispersed system. This behavior is likely to affect the performance of the material and this may influence the selection of the equipment needed to determine the state of dispersion.

Below approximately 100nm the situation becomes much more difficult to define as we approach the size of molecular aggregates or even the size of some of the larger proteins. The question must be raised as to whether or not the analyst at this conference should be concerned with particle size characterization at the molecular level of discrimination. This question is unlikely to be resolved readily but I should point out that submicrometre sizing methods currently available to us do, in fact, detect and discriminate particles down to about 1-5nm which will detect proteins with molecular weights in the 50-100kD region. A heavily hydrated protein molecule does not have a sharply definable interface, irrespective of the scale of scrutiny. Under the appropriate application of external forces, the molecule will move through space as a distinct entity although it will certainly drag a proportion of the continuum molecules with it as it moves. Under these conditions we can make the assumption that the particle-or molecule- is behaving as if it was inside a sphere of some material with a density intermediate between the true density and that of the continuum. It is this 'sphere of influence' that we can define and measure in terms of properties of matter. We can change the properties and influence, for example, the 'sphere of hydration' around the particle but providing the particle moves in a predictable way under the external forces we can measure a 'size' that has some relationship to the subsequent performance parameters of the product. Under these conditions the characterization of disperse systems is justified down into the sub-colloidal region that may include molecular dispersions and will certainly include systems without readily definable surfaces under some scales of scrutiny.

3 CURRENT METHODS AVAILABLE FOR SUB-MICROMETRE SIZING

The first method applied to the measurement of molecular weights of proteins was the ultracentrifuge in which forces in excess of 100,000g were applied to, generally, aqueous dispersions of the material under examination. Detection of material sedimenting or moving under the external force was originally limited to the measurement of Schlieren bands of different optical densities by photographic methods but methods of detection have improved considerably. Although less fashionable today, after more than 50 years of continuous development and application of these high speed machines, some slower centrifuges have found application where reduced forces are sufficient to make materials move(1,3). The upper limit of size measurement will be about 5µm, depending on the density of the material. It is, perhaps, time to re-evaluate the ultracentrifuge as a size characterization method because the methodology is thoroughly developed and it would be a relatively simple process to convert the older terminology of Svedberg units to terms involving diameters of particles of known densities that the modern analyst is familiar with.

More recent methodology is that of photon correlation spectroscopy (PCS) which depends on the detection of increased Brownian movement demonstrated by particles as they become smaller due to the influence of the kinetic energy of the dispersion medium. Nevertheless, the basic underlying equation upon which this principle depends involves a simple relationship between the particle 'diameter' and the diffusion coefficient or velocity of movement. Here the field is the environment surrounding the particle which is controlled by temperature and viscosity of the medium. The point of this is to note that, for both centrifugal and PCS methods, the 'particle' has to move as an independent entity, and this movement is detected either by application of a centrifugal field to overcome the randomness of orientation of the movement or by measurement of the rapidity of the random ('Brownian') movement using scattered radiation from an external source of monochromatic light. In neither case do we need to detect the particle by assuming that there is a sharp and distinct interface between the dispersed phase and the continuum. Effectively what we are admitting now is that we do not need to know the exact diameter of the particle substance but that what we are content to measure is the size or characteristic of a 'sphere of influence' which includes the particle of interest, irrespective of its chemical or physical nature. Philosophically or semantically this realization removes any doubts that we may have had about measuring particles which could be defined as molecules or molecular aggregates. The nature of the dispersed material becomes irrelevant, providing it behaves as if it was a distinct 'particle' from both a performance and an analytical point of view.

Other methods that detect a sphere of influence include those based on the Coulter principle which will also be reviewed at this conference. Here the data is reported in terms of a sphere of equivalent volume, irrespective of the shape or, in some situations, the state of the particulate interface. The method depends, essentially, on measuring the increase in resistance experienced between two electrodes as a particle passes between them and an essential requirement, therefore, is the presence of electrolyte in the measurement system. The method is realistically limited to particles down to about 1μm in diameter, and there is no practical upper limit to the principle. The presence of electrolyte in the environment is an advantage in some situations since the effect is to suppress charge effects at the particle interface and this simplifies the measurement of the 'size' of colloidal dispersions. Submicrometre dispersions can be measured but it should be noted that interference effects become more pronounced and there is less certainty about the magnitude of coincidence effects, quite apart from the intrinsic experimental difficulties of keeping orifices with diameters of less than 50um clean and operationally effective. Nevertheless, the Coulter principle has proved to be an invaluable technique for the detailed characterization of biological systems such as blood cells and, in some instances, bacterial suspensions.

Devices based on light extinction measurement or various forms of light scattering are becoming more widely available and are being utilized, especially in the area of particle detection when measuring contamination of liquids and air spaces. In general limited to particles with diameters larger than about 100nm-and more usually to larger than 2μm- these instruments have become widely employed for environmental monitoring in the pharmaceutical and electronic industries. In some cases the use of such devices is mandated by law (perhaps a unique American situation) but there has been some expression of discomfort with the sudden application of relatively untried and unproven instrumental principles. This is hardly a fair comment on the use of the HIAC Counter, for one example, since the device has been available in the market for at least 25 years and has been successfully applied in a number of different industries. However, the application of this instrument to the measurement of

particulate contamination in injection solutions in the United States
was mandated by law in 1985 and resulted in a necessary educational
period before operators began to understand and appreciate what the
instrument was capable of telling them (2).

The technique of photon correlation spectroscopy has become
much more widely available following resolution of some of the more
interesting computer algorithms. Using the appropriate laser
illumination intensity it is claimed by some manufacturers that
particles with diameters down to 1nm can be detected and measured,
and this is a size range that would comfortably encompass larger
protein molecules. The complexities of the methodology are by no
means totally resolved. The method would appear to be able to
provide a mean diameter if there is only a single species present,
with rather less certainty the width of the distribution around the
mean and, in a relatively minor number of cases, some resolution of
multimodal distributions if the system is mixed. Since many
manufacturers of this equipment have appeared and most have developed
their own software with different algorithms, a real need is becoming
evident for comparative trials of the available equipment to be
undertaken so that we can develop a better understanding of what we
are actually looking at. Nevertheless, as will be evident from this
conference alone, the technique overall is proving valuable, and
application to the molecular region of size is becoming a reality.

It will be evident that when we make a measurement we often
have no sensitivity to the appearance of the material we are making
the measurements on. This is not the case for the microscopy
measurements that are an often neglected but essential part of size
characterization methodology. Optical microscopy, using white light
to illuminate the particles, has become commonplace today and some
sophisticated scanning and image microscopes are available that
remove much of the fatigue associated with making adequate
measurements by this method. Many of the principles learned from
light microscopy have been applied to transmission and scanning
electron microscopy size characterization of materials. Surprisingly,
perhaps, some of the lessons learned from the older technique have
not always been applied to the more recent ones. There is still a
tendency for biologists to report the size of a biological structure
seen in the electron microscope based on one microscope field with no
attempt to validate the technique or the magnification of the field.
It is true that biological structures tend to have an intrinsic
variability which makes accuracy of the subsequent measurement
somewhat problematic. Nevertheless, an example of the fluctuation in
the estimate of dimensions of a microorganism due to sampling size
will be demonstrated, taken from recent work in our own laboratory
(5).

The main value of electron microscopy in recent years has been
to allow morphological features to be determined for small objects
and visualization of structures down to the molecular level. For
non-biologists, however, it should be pointed out that most electron
microscopes are operated in hard vacuum which can produce artifacts
due to desiccation. In addition, it is often necessary to stain
features with materials that are electron dense in order to provide
sufficient contrast in the field under examination, and this can
occasionally result in mis-identification of structures or
additional artifacts being produced in the specimen. Electron
microscopy is therefore a fascinating and specialized area with its
own challenges. However, here is another scientific discipline that
could benefit from the input of experienced particle size analysts in

order to develop the subject and, perhaps, moderate some of the more unrealistic claims over the years in the literature.

4   SOME EXAMPLES OF SPECIALIZED REQUIREMENTS OF PARTICLE SIZE ANALYSIS AND CHARACTERIZATION METHODOLOGIES IN BIOLOGICAL SYSTEMS PROTEINS AND OTHER COMPLEX (POLYMERIC) STRUCTURES

The biological activity of many proteins encountered in nature is exquisitely sensitive to conformational structure. This fact, literally of life, only became clear when some of the genetically engineered proteins were produced on a larger scale for evaluation in the laboratory. As in any polymeric system there is an average 'size' and a width or distribution around the mean. Chemical and manipulative skills have been applied to fix the type of terminal end groups on a linear polymer with a molecular weight that may vary between, say, 20-24kD in the case of human growth hormone. However, the biological activity is determined largely by the way the protein structure as a whole fits on its corresponding site of activity, whether on a cellular surface or in the aqueous medium between or inside cells. In other words, it is the 'sphere of influence' that controls the properties of the protein-and the particle size analyst has the skills and knowledge to determine the characteristics of that sphere. Here we see the direct connection between the analysis itself and the performance related parameters that should be driving the analysis. It is very likely that the size of the sphere of influence will be a major performance parameter in terms of receptor fit, and therefore the relative proportion of the available material that is biologically active becomes an important production consideration. In turn this has economic repercussions because of the significant cost of these materials. There is therefore some interest in being able to accurately determine the mean size and spread around the mean for different lots of large molecular weight proteins. Photon correlation spectroscopy and ultracentrifugation are methods currently available to the analyst in the region of 1nm and above but their successful application requires considerable development for the future.

<u>Cells</u>

As noted earlier, the Coulter principle has been effectively applied to the size characterization of blood cells and the generation of data has become extremely sophisticated from this one technique alone. Rather less effort has been applied to the characterization of bacterial cells but I will demonstrate with the aid of data obtained in our own investigations of a bacterial vaccine just how useful application of the Coulter principle can be, (5).

The Coulter has limited application to the determination of the characteristics of viral and micrococcoidal forms of life because they are so small, and these systems have received scant attention from size analysts, other than detailed electron microscopic evaluation by biologists. Indeed, there are examples in the literature where the size characteristics of a virus or bacteriophage may have been based on one electron micrograph.

<u>Aerosols</u>. Aerosol size characterization is a highly specialized science in its own right and has significance for drug delivery, environmental concerns-including production of pharmaceutical and electronic products- and areas such as chemical and biological warfare. Except to observe that many of the sizing

methods involve some form of light energy scattering (6) the subject
is not appropriate for review here.
   Drug Delivery. The biological activity of insoluble solid drug
particles is known to be influenced by the state of dispersion. Solid
particles are unlikely to be encountered with diameters much below
1μm because surface forces are released that make the handling of the
powder extremely difficult and encourages particle aggregation which
counteracts the advantages of reducing the particle size in the first
place.   These disadvantages are not encountered with liquid
dispersions and a great deal of scientific energy has been expended
in developing liposomes and emulsions as drug delivery systems.
Furthermore, newer methods for producing dispersions into the lower
nanometre size region have become available over the past two or
three years that seriously challenge the size analyst. Some attempts
to characterize liposomes by electron microscopy have been made but,
in general, most sizing of submicrometre systems has been attempted
using PCS. What is needed is performance-related sizing data, and a
review of current literature shows that there is a serious lack of
information at present in this field.   There are relatively few
liposomal drug systems on the market, in the main because of the
complexities of the production and stabilization of these complex
systems.   However, there is a great deal of optimism in this area
that, sooner or later, a major contribution to health care will be
made by drug delivery systems based on liposomes and emulsions. A
significant factor in this progress will undoubtably be the
contribution that particle size characterization will have made to
the subject since size and activity (performance) are both intimately
linked.
   Particulate   Contamination.   Although   environmental
contamination with particulate is of importance in a number of the
newer technologies (electronics, for example), I will confine my
remarks to the pharmaceutical arena since this is where I have
experience. Progress in this area was not made until instrumental
methodologies were introduced to overcome the subjectivity of the
visual inspection of injectable solutions that were used up to the
middle of the 1970s.   Initially based on the experience acquired
from the hydraulic industry, compendial inspection methods involved
the Coulter, the HIAC/Royco and filtration, with optical microscope
evaluation of the particles collected on the filter. Experience was
not, however, extensive and the whole industry went through a series
of learning experiences. Only now, for example, is the original
filtration/optical microscopy method receiving attention and re-
evaluation so that, ultimately, an improved method will be available
to the pharmaceutical analyst. In addition, the detection of sub-
threshold particulate has received attention, as has the meaning of
the word 'particle' because, as noted earlier, contaminating
particles consist of either solid or liquid materials. Bubbles of
gas cannot be avoided when handling injection solutions so common
sense suggests that these should be excluded from evaluation. Having
said that, obviously bubbles will be detected by instruments based on
the Coulter or HIAC so, experimentally, care has to be taken to avoid
their incorporation and detection. In addition, there is a better
understanding of the meaning of data obtained by different detection
principles, and there is some understanding of the nature of the
distribution of size when the system under examination consists of
random identities and random sizes in very low numbers per unit
volume—the situation encountered in a pharmaceutical solution that
has been repeatedly filtered down to the 200nm region in order to
remove all bacteria (7).

5   DISCUSSION

This Group, perhaps more than any other collection of scientists with widely differing backgrounds and interests, will be sensitive to the meaning of the words 'particle', 'size' and 'analysis'. I happen to think that we should incorporate the word 'characterization' into the title somewhere but our commonality involves a realization that the data we derive from some method of size analysis has to be related to the subsequent performance of the particulate in practice-no matter what the end point of the process is. In other words we are not necessarilly interested in just the collection of data per se. In the case of biological and pharmaceutical systems- and especially with the newer biotechnologically based generation of genetically-engineered drugs-there is a need to broaden the scientific approach to the subject in order to apply the methodologies developed earlier to systems that may not have sharp and defined surfaces between the continuum and the material being measured. Significant advances are being made in this area as we develop a better understanding of what the methods we have are telling us, and how the data can be interpreted. The next 25 years could be very exciting.

6   REFERENCES

1.   T. Allen, 'Particle Size Measurement',(Fourth Edition), Chapman and Hall, London, 1990.
2.   M.J.Groves, Proc. Int. Conf. on Particle Detection, Metrology and Control, Parenteral Drug Association, Washington, pps 82-102, 1990.
3.   M.J.Groves, in 'Modern Methods of Particle Size Analysis', (edited H.G.Barth), John Wiley, New York, pps 43-92, 1984.
4.   M.J.Groves, in 'Particle Size Assessment and Characterization', (edited T.Provder), American Chemical Society Monograph (in press, 1991)
5.   M.J.Groves, M.E.Klegerman, P.O.Devados, O.N.Singh and Y.Zong, Particle and Particle Systems Characterization, (in press, 1991)
6.   A.J.Hickey in 'Pharmaceutical Aerosol Technology', (edited A.J.Hickey), Marcel Dekker, New York, (in press, 1992).
7.   T.A.Barber and M.J.Groves, 'A Source Book on Particulate Contamination of Pharmaceuticals and Medical Devices', Interpharm Press, Buffalo Grove, Illinois, (in press, 1991).

Standard Powders for Particle Size Analysis

R. J. Gowland and B. Wilshire

DEPARTMENT OF MATERIALS ENGINEERING, UNIVERSITY COLLEGE, SWANSEA SA2 8PP, UK

1 INTRODUCTION

A variety of different routes for production of novel particulate materials are being investigated at Swansea. These research activities began using atomization-milling technologies to provide improved metallic flake materials for decorative printing, paints, etc. (1). The range of metallic materials being prepared and evaluated was then extended in a search for superior fine flake products for detection of latent fingerprints (2) and for obscuration of military targets from modern electro-optic weapon systems (3). The programme concerned with the procurement of new high-performance particulate obscurants also studied the effectiveness of various chemically-produced non-metallic materials, with the reaction conditions being controlled to obtain different particle sizes and shapes. In turn, this continuing programme on the attenuation of electro-magnetic radiation by air-borne particle dispersions focused on the need for uniformly-sized particles, with known controllable shapes, for assessment of the performance of aerosol characterization instruments (4).

In addition to the expanding range of instruments becoming available for aerosol characterization, many sophisticated methods have been and are being developed to meet the routine industrial requirement for reliable measurement of particle size distributions as a parameter of sale or of production control. However, different types of particle sizing equipment utilize different physical principles so that, even when measurements are completed for identical samples, different results are recorded because different parameters of the particles are quantified. For calibration of individual instruments and for comparison of different measurement techniques, it is therefore important to have access to known "standard" powder samples. In part, this requirement could be met by the provision of reproducible samples of chemically-inert spherical powders, with known particle size distributions, covering various size ranges, eg from 0.1 to 50µm, etc. Unfortunately, the problem of defining powder standards is complicated by the fact that most practical materials are not spherical. Consequently, several types of standard powder are ideally required, including mono-disperse particles of known size and shape. The availability of a broad range of standardized spherical and non-spherical powders could then lead to additional benefits. For instance, analysing the results obtained for a variety of standardized powders using several different types of instrument could identify combinations of measurement technique capable of quantifying not only particle size distributions but also particle shape functions. For these reasons, the present work describes the various production routes now being evaluated at Swansea for provision of fine powder products having controlled particle sizes and morphologies.

2 PRODUCTION OF POWDERS BY ATOMIZATION AND MILLING

A fine stream of molten metal can be disintegrated by high-velocity gas jets to create a cone of tiny droplets. The atomization conditions can be controlled, such that the droplets spheroidize under surface tension forces and then solidify in the gas flow to be collected as fine spherical powders. Figure 1 shows a sample of atomized copper powder, with the size distribution of the spherical particles evident from Figure 2. The particle sizes vary from about 0.1 to 50μm, giving a 500:1 dynamic range. However, the density of copper is 8.9 Mg.m⁻³. This value may be rather high for calibration of most particle sizing instruments, but this problem could be overcome by using atomized aluminium or aluminium alloy powders, with densities ranging from about 2.6 to 2.9 Mg.m⁻³. Similar density ranges could also be achieved with atomized silicate powders of selected compositions, if inorganic materials are preferred.

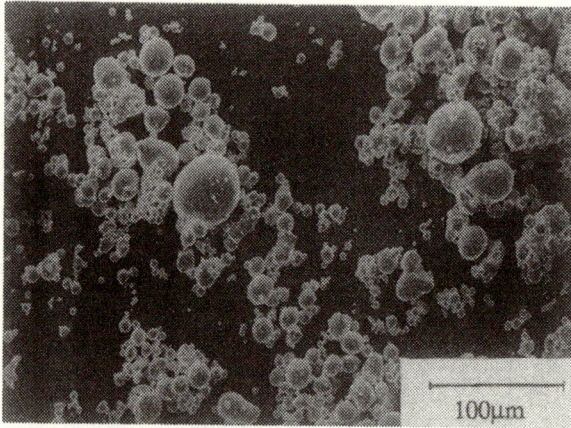

Figure 1   Scanning electron micrograph of atomized copper powder.

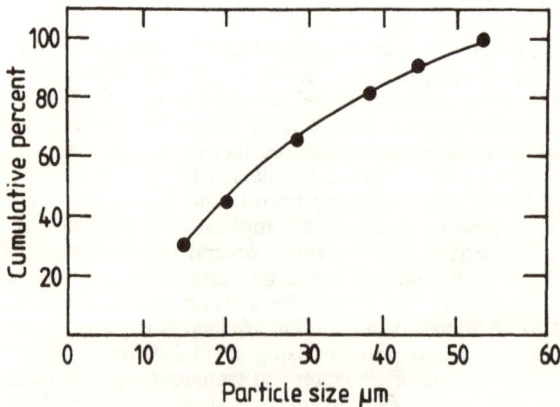

Figure 2   Typical particle size distribution of the atomized copper powder.

Ductile metal powders, produced either by gas or by water atomization, can be easily converted to flake form by standard ball milling operations. As the powder particles are trapped repeatedly between the impacting steel balls, the initially equiaxed powders are deformed to flake in a series of microforging steps, causing the diameter/thickness ratio to increase progressively until flake fracture occurs, as illustrated in Figure 3. The milling times and conditions can then be optimized so that flake within any desired size range can be provided. A variety of flake metal powders can be obtained using these procedures, to give particle density values of approximately 2.5 Mg.m⁻³ and above. However, in relation to the calibration of particle sizing instruments, two possible disadvantages may be experienced with this type of flake product. Firstly, the processes of microforging and flake fracture produce particles with ragged edge profiles, as evident particularly with the 0.5μm thick flake products shown in Figure 3. Secondly, the stearic acid added as a milling aid affects the ease with which the flake can be dispersed in water, but the stearic acid coating on the particles can be removed by soxhlet washing using hot acetone. Carefully optimized atomization-milling-classification procedures can then provide standardized flake products, with minimum thicknesses of about 0.5μm and with mean diameters controllable within the range 5 to 50μm.

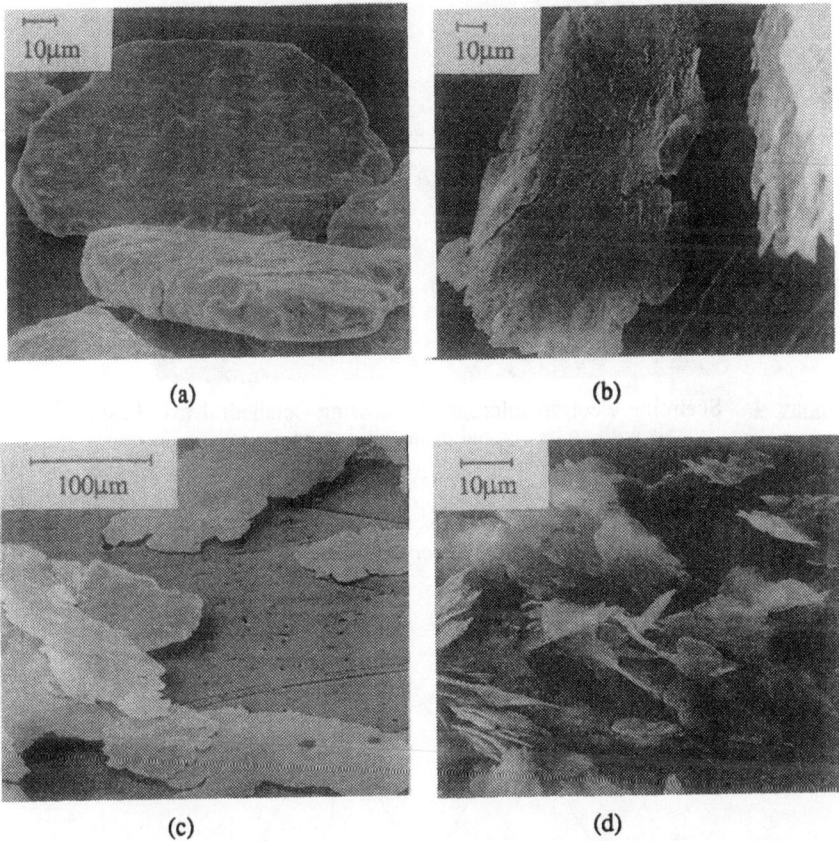

(a)

(b)

(c)

(d)

Figure 3   Scanning electron micrographs showing (a) the initial flattening of equiaxed brass powder during milling, (b) flake fracture as the diameter to thickness ratio increases, (c) brass flake with a minimum thickness of about 0.5μm and (d) aluminium flake with a minimum thickness of about 0.5μm.

3 PRODUCTION OF MONO-SIZE POWDERS WITH UNIFORM CRYSTAL SHAPES

While there is a need for standard powder products having a known distribution of particle sizes, a requirement also exists for mono-size particles having a uniform shape. Particles of constant composition, whose size and morphology can be controlled precisely, can be obtained by various chemical reaction routes. Specifically, several mono-disperse systems have been investigated, which rely on forced hydrolysis to provide various inorganic precipitates having a narrow size distribution in the range 1μm to 20μm. In this way, a number of uniformly sized particle types have been produced, each having a simple uniform crystal geometry. Thus, iron basic sulphate crystals can be obtained by heating iron nitrate solutions to 353K, in the presence of sodium sulphate (5). The crystals produced using this procedure have an octahedral morphology, with an hexagonal projected image, as shown in Figure 4. The size of the octahedral particles increases with reaction time, (Table I), confirming previously published data.

Figure 4   Scanning electron micrograph showing octahedral iron basic sulphate crystals with an hexagonal projected image

Table I. Variation in iron basic sulphate crystal size with reaction time

Time (h)	d_1(μm)	d_2(μm)	d_1/d_2
1.5	2.6	2.2	1.18
3.0	6.0	5.2	1.15
4.0	7.6	7.0	1.09
7.0	14.0	13.0	1.07

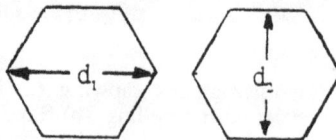

$d_1/d_2 = 1.15$ for a perfect hexagon

By heating to 373K, spindle-type iron oxide particles can be prepared from iron chloride dissolved in a 50% w/w ethanol-water solvent, (6), with the near-spherical iron oxide particles having a very narrow size distribution, Figure 5a. Moreover, by the addition of small quantities of sodium hypophosphite to the iron chloride solution, the particle morphology can be adjusted to ellipsoidal form, Figure 5b. With this approach, a continuous range of aspect ratios up to approximately three can be obtained by increasing the hypophosphite concentration.

While the spindle-type iron oxide crystals are already being used for assessment of light scattering equipment for aerosol characterization (7), it is interesting to note that iron oxide crystals with a disc-like morphology can also be obtained by forced hydrolysis. In this case, an iron (III) nitrate solution is heated to 523K in the presence of triethanolamine (TEA), sodium hydroxide and hydrogen peroxide. However, this procedure results in the simultaneous formation of very fine irregular particles of iron oxide (Figure 6), but the size distributions are such that the iron oxide discs can be separated by sieving.

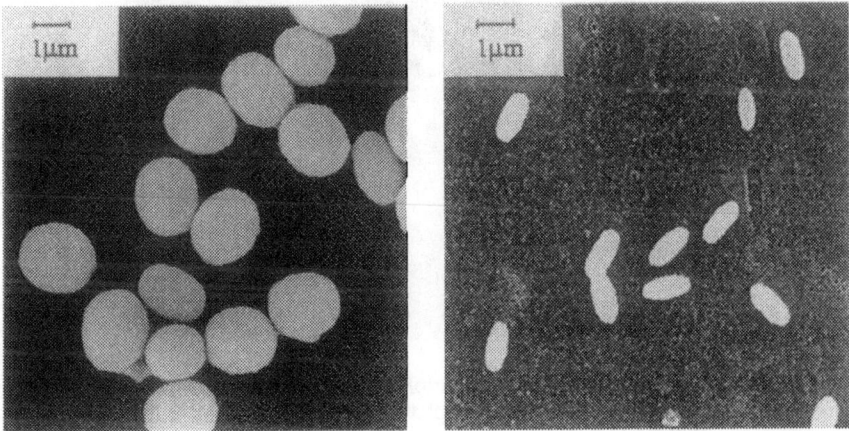

Figure 5   Scanning electron micrographs showing (a) near-spherical and (b) ellipsoidal iron oxide particles

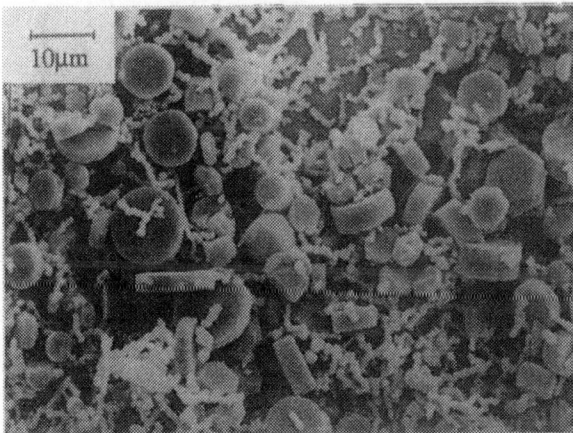

Figure 6   Scanning electron micrograph of iron oxide discs, together with the fine irregular particles formed simultaneously

A variety of alternative crystal shapes can be produced by forced hydrolysis of inorganic salts. For instance, copper basic sulphate crystals with a uniform leaf-like appearance (Figure 7) can be prepared by heating copper sulphate solutions in the presence of urea (8). Similarly, zinc oxide needles can be obtained by heating zinc nitrate solutions in the presence of triethanolamine (TEA), sodium hydroxide and hydrazine, Figure 8.

Unfortunately, while the leaf-like copper basic sulphate crystals and the rod-like zinc oxide crystals had uniform particle shapes, in both cases, the particle sizes varied appreciably, as evident from Figures 7 and 8. For this reason, experimental programmes are continuing with a view to identifying the reaction conditions which give uniformly-sized crystals.

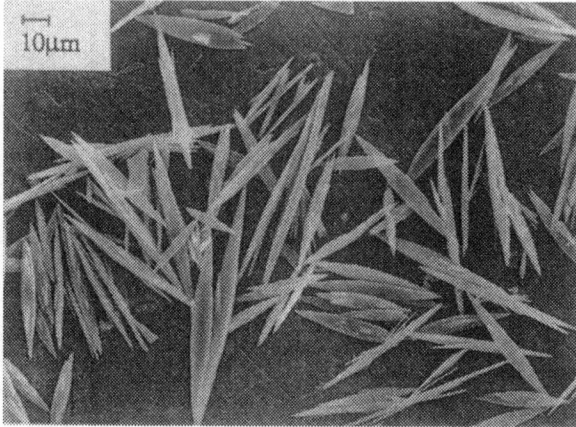

Figure 7   Scanning electron micrograph showing the size range distribution of the leaf-like copper basic sulphate crystals formed by forced hydrolysis of copper sulphate solutions at 373K

Figure 8   Scanning electron micrograph showing the size range distribution of the rod-like zinc oxide crystals formed by forced hydrolysis of dilute zinc nitrate solutions at 523K

4 PRODUCTION OF MONO-SIZE POWDERS WITH COMPLEX CRYSTAL SHAPES

In addition to the more usual crystal shapes shown in Figures 4 to 8 inclusive, less common crystal morphologies can also be produced by forced hydrolysis and other chemical reaction routes. For example, heating copper nitrate solutions to 373K in the presence of N-2 (hydroxyethyl) ethylenediaminetetramine (HEDTA) and sodium hydroxide (8) produces uniformly-sized star-like copper oxide particles. Yet, while the target particle shape was star-like, a substantial quantity of larger dendritic particles were formed simultaneously (Figure 9), but the size differences between the star-like and dendritic forms were sufficient to allow separation by sieving. Moreover, without changing the particle morphology, the star-like copper oxide particles could be reduced to copper by reduction with hydrogen at 673K.

Copper powders, having a variety of interesting particle shapes, can also be produced by the CQG method (9), a multi-step process involving gas/liquid and solid/liquid reactions. The process depends on the fact that copper can exist in two oxidation states. A copper (I) salt is first produced by reducing an aqueous copper (II) ammonium sulphate solution with sulphur dioxide as

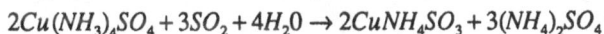

$$2Cu(NH_3)_4SO_4 + 3SO_2 + 4H_2O \rightarrow 2CuNH_4SO_3 + 3(NH_4)_2SO_4$$

This copper (I) salt is then reacted with sulphuric acid to yield particulate copper as

$$2CuNH_4SO_3 + 2H_2SO_4 \rightarrow Cu + (NH_4)_2SO_4 + 2SO_2 + 2H_2O + CuSO_4$$

Many factors affect the morphologies of the copper powders prepared in this way, including the reactant concentrations, reaction temperatures, etc. However, with all processing conditions studied, the powders invariably consisted of a complex mixture of particle morphologies, with separation of the individual particle types proving impractical because of their similarity in size, Figure 10. The possibility of achieving mono-size copper particles having a uniform crystal geometry is now being investigated by replacing the copper sulphate in the above reactions with solutions of other copper salts.

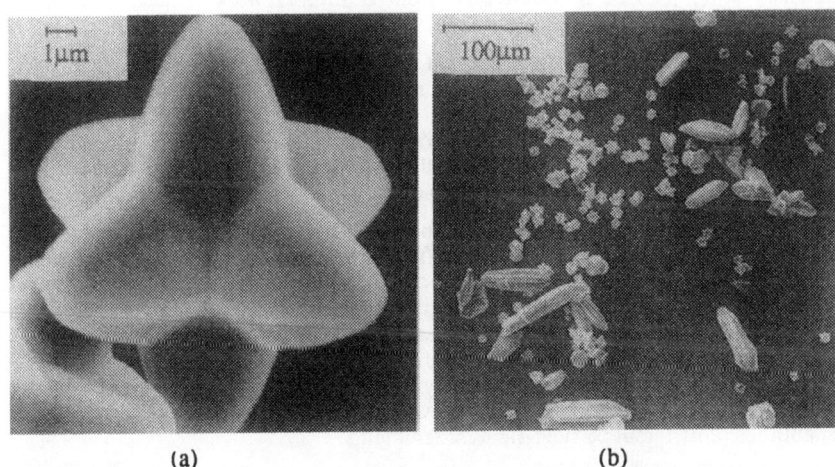

(a)   (b)

Figure 9   Scanning electron micrographs of copper oxide particles showing (a) the star-like target form and (b) the larger dendritic material formed simultaneously with the star-like crystals

Figure 10   Scanning electron micrographs of CQG copper powders having various
             mixed morphologies

5 CONCLUSIONS

(a)   Various chemical reaction routes based on forced hydrolysis of inorganic
salts can be used to manufacture mono-size powders of constant particle morphology and also uniformly-shaped powder products having a distribution of particle
sizes. In the search for mono-disperse powders, particular success has been
achieved with spindle-type iron oxide particles, which can be produced with aspect
ratios from one to approximately three. Other robust chemically-inert oxide
powders which can readily be dispersed in water include iron oxide discs and
rod-like zinc oxide crystals, although further studies are now underway to identify
the reaction conditions needed to supply these crystal forms as mono-size powders.
Similarly, mono-size octahedral iron basic sulphate crystals having an hexagonal
projected image can be prepared easily, but additional work is needed to obtain
leaf-like copper basic sulphate particles of uniform size.

(ii)   Atomization of liquid metals can produce spherical powders, with a size
distribution ranging typically from about 0.1 to 50μm. Using this technology,
particle densities varying from around 2.5 to 10 Mg.m^{-3} can be supplied by
controlling the melt composition. Moreover, using standard milling procedures,
flake products can be made available with minimum thicknesses of around 0.5μm
and with mean diameters between about 5 and 50μm.

(iii)   Atomization-milling procedures and various chemical reaction routes allow
a wide variety of particulate copper products to be manufactured reproducibly.
Thus, atomized copper powders can be produced having known particle size ranges
and these spherical powders can be converted to flake of varying thickness/diameter
ratios by milling. In addition, chemical reaction routes have been devised to supply
mono-size star-like copper particles and even copper powders with mixed complex
particle morphologies. While the density of copper may be inconveniently large
(8.9 Mg.m^{-3}), this range of copper powders could pose a severe test of the
capabilities of different particle-sizing instruments.

(vi)   Analysis of the differing results obtained using different particle sizing
equipment to characterize a broad range of standardized powder products could
identify combinations of measurement techniques capable of quantifying not only
particle size distribution but also particle shape functions.

6 ACKNOWLEDGEMENTS

The authors gratefully acknowledge the support funding provided by the Chemical Defence Establishment, Porton Down, which included a senior research assistantship for Dr. R.J. Gowland under the terms of MoD Contract Number D/ER1/9/4/2081/067/CDE and by the Science and Engineering Research Council which allowed purchase and construction of atomization and milling equipment under the terms of SERC Grant Number GR/F/03028.

REFERENCES

1. J.D. James, B. Wilshire and D. Cleaver, Powder Metallurgy, 1990, 33, 247.

2. J.D. James, C.A. Pounds and B. Wilshire, Powder Metallurgy, 1991, To be published.

3. R.J. Gowland, B. Wilshire and H. Edwards, Smokes/Obscurants XV, 1991, To be published.

4. R.J. Gowland, B. Wilshire and J.M. Clark, 5th Ann. Conf. of the Aerosol Society, 1991, 217.

5. I.A. Marshall, J.P. Mitchell and W.D. Griffiths, 4th Ann. Conf. of the Aerosol Society, 1990, 7.

6. M. Ozaki, S. Kratohvil and E. Matijevic, J. Colloid and Interface Sci., 1984, 1, 102.

7. J.M. Clark, K. Reid and P.H. Kaye, 5th Ann. Conf. of the Aerosol Society, 1991, 247.

8. E. Matijevic, Acc. Chem. Res., 1981, 14, 22.

9. E.J. Chupungco, J.A. Quintos and W.S. Godinez, Powder Met. Inter., 1982, 4, 14.

Instrumentation for Particle Size Analysis by Far Field Diffraction: Accuracy, Limitations and Future

B. J. Azzopardi

DEPARTMENT OF CHEMICAL ENGINEERING, UNIVERSITY OF NOTTINGHAM, NOTTINGHAM NG7 2RD, UK

1 INTRODUCTION

There have been significant developments in the instrumentation since the First Particle Size Analysis Conference in 1966. At that event, methods such as photo-microscopy and sedimentation were the focus of interest. A recent semi-automated instrument mentioned was the Ziess-Endter Particle Size Analyzer. This device consisted of a light spot whose size can be adjusted, by an iris, to match the size of each particle on enlarged microphotographs. When the spot is at the correct size, a button is pressed. This incremented the correct counter (out of, say, 64) and brought down a needle on an arm to pierce the particle image and so eliminate double counting. It is interesting to reflect that this instrument was a great step forward, decreasing the time to process the 1000-2000 particles necessary for a statistically meaningful analysis to hours instead of days for manual measurements. This brings into stark relief the ease of operation of modern diffraction based instruments which can obtain a result, based on several thousand particles, in seconds.

Analysis of particle size is important in a wide range of industries. When the particles are powders, the information is required for research and development as well as for process quality control. Allen[1] states that two thirds of the 3000 products of Dupont's products were sold as powders and that another fifteen percent contained powders. Whilst figures for other companies may be lower, this illustrates the importance of powders for the chemical process industry. If drops are being considered, the interest usually arises from research and development into sprays from nozzles or from confined two-phase flows. The former can have applications from combustion to crop protection. In two-phase flows, the size of drops can affect momentum, heat and mass transfer. In addition, drop sizes are essential for the design of compact gas/liquid separators.

In all cases fast, accurate methods are advantageous. However, because of problems such as coalescence, not all techniques or instruments are suitable for studies involving drops. Following a critical review of methods that had been used for drop size measurement, Azzopardi[2] concluded that non-intrusive optical techniques can be most suited for such applications.

This paper places the evolution of particle sizing by diffraction in historical context, examines development of modern instruments and examines the capability of instruments currently available. Consideration is then given to the accuracy of such instruments, their limitations are discussed and speculations on future developments of such instruments and their applications are made.

Historical Perspective

The scientific knowledge that the position of diffraction rings depended on particle size has been known for several centuries. Indeed, an instrument for sizing of fibres or particles was first reported more than 170 years ago. The next major step forward was the re-invention of the instrument by Pijper[3] in 1918. He used the technique for the study of the size of blood cells. A number of other applications in this field followed in the 20's and 30's. A commercial version of the instrument developed by Pijper was available from Carl Ziess, Jena - *Blutzellenprufer nach Dr Pijper*. This approach depended on the size distribution of the sample being narrow. The particles could be treated as essentially mono-disperse and particle size could be deduced from the position of the diffraction rings. Figure 1 illustrates the basis of a version of the instrument used for comparative studies.

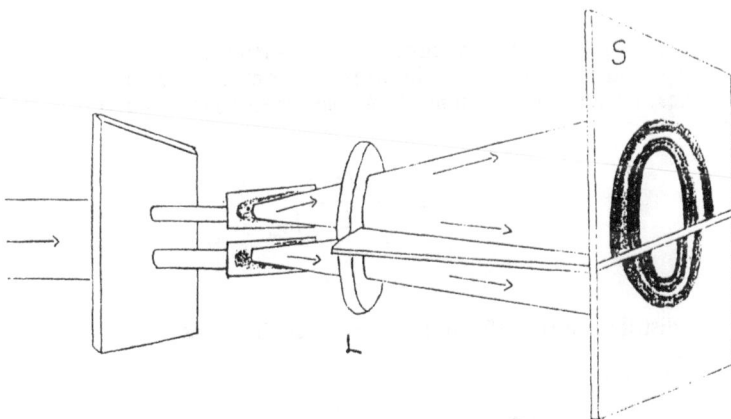

Figure 1 Schematic of Pijper Blood Cell Tester showing direct comparison of two blood films: normal blood on one, pernicious anaemia on the other[4].

Amongst the non-medical applications, an early paper is that of Lothian and Chappel[5]. The next step forward is the work of Sliepcevich and coworkers[6,7] who were originally motivated by studies on sprays. They developed the theoretical basis to relate the angular distribution of diffracted light to a particle size distribution and used it to deconvolute experimental data.

The first paper on this topic to appear in a PSA conference was that of Talbot[8] in 1970. In the seventies, papers such as that of Cornillault[9] put forward the use of light energy instead of intensity as having advantages in requiring less wide ranges of sensitivity. Advances in opto-electronic components led to the next series of developments with the work of Swithenbank et al[10] and Kopp et al[11] and Mullaney et al[12]. These used multi-element detectors developed for space studies. A paper by Butters and Wheatley[13] which described experience in the use of such an instrument appeared in PSA 1981. Modern instruments have developed from this type of work leading to configurations of the type shown in Figure 2.

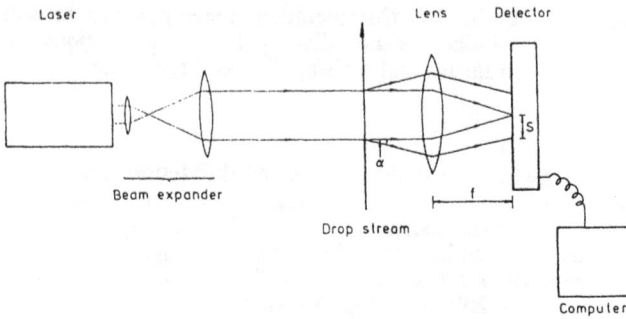

Figure 2 Main components of a modern diffraction based instrument for particle sizing

Theoretical Basis

In considering diffraction based instruments it is interesting to review the basic equations employed to abstract size information from scattered light data. For single particles, the angular variation of light scattered by diffraction is given by

$$I(\theta) = I_o \left(\frac{a J_1(\alpha\theta)}{\theta} \right)^2 \qquad (1)$$

Which for a distribution of different sized particles, becomes

$$I(\theta) = I_o \int_o^\infty f(a) \left(\frac{a J_1 \alpha\theta}{\theta} \right)^2 da \qquad (2)$$

Gumprecht and Sliepcevich[6] and Chin et al[7] inverted equation 2 to yield

$$a^2 f(a) = \frac{-4\pi^2}{\lambda} \int_o^\infty (\alpha\theta) J_1 (\alpha\theta) Y_1 (\alpha\theta) \frac{d}{d\theta} (\theta^3 I(\theta)) d\theta \qquad (3)$$

With this, the particle size distribution can be obtained from the angular variation of scattered light intensity. Knight et al[14] suggest an alternative form of equation 3

On the whole, the intensities to be measured cover a very wide range. In an alternative approach, Cornillault[9] and Swithenbank et al[10] considered the energy falling into an annular ring from a distribution of particles defined by their weight distribution.

$$a^2 f(a) = \frac{-4\pi^2}{\lambda} \left\{ \begin{array}{l} \displaystyle\sum_{n=1}^{N_{det}} In \int_{\theta_{in}^n}^{\theta_{out}^n} K(\alpha\theta) \frac{d}{d\theta} \left[A(\theta)\theta^3 \right] d\theta \; + \\[2em] \displaystyle\sum_{n=1}^{N_{det}-1} \int_{\theta_{out}^n}^{\theta_{in}^{n+1}} K(\alpha\theta) \frac{d}{d\theta} \left[A(\theta)\theta^3 I(\theta) \right] d\theta \end{array} \right\} \tag{4}$$

where $\quad K(\alpha\theta) = \alpha\theta \, J_1(\alpha\theta) \, Y_1 \, (\alpha\theta)$

$$A(\theta) = \left[1 - \frac{\theta^2}{\theta_{max}^2} \right]^2$$

$$E_{ij} = \int_{k=o}^{M} C'' \frac{W_k}{d_k} \left[\left(J_1^2(\alpha\theta) + J_o^2(\alpha\theta) \right)_{jk} - \left(J_1^2(\alpha\theta) + J_o^2(\alpha\theta) \right)_{ik} \right] \tag{5}$$

If the detector consists of a series of rings, a matrix equation is obtained

$$[E] = [T][W] \tag{6}$$

where [T] is the matrix of coefficients which define the energy distribution curve for each particle. The range of energies involved is much smaller than that for intensity and therefore easier to measure. Originally, the weight distribution was obtained from the energy distribution by selecting an equation to describe the particle size distribution guessing values of the constants and calculating the energy distribution. This was compared with the measured energy. The values of the constants were adjusted until the best fit between measured and calculated energy was achieved. The more direct approach of inversion was not used because of the large dynamic range of the coefficients which can result in considerable computational difficulties and non-physical solutions.

Caroon and Borman[15] followed Phillips[16] and Twoomey[17] in suggesting an inversion approach which overcame the problems discussed above.

$$[W] = \left([T]^* \, [T] + \gamma \, [H] \right)^{-1} [T]^* \, [E] \tag{7}$$

This inversion problem continues to receive attention, Bertero et al[18].

Recently, Boxman et al[19] have suggested that more information could be obtained from diffraction instruments if the fluctuations of the signals from each detector could be used together with the mean values. They define the standard deviation for each detector as

$$\sigma_{ii} = \left(\sum E_{ik}^2 - s\bar{E}_i\right) / s(s-1) \tag{8}$$

the standard deviations for signal and background are combined by

$$\sigma_{(ii)_t} = \sqrt{\sigma_{iis}^2 + 2\,\sigma_{iib}^2} \tag{9}$$

This is the used to weight the main signal from that detector element.

$$E_i = \bar{E}_i / \sigma_{iit} \tag{10}$$

Boxman et al[19] note that the approach can identify whether the inaccuracy is due to insufficient sampling of the detector array or imperfections in the optical model.

Development of Modern Instruments

The first of the modern diffraction based instruments for particle size analysis is probably that produced by CILAS (France). This is described by Cornillault[9] and used a moveable mask to detect light scattered at different angles. The Leeds and Northrup Microtrac dates from 1976, this was described by Wiess and Frock[20].

Developments in multi-element photo-detectors for space applications led to the next steps forward. In 1976 the Malvern Instruments ST1800 became available. This used an array consisting of 31 semi-annular detectors. The scattered light information was converted to a particle size distribution by a PDP8 computer using an electro-mechanical input/output device. Programme loading was by punched paper tape. Versions from this company using more flexible computers appeared in 1981 (Commodore PET) and 1985 (PC).

Instruments Currently Available

There are several instruments currently available for particle sizing based on far field diffraction. Details of many of them are given in Table 1. This is not intended to be comprehensive, but to illustrate the range of possibilities. The instruments are now easy to use. They provide guidance on correct particle concentration in the circulating dispersant, advice on the compatibility of the sample/instrument size ranges etc.

Table 1

Company	Model	Launch date	Particle Size Range μm	Control/ output	Low Size Method
Cilas	715 850	(1972)	1-192 0.1-600	MP	
Coulter Electronics	LS100 LS130	1989	0.4-850	PC	P.I.D.S. Double lens train
Fritsch	Analysette 22	1987	0.16-1160	PC	
Horiba Instruments Ltd.	LA500	1989	0.1-200	MP	Backscatter
Leeds & Northrup	Microtrac SRA FRA OPA	(1976) 1989 1990 1990	0.7-700 0.1-700 0.005-3	earlier MP Now PC	Highangle Brownish
Malvern Instruments Ltd.	ST1800 2200 2600 3600 Mastersizer	1976 1981 1982/85 1983 1987	1.2-1880 0.5-1880 0.1-600	PDP8/ teletype PET }PC } } }	}Anomalous }Mie
Shimadzu	SALD 1100	1989	0.1-45 1-150 5-500	MP	Mie sidescatter

Almost all instruments have the following sample presentation options:-

Dispersion in liquid; Ultrasonic dispersion; Stirrer; Circulation pump; Cell; Dry powder dispersion.

2 ACCURACY

Definition

It is obvious that in any system where quantitative measures are used, accurate measurements are essential. Allen[1] points out that absolute accuracy is more important for research and development functions whilst for process control, the important criterion is reproducibility or precision. However, modern developments place extra demands on production processes. The increasing importance of Quality Assurance requires knowledge of measurement systems and an ability to trace back the attribution of accuracy to an acceptable standard.

Accuracy has to be carefully defined, particularly for a (drop or particle) population containing a range of sizes. The simplest description is via some appropriate mean size such as the Sauter Mean Diameter (d_{32}, particle with the same volume /surface area ratio as ensemble of particles, sphere equivalent) or the mass median diameter (d_{50}, size below which lie 50% by weight of the sample). The next step is to consider a second parameter describing the width of the distribution. A simple definition is a dimensionless span,

$$S = \frac{d_{90} - d_{10}}{d_{50}} \tag{11}$$

Alternative width parameters are standard deviations for ensembles described by normal or log-normal distributions or the parameter N in the Rosin Rammler equation. These, of course, require the ensemble to be a good fit to the appropriate equation. More detailed comparisons are possible by examining the cumulative volume undersize curves, but this is a graphical comparison.

In an attempt to reduce this comparison to a single number, Allen[1] ,suggests a measure of how closely a sample reflects a standard. This accuracy, A, is defined as

$$A = 100 \frac{(x-x_s)}{x_s} \tag{12}$$

where x and x_s are the measured and standard sizes at the percentiles. He further suggests that since the differences between the measured and standard distributions are difficult to quantify at the extremes of a distribution, a standard error as the average value of A from 10% to 90%

$$\bar{A} = \sum_{10}^{90} \frac{|A|\Delta P}{80} \tag{13}$$

However, this approach favours differences at the small size end of the distribution. An alternative approach is to consider the differences in volume fraction over successive size intervals. These intervals could be 1/10 of the range of particle present.

For reproducibility, Allen[1] suggests an average standard deviation

$$\bar{S} = \sum_{0}^{100} \frac{S\Delta P}{100} \tag{14}$$

Early Testing

The earliest test of diffraction based particle sizing instruments was published by Haden[21] in 1938. He considered the Emmon Eriometer, the Bock Erythometer and Dr Pijper's blood cell tester. These instruments depended on near mono-disperse samples. The position of the diffraction rings was used to determine particle size. Haden compared these diffraction methods to direct microscopic measurements. He concluded that diffraction methods are satisfactory for measuring mean sizes and recommended specific instruments for different types of tests.

For a far field diffraction instrument (Malvern ST1800), amongst the first to publish results of testing were Azzopardi[22] and Hammond[23]. The former used photography and sedimentation to determine the size distribution of samples of glass spheres. Hammond employed commercially available "standard" distributions of polystyrene spheres. The result of the comparison is shown in Figure 3 in terms of the Sauter Mean Diameter.

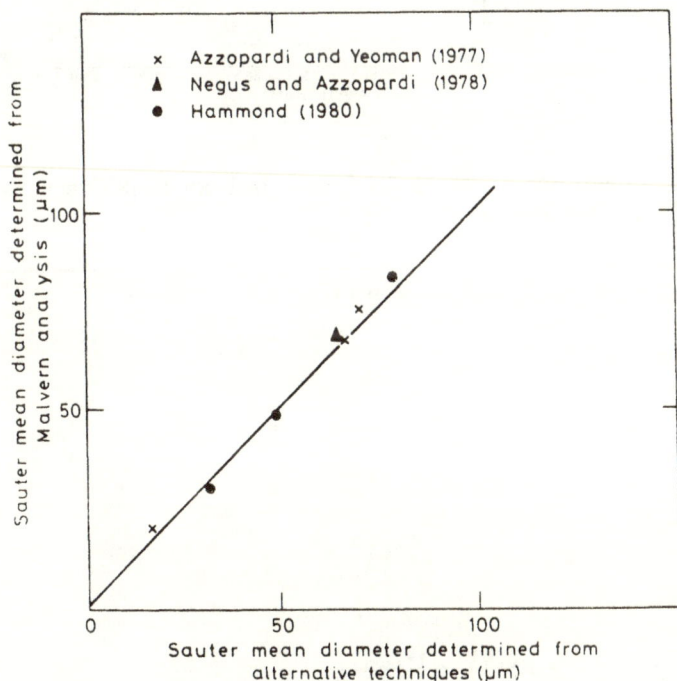

Figure 3 Testing of Malvern ST1800 - Sauter Mean Diameters (Azzopardi[22])

This indicates good agreement between the diffraction approach and the alternative measuring methods. However, examination of the distributions, Figures 4, 5 and 6, shows that for some cases there is good agreement throughout the distribution but in others there are noticeable differences. These differences are not necessarily an indication of inaccuracy in the diffraction instrument but could be due to difficulties in the sampling/photography technique.

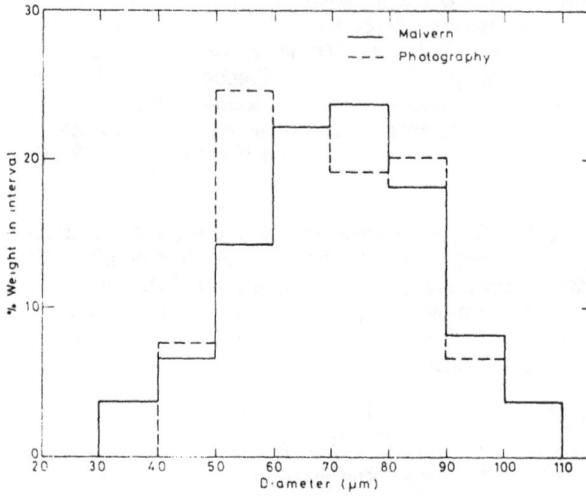

Figure 4 Comparison of the results from a Malvern ST1800 with those from photography (Azzopardi[22])

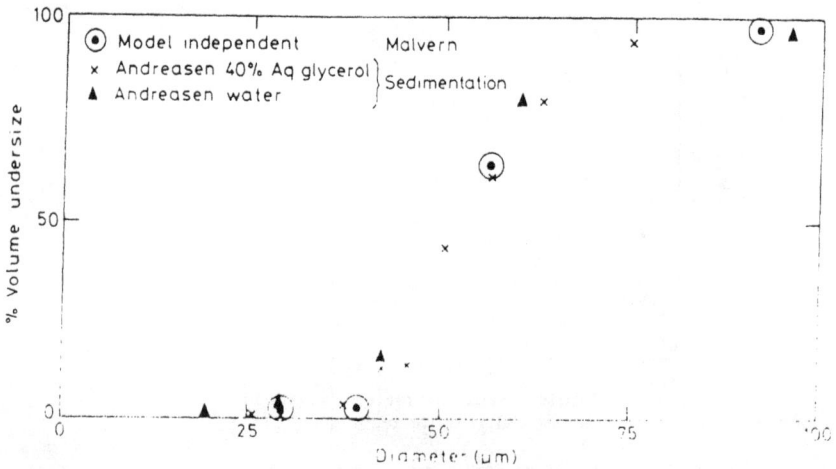

Figure 5 Comparison of the results from a Malvern ST1800 with those from sedimentation (Azzopardi[22])

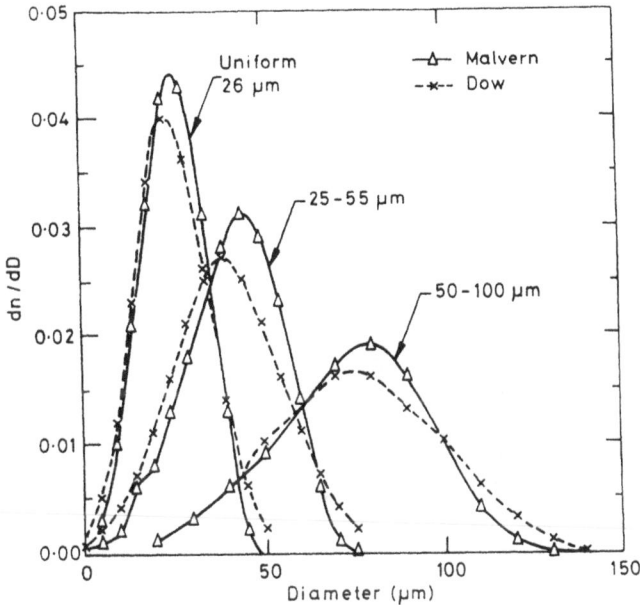

<u>Figure 6</u> Comparison of the results from a Malvern ST1800 with supplier specification for latex spheres (Hammond[23])

An Alternative Standard

A most useful approach has been published by Hirleman et al[24]. This consisted of an artificial "aerosol" made up of an array of chrome thin-film circles on a transparent glass substrate. This calibration reticle[*] contains more than 10000 of these particles randomly positioned in an 8 mm circle to simulate a Rosin Rammler distribution of spherical particles. Versions are available with different Rosin Rammler parameters X and N, and at least two levels of obscuration. The continuous Rosin Rammler size distribution is approximated on the reticle by 24 discrete particle sizes ranging from 2 to 105 μm depending on the individual reticle. Each reticle is individually certified. An example of the goodness of fit to the Rosin Rammler distribution is shown in Figure 7. Hirleman et al use the reticle to check the performance of a Malvern 2200 model.

---

[*]Laser Electro-optics, Tempe, Arizona, USA.

RR-50-3.0-0.08-102-CF-#111

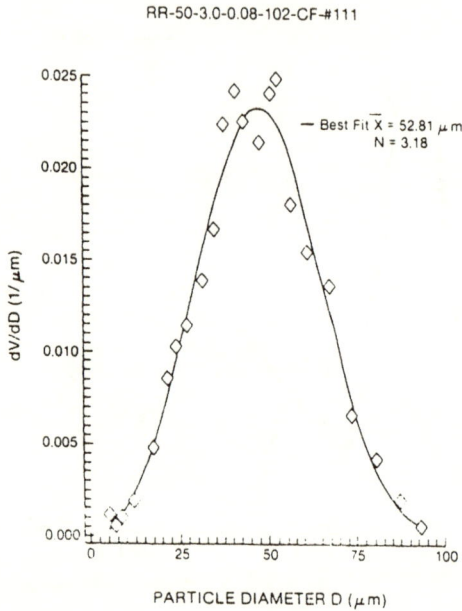

PARTICLE DIAMETER D (μm)

Figure 7 Goodness of fit of Rosin Rammler equation of 'particle' population on reticle, Hirleman et al[24]

Recent Testing

A systematic test of particle sizing instruments has been carried out by Allen of Dupont. This covered a wide range of instruments, including those which do not use far field diffraction. Standard BCR silica was used in this exercise. This is supplied by the Community Bureau of Reference in Brussels (part of the CEC) and is available in different median sizes from about 1 to 50 μm. At the 2nd International Congress on Optical Particle Sizing, Allen[1] described a series of tests with BCR 66 (particle size around 1 μm). A summary of the accuracy and reproducibility for far field diffraction instruments is shown in Figure 8. Here equations 13 and 14 are used to define accuracy and reproducibility respectively. It is noted that because of the size of the particles, it is not the diffraction aspect of the instrument that is being tested. This probably explains the apparent poor accuracy of some instruments.

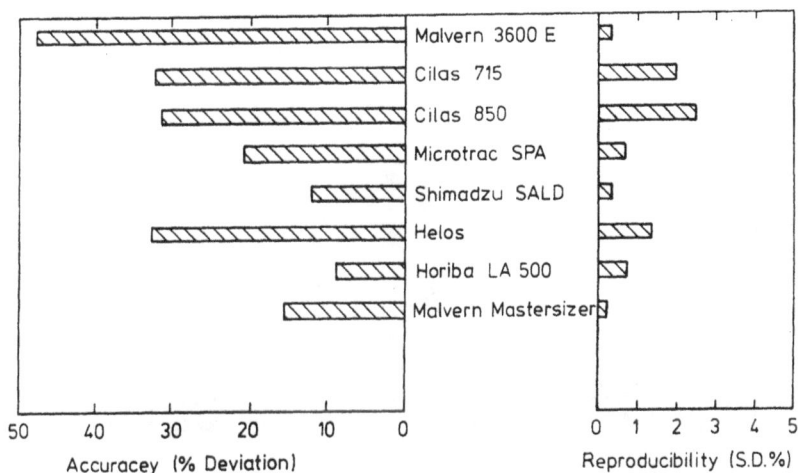

Figure 8 Comparison of accuracy and reproducibility of instruments, after Allen[1]

Within ICI, Bell[25] has been carrying out a comparative study on diffraction based instruments. Part of the exercise uses a mixture of standard silica (20% BCR66/80% BCR96). Figure 9 shows both the differential and cumulative undersize distributions as measured by one instrument. The variations in results produced by five different instruments is illustrated in Figure 10, the line marked calculated has been produced from data supplied by BCR for their reference materials. This information has been collected over a number of years and it is noted that in some cases improvements have been made to the algorithms used by manufacturers since the tests were carried out. In order to particularly test the accuracy of instruments at the lower end of the size spectrum, Bell has identified the cumulative percentage in sizes less than 10 μm, the split point between the two fractions making up the mixture. Figure 11 shows the wide range of results produced by different instruments. Where instruments use the Fraunhofer diffraction model, the data is marked with an 'F'. The other data refer to instruments which employ more elaborate optical models demanding knowledge of the optical properties of the sample.

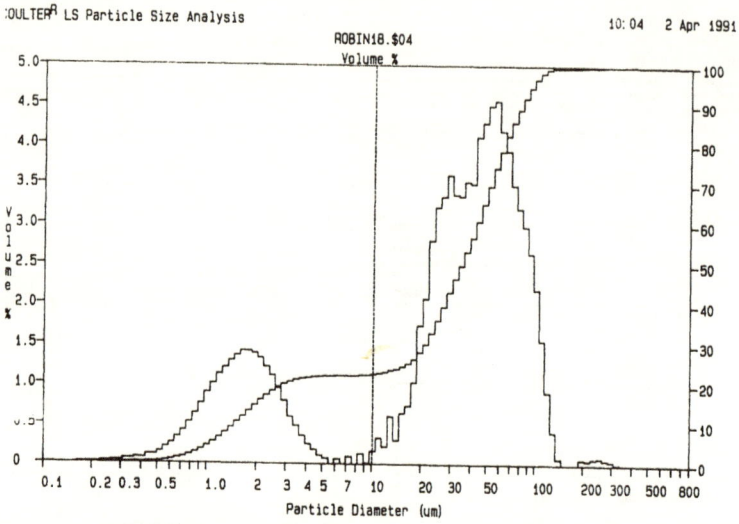

Figure 9 Mixture of BCR66 and BCR69 used by Bell[25]

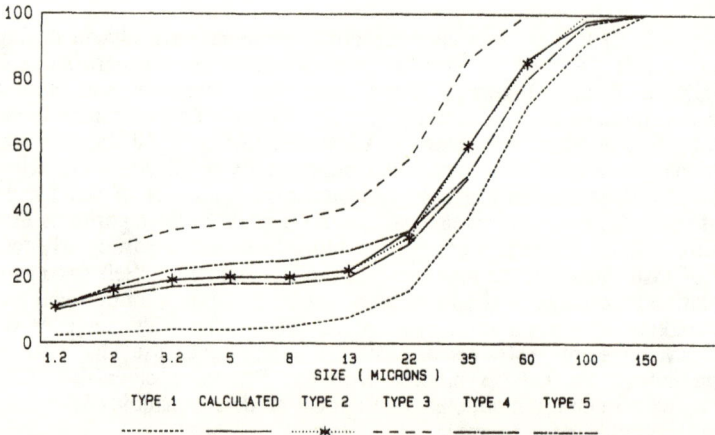

Figure 10 Comparison of measurements of different diffraction based instruments, sample is 20% BCR66/80% BCR69, Bell[25]

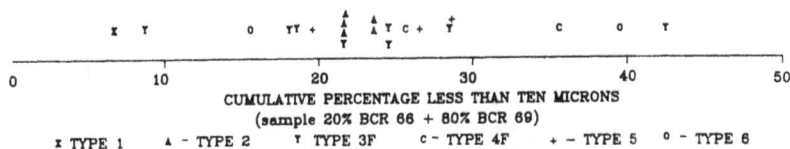

CUMULATIVE PERCENTAGE LESS THAN TEN MICRONS
(sample 20% BCR 66 + 80% BCR 69)

x TYPE 1   ▲ - TYPE 2   ▼ TYPE 3F   c - TYPE 4F   + - TYPE 5   ° - TYPE 6

<u>Figure 11</u> Instrument performance by type, Bell[25]

Present Exercise

As part of the preparation for this paper a short comparative study has been carried out. Its purpose was to gain familiarity with as wide a range of instruments as possible in the time available and particularly to learn from operator's experiences. Four instruments were involved: (i) a Malvern 2600; (ii) an early Microtrac; (iii) a Horiba LA500 and (iv) a Coulter LS130. Four samples were selected for the exercise: (a) a narrow sieve cut of glass spheres; (b) precipitated calcium carbonate; (c) barium sulphate and (d) a reticle containing 13252 circles whose sizes corresponded to a Rosin Rammler distribution with parameters 50 and 2.0 (ref no. #69). It was recognised from the outset that some instruments would not cope well with some samples as (b) and (c) contained particles smaller than the bottom limit of some of the instruments. Results were obtained for all samples on all instruments. This was particularly pleasing as most cases the visits to outside laboratories only occupied part of a day. The results obtained are shown in Figures 12 to 15 .

These samples were selected for their availability but also because they could illustrate specific points. They needed to be, where appropriate, dispersable in water. The reticle was taken as an absolute standard whilst the glass beads were in the size range where the defraction theory is known to be appropriate. Moreover, they were spherical. The two insoluble salts ($CaCo_3$, $BaSO_4$) were in the micron and sub-micron range, extending beyond the limits of some instruments and where diffraction theory by itself was insufficient.

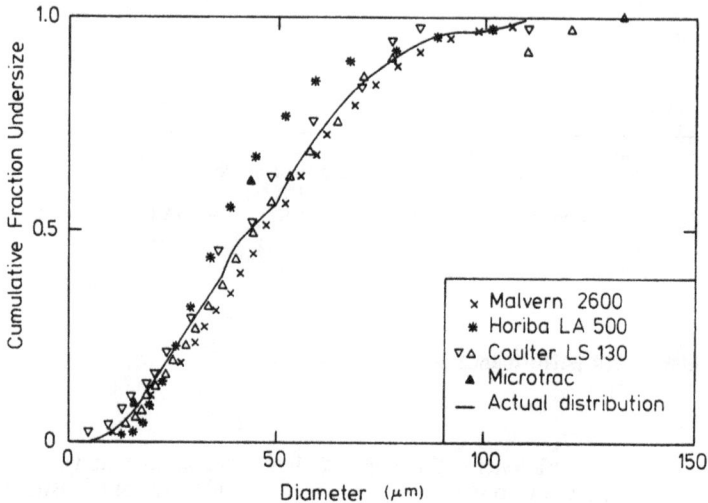

Figure 12 Comparison of output of different instruments - reticle

The results obtained with the reticle show that the Malvern and Coulter instruments give accurate measurements. In contrast, the results obtained with the Microtrac and Horiba show noticeable deviations from the standard. However, it is believed that this is not inaccuracy in the instrument but due more to a problem of range. The reticle has 'particles' in the range 2-110 μm. the particular Microtrac model has a specified range of 0.7-125 μm whilst for the Horiba it is 0.1-200 μm. At first glance the reticle and instrument ranges appear compatible but in most instruments sub-ranges are not uniform subdivisions, they tend to have an approximately logarithmic increase towards the top end. This is usually related to differences in the size of detectors. Therefore, unless the peak in the size spectrum corresponds to one of the detectors in the middle of the range, a significant amount of information will go unseen. this effect is best illustrated by the results shown in Table 2. These were taken using a sample of the glass beads used above, together with a second narrow sieve cut at around 150 μm. These samples were held in separate cells and measurements were made with 63, 100 and 300 mm focal length lenses using a Malvern 2600. Inspection of the ranges and sub-ranges would normally have led to the use of the 300 mm lens. Interestingly, the other (smaller focal length) lenses give values of the mass median lower than that obtained with the most appropriate lens. This is the same deviation as was seen with the reticle.

Table 2 Effect of focal length of Fourier Transform Lens on measured particle size

Focal length of lens (mm)	Apparent Size Range (μm)	Larger Mass Median Diameter eq. (μm)	Span Eq. (11)	Smaller Mass Median Diameter (μm)	Span Eq. (11)
63	1.18-118	84.67	0.63	52.84	1.09
100	1.88-188	137.09	0.58	70.3	0.46
300	5.64-564	139.16	0.34	76.4	0.39

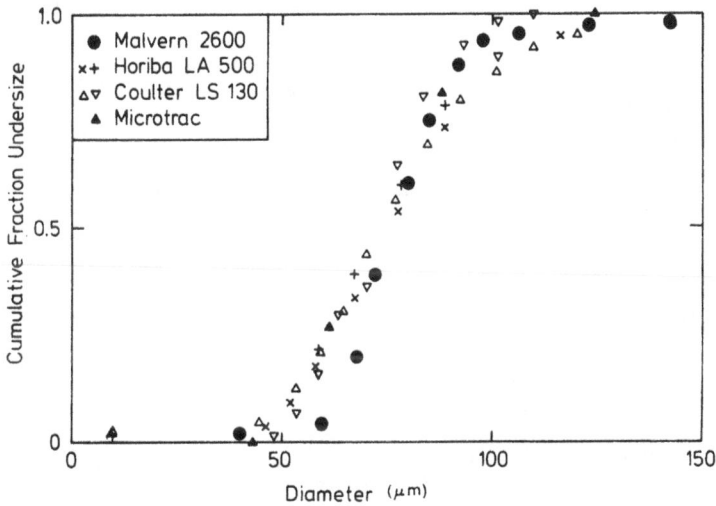

Figure 13 Comparison of output of different instruments - glass beads 77-89 μm nominal

In the case of the glass beads, there is good agreement from all instruments.

The two powders ($CaCo_3$, $BaSO_4$) illustrate the need to have an instrument with a size range that encompasses that of the sample. The Horiba and the Coulter have suitable ranges, but the particular Microtrac and Malvern models do not go down to small enough sizes.

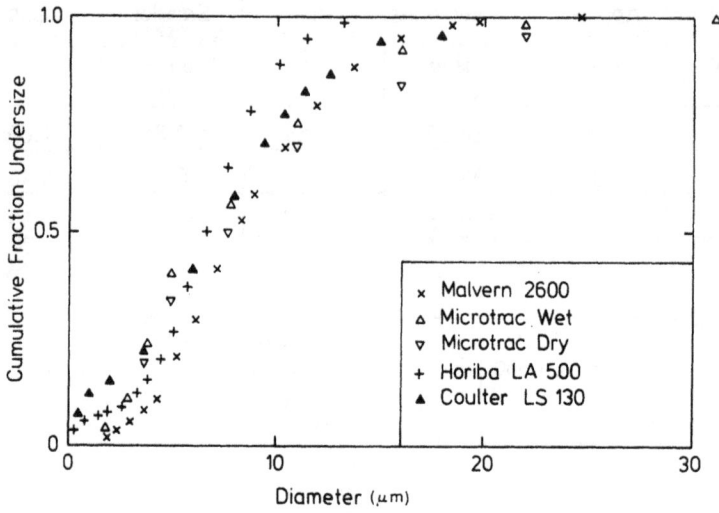

Figure 14 Comparison of output of different instruments - calcium carbonate

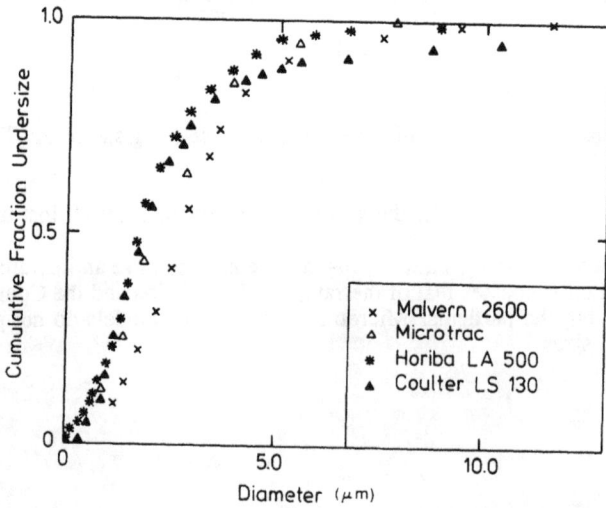

Figure 15 Comparison of output of different instruments - barium sulphate

3 LIMITATIONS

Multiple Diffraction

Most modern instruments specifically designed for the sizing of powders provide an automatic or semi-automatic control on the concentration of particles in the dispersion on which measurements are being made. This ensures that there is sufficient scattered light to provide a good signal to noise ratio whilst keeping out of the region where there is significant re-scattering of the scattered light by other particles, an effect illustrated in Figure 16. Obviously, it is not feasible to effect this type of control in sprays and other drops systems.

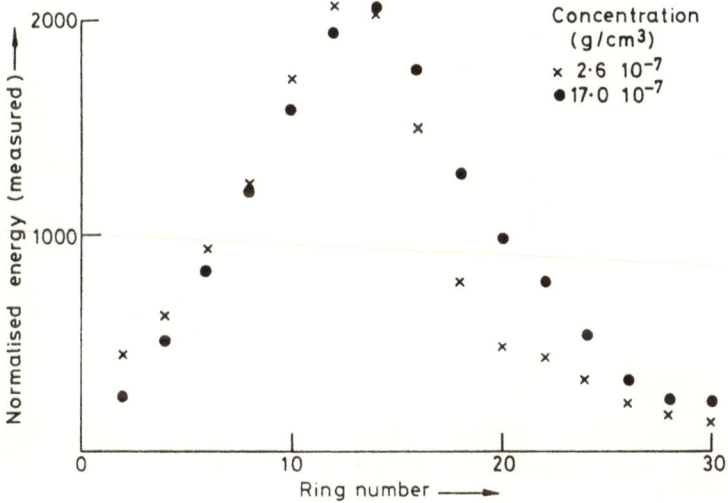

Figure 16 Measured normalised energies - Effects of concentration (Azzopardi[22])

A number of workers have put forward methods to correct for multiple diffraction. Felton et al[26] provided correction factors for those cases where the distribution information was provided in the form of a two parameter equation such as Rosin Rammler, normal or log-normal. Hamidi and Swithenbank[27] published a method which was not linked to specific distribution equations. Figure 17 shows the efficacy of this method. These workers divided the spray into slices along the laser beam and calculated the cumulative effect.

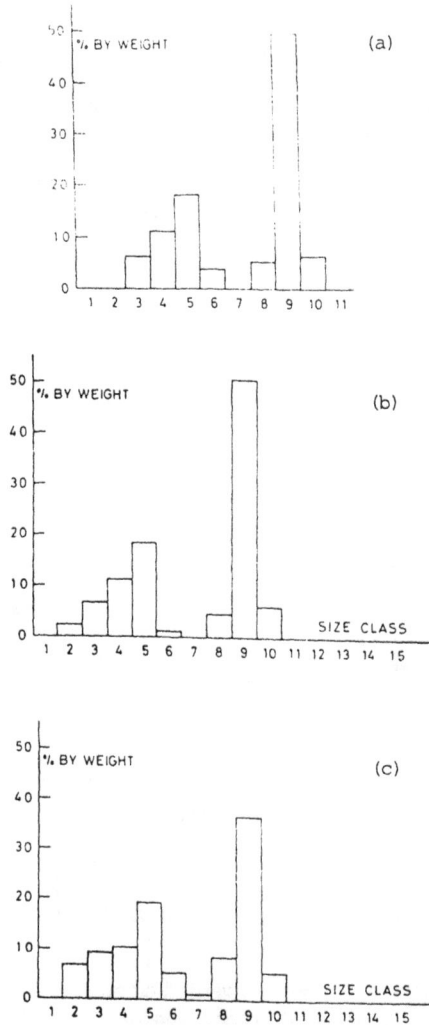

Figure 17 Effect of multiple scattering and correction: (a) sample at low concentration; (b) apparent size distribution at high concentration; (c) corrected high concentration data (Hamidi and Swithenbank[27])

Cao et al[28] extended the approach by paying special attention to scattering events which occurred off the optical axis and give a diffraction pattern the centre of which is displaced from the centre of the light energy detector plane. Figure 18 shows that the corrected diffraction data agrees well with phase doppler measurements whilst the uncorrected data (single model) is significantly different. Hirleman[29] has also given attention to this problem.

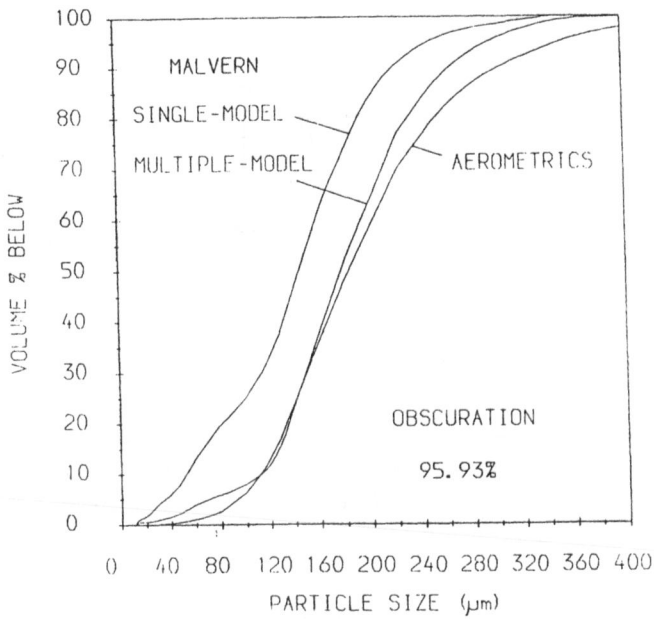

Figure 18 Comparison between Diffraction and phase doppler particle size measurement (Cao et al[28])

Particle/Medium Refractive Index Ratio

The basic diffraction theory assumes that the particle is to a good approximation opaque. However, for many particles of industrial interest, this is not a reasonable assumption. Consideration has then to be given to the ratio of refractive indices of the particle and the medium in which it is dispersed. Brown et al[30] have studied the effect of changing the refractive index of the medium. Figure 19 shows that when the particle and liquid refractive indices are equal there can be large error in the particle size determined using diffraction, even with large particles where diffraction theory would normally be expected to hold.

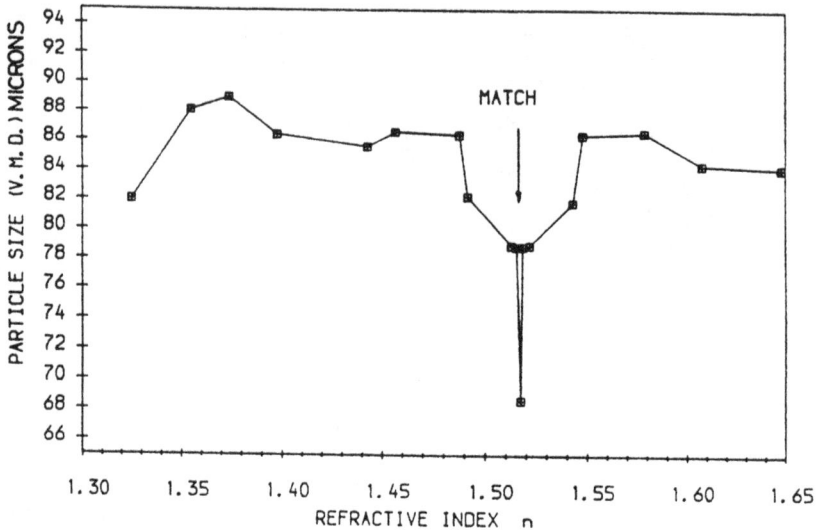

Figure 19 Change of apparent particle size with refractive index

Lightfoot and Watson[31] argue that full Mie scattering theory needs to be used for these difficult cases.

Other Limitations

There are limits to the size of particles whose size can be determined from diffraction theory. Jones[32] gave some error contour charts for the lower size end. However, most modern instruments include methods, over and above the diffraction theory, for this end of the spectrum. Some measure the scattered light distribution over angles of up to 50 degrees. Others have extra detectors at the side. Full Mie theory and approaches such as the ratio of light of different polarities scattered are employed. Cornillault[33] gives details of the method used in one instrument.

Another property of powders that could affect the results obtained with diffraction based instruments is the shape of the particles. As many particles are approximately spherical or at least circular in projection this might not be a big problem. Needle shaped particles or cylinders pose a completely different problem. Swithenbank et al[34] have considered the effect of cylinders. Obviously, a cylinder, like a slit has a linear not a circular diffraction pattern. However, for randomly orientated cylinders an equivalent circular diffraction pattern was produced which gave a dimension 12% smaller than the cylinder diameter. The scattered light is insensitive to the cylinder length (assuming $L/D > 3$).

4 FUTURE

It is much more difficult to write about future developments than about accuracy or limitations on present instruments. Some trends are evident. The increased emphasis on Quality Assurance or Total Quality Management will produce a requirement for regular testing of instruments against standards. With increasing computing power it should be very easy to store the last result from the standard and output the differences between the two tests (one instrument already has this capability). With the growth in packaged instruments with built in suspension circulation systems, cleanliness will probably become of increasing importance with recording of cleaning (and checking with standards) the order of the day. There might even be a requirement for instrument manufacturers to make available response curves for the detector elements. Self checking of instruments should also be possible. An approach such as that of Boxman et al[19], examining the fluctuating component of the signal from each detector might identify whether it is the signal quality or the optical model used for data abstraction that is the weak link. Figure 20 shows data treated in this manner. The important trends in the energy distribution can be seen to emerge from the noise as the number of scans is increased. The approach might allow an operator to assess when more scans would be useful.

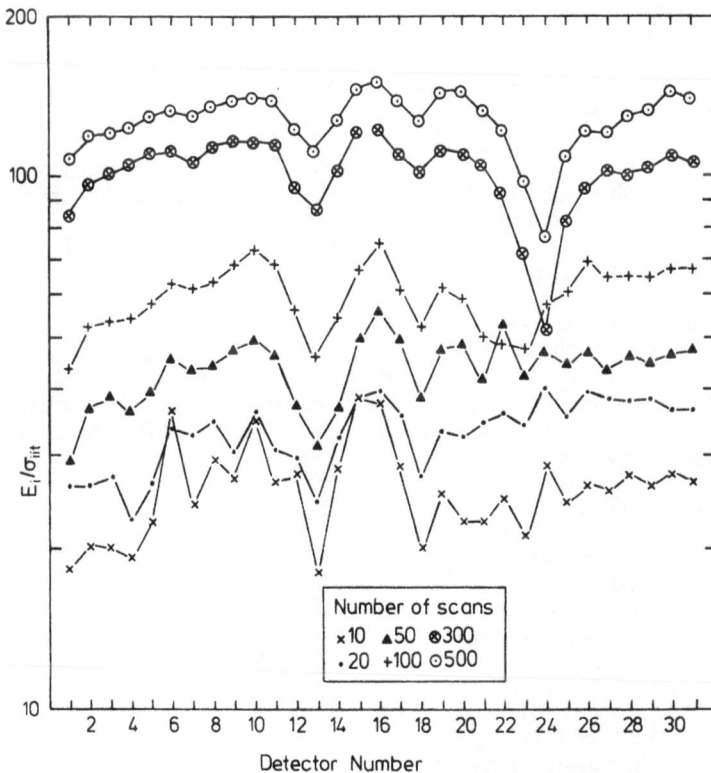

Figure 20 Effect of number of scans on light energy distributions, analysis utilising signal fluctuations.

Another trend that can also be seen is for instruments to have a bottom limit well into the sub-micron region.   This will probably become more wide spread.

On line measurement has long been considered by some people as a 'holy grail'.   However, because of the long path-lengths involved across most industrial ducts and the problems of keeping windows clean, it will always be unattainable. Only when there is no alternative, such as with gas/liquid flows in pipes, will this approach be used.   With 'near' on-line systems, the method of sampling must be designed with great care.   It is possible that extracting a sample from a side tube will introduce errors.   Such an arrangement is a form of T junction, a geometry that is notorious for separating phases.   Indeed, Nasr-el-Din et al[35] have shown that such an arrangement can also lead to partition by size.

5 CONCLUSIONS

There is no doubt that following the availability of lasers, opto-electronic detectors and microprocessors, great steps forward have been taken in diffraction based instruments for particle size analysis.   The range of sizes that the instruments can handle and their facility to advise on optimal concentrations of dispersions are particularly helpful.   However, care must still be taken in applications in ensuring that the particle size range is correctly within the range of the instrument.   It is insufficient to be within the stated size range, the peak of the distribution should be within the central sub-ranges.

The accuracy of instruments needs to be defined with care.   Additionally, because of the demands of Quality Assurance, manufacturers will probably need to justify the results produced by their instruments.   It might be that certified data on the linearity and response of detectors etc will need to be provided with each instrument together with a standard test method (such as the reticle described above) and calibration check procedures.

ACKNOWLEDGEMENTS

The author would like to thank Messrs R. Crouch (Coulter Electronics Ltd), A. Howells (Chroda Applications Chemicals), R. Ward (Frodingham Cement Co.), Ms D. Rudland (Emca-Remex Products) and Mr B. Wilson (Department of Chemical Engineering, University of Nottingham) for their assistance in obtaining the results described above.   He would also like to thank Miss S. Close for her assistance in preparing the paper.

NOMENCLATURE

a	particle radius
[E]	light energy matrix
[H]	smoothign matrix
I	sattered light intensity
I_o	incident light intensity
J_o, J_1, Y_1	Bessel functions
[T]	Matrix of coefficients
[T]*	inverse of [T]
[W]	weight distribution matrix
α	$2\pi a/\lambda$
λ	wavelength of incident light
θ	angle from forward direction

REFERENCES

1. T. Allen, Proc. 2nd Int. Congress on Optical Particle Sizing, Tempe, Arizona, 1989, 359.

2. B.J. Azzopardi, Int.J.HeatMassTrans., 1979, 22, 1245.

3. A. Pijper, Med.J.S.Africa, 1918, 14, 211.

4. A. Pijper, J.Lab.Clin.Med., 1947, 32, 857.

5. G.F. Lothian and F.P. Chappel, J.AppliedChem., 1951, 1 475.

6. R.O. Gumprecht and C.M. Sliepcevich, J.Phys.Chem., 1953, 57, 90 and 95.

7. J.H. Chin, C.M. Sliepcevich and M. Tribus, J.Phys.Chem., 1955, 59, 841 and 845.

8. J.H. Talbot, Particle Size Analysis 1970, Soc. Analytical Chem.

9. J. Cornillault, Appl.Opt., 1982, 11, 265.

10. J.Swithenbank, J. Beer, D.S. Taylor, D. Abbot and C.G. McCreath, AIAA Paper No. 76-69, 1976.

11. R.E. Kopp, J.Lisa, J. Mendelsohn, B. Pernick, H. Stone and R. Wohlers, J.Histochem.Cytochem., 1976, 24, 123.

12. P.F. Mullaney, J.M. Crowell, G.C. Salzman,J.C. Martin, R.D. Hiebert and C.A. Goad, J.Histochem.Cytochem., 1976, 24, 298.

13. G. Butters and A.L. Wheatley, Particle Size Analysis 1981, Wiley-Heyden Ltd.

14. J.C. Knight, D. Ball and G.N. Robertson, Proc. 2nd Int. Congress on Optical Particle Sizing, Tempe, Arizona, 1989, 580.

15. T.A. Caroon and G.L. Borman, Combust.Sci.Tech., 1979, 19, 255.

16. B.L. Phillips, J.Assoc.Comp.Mach., 1962, 9, 84.

17. S. Twoomey, J.Assoc.Comp.Mach., 1963, 10, 97.

18. M. Bertero, P. Boccacci and E.R. Pike, Inverse Problems, 1985, 1, 111.

19. A. Boxman, H.G. Merkus, P.J.T. Verheijen and B. Scarlett, Proc. 2nd Int. Congress on Optical Particle Sizing, Tempe, Arizona, 1989, 178.

20. E.L. Wiess and H.N. Frock, Powd.Tech., 1976, 14, 287.

21. R.L. Haden, J.Lab.Clin.Med., 1938, 23, 508.

22. B.J. Azzopardi, Filt.&Sep., 1984, 21, 415.

23. D.C. Hammond, General Motors Report, GMR 3195.

24.   E.D. Hirleman,V. Oechsle and N.A. Chigier, Opt.Eng., 1984, 23, 610.

25.   J. Bell, RSC Meeting on Particles Today, Luton, 16th April 1991.

26.   P.G Felton, A.A. Hamidi and A.K. Aigal, Proc 3rd ICLASS, Institute of Energy, London, 1985.

27.   A.A. Hamidi and J. Swithenbank, J.Inst.Energy, 1986, 59, 101.

28.   J. Cao, D.J. Brown and A.G. Rennie, J.Inst.Energy, 1991, 64, to be published.

29.   E.D. Hirleman, Proc. 2nd Int. Congress on Optical Particle Sizing, Tempe, Arizona, 1989, 159.

30.   D.J. Brown, K. Alexander and J. Cao, Proc. 2nd Int. Congress on Optical Particle Sizing, Tempe, Arizona, 1989, 451.

31.   N.S. Lightfoot and D.J. Watson, Proc. 2nd Int. Congress on Optical Particle Sizing, Tempe, Arizona, 1989, 511.

32.   A.R. Jones, J.Phys.D: Appl.Phys., 1977, 10, L163.

33.   J. Cornillault, 17th Annual meeting of the Fine Particle Society, San Francisco, July 28 -August 2, 1986.

34.   J. Swithenbank, J. Cao, P. Wild, F. Boysan and A.A. Hamidi, Conf. on Spray Combustion, Rouen, France, 28-29 March, 1988.

35.   H. Nasr-el-Din, A. Afacan and J.H. Masliyah, Chem.Eng.Comm., 1989, 82, 203.

Hydrophobic Drug Substances: The Use of Laser Diffraction Particle Size Analysis and Dissolution to Characterize Surfactant Stabilized Suspensions

T. W. Atkinson and S. White

PHARMACEUTICAL SCIENCES, SMITHKLINE BEECHAM PHARMACEUTICALS,
CLARENDON ROAD, WORTHING, UK

1 ABSTRACT

Pre-clinical studies are often performed with aqueous suspensions of the drug compound. the inclusion of a surfactant aids powder wetting and ensures good particle distribution. Tween 80[TM] (polyoxyethylene 20 sorbitan mono-oleate) is widely established in the pharmaceutical industry as a suitable agent for this use. However, it was found unsuitable for producing a stable suspension for a particularly hydrophobic drug substance.

Seven surfactants were selected with differing HLB (Hydrophile Lipophile Balance) numbers and used at concentrations of 0.5 - 2.5% w/v in aqueous solution. An intermediate concentrate was prepared by high shear mixing and subsequently diluted with 1% methylcellulose solution to give a final drug concentration of 10% w/v. Methods of analysis included laser diffraction particle size determination, optical microscopy and dissolution into an aqueous medium. The suspensions were analysed immediately after preparation and after 4 and 96 hours.

The most stable particle size distributions and dissolution profiles were obtained using Span 20[TM] (Sorbitan monolaurate) and Pluronic L62[TM] (Poloxamer 181). These agents have HLB values of 7 to 9 which is the approximate range recommended by HLB theory (7-10) for hydrophobic powder wetting[1].

2 INTRODUCTION

Hydrophobic drug substances are often presented as simple aqueous suspensions for use in pre-clinical studies. The resulting suspensions must have a stable particle size distribution to ensure consistent drug dissolution and subsequent bioavailability.

To enable formulation of such a suspension, a suspending agent and a powder wetting agent (in this study only surfactants were considered) are normally required. The pharmaceutical industry is conservative in its use of excipients due to the high costs incurred in obtaining acceptance by the regulatory authorities. Therefore, any surfactant or suspending agent used must comply with Pharmacopoeial requirements, even though the properties may not be ideal for the material concerned.

Scientific literature has indicated the wide acceptability within the pharmaceutical industry of Tween 80^{TM} (polyoxyethylene sorbitan mono-oleate) and Methylcellulose for use in the preparation of such a suspension. However, a more suitable surfactant was required to pre-wet the hydrophobic substance under test.

The purpose of this study was to find a suitable wetting agent and identify a method of preparing a stable suspension of this extremely hydrophobic drug substance with respect to its particle size distribution and aqueous dissolution properties. As a consequence, suitable methods of assessing these properties also needed to be developed.

Two main factors led to the selection of the various surfactants used in this study:

1. Previous acceptability in pharmaceutical preparations.

2. Physical properties of the surfactants (specifically the Hydrophile Lipophile Balance (HLB) number).

A comprehensive review of Pharmacopoeial surfactants was undertaken to ensure that the wetting agents selected were acceptable. Table 1 lists these surfactants together with their respective HLB numbers.

Table 1 HLB of the surfactants used.

Surfactant Type	HLB Number
Pluronic L62^{TM}	7 to 9
Span 20^{TM}	9
Tween 80^{TM}	15
Cetomacrogol 1000^{TM}	16
Tween 20^{TM}	17
Pluronic F88^{TM}	28
Sodium Lauryl Sulphate	40

The drug is very hydrophobic in nature and HLB theory suggests the use of a lipophilic wetting agent of HLB approximately 7 to 10[1]. Span 20^{TM} and Pluronic L62^{TM} have 'ideal' HLB values for this purpose, however a representative range of surfactants was examined.

Methylcellulose solution was chosen as the suspending agent and was used exclusively throughout the study.

All the surfactants selected are classed as being non or slightly toxic apart from SLS which is now considered to be an irritant.

3 MATERIALS

The present study compares the ability of seven surfactants to pre-wet a hydrophobic drug substance: Polyoxyethylene 20 sorbitan mono-oleate (Tween 80[TM], I.C.I. Pharmaceuticals, U.K.), Polyoxyethylene 20 sorbitan monolaurate (Tween 20[TM], I.C.I. Pharmaceuticals, U.K.), Sorbitan monolaurate (Span 20[TM], I.C.I. Pharmaceuticals, U.K.), Polyoxyethylene – Polyoxypropylene block copolymers (Pluronic L62[TM] and Pluronic F88[TM], Blagden Chemicals Ltd., Surrey, U.K.), Polyoxyethylene 20 cetyl ether (Cetomacrogol 1000, I.C.I. Pharmaceuticals, U.K.) and Sodium Lauryl Sulphate (Fisons PLC, Loughborough, U.K.). The suspending agent used for all experiments was Methylcellulose (2%, 4000cps, Sigma Chemical Co., St.Louis, U.S.A.), prepared as a 1%w/v aqueous solution.

4 METHODS

(a) Suspension Preparation

The suspensions were prepared using a mortar and pestle for all mixing operations. The powder was pre-wetted by adding the surfactant solution dropwise whilst mixing, until a smooth paste was obtained. The methylcellulose (1%w/v) solution was then added slowly to obtain a workable suspension. The resulting suspension was stored in a syringe for ease of dispensing into the analytical equipment.

The formula of each suspension was as follows:

Hydrophobic Drug Substance	10% w/v
Surfactant Solution	20% v/v
1% Methylcellulose Solution	to 100 %

The surfactant solution concentration for the pre-wetting stage was taken from the literature[1,2,3,4] and listed below:

Sodium Lauryl Sulphate	0.5% w/v
Cetomacrogol 1000[TM]	0.5% w/v
All other agents	2.5% w/v

(b) Particle Size Measurement

All particle size data was generated using a Malvern MasterSizer (300μm lens) fitted with a small volume sample presentation unit, the suspensions being dispersed in water, saturated with the drug substance. Methylcellulose solution was shown not to interfere with the determination.

All suspensions were analysed at the time of preparation and after 4 and 96 hours. The particle size distribution of the drug substance raw material was also measured (essentially dispersed to its primary particle size in iso-octane) for reference.

(c) Optical Microscopy

Each suspension was viewed under a microscope about 24 hours after preparation and photographed to provide a visual record of the particle size distribution and degree of aggregation exhibited.

(d) In-Vitro Dissolution

The dissolution method used involved a continuous flow system for sample removal as shown in figure 1. 0.5ml of the suspension was introduced into 1000ml of buffered aqueous media at 37°C. The dissolution medium (stirred at 100rpm by a paddle to conform with the BP 1988 Paddle Method) was cycled over an autosampler weir system to ensure continuous flow (cycle time approximately 1 min.). Samples were removed from above the weir at 10 minute intervals and assayed by HPLC. Results are displayed graphically as mg released against time (figure 5).

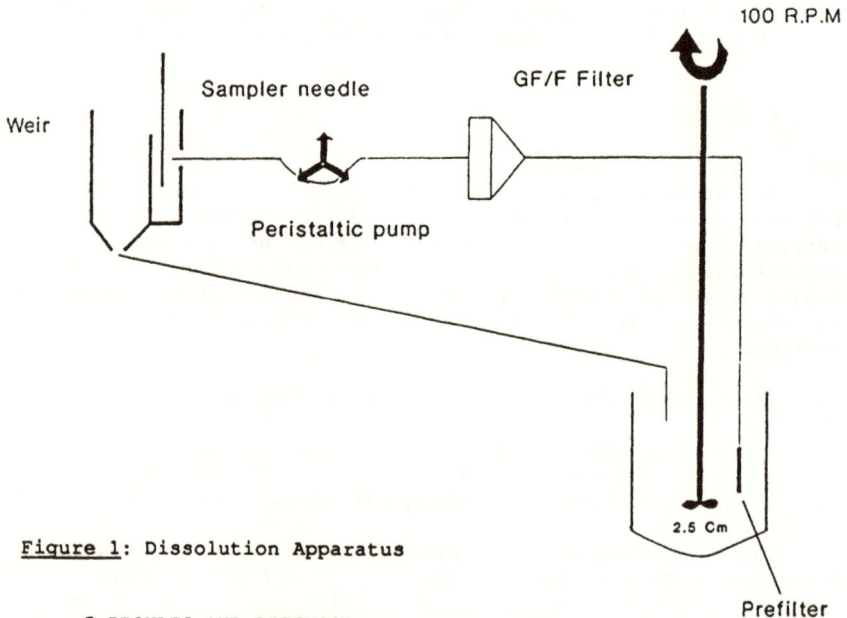

Figure 1: Dissolution Apparatus

5 RESULTS AND DISCUSSION

The particle size distributions of each suspension at 0, 4 and 96 hours are displayed in figures 2, 3 and 4 respectively. The raw material data is displayed in figure 2 for reference. The results clearly show the size distributions in the presence of Span 20[TM] and Pluronic L62[TM] as being the most consistent over the course of the experiment with median particle sizes of 10-15μm and little evidence of aggregation. Cetomacrogol 1000[TM] gives a median particle size of 7-9μm but shows a much higher degree of aggregation. All other particle size distributions show high degrees of aggregation on standing. Figures 3 and 4 illustrate the particle size distributions 4 and 96 hours after preparation.

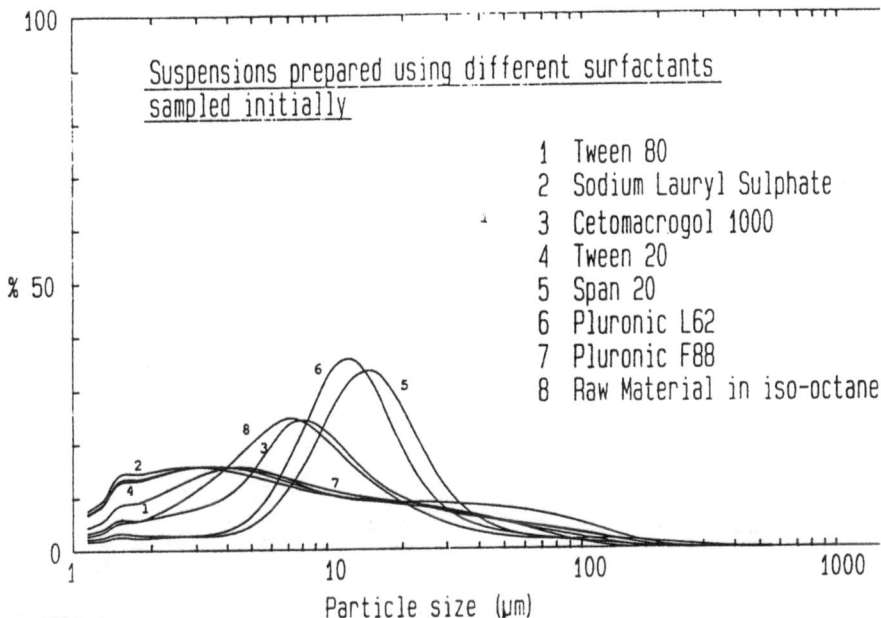

Figure 2: Suspensions prepared using different surfactants, sampled initially

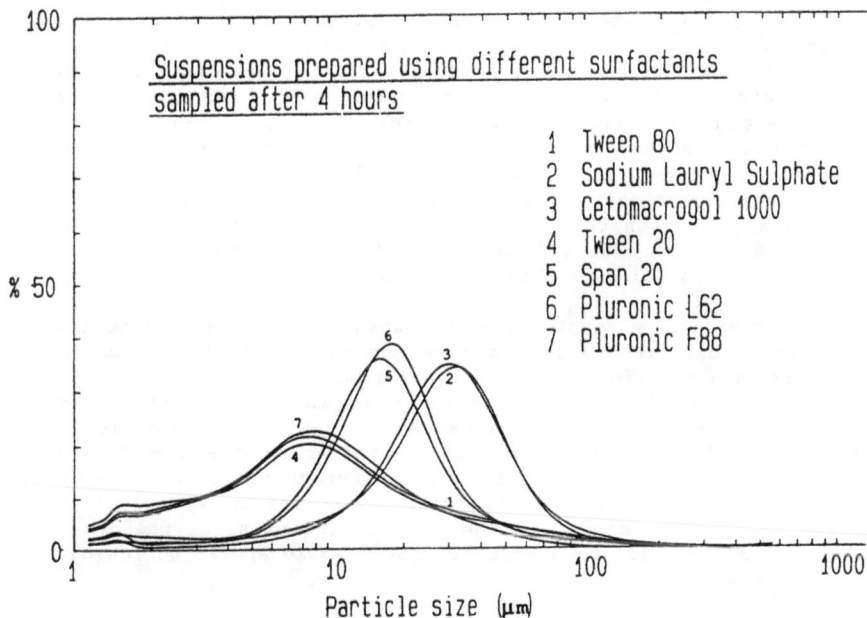

Figure 3: Suspensions prepared using different surfactants sampled after 4 hours

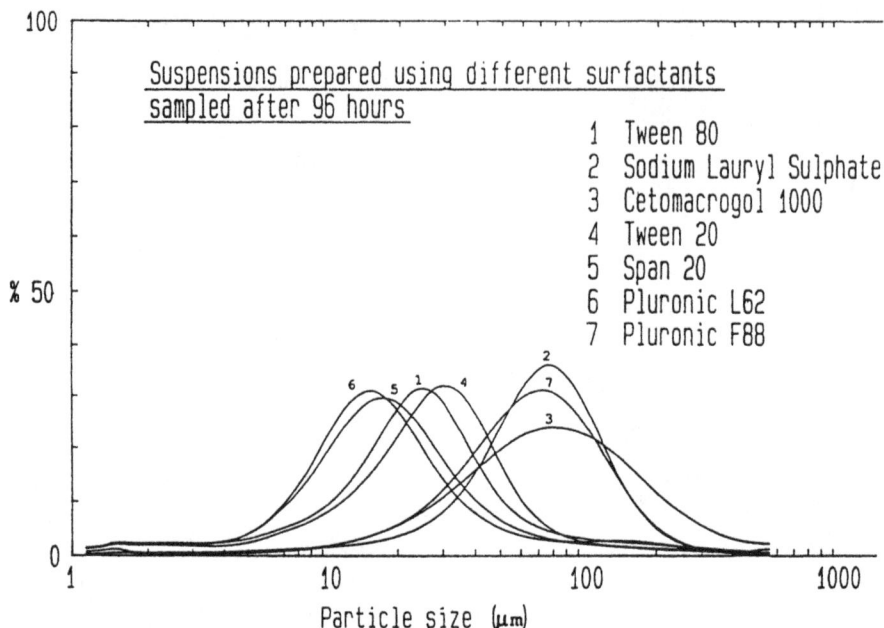

Figure 4: Suspensions prepared using different surfactants sampled after 96 hours

The significant tendency for all the suspensions to aggregate is clearly shown except for those incorporating Span 20[TM] and Pluronic L62[TM]. The suspensions prepared using these two surfactants have comparatively stable particle size distributions for at least 4 days. All other surfactants tested exhibited very unstable particle size distributions and would therefore be unsuitable for this particular drug substance. Furthermore, suspensions prepared using Cetomacrogol 1000[TM] and SLS became very viscous and almost solidified over time. Using this data alone, it can clearly be seen that the two surfactants with 'ideal' HLB (i.e. HLB values 7 to 10) have the desired properties for a suspension preparation of this drug substance.

The dissolution results for the seven surfactants in figure 5 show marked differences in the rate and extent at which drug is released into solution. This data would be expected to correlate with the particle size distributions and indeed the best dissolution profiles were obtained using Span 20[TM] and Pluronic L62[TM]. Tween 80[TM] and SLS also demonstrated acceptable dissolution profiles. However, the particle size distributions show evidence of suspension instability and the dissolution results could differ significantly dependent upon the time of analysis. Of the other suspensions analysed, Cetomacrogol 1000[TM] would appear to inhibit drug release.

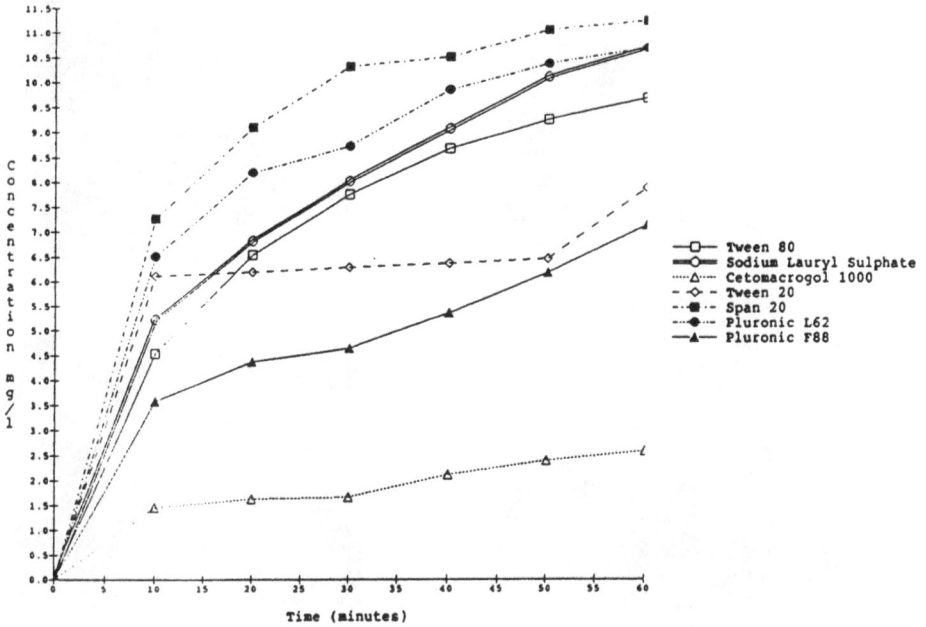

Legend:
- □ — Tween 80
- ○ — Sodium Lauryl Sulphate
- △ — Cetomacrogol 1000
- ◇ — Tween 20
- ■ — Span 20
- ● — Pluronic L62
- ▲ — Pluronic F88

x-axis: Time (minutes)
y-axis: Concentration mg/l

Figure 5: Dissolution results: Drug release into solution against time

The photomicrographs displayed in figure 6 clearly show the expected differences (with reference to the particle size distributions and dissolution profiles) between the suspensions 24 hours after preparation. The photomicrographs of suspensions prepared using Span 20™ and Pluronic L62™ show superior particle dispersion and reduced aggregation to the other suspensions. SLS and Pluronic F88 show very high levels of aggregation (in comparison to the other surfactants) as would be expected from the other data presented.

Figure 6: Photomicrographs of suspensions prepared using each of the surfactants in the study

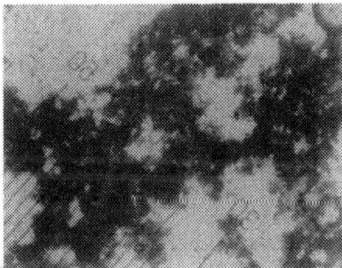

(a) Tween 80™

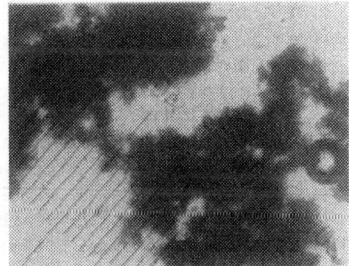

(b) S.L.S.

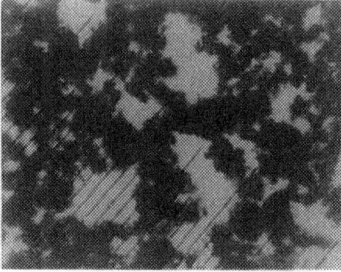

(c) Cetomacrogol 1000^{TM}

(d) Tween 20^{TM}

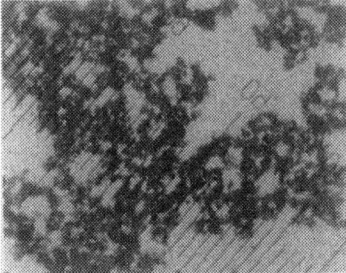

(e) Span 20^{TM}

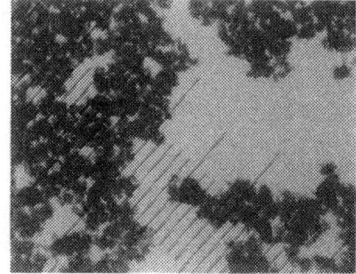

(f) Pluronic L62^{TM}

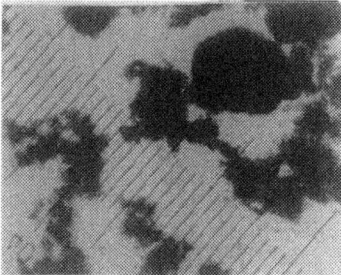

(g) Pluronic F88^{TM}

It was apparent that only Span 20^{TM} and Pluronic L62^{TM} were acceptable for suspension preparation and as such, the concentrations used were lowered in an attempt to assess the effect on the particle size distribution of the drug substance. Surfactant concentration was reduced from 2.5% to 0.5%w/v and three suspensions were prepared using Span 20^{TM} and Pluronic L62^{TM}, with Tween 80^{TM} for reference. Results are displayed in figure 7. The resulting suspensions show considerably different particle size distributions compared to the more concentrated material (figure 2). The effect of reducing the surfactant concentration to this level would appear to be detrimental to the resulting suspension. Therefore, the concentrations of Span 20^{TM} and Pluronic L62^{TM} cannot be reduced. This finding is consistent with the literature which suggests a 'lower' limit for Pluronic L62^{TM} of 2.4%w/v[4].

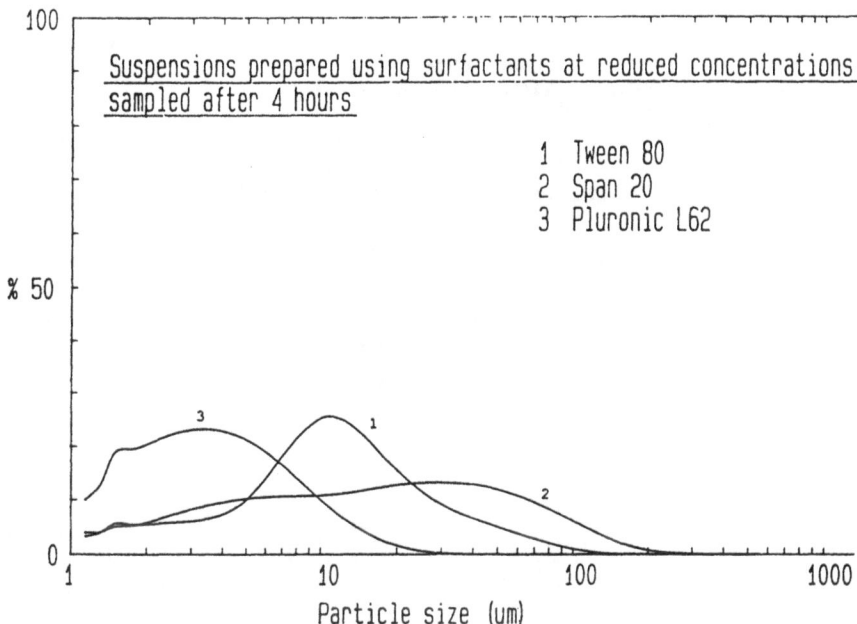

Figure 7: Suspensions prepared using surfactants at reduced concentrations sampled after 4 hours

6 CONCLUSION

Of the various surfactants tested, a suitable suspension with respect to particle size distribution and dissolution profile can be obtained by pre-wetting this particularly hydrophobic drug substance with Span 20^{TM} and Pluronic L62^{TM} at a concentration of 2.5% w/v, resulting in a final suspension concentration of 0.5% w/v.

Other surfactants tested including Tween 80^{TM}, Tween 20^{TM}, Cetomacrogol 1000^{TM}, Pluronic F88^{TM} and SLS were all found to be unsuitable owing to the resulting poor particle size distributions on storage. Tween 20^{TM}, Pluronic F88^{TM} and Cetomacrogol 1000^{TM} were shown to produce suspensions with inferior dissolution profiles (reduced drug release into aqueous media). All photomicrographs clearly supported the conclusions drawn from the particle size and dissolution data, showing the aggregation tendencies of specific surfactants used.

The findings of this study clearly vindicate the HLB approach to the selection of a surfactant for the preparation of suspensions of drug compounds.

It is also apparent that a standard surfactant used in the preparation of pre-clinical studies is not always suitable for individual drug substances.

7 REFERENCES

1. I.C.I. "Atlas" Surfactant range literature.
2. B.A.S.F. "Pluronic" Surfactant range literature.
3. Martindale 29th Ed. pp1245.
4. K.N. Prasad et al. "Surface activity and association of ABA Polyoxyethylene - Polyoxypropylene Block Co polymers in aqueous solution". J.Coll-Interface Sci., 1979, 69, 225-232.

8 ACKNOWLEDGMENTS

    The contributions to the practical work involved in this study made by Messrs. N.Simpson and A.J.Smith are gratefully acknowledged.

The Use of Laser Diffraction Particle Size Analysis to Predict the Dispersibility of a Medicament in a Paraffin Based Ointment

T. W. Atkinson, M. J. Greenway, S. J. Holland, D. R. Merrifield and H. P. Scott

PHARMACEUTICAL SCIENCES, SMITHKLINE BEECHAM PHARMACEUTICALS, CLARENDON ROAD, WORTHING, UK

1 ABSTRACT

The particle size of active constituents suspended in pharmaceutical ointments for topical and particularly ophthalmic use needs to be rigorously controlled. Often sufficiently fine material can only be obtained by micronisation (fluid energy milling). Whilst this process achieves a very fine particle size (<5μm), it can also lead to serious aggregation of the active. Adequate comminution and subsequent successful dispersion of the micronised drug will depend on its physical properties, the nature of the ointment base used, and the energy imparted during the dispersion process.

It is important that the particle size analysis accurately reflects the actual size of particles and aggregates in the final product. Various physical characterisation methods were used to study the effects of modifying the precipitation conditions used in the preparation of the drug. In particular, analysis of suspensions in liquid paraffin by laser diffraction is described. This is considered to have especial value in predicting the dispersibility of the active in paraffin based ointments.

2 INTRODUCTION

All pharmaceutical dosage forms are required to meet stringent standards of drug content uniformity and quality. This is to ensure that the product will consistently deliver the intended dose, thereby achieving the desired efficacy and avoiding the adverse effects of inaccurate dosage [1,2]. This equally applies to drugs intended for topical application where the active compound is dispersed as a suspension in an ointment base, such as semi-solid White Soft Paraffin BP. It is necessary to comminute the active material to a very fine particle size so that a sufficiently well dispersed system can be achieved. This will help to ensure bioavailability and hence efficacy of the medicinal product. Moreover for ointment intended for ophthalmic use it is important to achieve a very fine product texture to avoid irritation of the very delicate surfaces of the eye that would arise from the presence of large abrasive particles. Pharmacopoeial standards include tight specifications for the control of particle size of drug materials in ophthalmic presentations. These requirements are discussed fully below, but essentially all particles must be smaller than 25μm. The method of assessment specified for this test is microscopic examination and analysis. This is a slow and labour intensive method.

It is rarely possible that a drug substance may be incorporated directly into a suspension formulation without some intermediate treatment such as hammer milling or micronisation to ensure a sufficiently fine particle size. The latter process is particularly time consuming and labour intensive, as well as requiring considerable energy in respect of the volume of compressed gas used in the grinding process. Micronisation often becomes a rate determining step in the manufacture of such products.

Having achieved the required degree of fineness in the powder it is then necessary to ensure that the powder is satisfactorily dispersed in the ointment [3]. Finely divided powders have a marked tendency to reaggregate [4]. Dispersion would typically be done using high shear homogenisers of the shrouded turbine type, at a further energy cost.

This work forms part of a larger programme to define those characteristics of the raw material drug substance that would enable rapid and effective dispersion in the ointment base. An ultimate objective is to be able to prepare these dispersions with an acceptable particle size without the need for prior micronisation.

To succeed in this it was considered worthwhile developing a particle sizing technique that would be able to predict the likelihood of achieving an acceptable particle size distribution in the ointment.

Attempts were made to change the physical characteristics of the drug substance by varying those precipitation conditions most likely to affect the particle size and robustness of the precipitate. A number of batches of this drug substance were prepared under various precipitation conditions. These were characterised as milled and micronised powders.

Ointments were prepared from the powders and tested against pharmacopoeial standards. A major requirement of this characterisation was a knowledge of the particle size of the powder as first precipitated, after comminution, and as present once incorporated into the ointment base, or vehicle. Whilst various techniques could be used to determine the size distribution of the powder samples, it was considered that these would fail to account for any particle-vehicle interactions that were likely to occur in the paraffin based product. The size and dispersion quality of the drug substance in the ointment base may be assessed microscopically using established pharmacopoeial techniques. These are however labour intensive, and it was considered that a rapid, and preferably automated technique should be developed that would predict the likely performance of the drug substance in the ointment base. It was not possible to use the intended vehicle itself, largely due to the practical difficulties arising from its viscosity, so a method was developed employing a low viscosity liquid paraffin rather than the semisolid grade. The conditions of the analysis were selected so that they would closely match the conditions of preparation of the ointment (in that a high shear mixer was employed to form the dispersion to be analysed) and that this dispersion would be of a similar concentration to that of the formulated ointment. This approach to assessment of particle size should give a more meaningful and reliable indication of the propensity of the drug substance to disperse in the intended vehicle.

2 EXPERIMENTAL

Ten small scale (100g) batches of the drug substance were prepared. The conditions of precipitation were varied in respect of temperature and solvent composition for each batch. The precipitate was isolated and dried by a standard method. The nature of the powders so obtained was characterised as described below.

The particle size of the drug substance was reduced using a hammer mill (Apex 314S) operating with hammers forward at 7980 rpm.,passing the material through a 0.027" screen. Samples of this material were taken for particle size analysis and other physical and chemical characterisation tests. The milled material was incorporated into paraffin based ointments.

Particle size was further reduced by fluid energy milling, using a Berk 2" microniser operating at 85-95 psi feed pressure and 35-45 psi grind pressure. Micronised materials were collected and again sampled for characterisation, and incorporation into ointments.

Ointments were prepared by high shear incorporation and homogenisation of the drug substance into White Soft Paraffin, followed by triple roller milling of the resulting suspension using a Pascall Model I roller mill (Pascall Ltd., Crawley, UK.). The quality of the dispersion and the particle size of the drug substance were assessed by optical microscopy. In particular, particle size was measured using the method of the British Pharmacopoeia 1988. This states that a small, known quantity of ophthalmic ointment is gently spread as a thin layer onto a microscope slide. An area corresponding to a solid phase content of 10μg is then scanned for the assessment of particle size. The numbers of particles observed to have a maximum Feret dimension exceeding 25, 50 and 90μm are recorded. The criteria for a pass on this test are given in both the British and the European Pharmacopoeias [5,6] as being not more than 20 particles exceeding 25μm, not more than 2 exceeding 50μm, and none exceeding 90μm. The results of the evaluations are given in Table 1.

Characterising tests were applied to the powder samples.
Optical microscopy and Scanning Electron Microscopy was performed
on all samples. Aerated and tapped densities were determined for
the powders. The aerated density was measured after passing the
powder through a 20# sieve, and allowing it to fall into a weighed
measuring cup, and reweighing. The tapped density was obtained
after tapping until a maximum or equilibrium density value was
obtained. Specific surface was determined by nitrogen gas
adsorption (BET technique) using the dynamic flow technique on the
Quantasorb Junior analyser (Quantachrome Corp.,Syosset, NY.,US.).
Samples of c.0.7g. were outgassed at 50^0C under a helium gas flow
until there was no detectable change in the conductivity of the
gas. Triple point analyses were conducted at 0.1,0.2 and 0.3
partial pressure of nitrogen in helium, with the gas mixes being
prepared by use of the Quantachrome dynamic gas flow mixer. Each
analysis made use of the desorption peak arising from the rapid
warming of the sample to room temperature, following the
adsorption of gaseous nitrogen occurring at liquid nitrogen
temperature. The instrument was calibrated by measuring the peak
due to direct injection of a known volume of nitrogen gas.

Particle size distributions were determined by the electrical
sensing zone technique using the Coulter Multisizer (Coulter
Electronics, Luton, UK.). The suspension medium consisted of 1%
w/v sodium chloride dissolved in 5% propan-2-ol/95% water
presaturated with the drug substance. The medium was filtered
through a 0.22µm filter prior to use, and a background count
performed on it. Suspensions of the powder samples were counted on
passing through an appropriate size orifice. The orifice tubes
were calibrated using polymer latex particles. Size distributions
were obtained as frequency plots and median and upper and lower
quartile values. The results are shown in Table 3 and 4.

Particle size analysis of the drug substance was also
performed by the laser diffraction technique (Malvern MasterSizer,
Malvern, UK.). A quantity of the powder sample was dispersed in
liquid paraffin (water white liquid paraffin, Fisons,
Loughborough, UK.) S.G. 0.83 - 0.86. This dispersion was
achieved by adding 20 ml. of the liquid paraffin to c. 500 mg. of
the powder in a 50 ml. glass beaker. The high shear mixer
(Silverson bench top fitted with a $^5/_8$" micro unit general purpose
disintegrating head) was run for 5 minutes operating at 3000 rpm.
Small aliquots of this suspension were sampled using a pipette and
added to the small volume sampler of the Malvern MasterSizer,
which also contained liquid paraffin. Sufficient suspension was
added until an obscuration of 0.25 was obtained, after which the
sample was analysed. The results were obtained as frequency plots,
% of particles (by weight) below 11 and 55µm, as medians, and as
upper and lower deciles.

The robustness and reproducibility of this method were
evaluated by varying the dispersion time of the drug substance in
the liquid paraffin, by checking replicate samples of individual
batches and by conducting similar analyses on separate occasions.

3   RESULTS

Table 1   Dispersion quality and assessment of ointments, prepared from micronised drug substance.

Ointments prepared from all batches of hammer milled drug substance were assessed as being unacceptable as topical ointments.

All ointments prepared from micronised material were acceptable as topical ointments. The assessment given below applies to the ophthalmic test.

Batch	Numbers of particles exceeding size limits, and assessment of ointments			Quality of dispersion (topical)
	> 25 μm	> 50 μm	Assessment	
1	6	2	pass	good
2	9	2	pass	good
3	4	1	pass	good
4	3	0	pass	very good
5	3	0	pass	very good
6	4	0	pass	good
7	12	3	fail	poor
8	2	0	pass	very good
9	5	0	pass	good
10	17	4	fail	poor

<u>Table 2</u>   Particle size analysis of micronised material by laser diffraction technique (Malvern)

	> 25 μm[1]		> 50 μm [2]		lower decile [3]		median μm		upper decile [4]	
1	10.4	11.6	3.3	3.6	2.2	2.2	7.0	7.2	26.3	28.2
2	8.9	6.7	2.6	1.3	1.5	1.5	4.8	4.4	23.4	19.9
3	0.2	0.1	0.0	0.0	1.2	1.2	2.8	2.8	7.6	7.1
4	8.4	10.0	2.6	5.2	1.5	1.6	4.2	4.4	21.8	25.5
5	3.8	0.1	2.4	0.0	1.6	1.5	4.1	3.9	10.9	8.3
6	4.1	0.7	1.5	0.0	2.2	2.2	5.7	5.5	15.0	12.0
7	19.7	16.8	8.4	7.8	2.6	2.6	8.1	7.8	45.6	41.4
8	0.0	0.0	0.0	0.0	1.9	1.9	4.6	4.6	9.9	9.9
9	1.0	1.1	0.1	0.0	2.4	2.3	5.9	5.8	13.0	13.1
10	12.0	10.2	3.4	2.4	2.5	2.4	7.1	7.0	42.0	22.0

Notes
1)   % of particles by weight (volume) which exceed 25μm.
2)   % of particles by weight (volume) which exceed 50μm.
3)   The particle size in microns than which 10% by weight are smaller.
4)   The particle size in microns which 10% by weight exceed.

Table 3 Characterisation of material comminuted by hammer milling

	Density gcm^{-3}		Particle size analysis (Coulter)				
	aerated	tapped	>11µm	>55µm	lower quartile	median	upper quartile
1	0.30	0.43	62	4	6.4	18.6	33.4
2	0.26	0.40	68	35	6.9	34.0	71.3
3	0.15	0.25	39	5	5.2	7.2	23.6
4	0.28	0.44	69	17	7.4	25.0	46.2
5	0.26	0.40	48	2	5.7	10.1	23.6
6	0.33	0.48	26	1	4.7	6.0	11.6
7	0.29	0.42	60	2	6.2	16.9	33.2
8	0.29	0.43	72	33	9.3	29.9	69.2
9	0.34	0.50	63	18	6.8	21.4	45.2
10	0.34	0.49	70	29	8.6	24.2	64.9

Table 4 Characterisation of material comminuted by micronisation

	Density gcm^{-3}		Particle size analysis (Coulter)					Surface Area m^2g^{-1}
	aerated	tapped	>11µm	>55µm	lower quartile	median	upper quartile	
1)	0.18	0.25	13.1	0	4.8	5.9	7.7	11.4
2)	0.10	0.15	7.1	0	4.1	4.9	5.8	11.2
3)	0.07	0.11	0.8	0	4.3	5.1	5.8	11.3
4)	0.06	0.11	0.4	0	4.0	4.8	5.5	12.4
5)	0.11	0.18	1.3	0	4.8	5.7	6.7	11.8
6)	0.13	0.20	6.1	0	4.4	5.5	7.0	11.1
7)	0.14	0.20	9.2	0	3.4	4.4	6.3	17.3
8)	0.13	0.19	4.1	0	3.7	4.7	6.3	12.6
9)	0.16	0.24	9.5	0	4.4	5.6	7.5	12.3
10)	0.21	0.28	21.4	3.8	5.5	7.0	9.9	8.4

4 DISCUSSION

The assessments of ointments prepared from hammer milled drug
substance clearly indicate that a sufficiently fine particle size
cannot be achieved. The micronisation stage is still required. The
chemical modifications to the isolation stage that were studied
have not enabled the preparation of a form of the drug substance
that can be directly incorporated into the ointment base.

A comprehensive characterisation of the batches was conducted
in an attempt to establish which physical properties correlated
with good dispersion. A cross correlation analysis was performed
on the data. This only demonstrated correlation where it may have
been expected, notably between sets of particle size data. There
was some association between the surface area data of the batches
and their appearance as assessed by Scanning Electron Microscopy.
Otherwise the surface area data did not correlate with either the
particle size data or with the ability to form acceptable
ointments. The surface area data reflects the porous nature of the
material, since a reduction in particle size does not markedly
affect the specific surface.

A method of particle size analysis of the drug substance
that may be predictive of ultimate ointment quality has been
developed. This is shown by the data in Tables 1 and 2. The
laser diffraction method is able to establish which micronised
batches are likely to give acceptable ointments. This method,
which is carried out in liquid paraffin following dispersion with
a high shear mixer, is much more representative of actual
processing conditions than the electrical sensing zone method
which is conducted in aqueous electrolyte.

Specifically, data from the laser diffraction method
identifies batches 7,10, and 1 as having the largest median
diameters (8.0, 7.1 and 7.1 µm) and upper deciles (43.5, 32.0
and 27.3 µm) respectively. At the lower end of the distribution,
although the differences are small, batches 7 and 10 also show the
greatest values for the lower decile (2.6 and 2.5µm
respectively). Considering the distributions in another way,
these batches show higher percentages of particles exceeding 25 µm
and 50 µm, where batch 7 is considerably worse than the rest while
batch 10 is also poor. The electrical sensing zone technique is
not as discriminating. In this analysis batch 10 shows the
greatest median size (7.0 µm) but batch 7 the lowest (4.4 µm).
Only batch 10 displays a presence of particles that exceed 55 µm
(3.8%), and there is no indication that batch 7 has a greater
proportion of these large particles than the other batches.
Indeed, in respect of particles in the lower quartile, it would
seem finer than most. Both batches 7 and 10 give rise to
unacceptable ophthalmic ointments, containing high levels of large
particles. These fail the pharmacopoeial specifications for
ophthalmic ointments.

The next worse batches as described by the particle size
limit test for ophthalmic ointments are batches 1 and 2. By the
laser diffraction technique both of these give high values for the

upper decile (27.3, 21.7 µm) and the numbers of particles exceeding 50 µm (3.5, 1.9%) respectively. Additionally a high median value is obtained for batch 1, although batch 2 is not exceptional in this respect. The electrical sensing zone method does not pick out these batches as having significant levels of coarser particles. It does not discriminate between these batches and the remainder.It was not able to predict the considerably inferior product obtained from batches 7 and 10.

The laser diffraction technique shows a consistent picture for those batches giving rise to very good ointments. A fine particle size is seen for batches 4, 5 and especially 8. Whilst there is nothing striking about the particle size distribution of 4, both 5 and 8 give low particle size for material in the upper decile (9.6 and 9.9 µm) and give low medians (4.0 and 4.6 µm).

It is considered that whilst the information given by the electrical sensing zone method may well be representative of the nature of the comminuted material, it does not accurately portray the effective size likely to be achieved in the ointment. It is not possible to analyse the ointment other than by microscopy or image analysing techniques. The analysis of the drug substance suspended in liquid paraffin is a relatively good predictor of ointment quality, using a convenient automated technique. In this case the ability to size the drug substance in a medium of similar chemical nature and polarity to that of the pharmaceutical vehicle,at a relevant concentration, using a dispersion technique similar to that employed in the manufacturing process, along with the rapidity of the analyses, offers major advantages over the electrical sensing zone method.

It is considered that this approach, which attempts to measure the effective particle size achieved in a close match to the vehicle used, by a simulated dispersion technique, is the correct approach to determining particle size in industrial and manufacturing applications.

5 CONCLUSIONS

A method of particle size analysis for the drug substance using the laser diffraction technique has been developed. This method attempts to simulate the actual conditions of manufacture of the ointment.

The method offers important advantages over other sizing techniques in that it is conducted at a relevant concentration, in a medium of like chemical nature and polarity, and it employs a similar dispersion technique.

As it is an automated instrumental technique, it is both convenient and rapid. It is considered to be particularly useful in assessing the ability of batches to form acceptable ointments.

6   REFERENCES

1   P.A.Hartley, G.D.Parfitt and L.B.Pollack, <u>Powder</u> <u>Technol.</u>
     1985, <u>42</u>, 35

2   N.A.Orr, E.A.Hill,and J.F.Smith, <u>Int.J.Pharm.Tech.&</u>
     <u>Prod.Mfr.</u> 1980, <u>1</u>, 4

3   P.M.Heertjes and W.C.Witvoet, <u>Powder</u> <u>Technol.</u> 1969/70, <u>3</u>,
     339

4   J.A.Hersey and P.C.Cook, <u>J.Pharm.Pharmacol.</u> 1973, 25
     Supplement 21P

5   Unguenta Ophthalmica, European Pharmacopeia 1980, 61

6   Eye Ointments, British Pharmaceopia 1988, 686.

7 ACKNOWLEDGMENTS

     The contribution to the practical work involved in this
study made by Messrs. J.Mackew, T.Hamilton and S.W.Campbell is
gratefully acknowledged.

Optical Particle Sizing of Coarse-dispersed Aerosols Under Multiple Light Scattering in a Medium*

E. P. Zege and A. A. Kokhanovsky

INSTITUTE OF PHYSICS, BSSR ACADEMY OF SCIENCES, 220602, MINSK, CIS

1 INTRODUCTION

Determination of the particle size distribution function (DF) using a radiation scattering phenomenon is the field which incorporates now not only classical methods (of a full- and small-angle indicatrix, of spectral transparency), but also such modern trends as dynamic spectroscopy, multiwave laser sounding, etc. All these methods rely on measuring angular or spectral characteristics of single-scattering radiation. In this case, DF or its integral parameters are reconstructed by the Mie theory.

It is frequently difficult (if possible at all) to provide the condition of light single scattering in a test medium, and only multiple scattering characteristics can be measured. It is the situation which appears in studying geophysical objects (clouds, mists), in all situations when nondestructive control of finished samples is required, etc. In these cases, the problem of estimating particle DF by their sizes becomes somewhat involved.

This work is concerned with analyzing some possible approaches to optical sizing of coarse particles ($\rho = ka \gg 1$, $k = 2\pi/\lambda$, a is the particle radius, λ is the used radiation wavelength) under multiple scattering. Remember that for determining coarse particle sizes at the extinction coefficient ε independent of the wavelength spectral methods cannot be applied.

2 DETERMINATION OF PARTICLE SIZES BY MEASURING ANGULAR DISTRIBUTION OF TRANSMITTED RADIANCE

Let us consider a scattering coarse particles slab, with normal radiation incident on one of its surfaces. We measure angular distribution of transmitted radiance $I(\vartheta)$, where ϑ is the angle between the beam direction and normal to the slab. The scattering indicatrix of

* Not presented at PSA 91 Conference

coarse particles $x(\beta)$ (β being the scattering angle) is prominently forward-extended. Then, for a moderate optical thickness τ, we have[1]:

$$I(r) = \frac{1}{E_0} \int_0^\omega I(\beta) \; J_0(p\beta) \; \beta \; d\beta, \qquad (1)$$

$$I(p) = \exp(-\tau + \Lambda \tau x(p)), \qquad (2)$$

$$x(p) = \frac{1}{2} \int_0^\omega x(\beta) \; J_0(p\beta) \; \beta \; d\beta. \qquad (3)$$

Here E_0 is the irradiance induced by an incident beam on the upper slab surface, p is the angular frequency, $\Lambda = \sigma/\varepsilon$ is the single-scattering albedo (σ is the scattering coefficient of scattering medium), $J_0(\cdot)$ is the zero-order Bessel function.

For large spherical particles of a radius, the scattering indicatrix for a small-angle region, is of the form[2]

$$x(\beta,\rho) = \frac{4 \; J_1^2(\beta\rho)}{\beta^2}, \qquad \frac{1}{2} \int_0^\infty x(\beta,\rho) \; \beta \; d\beta = 1, \qquad (4)$$

where $J_1(\cdot)$ is the first-order Bessel function.

Regarding Eq. 4, from Eq. 3 we have:

$$x(p,\rho) = 2 \int_0^\infty J_1^2(\beta p) \; J_0(p\beta) \; \beta^{-1} \; d\beta. \qquad (5)$$

Integral Eq. 5 has been carefully studied in Fourier optics, and determines the optical transfer function (OTF) for the round orifice of a radius[3]. It appears here not by chance, since in a small angle region, radiation scattering on a particle (opaque or with an essential phase shift) is equivalent to optical radiation diffraction on a disc or orifice of similar shape and area[2]. Therefore, in the case under study, particles are replaced by equivalent opaque screens, which essentially simplifies the analysis. An analytical expression for $x(p)$ is known[3]:

$$x(p,\rho) = \begin{cases} 2\left[\cos^{-1}\xi - \xi(1-\xi^2)^{1/2}\right]/\pi, & \xi \le 1, \\ 0, & \xi > 1 \end{cases}, \qquad (6)$$

where $\xi = p/2\rho$.

For dispersed media with a particle-size DF $f(a)$, instead of Eq. 6, we have:

$$\bar{x}(p) = \frac{1}{S} \int_0^\infty a^2\, x(p,ka)\, f(a)\, da, \quad S = \int_0^\infty a^2\, f(a)\, da,$$

$$\bar{X}(\beta) = \frac{1}{S} \int_0^\infty a^2\, x(\beta,ka)\, f(a)\, da, \quad \int_0^\infty f(a)\, da = 1.$$

(7)

Relations Eq. 6 and Eq. 7 can be used to determine the values of a or $f(a)$ from $x(p)$ and $\bar{x}(p)$ measurements. In so doing, the small-angle $I(\vartheta)$ part should be measured and Fourier-Bessel transformation Eq. 1 performed. The value of $x(p)$ is found from Eq. 2. Also, an optical thickness $\tau = -\ln T$, where T is the transmission coefficient of unscattered radiation, should be measured.

Indeed, it is the small-angle method[2] including multiple scattering. The regularizing algorithm from integral equation Eq. 7 solution to reconstruct $f(a)$ was discussed[4]. However, realizing of the method proposed in the paper[4] requires measuring the function of spatial radiance correlation to calculate then function $x(p)$. Here we propose to measure an angular radiance distribution $I(\vartheta)$ which is in fact more convenient than measuring the coherence function or the function of a spatial radiance correlation. These methods may be regarded to be two different approaches to experimental determination of one and the same quantity, since in this case, an angular radiance distribution can be considered as an angular spectrum of the coherence function[1]. Thus, knowing $I(\vartheta)$, the functions $f(a)$ can be reconstructed. However, rather a high measurement accuracy and intricate mathematical processing of experimental data are required. Below, we present a more simple and convenient method which, however, provides less comprehensive DF information. It is based on measuring an OTF of a scattering layer.

3 OTF OF A COARSE DISPERSED PARTICLES LAYER

For coarse particles and not very optically thick layers, the OTF $S(\omega)$ ($\omega = \nu l$, where ν is the spatial frequency and l the geometric layer thickness) is of the form[1]:

$$S(\omega) = \exp(-\tau + \Lambda\tau\psi(\omega)),$$

(8)

where

$$\psi(\omega) = \int_0^1 x(\omega y)\; dy, \tag{9}$$

$$x(\omega y) = \frac{1}{2} \int_0^\infty x(\beta)\; J_0(\omega y \beta)\; \beta\; d\beta. \tag{10}$$

Considering Eq. 4, for $\psi(\omega)$, we easily arrive at

$$\psi(\omega) = \begin{cases} \psi_1 = \dfrac{4}{3\pi b}\left[1 - \dfrac{2+b^2}{2}\,(1-b^2)^{1/2} \right] - \dfrac{2}{\pi}\cos^{-1}b,\; b \le 1, \\[2mm] \psi_2 = \dfrac{4}{3\pi b},\qquad b > 1. \end{cases} \tag{11}$$

where $b = \omega/2\rho$. At $b \to 0$, from Eq. 11, we have:

$$\psi(\omega) = 1 - \frac{2b}{\pi}\left[1 - \frac{b^2}{12} \right] + O(b^5). \tag{12}$$

As a special analysis shows, Eq. 11 is applicable at $b \le 1$ with a relative error of 2% and less. Here, substitution of Eq. 11 or Eq. 12 into Eq. 8 gives a simple OTF $S(\omega)$ relation to a particle a radius which can be used for its determining.

Eq. 11 is simply extended to the case of dispersed media with different diameter particles:

$$\overline{\psi}(\omega) = \frac{4}{3\pi S\alpha} \int_0^\alpha a^3 f(a)\; da + \frac{1}{S} \int_\alpha^\infty a^2 f(a)\; \psi_1(a)\; da, \tag{13}$$

where $\alpha = \omega/2k$.

At $\alpha \to 0$, from Eq. 13, we have:

$$\overline{\psi} = 1 - \frac{\omega}{\pi k a_{21}}, \tag{14}$$

while at $\alpha \to \infty$:

$$\overline{\psi} = \frac{8k a_{32}}{3\pi\omega}, \tag{15}$$

where the mean-volume particle radius a_{32} and parameter a_{21} are the DF moments ratios

$$a_{32} = \frac{\displaystyle\int_0^\infty a^3 f(a)\, da}{\displaystyle\int_0^\infty a^2 f(a)\, da} \qquad\qquad a_{21} = \frac{\displaystyle\int_0^\infty a^2 f(a)\, da}{\displaystyle\int_0^\infty a\, f(a)\, da} \qquad (16)$$

The general DF form depends on the physics of particle formation (coagulation, breaking, etc.) and frequently allows an approximation of the form:

$$f(a) = B\, a^\mu \exp(-\mu a/a_0), \qquad (17)$$

where B is the normalization constant, a_0 is the modal radius, μ is the DF half-width parameter. From Eq. 13 and regarding to Eq. 11, Eq. 12 and Eq. 17 may yield

$$\overline{\psi}(\omega) = 1 - P(\mu+3,y) + \frac{4(\mu+3)}{3\pi y} P(\mu+4,y) -$$

$$- \frac{2y(1-P(\mu+2,y))}{\pi(\mu+2)} + \frac{y^3(1-P(\mu,y))}{6\pi\mu(\mu+1)(\mu+2)}, \qquad (18)$$

where $y = \mu\omega/2\rho_0$, $\rho_0 = ka_0$, $P(z,y)$ is the incomplete

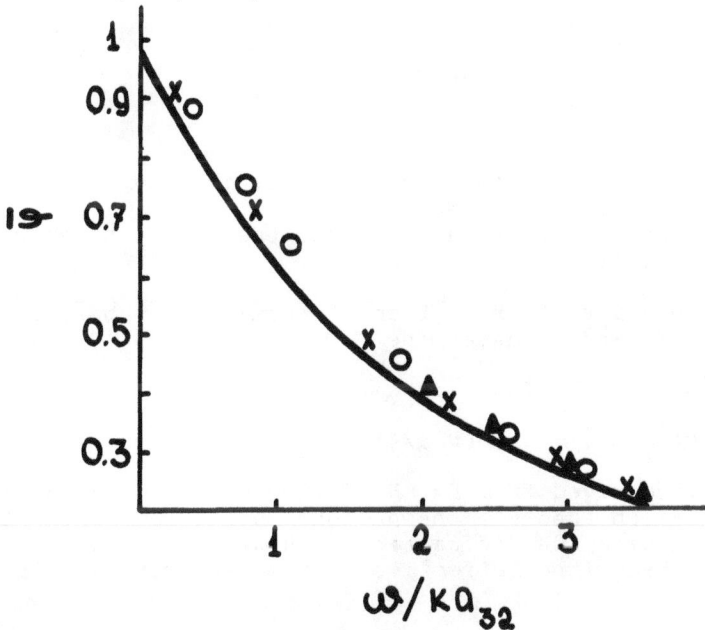

Figure 1   Function $\overline{\psi}(\omega/ka_{32})$ at $\mu = 1$ (solid line), $\mu = 6$ (crosses), $\mu = 30$ (circles). Triangles calculated using Eq. 15.

Γ-function. At integer z, we have:

$$P(z,y) = 1 - \exp(-y) \sum_{j=0}^{z-1} x^j / j! .$$

It should be noted that the last term in Eq. 18 is small and can be neglected in the majority of cases. The function $\bar{\psi}$ from Eq. 18 is plotted in Figure 1 vs ω/ka_{32} at different μ. It is seen to weakly depend on μ variations at fixed a_{32}, asymptote Eq. 15 being achieved at $\omega/ka_{23} \approx 2 \div 3$.

4   CONCENTRATION AND MEAN-VOLUME PARTICLE RADIUS MEASUREMENT TECHNIQUE

Eq. 8 and Eq. 15 can be used as a basis for a new method of determining sizes of coarse particles in multiple scattering media. Measuring a_{32} and volume particle concentration C_v requires:

•To measure a point spread function (PSF) of a medium, i.e. radial irradiance distribution of a transmitted light, when a scattering slab is normally illuminated by a monodirected point source. Then OTF can be obtained using the Hankel transform of a PSF $E(r)$[1]:

$$S(\nu) = \frac{2\pi}{W_0} \int_0^\infty E(r) \, J_0(\nu r) \, r \, dr, \qquad (19)$$

where W_0 is the source power. A radiation wavelength is arbitrary, but $\lambda \ll a_0$. Naturally, any other direct OTF measurement procedure can be used, the one with special test objects[1].

•To measure direct radiation transmission coefficient $T = \exp(-\tau)$ and to determine τ.

•To determine the function

$$\psi(\nu) = 2(1 + \ln(S(\nu))/\tau). \qquad (20)$$

Eq. 20 is derived from Eq. 8. Following the small-angle approximation theory[1], the photons scattered into the angles outside the diffraction peak were considered to be absorbed (the effective single-scattering albedo $\Lambda = 0.5$).

•To find a mean-volume particle radius a_{32} using formula (see Eq. 15)

$$a_{32} = \frac{3\pi \nu \, l \, \overline{\psi}(\nu l)}{8k} \tag{21}$$

or by the slope of the line $\overline{\psi}(\nu^{-1})$. The rate of ν changes should be chosen within a linear function $\overline{\psi}(\nu^{-1})$, i.e. at $\overline{\psi} \le 0.3 + 0.4$ (see Figure 1). Here, the error of a_{32} reconstruction is equal to the measurement error of a function $\overline{\psi}$.

• To determine volume particle concentration C_V using the formula[1]:

$$\tau = 1.5 C_V l / a_{32}, \tag{22}$$

which hold for large scatterers.

5 ALLOWANCE FOR A GEOMETROOPTICAL PART OF A SCATTERED FIELD

The proposed methods imply prescribing the scattering indicatrix in form Eq. 4. In practice, the indicatrix $x(\beta)$ of coarse particles consists of a diffraction $x_1(\beta)$ (see Eq. 4) and geometrooptical $x_2(\beta)$ components:

$$x(\beta) = \frac{\sigma_1 x_1(\beta) + \sigma_2 x_2(\beta)}{\sigma_1 + \sigma_2}, \tag{23}$$

where σ_1 (σ_2) is the scattering cross section related to diffraction (reflection and refraction). The function $x_2(\beta)$ at $\beta \le \pi/4$, $n = 1.34$, $\chi = 0$ ($m = n - i\chi$ is the refraction index) can be obtained by approximating tabular data[5]

$$x_2(\beta) = 2 \, \alpha^2 \, \exp(-\alpha\beta), \tag{24}$$

where $\alpha = 3.3$. The quantity α is a weak function of optical particle constants n and χ. Thus, the conclusions made below are valid for various dispersed substances.

Substitute Eq. 24 into Eq. 9 and Eq. 10 to obtain

$$\psi = \frac{\sigma_1 \psi_1^* + \sigma_2 \psi_2^*}{\sigma_1 + \sigma_2}, \tag{25}$$

$$\psi_1^* = \frac{4}{3\pi b} \left[1 - \frac{2 + b^2}{2} (1 - b^2)^{1/2} \right] - \frac{2}{\pi} \cos^{-1} b, \tag{26}$$

$$\psi_2^* = \left[1 + (\omega/\alpha)^2 \right]^{-1/2} \tag{27}$$

At $b > 1$, we have $\psi_1^* = 4/3\pi b$ (see Eq. 11) instead of Eq. 26. Special calculations show, $\psi_1 \gg \psi_2$ for all ω except for a narrow band around zero frequency. This proves the validity of Eq. 4 used in our calculations for the media with any absorption and not only for black screen, as it is usually meant when applying Eq. 4 and Eq. 6.

Figure 2 gives a comparison between the functions $S(\omega)$ (Eq. 8) with the scattering coefficients and indicatrices specified in a different way: from the Mie formulae[6] and from approximate solution (Eq. 4, Eq. 7), described only the small-angle part of scattering indicatrix. The data are represented for $\tau = 1$ layer containing scattering particles of $m = 1.33 + i0$ and particle size distribution (Eq. 17) with $\mu = 6$, $\rho_0 = 36$. As seen, in this case both curves coincide at $\omega_0 = 25$. The value of ω_0 decreases with growing n, χ, and ρ_0. Thus, at high spatial frequencies, the function $S(\omega)$ is determined only by a small-angle part of a scattering indicatrix Eq. 4. It is this fact which underlies the proposed approach to determining the mean-volume radius and concentration of dispersed particles.

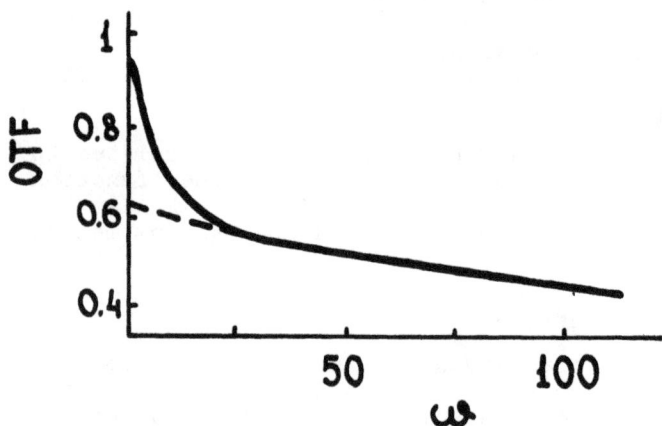

Figure 2   OTF $S(\omega)$ by Eq. 8 with $x(\beta)$ derived from Mie theory (solid line) and calculated from Eq. 4, Eq. 7 (dashed line).

6   MEASUREMENT OF COARSE PARTICLE SIZES IN OPTICALLY THICK WEAK-ABSORBING MEDIA

We have discussed possible measurement of particle sizes at moderate optical thickness of samples. Now consider one more case, when $\tau \gg 1$ and weak radiation absorption in a medium is realized, i.e. from[1]:

$$1 - \Lambda \leq 0.2(1 - g), \tag{28}$$

where

$$g = \frac{1}{2} \int_{-1}^{1} x(\cos\beta) \cos\beta \; d(\cos\beta)$$

is the scattering asymmetry factor. Then the diffusion reflection $r(\mu_0)$ and transmission $t(\mu_0)$ coefficients for the sample illuminated by parallel beams are described by the formulas[1]:

$$r(\mu_0) = \frac{shZ_1}{sh(Z_1 + Z_2)}, \qquad t(\mu_0) = \frac{shZ_2}{sh(Z_1 + Z_2)}, \qquad (29)$$

where $Z_1 = \gamma[\tau + 4q(1 - g(\mu_0))]$, $Z_2 = 4q\gamma g(\mu_0)$, $q = 1/3(1 - g)$, $\gamma = ((1 - \Lambda)/q)^{1/2}$, $\mu_0 = \cos\vartheta_0$, $g(\mu_0) = 3(1 + 2\mu_0)/7$, ϑ_0 is the angle between the normal and incident radiation. Upon measuring r and t from Eq. 29, we easily arrive[7] at the parameters Z_1 and Z_2:

$$Z_1 = \text{arch}(Q/2r), \qquad Z_2 = \text{arch}(Q/2t),$$

$$Q = \sqrt{(1 + r^2 + t^2) - 4r^2}, \qquad (30)$$

which are simply related to the medium microstructure parameters. For example, for coarse particle[8]:

$$\tau^* = \frac{4Z_1}{3Z_2} = G_0(n)\tau, \qquad \tau_a = \frac{Z_1Z_2}{4} = V(n)KC_v l,$$

$$(31)$$

$$\tau = \frac{1.5C_v l}{a_{32}},$$

where $K = 4\pi\chi/\lambda$. Essentially, the general form of Eq. 31 holds for nonspherical scatterers. For spheres[8]:

$$V(n) = n^2 \left[1 - (1 - n^2)^{3/2} \right],$$

$$(32)$$

$$G_0(n) = c(n - 1)^2 + d(n - 1) + h,$$

the latter being derived by approximating tabular data[8] at $n \in [1.1, 1.65]$ within a relative error less than 2% ($h = -0.01684$, $d = 0.4465$, $c = -0.1459$).

Eq. 31 and Eq. 32 allow the mean-volume particle radius a_{32} and volume concentration C_v to be determined by measure $r(\mu_0)$ and $t(\mu_0)$ knowing the constants n and χ. It is impossible to reconstruct other DF parameters. For Mie scattering, a similar approach was earlier used to determine the sizes of erythrocytes in blood[9,10] and of cloud droplets[11]. It is essential that the value of τ (and, hence, of a_{32}) should be obtained by measuring only reflected light characteristics. For example, for nonabsorping optically thick radiation scattering layers, we have[1]:

$$\tau = \frac{4(r(\mu_0) + g(\mu_0) - 1)}{3G_0(n)(1 - r(\mu_0))} . \tag{33}$$

Application of Eq. 31 and Eq. 33 implies knowledge of optical constants n and χ. On the other hand, given C_v and a_{32}, a complex particle refraction index $m = n - i\chi$ can be found from Eq. 31.

7 CONCLUSION

It is shown possible to determine the particle size distribution function $f(a)$ by measuring the angular distribution $I(\vartheta)$ of transmitted radiance (i) as well as the DF parameter a_{32} by measuring PSF (OTF) (ii) or reflection and transmission coefficients (iii) under multiple light scattering in a medium. Simultaneously, concentration of particles is estimated. The first two methods can be applied for dispersed media at $\Lambda\tau(1 - g) \leq 0.5$ (knowing the optical constants of particle material is not required); the third is used for $\Lambda\tau(1 - g) \geq 2$ and the known optical constants of particles. It is important that all these approaches use traditional measurement techniques and can be easily realized.

References

1. E.P. Zege, I.L. Katsev, A.P. Ivanov, 'Image Transfer Through a Scattering Medium', Springer-Verlag, Berlin, 1991.
2. K.S. Shifrin, 'Introduction in Ocean Optics', Gidrometeoizdat, Leningrad, 1983.
3. J.W. Goodmen, 'Introduction to Fourier Optics', McGraw-Hill, New York, 1968.
4. A.G. Borovoi, N.I. Vagin, V.V Veretennikov, Opt. Spectr., 1986, 61, 1326.
5. K.S. Shifrin, 'Scattering of Light in a Turbid Medium', NASA Report No.TTF-447, Washington, 1951.
6. A.S. Drofa, A.L. Usachev, Izv. AN SSSR, FAO, 1980, 16, 933.
7. E.P. Zege, M.P. Znachenok, I.L. Katsev, Zh. Prikl. Spectr., 1980, 33, 735.
8. E.P. Zege, A.A. Kokhanovsky, Izv. AN SSSR, FAO, 1988, 23, 691.
9. A.Ya. Khairullina. Zh. Prikl. Spectr., 1987, 46, 1000.
10. E.K. Naumenko, A.Ya. Khairullina. Zh. Prikl. Spectr., 1990, 52, 654.
11. T. Nakajama, M.D. King, J. Atm. Sci., 1990, 47, 1878.

Particle Size Distribution From Light Scattering*

H. T. Sommer, C. F. Harrison and C. E. Montague

HIAC/ROYCO DIVISION OF PACIFIC SCIENTIFIC, SILVER SPRING, MARYLAND, USA

1   SUMMARY

Particle sizing instruments can be categorized by the method applied to determine a size distribution. Integral optical methods like Fraunhofer Diffraction or Dynamic Light Scattering require extensive mathematics to extract the particle size distribution from samples. The general shape of this distribution is ordinarily assumed and a few parameters of this assumed distribution are adjusted to represent features of the data. Because of these restrictions, integral optical methods are not able to resolve all fine details of a particle size distribution.

Optical particle counters have been used extensively to detect, classify, and count individual particles in contamination monitoring of many liquids and gases. These sensors automate the laborious task of single particle size classification by microscopy. The paper describes an instrument suitable of single particle sizing and counting and presents a comparison of size distributions obtained with a Fraunhofer diffraction instrument and an optical single particle sizing instrument from the same sample. Methods and techniques applied to overcome the limitations of Optical Single Particle Sizing are described and the results of the comparison are critically interpreted.

2   INTRODUCTION

Optical Particle Counters (OPC) possess many favorable characteristics such as high size resolution and speed in counting and measuring distributions of particles in suspensions. Global optical sizing methods [1-7] like Fraunhofer Diffraction are also used to measure particle size distributions of wide size range. These methods determine the particle size distribution indirectly by assuming the shape of the distribution function and fitting two or more parameters, relating the distribution to the light scattering from many particles. In many cases, the presumed distribution does not include details of the actual distribution as a larger number of fitting parameters is need to describe these details than global methods can provide. OPC's are able to size and count individual particles with high resolution [8]. Two optical methods have been proven very effective: particle sizing by light scattering and light extinction. Scattering techniques [9,10, 11] are commonly used to detect and size particles in the submicron range, $(0.1\mu m - 1.0\mu m)$ and extinction methods [12,13,14] are suitable to determine the size of larger particles (> $1.0\mu m$).

In the following, an instrument will be described that performs sizing and counting of particles in highly concentrated suspensions. The instrument consists of an optical sensor, an electronic counter, a sample preparation and delivery system, and a computer program for data display and statistical manipulation (Fig 1).

* Presented by P. Rossi

**PDAS Particle
Sizing Software**

Sensor

Model 9064 Counter

HIAC/ROYCO

Sample Delivery System

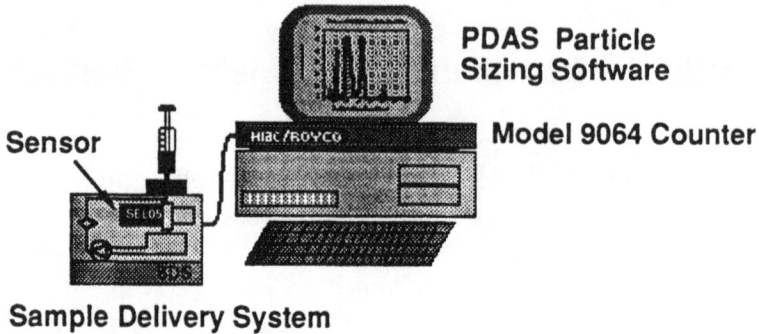

Fig. 1 HIAC/ROYCO Optisizer™ components

3   SENSOR

The heart of the instrument is the wide dynamic range, single particle sensor. It operates on the principles of light scattering and extinction simultaneously. The sensor delivers two signals, a scattering signal and an extinction signal to the electronic counter. The counter performs the sizing and counting operation. Forward scattering is applied in the model MicroCount™05 sensor to detect and size particles in the size range from 0.5μm to 1.63μm. Because of the strong dependency of the light scattering signal from the particle size (signal ~ d^6 for particles smaller than the wavelength of the laser beam), the dynamic range of the scattering signal channel is limited. The scattering size range is overlapped in its upper part by the extinction size range. For light extinction, the size dependency of the signal reduces to square power and for larger particles (> 40μm) this dependency becomes linear. Therefore, the extinction channel covers the larger part of the size range from about 1.53μm to 350μm. The upper limit is determined by the dimensions of the measurement cell of the sensor and might vary with the sensor model.

The sensor (Fig. 2) combines the measurement principles of light scattering and extinction in a single unit. The particles follow the suspending fluid flow in the measurement cell and pass through the optical measurement volume. This volume is defined by the flow passage cross section and the height of the focused laser diode beam. The laser beam is shaped to illuminate the entire flow passage uniformly at the location of the measurement volume. Sizing resolution in both signal channels, scattering and extinction, strongly depends on the uniformity of illumination in the measurement volume.

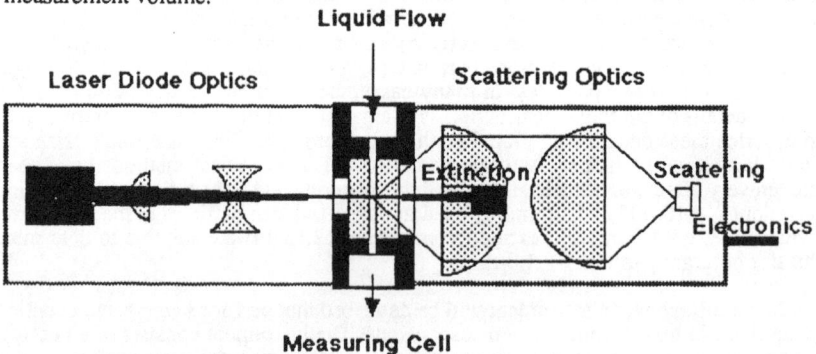

Liquid Flow

Laser Diode Optics   **Scattering Optics**

Extinction   Scattering

Electronics

Measuring Cell

Fig. 2 HIAC/ROYCO MicroCount™05
(combined light scattering and extinction, single particle sizing sensor)

Typical dimensions of the flow passage are: width $W = 0.5$mm and depth $D = 1$mm. The laser beam is focused to approximately 40μm in the center of the measurement volume. Since the local intensity of a focused laser beam does change along the beam with the highest intensity in the focal plane, it is important to adapt the beam shaping optics not only for uniformity of illumination across the beam, but also along the depth of the flow passage. For a detailed analysis of the interaction of a focused laser beam with a sample flow, see [15].

Particles that are carried by the sample flow through the optical measurement volume scatter light in all directions. The scattered light is registered by the extinction photo detector as an intensity defect of the laser beam. The fraction of the light scattered in the direction of the collection optics is focused on the scattering photo detector. Both detectors convert the light signals into electrical signals for further processing by the counter.

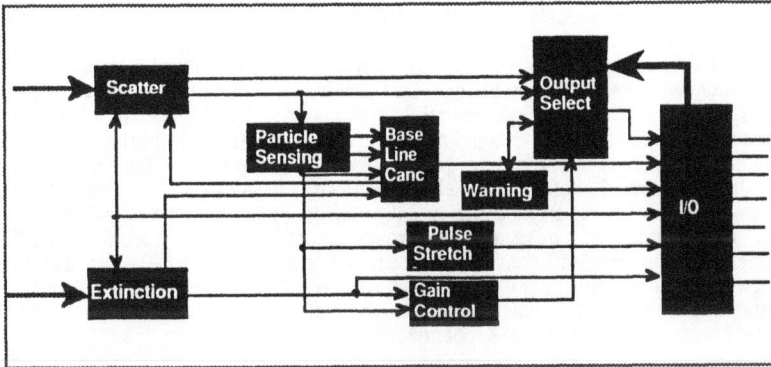

Fig. 3 Analog signal block diagram of multi-mode sensor

Fig.3 shows the block diagram of the analog signal treatment of the sensor electronics. Going into the details of the circuit would exceed the objective of this paper. The electronics in the sensor are kept to a minimum. Most of the signal processing is conducted by the counter.

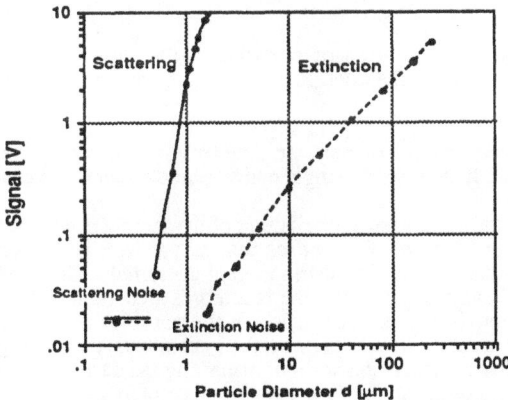

Fig. 4 Calibration curve of a combined Scattering-Extinction sensor
HIAC/ROYCO model MicroCount™05

Fig.4 shows the typical calibration curves of a dual mode sensor. The first two data points of the curve for light extinction overlap the last two data points of the scattering curve. The smooth connection of the two curves is important for accurate measurement performance in this size range. Since the size resolution of particle measurements is dependent on the slope of the calibration curve, both curves should show comparable slope values at the connection point. The slope values in the logarithmic plot of Fig.4 are 0.005 for scattering and 0.0042 for extinction. The extinction curve exhibits a strong change in slope in the size range from $2\mu m$ to $4\mu m$. This behavior of the extinction calibration curve can be explained with the extinction paradoxon [16,17], that locates the first minimum of the scattering intensity for the given arrangement of collection optics in this size range.

4   COUNTER

The counter continuously monitors the first signal line. This signal is the highest amplified (factor 100) scattering signal in which particles near the sensitivity limit of the sensor will appear ($\sim 0.5\mu m$- $0.7\mu m$). When this signal saturates at about 9.5V, the counter switches to the next channel that provides an unamplified scattering signal. This signal covers the size range from $0.7\mu m$ to about $1.63\mu m$. If the signal pulse also saturates this signal channel, channel number three, which is the amplified (factor 100) extinction channel , is monitored. In the case that the signal range of this channel is also exceeded ($1.63\mu m$ - $25.0\mu m$), channel number four is monitored. This channel covers the remaining size range from $25\mu m$ to $350\mu m$. Signals from particles within the size range of the sensor will have a voltage reading that is detected by one of the four channels. The counter will convert this reading into a number with 14 bit resolution. The two remaining bits of the 16 bit digital representation are used to identify the channel and amplification that transmitted the signal. The microprocessor sorts the digitized voltage signal into 64000 channels.

5   SAMPLE DELIVERY SYSTEM

Optical particle counters require dilution of highly concentrated particle suspensions to a concentration level below the concentration limit of the sensor. The low "overconcentration" limit of a OPC was in the past the major disadvantage, limiting the use of optical single particle counters. The main objectives of the sample delivery procedure are:

1) To reduce the concentration of particles in the flow through the sensor to a concentration level below the concentration limit of the sensor by adjusting dilution automatically.

2) To reduce manual sample preparation time by allowing highly concentrated "raw" samples to be directly used with a single optical particle sizing and counting sensor.

3) To ensure that the particle size distribution of the sample is not artificially altered by the sample delivery system. This means that all particles of the sample have to be analyzed to build the particle size distribution of the sample. The user has to make sure that the analyzed sample in the syringe is a representative sample of the suspension under investigation. The size distribution of the particles should not change during dilution. In order to meet this requirement, the system must dilute a highly concentrated sample of particles in suspension without disturbing the particle size distribution. The method must avoid segregation of particles based on their sizes.

4) To increase the size resolution of measurement by injecting the sample of highly concentrated particulate matter in the center of a particle free constant diluent flow, thus forcing the particles always through the same part of the optical measurement volume that is most uniformly illuminated.

Depending on the fluid used in suspending the particulate matter (powder) of interest, a closed-loop or open diluent flow system can be applied to carry the particles through the sensor cell. Typically, a constant flow (between 20 - 100ml/min) of clean filtered diluent is generated through a closed-loop pump system or an open diluent supply system with a sufficient pressure head that is controlled to maintain a constant flow through the sensor. Fig. 5a illustrates the closed-loop flow system. The sample of highly concentrated particulate matter is introduced through a syringe into the constant flow of clean carrier fluid (diluent) just above the optical measurement volume. The needle of the syringe is positioned in the center of the flow. The flow of sample and diluent is filtered by a replaceable filter and collected in a reservoir. The pump moves the diluent at a preset constant flow rate through another fine filter to the sensor. This closes the flow cycle. The pump speed is controlled by a flow controller to maintain constant flow. The flow controller is connected to a personal computer that provides the set point for the flow controller presetting the flow rate depending on sensor model.

Highly concentrated particulate matter is sampled in a small (0.01 - 5 ml) syringe. The user has to make sure that the sample in the syringe is representative of the larger sample on hand or the analysis of a statistically representative number of "small" samples has to be conducted to represent an "average" measurement. The syringe is mounted in a computer controlled syringe pump that advances the piston of the syringe at a controlled speed. During sample analysis, the syringe pump will start slowly to advance the piston at a constant acceleration. During this process sizing and counting of particles in the sensor takes place. When the counting rate of the sensor reaches a number close, but safely below the count rate limit (concentration limit) of the sensor, the syringe pump stops accelerating and maintains the current piston velocity constant until the sample is exhausted.

Since all the sample volume is analyzed, a true particle size distribution of the sample in the syringe is obtained. Size segregation in the syringe and settling during the analysis is not a problem because the total sample is analyzed. The dilution ratio is determined by the system itself during the initial phase and the computer can calculate a dilution factor as a function of analysis time from the diluent flow rate and the sample flow rate. Dilution factors of 10,000:1 are possible using 100ml/min diluent flow and 0.01ml/min sample flow.

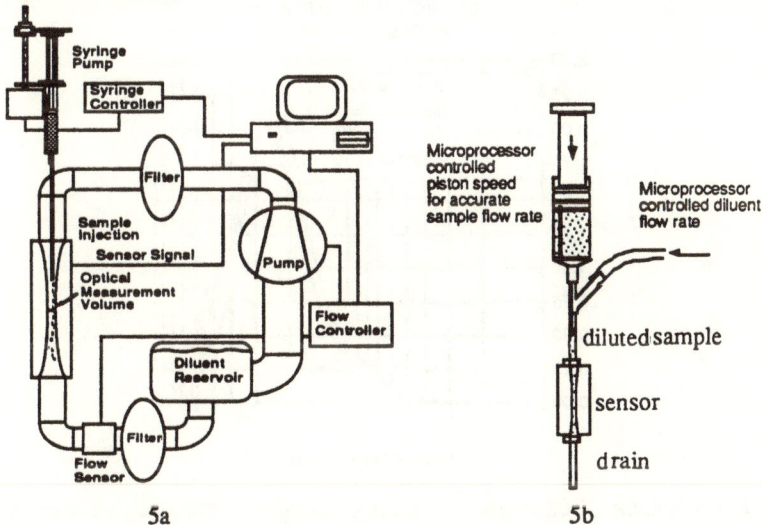

Fig. 5a Sample Delivery System (closed loop)
5b Detail of sample injection (open system)

Fig.5a illustrates the sample delivery system. The integrated control of sample injection, sensor count rate, and diluent flow allows easy particle size distribution analysis. Fig. 5b shows the details of sample injection. The exit of the injection needle is located in the center of the diluent flow, placing all particles at the same location in the flow. The laminar diluent flow carries each individual particle through the center of the measurement volume. This greatly enhances size resolution of the sensor since all particles pass through the same part of the measurement volume.

6   STATISTICAL ANALYSIS SOFTWARE AND USER INTERFACE

The Particle Distribution Analysis Software (PDAS) is the link between the user and the data generated by the single particle sizing system. This program requires an IBM-AT-type computer. It contains all manipulation tools to perform useful statistical analysis of the measured data including mass, volume, and surface weighted distribution. Display of multiple distributions in one graph and algebraic operations between distributions (background subtraction) is executed at a keystroke. The software gathers the data from the digital counter and displays the particle distribution in a number of size bins selected by the user. The largest number of size categories is 512. At this value, the contents of 125 size channels of the counter are combined in one size bin and displayed on the computer screen.

In addition to data manipulation and display, the software conducts control functions for the sample delivery system. The acceleration control of the sample injection pump is determined from the particle count rate and the count rate limit of the sensor and directions are sent to the syringe pump controller limiting the count rate to a value below the maximum count rate of the sensor. PDAS does the actual conversion from measured signal voltages to particle size using the sensor calibration curve.

7   RESULTS AND CONCLUSION

Experimental results have been obtained from the sensor, counter, and software described here. The measurement results from a set of four similar silica gel samples are shown in Fig. 6.

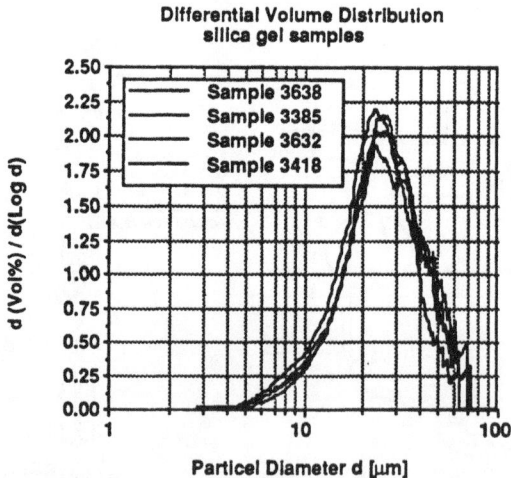

Fig. 6   Volume Distribution of four silica gel samples (Optical Single Particle Sizer)

In these tests the diluent was filtered water circulating at 25 mL/min. The test particles were amorphous silica gel beads of varying shapes with sizes in the 3μm to 100μm range. About 10mg of the dry material was suspended in water and added to the continuous flowing water through the MicroCount 05 sensor. The sample was measured in approximately 45 seconds. The total number of particles measured in this period was typically 50,000. Data were processed through the counter and software in 512 size bins.

In single-particle counting, the number count is the directly measured result. Fig.7 shows the differential number count fraction (normalized by the log of diameter) from four samples. The resolution and sensitivity of the method is apparent. Three samples show a prominent peak near 7μm, while the fourth shows a broader and larger-diameter distribution. Subtle differences can be resolved among the samples. The degree of repeatability is shown in Fig.8, where two samples of the same material were run at different concentration; total counts were 32000 and 62000, with excellent consistency in distribution shape.

Differential Count Distribution
silica gel samples

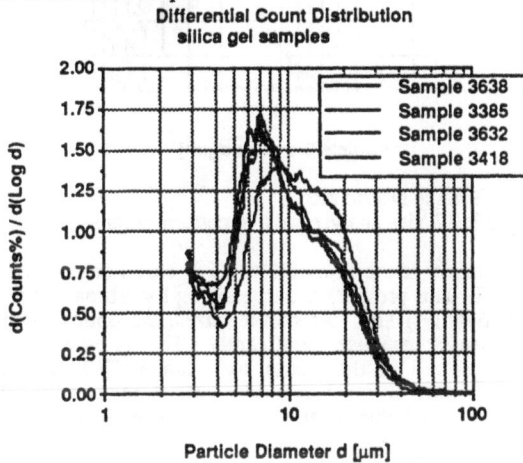

Fig. 7 Differential Count Distribution of four silica gel samples
(Optical Single Particle Sizer)

Differential Count Distribution
silica gel sample 3632

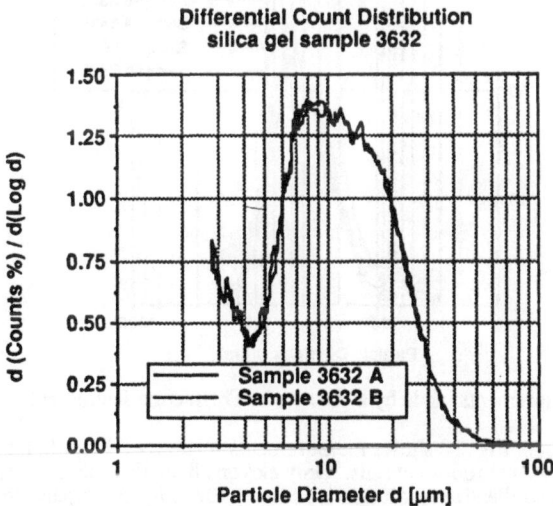

Fig. 8 Repeatability of Number Counts:
Sample 3632, size analyzed at two concentrations

Differential Volume Distribution
silica gel samples

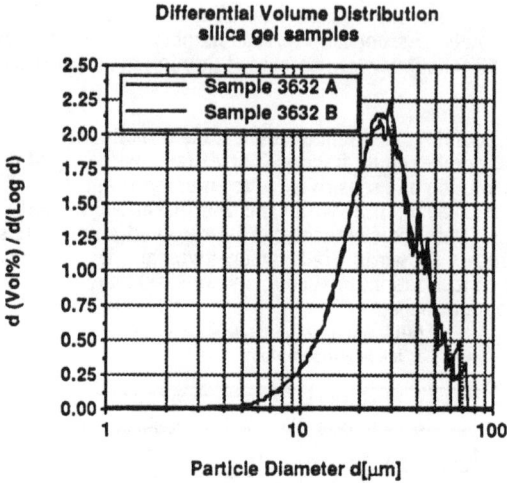

Fig. 9 Repeatability of Volume Distribution:
Sample 3632, size analyzed at two concentrations

The standard calibration of the MicroCount 05 is based on uniform polystyrene beads traceable to NIST. This provides accurate sizing for a broad range of materials in the ranges above roughly 10μm. At smaller sizes, the optical properties of the sample influence the sensor response significantly. This is especially true for the smallest particles measured in the scattering channel. When calibrated with polystyrene, the sensor may be used as a comparison instrument. If additional applications information is available, a calibration curve specific to the material can be developed.

Differential Volume Distribution
silica gel samples

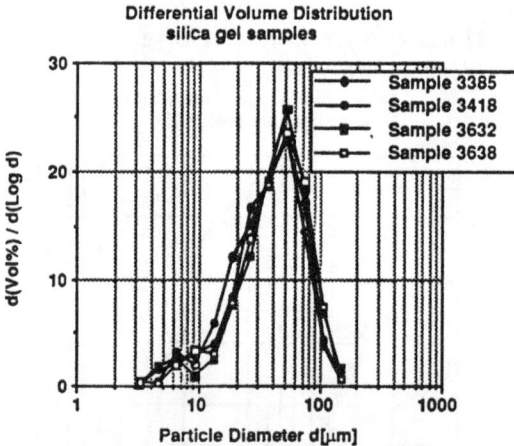

Fig. 10 Measurements made by a Fraunhofer Diffraction instrument

For comparison, Fig. 10 shows measurements made by a Fraunhofer Diffraction instrument on the same four materials. Only eleven channels of data are registered on this distribution, so the size resolution is limited. Even for large particles, the direct measurement of concentration in this type of instrument is affected by the scattering efficiency of the materials, so the volume and number counts must be computed indirectly based on material assumptions. The Fraunhofer Diffraction instrument showed a volume peak substantially larger than single-particle counting. The cause of this disagreement is uncertain, although sample preparation and instrument adjustments are possible sources of error.

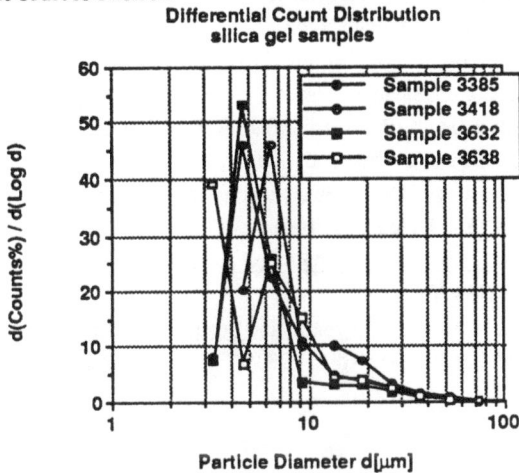

**Differential Count Distribution
silica gel samples**

Fig. 11   Fraunhofer data as differential counts

In integral optical methods, the energy scattered from the larger particles is much greater than that from the small ones, which can lead to poor accuracy at the lower-diameter tail of the distribution. This effect is pronounced when the above Fraunhofer data are presented as differential counts (Fig. 11); the wide variation in counts near 5 μm is artifact.

8 REFERENCES

[1] G. B. J. de Boer, C. deWeerd, D. Thoenes, H. W. J. Goosens,
    Particle Characterization, vol. 4, no. 1, 1987
[2] Suezou Nakadate, Hiroyoshi Saito,
    Optics Letters, vol. 8, no. 11, Nov. 1983
[3] Philippe Herve and Nathalie Balu, Proceedings :
    Fine Particle Society Annual Meeting , 10/9-13, 1989, Reno, NV
[4] R. Pecora, Pure & Appl. Chem., vol. 56, no. 10, 1984
[5] R. S. Stock, W. Harmon Ray,
    Polymeric Materials Science And Engineering, vol. 53, 1985
[6] A. M. Ganz, Rev. Sci. Instrum., vol. 56, no. 11, 1985
[7] H. Auwetter, D. Horn,
    Journal of Colloid and Interface Science, vol. 105, no. 2, 1985
[8] J. K. Dhont, C. G. De Kruif, A. Vrij,
    Journal Of Colloid and Interface Science, vol 105, no. 2, 1985
[9] Esin Gulari, Erdogan Gulari,
    Polymeric Materials Science and Engineering, vol. 53, 1985

[10] H. W. Schrader, W. G. Eisert, Applied Optics, vol. 25, no. 23, 1986

[11] H.T Sommer, Proceddings: 36th Annual Meeting of the Institute of Environmental Sciences, New Orleans, April 1990

[12] A.D.Bates "Performance Limit of an Obscuration Type Particle Counter with Tungsten Source" Private communication

[13] Julius Z. Knapp "A Critical Examination of Light Obscuration Detection Measurements" Private Communication

[14] H.T. Sommer, Proceedings: 10th International Symposium on Contamination Control, Zürich, Switzerland, September 1990

[15] H.T Sommer, Proceedings: 37th Annual Meeting of the Institute of Environmental Sciences, San Diego, May 1991

[16] H.C. Van De Hulst ,Light Scattering by Small Particles, John Wiley & Sons New York 1957

[17] C.F. Bohren, D.R. Huffman, Absorption and Scattering of Light by Small Particles, John Wiley and Sons, New York, 1983

Twenty Seven Years of QELS: A Review of the Advantages and Disadvantages of Particle Sizing with QELS

B. B. Weiner

BROOKHAVEN INSTRUMENTS CORPORATION, 750 BLUE POINT ROAD, HOLTSVILLE, NEW YORK 11742, USA

1 INTRODUCTION

It is well known that the existence of light is mentioned at the beginning of the Bible. Less well known is a footnote in the bibliography to the following effects: let everything scatter light; let the relationship between particle size and scattering change dramatically over the size range of interest; and let ill-conditioned Laplace transforms run rampant over the whole field. So much for the religious aspects of particle sizing with light, though anyone who stays in particle sizing long enough must wonder if some of our efforts are not theological in nature.

QELS stands for quasielastic light scattering. It was the name first given to the technique because, when photons are scattered by mobile particles, the process is quasielastic. QELS measurements yield information on the dynamics of the scatterer; this gave rise to the acronym DLS, which stands for dynamic light scattering. Finally, since most measurements are made with a digital correlator, the acronym PCS, which stands for photon correlation spectroscopy, is widely used.

QELS has something in common with many modern techniques for particle sizing: it was not developed for that purpose. In fact, it was not commercially available for almost 7 years after the first measurements were made in 1964[1]--and then only in a form suitable for experts. The majority of practitioners built their own equipment. Slowly, through the 1970's it gained wide acceptance among experts in light scattering.

Physicists and physical chemists developed the technique for three purposes: curiosity about the statistical nature of light;[2] as a tool to study critical phenomena;[3] and as a way to probe the dynamics of polymer solutions.[4] Early on it was used to measure the diffusion coefficient of macromolecules, from which a hydrodynamic size was calculated.[5] A few industrial users tried it for submicron particle sizing, mostly to replace TEM measurements in QC applications. The approach proved rather awkward in those early days.

By the early 1970's the fundamentals had been laid down and tested. The second half of the 1970's consisted of: the improvement of digital autocorrelators; the introduction of several algorithms for analyzing linewidth distributions--later these would have interesting consequences for particle sizing; and the automation/integration of instrumentation, making the technique more suitable for use in particle sizing. These included microprocessor control, stepping motor control of angular measurements, and the use of small, stable lasers, most notably the 5mW HeNe. Measurements that had taken months to setup and hours to make, were now reduced to minutes.

This author remembers with some nostalgia that many current users of PCS had never heard of it prior to the early 1980's; much time was taken up explaining the fundamentals. Though the limitations of the technique were already known, they had not been presented in a form suitable for digestion by particle size analysts. With hindsight, it was predictable that this situation would lead to misunderstanding and to the reinvention of the wheel.

After a brief review of the fundamentals, the current boundaries of the technique as they apply to particle sizing will be illustrated with examples. The paper concludes with a brief discussion of new developments in PCS instrumentation.

2 FUNDAMENTALS

QELS theory is built upon the earlier foundation of classical light scattering theory. The latter is usually dated by Rayleigh's papers in 1871 on the scattering from a single particle small compared to the wavelength of light. Scattering from larger particles was added later; it is often referred to as Mie theory after Gustav Mie who gave the complete solution for spherical particles of any size.

As early as 1908 the temporal fluctuations about the average scattered light intensity were identified with the motion of the particles and their diffusion coefficients.[6] Since Einstein had also published the fundamental relationship between diffusion and size,[7] the way was open for the birth of a new particle sizing technique. The gestation period was over 60 years: lasers, high-speed digital circuits, and light-beating spectroscopy had to come first.

The fundamentals have been written about so often and in so many places that it seems only fitting here to quickly come to the important parts necessary for the particle analyst. The references should be consulted for a more complete treatment.[8-12]

Colloidal-sized particles in a liquid undergo random motion due to multiple collisions with the thermally driven molecules of the liquid. Light scattered by these particles will fluctuate in time, the characteristic dependence describable in terms of the diffusion coefficient of the particle. Like the sedimentation coefficient in a Stokes' Law device, the diffusion coefficient can be related to the particle size.

The time dependence of the scattered photons are most efficiently analyzed by measuring the second-order autocorrelation function (ACF) of the photon pulse train,

$$C(\tau) = <n(t) \bullet n(t-\tau)> \qquad , \qquad (1)$$

where n(t) is the number of pulses at time t during a short sampling time interval, and n(t-τ) is the number in the same sampling time interval but delayed by τ. For dilute, rigid, globular particles, measured in the self-beat mode, the ACF is given by

$$C(\tau) = B\{1 + f \mid g^1(\tau) \mid^2\} \qquad . \qquad (2)$$

For a monodisperse sample,

$$| g^1(\tau) | = \exp(-\Gamma\tau) \qquad , \qquad (3)$$

and

$$\Gamma = D_T \cdot q^2 \qquad , \qquad (4)$$

where Γ is called the linewidth after the broadening of the frequency distribution of scattered light. The inverse of Γ is the characteristic relaxation time of the fluctuations in the scattered light.

The amplitude of the scattering wave vector, q, is given by

$$q = (4\pi n/\lambda)\sin(\Theta/2) \qquad , \qquad (5)$$

where n is the refractive index of the liquid, λ is the laser wavelength, and Θ is the scattering angle.

A particle size is calculated from the translational diffusion coefficient D_T once a shape is assumed; by far the most common choice is the sphere. For spheres the Stokes-Einstein equation is

$$D_T = k_B T/(3\pi\eta d_h) \qquad , \qquad (6)$$

where k_B is Boltzmann's constant, T is the absolute temperature, η is the viscosity of the suspending liquid, and d_h is the hydrodynamic diameter. Except for very small, highly charged particles, d_h is practically equal to the geometric diameter.

For a polydisperse sample $| g^1(\tau) |$ is an intensity-weighted sum or integral over all the species contributing to the scattering. It can be written as

$$| g^1(\tau) | = \int G(\Gamma)\exp(-\Gamma\tau)d\Gamma \qquad , \qquad (7)$$

where $G(\Gamma)$ is the relative intensity for the particle that gave rise to linewidth Γ.

The most common approach is to expand the exponential in equation (7) about an average, and then integrate term-by-term. The result, called cumulant analysis, is:

$$\ln | g^1(\tau) | = -<\Gamma>\tau + (\mu_2\tau^2)/2! - (\mu_3\tau^3)/3! + ... \qquad . \qquad (8)$$

An average diffusion coefficient, from which an average particle size can be calculated, is obtained from $<\Gamma>$; the relative second moment, $\mu_2/<\Gamma>^2$, a dimensionless quantity, is a measure of polydispersity. It is the intensity-weighted variance divided by the square of the intenstiy-weighted average of the diffusion coefficient distribution. The relative second moment is also called poly or the polydispersity index.

3 SOME VICTORIES

<u>Narrow Distributions</u>

It is difficult to imagine a faster, more repeatable, simpler, more accurate technique than PCS for narrow size distributions, provided the size is appropriate for the instrument configuration. Here is one example of many.

Samples of narrow polystyrene (PS) latex, steam-stripped to remove excess, unreacted monomer, were purchased from a supplier in order to test the calculation of extinction efficiencies for use in correcting data from a disc centrifuge photosedimentometer.[13] It was crucial that the sizes were known with an accuracy of ±1%, independent of the DCP measurement. TEM values were given by the manufacturer. In addition PCS and DCP measurements were made. Table 1 shows the results.

Table 1 Comparison of average diameter of latex standards using different techniques.

Technique	Std.1 (nm)	Std.2 (nm)
TEM (labeled)	300±1	121±1
PCS	313±1	151±1
DCP	316±2	151±2

The accuracy of any Stoke's Law device is, of course, subject to the uncertainty in the density difference. PS in water represents a case where this error is unavoidably high. The PCS and DCP results, however, agree within ±1%. Subsequent TEM measurements by the manufacturer on Std.2 revealed an error in the original calibration. The new TEM result was given as 149 nm, in excellent agreement with the DCP and PCS measurements.

It is known that commercially available standards are not always labeled properly. That is not the point here. The point is that accurate results from the PCS measurements took only a few minutes without any need to calibrate or correct for particle properties like density or refractive index. Also, the polydispersity index was <0.02 for both standards, implying narrow distributions.

Fortunately there is a continuing need to measure narrow distributions in the submicron range. Narrow standards, liposomes, and specialty latexes are examples, just to name a few.

Measurements of Very Small Particles

Figure 1 shows the results of two different PCS measurements on tetrapropylammonium bromide, $(CH_3CH_2CH_2)_4NBr$, in pure water at 25 °C. The data were collected at 30° scattering angle; 300 mW of Argon laser power was necessary to achieve good results in a reasonable time.[14]

Curve 1 is not a straight line at all, indicating polydispersity. The average size calculated is ~2 nm. Yet the tetrapropylammonium ion is monodisperse; its size, estimated from bond lengths and angles, is ~1 nm. Curve 1 was obtained using standard cleaning techniques: all syringes and glassware were oven dried after rinsing with distilled and filtered benzene until no dust flashes could be seen in the benzene with a laser; water was freshly distilled; a $0.2\,\mu$ filter was pre-rinsed several times before filtering the solution into the cell. Most likely recontamination by dust has occurred.

Curve 2 is a straight line, indicating a monodisperse sample. The average size calculated is 0.93 nm, in excellent agreement with theory. Curve 2 was obtained using a completely enclosed, recirculating filtration system. Recontamination by dust has not occurred, allowing an accurate measurement.

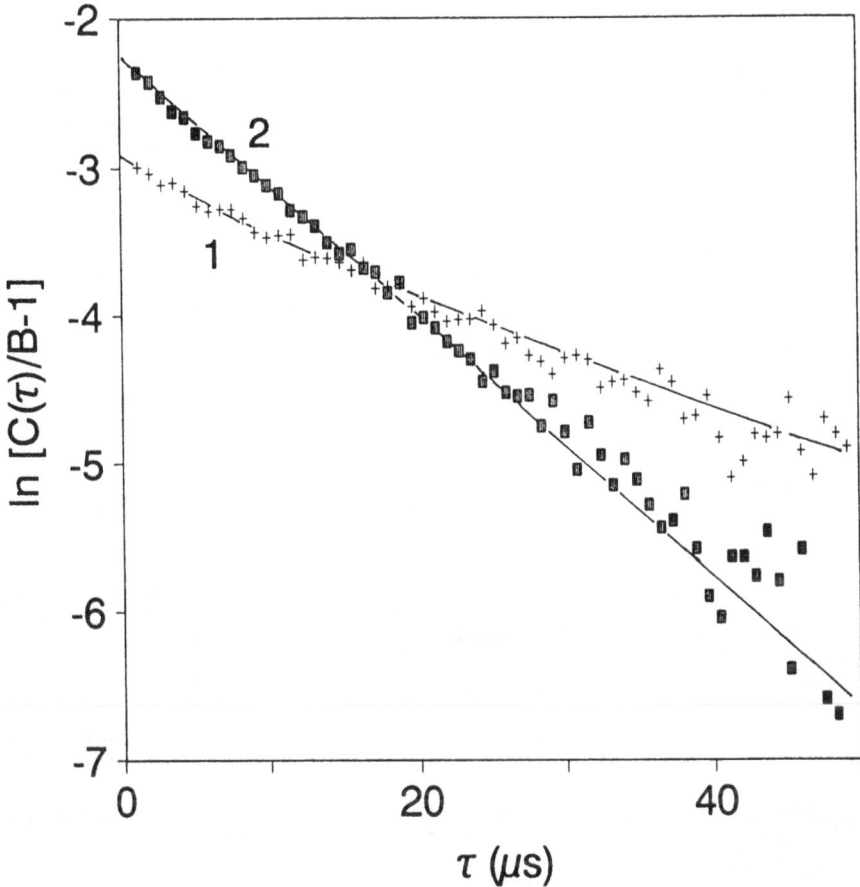

<u>Figure 1</u>   Semi-log plot of the ACF for $(CH_3CH_2CH_2)_4NBr$ in water. Curves 1 and 2 correspond to different dust removal techniques.

This is an example of several features of PCS. First, one can measure down to 1 or 2 nm with PCS. Second, one cannot do it, generally, with a 5 mW HeNe laser. Third, the average size and distribution obtained is dramatically dependent on proper precautions to preclude dust.

Testing of Colloidal Theories or Particle Sizing with a Pinch of Salt

When particle-particle interactions are not negligible, then the diffusion coefficient from a PCS measurement is concentration dependent. Particle size should then be calculated from the diffusion coefficient extrapolated to zero concentration. Fortunately, in many practical applications, extrapolation is not necessary because sufficient electrolyte exists in solution to shield the particles. Dilution in filtered tap water of charge stabilized dispersions is often sufficient to maintain this shielding. Occasionally one runs into trouble. Here is an example.

A biotechnology company was required to validate their PCS measurements on liposomes by showing agreement with an NIST traceable standard, in this case 96 ± 3.9 nm polystyrene. The concentration of the stock suspension was

5-10%, and it was charged stabilized with $\sim 0.2\%$ of sodium dodecyl sulfate (SDS), a 1:2 electrolyte. When diluted for PCS measurements, the particle and SDS concentrations were $\sim$ 1E-04 g/mL and $\sim$ 4E-3 g/L ($\sim$ 1.3E-5 mol/L), respectively. Deionized water was used. Particle sizing results were repeatable to 1%, but the average size was $\sim 8\%$ too high. By adding a pinch of salt (NaCl, a 1:1 electrolyte) the average size dropped to 97 nm, well within specification.

A colloid chemist will be familiar with the problem and the answer. Charged particles in suspension are surrounded by an electrical double layer. The double layer influences the distance over which the electrostatic potential of one particle can be felt by other particles. Estimates of the double layer thickness, also called the Debye length or K^{-1} are shown in Table 2.

Table 2: Double layer thickness, K^{-1}, at 25°C in water as a function of electrolyte concentrations.

Molarity	K^{-1}, 1:1 (nm)	K^{-1}, 1:2 (nm)
1E-7 (Pure H_2O)	960	560
1E-5	96	56
1E-3	9.6	5.6
1E-1	0.96	0.56

The double layer thickness is a measure of the distance over which repulsive forces that keep charged particles stable is important. At high salt concentration K^{-1} is low; aggregation can occur. At low salt concentration K^{-1} is high; particles may not be freely diffusing.

It has been shown theoretically by Oshima[15] and experimentally by Schumacher and van de Ven[16] that when $K \cdot r \sim 1$, the apparent diffusion coefficient is less than the true one by as much as 10%. Here r is the radius of the particle. In this case r = 48 nm, and, fortuitously, this is about equal to K^{-1} under the initial experimental conditions. That is why the apparent particle size was roughly 10% too high.

To solve the problem a pinch of salt was added. This amounted to 2 or 3 grains of salt in $\sim$ 3 mL of suspension, about 1E-3 molar. The double layer collapses, though not enough to cause aggregation; $K \cdot r$ rises; particle-particle interactions are shielded; and the measured particle size agrees with the standard value.

High Throughput Testing of Sunscreens

We are all familiar with the concerns over the destruction of the ozone layer, the increase in UV light, and the predicted rise in skin cancer. Consequently, the use of products that block harmful UV rays are on the rise. Active ingredients in sunscreens are of two types: organics, para aminobenzoic acid (PABA), is a common example; and inorganics such as TiO_2, ZnO, and even Mica.

Some allergic reactions occur with the organic sunscreens. Recent evidence suggests that the organics, which are known to work best when absorbed into the epidermis, may cause health problems if used daily in large amounts.[17] As a result, the inorganics are being considered for a greater role. Currently both TiO_2 and ZnO are found in several commercial sunscreens. Consumer research has shown that opaque creams will not be accepted if worn throughout the day, except, perhaps, at the seashore.

The problem is to find the right combination of scattering, absorption, and cosmetic acceptance. Clearly particle size has a part to play in all of this. Figure 2 shows the weight distribution for two different samples of ZnO. Notice that both show strong peaks at $\sim 0.2 \mu$, and one shows a tail up into the micron range. PCS results on these same samples shows the following: the narrower distribution in Figure 2 has an effective diameter and poly of 249 nm and 0.15, respectively; whereas, the broader distribution results are 596 nm and 0.65, respectively. It is not currently known which, if any, of these two samples is clinically and cosmetically superior, though a first guess would indicate it is the one with the narrower, smaller size distribution. It is, however, interesting to note that the major trends in the particle size distributions are quite apparent in the shift in the PCS diameter and polydispersity index.

Since the PCS measurement is faster and simpler to make, it is suggested that high-throughput, QC testing of these types of materials lend themselves to PCS measurements.

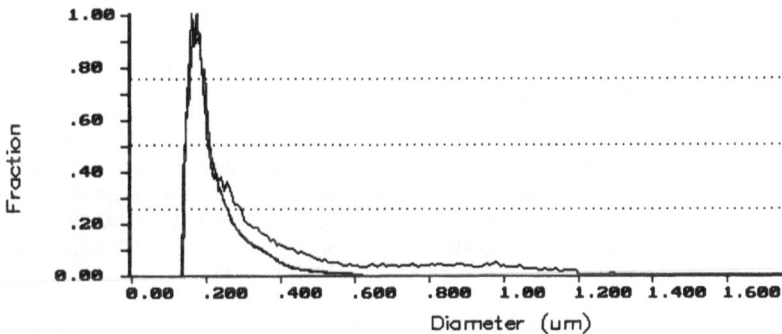

Figure 2   Weight distribution for two ZnO samples measured on a DCP.

4 A DEFEAT

Equation (7) is an ill-conditioned Laplace transform. As such it is known that no unique solution exists. A number of formally different distribution shapes, including physically unreasonable ones, will fit the measured ACF within experimental error. This is a fundamentally limiting feature of PCS. And it is the primary reason why PCS does not produce, reliably, well resolved size distributions, except for computer simulations.

Several methods have been proposed over the years for extracting the maximum information from equation (7); these are reviewed in Chu's recent book.[8] All require data of high precision, at least 0.1%; all require that the baseline B in equation (2) be determined to within 0.1%. Often neither of these prerequisites are achieved in practice with routine PCS instruments. Furthermore, the most sophisticated algorithms require user interaction with the data and with the intermediate results.

The foregoing does not mean that acceptable size distributions have not been determined from PCS measurements. They have. It means that either the reproducibility of the distribution is poor or an algorithm that worked well for one type of distribution did not work well for another. This is the reliability problem.

Nor should the foregoing be misinterpreted. The mean and variance of a distribution may be quite repeatable even if the overall shape is not reliably reproduced. In addition, cumulants analysis is always repeatable, though for $\mu_2/<\Gamma>^2 \geq 0.3$, the polydispersity index is increasingly less sensitive to variations in the width of the true distribution.

5 SOME MISUNDERSTANDINGS

Measurements near 1 Micron

PCS measurements are more difficult around $1\,\mu$. There are many reasons for this. The particle may be so dense and large that it sediments before a proper measurement can be made. Diffusion slows down inversely with size; what was a repeatable 2 minute experiment at $0.1\,\mu$ is a 20 minute experiment at $1\,\mu$, provided the intensity remained constant. It doesn't. Large particles scatter more in the forward direction. There is, however, a more subtle effect that limits measurements on large particles.

Equation 2 has another underlying assumption: The number, N, of particles in the scattering volume should satisfy the criterion $N^2 >> N$. This ensures that the scattered field is a Gaussian random variable since it is then the sum over the fields scattered by many particles. If this is not true, number fluctuations become important;[18] the baseline to which the ACF decays is more difficult to determine properly; and small errors in the baseline give rise to large errors in the calculated distribution parameters. While this number fluctuation effect has some interesting uses in studying the mean-free path of motile microorganisms, it is a nuisance in particle sizing. A reasonable rule is $N > 100$.

Typical volume fractions used in PCS measurements are in the range 1E-5 to 1E-3. A common value for the scattering volume is $\sim$ 1E-6 cc at 90°; at 30° it is twice this value. Given these numbers one can calculate that at the lower end of the concentration range $N < 100$ for $1\,\mu$ particles. At the higher end, particle-particle interactions and multiple scattering become problems. It is, however, very easy to arrange for $N > 100$ for sizes less than $\sim 0.5\,\mu$.

Though it is possible to change the optical configuration such that $N > 100$ even for particles up to a few microns, it is still difficult to find a particle and electrolyte concentration at which particle-particle interactions are negligible. Thus, measurements are much easier below $\sim 0.5\mu$. This is one of the reasons that the vast majority of PCS measurements are made on particles of a few hundred nanometers and less.

The size range below, roughly, 0.5μ has been called the deep submicron,[19] and by analogy this author proposes the term shallow submicron for the size range from 0.5 to 1μ. In the shallow submicron range PCS looses many of its advantages: speed, simplicity, and relative independence from concentration effects. Fortunately, in this size range there are a number of alternative techniques for particle sizing.

Light Scattering Corrections

Mie scattering corrections are not the limiting factor in PCS measurements of broad distributions. While it is true that intensity transformations may be difficult to apply in some cases, the limiting factor is the instability of the result to small amounts of noise on the ACF. It is the ill-conditioning that limits the information. Typically this yields an apparently broader size distribution in intensity space. Subsequent application of Mie theory distorts the results even further from the truth.

Low Angle Measurements

Theoretically, results from low angle measurements contain more information on larger particles. But in practice the opposite may be true, especially in aqueous dispersions. Routine filtration techniques that yielded good answers at high angles fail as the angle decreases because the contribution from dust to the total intensity increases. Increasing the particle concentration to compensate often results in apparently lower average sizes. This is typical of charge-stabilized, repulsive interactions between particles.[20]

Averaging Process

It is sometimes claimed that the average diameter in a PCS measurement is the z-average. This is incorrect, as can be seen from the following arguments. In the Rayleigh range, where the intensity is independent of angle, the average size is calculated from a z-average diffusion coefficient. The diffusion coefficient is inversely proportional to size. Since the inverse of an average is not equal to the average of the inverse, it is never correct to say that PCS yields a z-average diameter.

6 SOME COMMON THREADS

Fraunhofer diffraction (FD), PCS, and high angle intensity measurements share many common attributes as techniques for particle sizing even though PCS does not depend on interpreting an intensity pattern as the other techniques do.

Both particle size and amount are determined from the scattered light. In all of these techniques the results are from an ensemble average over all the particles in the scattering volume during the measurement period.

Measurements are very fast. Assuming sufficient scattering, the maximum information is obtained within a few minutes at most. Longer runs yield diminishing returns as a result of the square root relationship between noise and experiment duration.

Measurements are highly precise. Averages are repeatable to better than $\pm 1\%$ when the instruments are used in the proper size ranges.

Measurement resolution is low compared to other techniques. Resolution is here defined as the minimum ratio of two monodisperse distributions that can be separated. High resolution instruments such as electrozone counters, linestart DCPs, CHDFs, and SdFFFs can claim resolution of 10% or better. Resolution in light scattering instruments is very roughly 100%, unless *a priori* information is used during data analysis.

Particle density is not important. Theoretically this is true. Neither the Stokes-Einstein equation (PCS) nor the Airy function (FD) contain density as a parameter. Practically, however, particle density can affect the final results. Large and dense particles may never reach the sensing zone in the case of an FD instrument; they may sediment below the beam before or during a PCS measurement. These are examples of sampling errors.

Measurements do not depend on particle refractive index (RI), at least to first order. Yet RI does play a central role in defining limits in both techniques, though for different reasons and in fundamentally different ways.

A Fraunhofer particle can be defined as one where the extinction efficiency, Q, is independent of particle size. Figure 3 shows two examples. These were calculated for a tungsten-halogen source. Results for a laser would show even higher amplitude oscillations. For low relative RI's, poly(methylmethacrylate) in water for example, Q is not constant until d is greater than ~ 10μ; for high relative RI's, Si_3N_4 in water for example, Q displays both interference and ripple structure. For broad distributions and for QC, these affects may be ignored. For quantitative work, however, it is difficult to see how they can be.

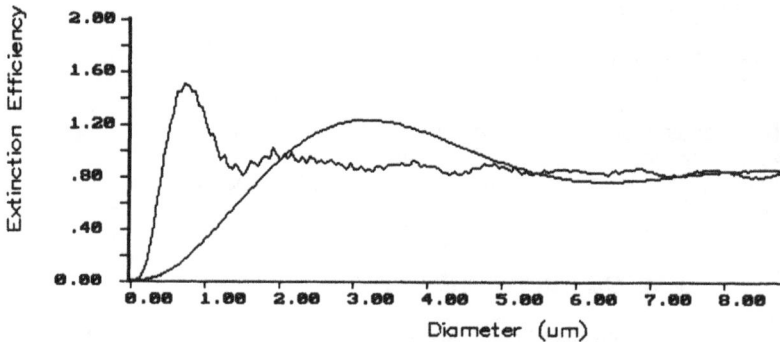

<u>Figure 3</u>   Integrated extinction efficiency vs. particle diameter for PMMA (smooth) and Si_3N_4 (rough) in water.

PCS results are independent of RI for Rayleigh scatterers and monodisperse samples; its effect on narrow distributions of larger particles is easy to predict. For broader distributions in the Mie region the RI is needed to transform from intensity-weighted into volume- and number-weighted results. As stated earlier, most algorithms overestimate the width of the intensity-weighted particle size distribution, and further massaging with Mie corrections often leads to even more spurious results.

Particle shape is not measured directly, though indirectly it affects the results. For an ellipsoid of revolution with an aspect ratio of 2, the equivalent spherical diameter from a PCS measurement is roughly 50% of the length of the ellipsoid.[21] For an aspect ratio of 1.2, the length is underestimated by only 12%. So for globular particles, PCS equivalent diameters are not much effected.

In some ways a measurement on submicron particles with a PCS instrument is similar to a measurement above a micron with an optical microscope. The similarity is this: relatively quickly some useful information can be obtained. For this reason, a PCS measurement--given its speed and relative simplicity compared to other submicron techniques--is as desireable a first measurement as the old dictum that all powders should be examined with a microscope before more elaborate techniques are employed.

7   FUTURE DEVELOPMENTS

In the opinion of this author it is unlikely that the fundamental resolution and distribution capabilities of PCS measurements will be dramatically increased, though it has been proposed that simultaneous data reduction of multiangle ACFs will do just that.[22] To date, multiangle measurements have yielded incremental advantages in special situations without pushing the resolution of the technique into the realm of its competitors: disc centrifugation, column hydrodynamic fractionation, sedimentation field flow fractionation, and electron microscopy.

More promising is the continual downsizing of the technique by virtue of the downsizing of the components that are available today. These include single-board correlators, laser diodes, optical fibers, and diode detectors. Smaller more rugged instrumentation should lead to wider applications of the technique.

Efficient, first generation correlators were multiboard hardware instruments with traditional front panels; the data were transferred to remote computers for analysis. Second generation correlators were microprocessor controlled and easily attached to an external computer, sometimes a desktop, though these rarely had sufficient computing power. With the widespread use of the AT-compatible both the hardware and the data analysis came under the control of one CPU in these third generation correlators; rarely was more computing power necessary. The fourth generation consists of single boards mounted directly inside the computer. For special applications, single chip correlators are also available.

Brown and Grant were the first to thoroughly report on laser diodes for use in PCS.[23] Though the power output stability was, at that time, marginal at 1%, no excess correlations were found in the output on the time scale of interest for most particle sizing applications. The lasers examined operated between 750 to 780 nm, making it difficult to align a system. Lasers diodes operating at 675 nm with polarized outputs around 5 mW are now available for half the cost of a comparable HeNe. With a length about one quarter and a diameter about one half of the traditional HeNe they replace, the size of the laser is no longer a limiting design consideration, at least for particle sizes above a few tens of nanometers. The small size also eliminates the need to fold the optics into a convenient package. Fewer optical elements improve reliability.

Launching of the beam into the scattering volume requires a polarization-preserving, single-mode fiber optic to maintain the proper coherence length. Multimode fibers are sufficient on the detection side. A fiber optic backscatter system allows the use of the technique *in situ*, though major hurdles remain in the interpretation of PCS results in concentrated dispersions.[24] Yet, for QC applications this may be sufficient. This work was pioneered by several groups.[25-27] Instruments are commercially available.

Early on diodes were tested for use with PCS. Smaller and more robust than a photomultiplier tube (PMT), diodes are also less costly. Unfortunately, no diodes were sufficiently noise-free to be used in the single photon counting mode with success, until recently.

Brown and coworkers have tested avalanche photodiodes (APD) as replacements for PMTs.[28] Preliminary tests were encouraging. Single photon counting was possible, though dead-time effects in the range of ~ 1-2 μs limited the maximum count rate. Special active quenching circuitry has reduced this dead time to below 100 ns with acceptably low afterpulsing, a form of inherent correlation in detectors.

In addition to their small size, the most intriguing aspect of an APD is its quantum efficiency compared to a PMT. Quantum efficiency around 8% in the red is typical of S20 photocathodes in selected PMTs suitable for use in PCS. The quantum efficiency of an APD in the red is around 8 times higher. Thus, the combination of a 5 mW, 675 nm laser diode with an APD yields roughly the same sensitivity as a 35 mW HeNe gas laser and a traditional PMT. Considering that the 35 mW HeNe is over a meter long and that the combined cost of the HeNe/PMT is ~ 4 times the cost of the laser diode/APD, one can look forward to high sensitivity instruments in very small packages.

There are still problems with APDs. Measurements performed over the last 2 years in our lab have shown that, sporadically, the polydispersity of the size distribution obtained with an APD is unacceptably too large. This effect was traced to occasional instabilities in the gain of the APD, though the average size agrees with that measured with a PMT. Brown noted this same effect last year.[29]

It can be expected that today's desk-top PCS instruments will become smaller, perhaps even truly portable. Given the simplicity of operation, this could lead to even more widespread use in the latex industry as well as in the measurement of liposomes and other applications where the product performance can often and usefully be described with one or two parameters of the size distribution.

8 CONCLUSION

Like all other particle sizing techniques, PCS has found its place. No longer a curiosity, PCS continues to serve a significant need in the submicron range for fast, simple, repeatable measurements. In many applications it is the instrument of choice. Examples include--very narrow distributions, distributions below ~ 20 nm, macromolecular distributions, micelles/microemulsions, vesicles/liposomes, and monoclonal antibodies to name a few. In other applications where only an average size and distribution width are sufficient, PCS is an excellent choice, especially for QC applications.

REFERENCES

1. H.Z. Cummins, N. Knable and Y.Yeh, Phys.Rev.Lett., 1964, 12, 150.
2. E. Jakeman and E.R. Pike, J.Phys. A, 1968, 1, 128.
3. H.L. Swinney, Critical Phenomena in Fluids, in "Photon Correlation and Light Beating Spectroscopy", edited by H.Z. Cummins and E.R. Pike, Plenum, 1974, Vol. 3, 331.
4. R. Pecora, J.Chem.Phys., 1964, 40, 1604.
5. R. Foord, et. al., Nature, 1970, 227, 242.
6. M. Smoluchowski, Ann.Phys., 1908, 25, 205.
7. A. Einstein, Ann.Phys., 1905, 17, 549.
8. B. Chu, "Laser Light Scattering: Basic Principles and Practice", 2nd edition, Academic, 1991.
9. K.S. Schmitz, "An Introduction to Dynamic Light Scattering by Macromolecules", Academic, 1990.
10. B. Dahneke, editor, "Measurement of Suspended Particles by Quasielastic Light Scattering", Wiley-Interscience, 1983.

11. B.B. Weiner, <u>Particle Sizing Using Photon Correlation Spectroscopy</u>, in "Modern Methods of Particle Sizing", H. Barth editor, Wiley-Interscience, 1984, Chapter 3, 93.
12. B.B. Weiner and W.W. Tscharnuter, <u>Uses and Abuses of Photon Correlation Spectroscopy in Particle Sizing</u>, in "Particle Size Distribution: Assessment and Characterization", T. Provder editor, ACS Symposium Series 332, ACS Washington, 1987, Chapter 3, 48.
13. W.W. Tscharnuter, B.B. Weiner and D. Fairhurst, <u>Particle Size Analysis with The Disc Centrifuge: The Importance of The Extinction Efficiency</u>, in "Particle Size Assessment and Characterization", T. Provder editor, ACS Symposium Series 472, ACS Washington, 1991.
14. R.D. Neuman and Z-J. Yu, <u>A New Optical Cleaning Technique for Light Scattering Measurements</u>, to be published.
15. H.Ohshima, et. al., <u>J.Chem.Soc.Faraday Trans.II</u>, 1984, <u>80</u>, 1299.
16. G.A. Schumacher and T.G.M. van de Ven, <u>Brownian Motion of charged Colloidal Particles Surrounded by Electric Double Layers</u>", PGRL Report #382, Pulp and Paper Reserach Institute of Canada, Pointe Claire, Quebec, 1987.
17. C.R. Taylor, et.al., <u>J.Amer.Acad.Dermatology</u>, 1990,<u>22</u>, 1.
18. B.J. Berne and R. Pecora, "Dynamic Light Scattering with Applications to Chemistry, Biology & Physics", Wiley-Interscience, 1976, Chapter 5, p. 65.
19. D.F. Nicoli, private communication.
20. R. Finsy, <u>Particle Characterization</u>, 1990, <u>1</u>, 238.
21. See Ref. 18, Chapter 7, p. 143.
22. P.G. Cummins and E.J. Staples, <u>Langmuir</u>, 1987, <u>3</u>, 1109.
23. R.G.W. Brown and R. Grant, <u>Rev.Sci.Instrum.</u>, 1987, <u>58</u>, 928.
24. D.J. Pine, et.al., <u>J.Phys.(Paris)</u>, 1990, <u>51</u>, 2101.
25. D.A. Ross, H.S. Dhadwal and R.B. Dyott, <u>J.Colloid Interface Sci.</u>, 1978, <u>64</u>, 533.
26. H. Auweter and D. Horn, <u>J.Colloid Interface Sci.</u>, 1985, <u>105</u>, 399.
27. J.C. Thomas and V. Dimonie, <u>Appl. Opt.</u>, 1990, <u>29</u>, 5332.
28. R.G.W. Brown, R. Jones, J.G. Rarity and K.D. Ridley, <u>Appl. Opt.</u>, 1987, <u>26</u>, 2383.
29. R.G.W. Brown, J.G. Burnett, J. Mansbridge and C.I. Moir, <u>Appl. Opt.</u>, 1990, <u>29</u>, 4159.

Singular Value Analysis and Reconstruction of Single- and Multi-angle Photon Correlation Data

R. Finsy[1], P. De Groen[2], L. Deriemaeker[1], E. Geladé[3] and J. Joosten[3]

[1] THEORETICAL PHYSICAL CHEMISTRY – VRIJE UNIVERSITEIT BRUSSEL, PLEINLAAN 2, B – 1050 BRUSSEL, BELGIUM

[2] DEPARTMENT OF MATHEMATICAL AND COMPUTER SCIENCE – VRIJE UNIVERSITEIT BRUSSEL, PLEINLAAN 2, B – 1050 BRUSSEL, BELGIUM

[3] DEPARTMENT OF PHYSICAL AND ANALYTICAL CHEMISTRY DSM-RESEARCH, P.O. BOX 18, N – 6160 MD GELEEN, THE NETHERLANDS

ABSTRACT.
Particle sizing in colloidal dispersions is performed nowadays on a routine basis with Photon Correlation Spectroscopy (PCS). The inversion of PCS data for particle size distributions is however ill-conditioned. Several methods, such as Contin and Maximum Entropy, dealing with the ill-conditioned nature of the inversion problem are used. Their main drawback is that they require prior knowledge, e.g. about the expected range of particle sizes. Singular value analysis and reconstruction (SVR) on the other hand does not require any prior knowledge. In SVR the inversion problem is reduced to well conditioned problems which are solved by reliable and fast algorithms. SVR also provides a practical criterion that allows to separate the information content of the data from the noise. Some practical performances of SVR in analyzing data collected at single angles and for the collective analysis of data registered at several scattering angles are presented. A further asset of SVR in analyzing multi-angle data-sets is that no prior knowledge on the angular dependence of scattering power on particle size (e.g. Mie scattering) is required.

1 INTRODUCTION

Particle sizing in colloidal dispersions is performed nowadays on a routine basis with Photon Correlation Spectroscopy (PCS) [1]. The increasing success of PCS is mainly based on the fact that it provides absolute estimates of particle sizes in a very short measuring time without severe sample preparation procedures and that easy to use commercial equipment is available. It is well known that PCS yields reliable results for monodisperse samples with particle sizes in the submicron range [2]. In the last decade many efforts have been spent for determinations of particle size distributions by PCS. It appeared that the interpretation of PCS measurements is considerably more difficult for polydisperse samples due to the ill-conditioned nature of the data inversion. The ill-conditioning is reflected in the fact that the resulting size distributions depend very sensitively on the data and on the inversion method [3]. These ambiguities in the answers for the particle size distribution are partially due to the fact that most inversion methods(e.g.Contin [4] and the Maximum Entropy Method [5], MEM) require prior knowledge such as the expected range of particle sizes. Since the scattering power of submicron and micron size particles is strongly dependent on scattering angle the analysis of PCS measurements recorded at different scattering angles may improve the conditioning of the inversion procedure. Some evidence supporting this argument has been published [6]. However, since the angular dependence of particle scattering power is used as a constraint, the published method can only be used if prior knowledge, such as particle shape, composition, refractive index and absorption coefficient, is available.

The singular value analysis and reconstruction (SVR) method [7] on the other hand does not require any prior knowledge about the expected range of particle sizes. With SVR the inversion problem is reduced to well conditioned problems which are solved by reliable and fast algorithms. SVR provides also a practical criterion that allows separation of the information content of the data from the noise. Furthermore SVR can be used without prior knowledge, for the collective analysis of PCS data collected at several scattering angles.

In this contribution the application of SVR for the interpretation of some single-angle and multi-angle data sets is presented.

2 INTERPRETATION OF PCS MEASUREMENTS

General outline

In most PCS experiments the intensity autocorrelation function $G_2(\theta,\tau)$ is measured at one or several scattering angles θ as a function of time delay τ. In the first step of the interpretation procedure $G_2(\theta,\tau)$ is related to the modulus of the normalized field autocorrelation function $|g_1(\theta,\tau)|$ by a Siegert relation :

$$G_2(\theta,\tau) = A + B|g_1(\theta,\tau)|^2 \qquad (1)$$

In eq. (1) A is a, in principle constant, background signal and B is an instrumental factor. Note that eq. (1) applies only to scattered fields with Gaussian statistics an hypothesis which is not always fulfilled experimentally. Especially for particles larger than roughly 0.5 to 1 μm additional time delay dependent factors can be distinguished in eq. (1) [8]. In a second step the time decay of the field auto-correlation function is related to the particles Brownian motion. Thereby it is assumed that the particles scatter independently. In particular for monodisperse samples $|g_1(\theta,\tau)|$ is an exponentially decaying function :

$$|g_1(\theta,\tau)| = \exp\left[-\Gamma(\theta)\tau\right] \qquad (2)$$

where the decay rate $\Gamma(\theta)$ is connected to the particles diffusion coefficient D by $\Gamma = Dq^2$, where q is the modulus of the scattering vector

$$q = \frac{4\pi m_1}{\lambda_0} \sin(\theta/2) \qquad (3)$$

In eq. (3) m_1 is the index of refraction of the solution, λ_0 the wavelength in vacuo of the incident light.

For polydisperse samples the generalization of eq. (2) is a sum of exponentials :

$$|g_1(\theta,\tau)| = \sum_{i=1}^{n} c_i(\theta) \exp\left[-\Gamma_i(\theta)\tau\right] \qquad (4)$$

The coefficients $\{c_i(\theta), i = 1 \dots n\}$ represent the intensity weights of the different particles with diffusion coefficients $\{D_i = \Gamma_i/q^2, i = 1 \dots n\}$ at scattering angle θ. In the third step the set diffusion coefficients $\{D_i\}$ are related to the set particle sizes $\{d_i\}$. To this end the Stokes-Einstein expression for the diffusion coefficient is used

$$D_i = \frac{kT}{3\pi\eta d_i} \qquad (5)$$

In eq. (5) k is Boltzmann's constant, T the absolute temperature, η the viscosity of the dispersion medium and d the particle diameter. Note that the use of eq. (5) implies the assumption of non-interacting spherical particles [9]. In a last step the intensity weights $\{c_i\}$ can be converted into number weights $\{n_i\}$ or volume or mass weights $\{w_i\}$. Thereby prior knowledge of the dependence of the scattering power on particle diameter (and refractive index) must be used.

When inverting experimental data a first and severe complication arises due to the experimental uncertainties and noise. A second experimental complication arises from the fact that, due to unavoidable minor drifts in incident power, the background term is not a constant rather it is a slowly decaying function of time delay τ. Hence eq. (1) should be written as

$$G_2(\theta,\tau) = A(\tau) + B|g_1(\theta,\tau)|^2 + \varepsilon(\tau) \qquad (6)$$

where $\varepsilon(\tau)$ are the (unknown) experimental uncertainties. The main difference between eq. (1) and eq. (6) is that eq. (6) is strictly not invertable for the modulus of the field autocorrelation function without unknown bias. Since the second step, i.e. the Laplace inversion of the field autocorrelation function for the intensity weights $\{c_i\}$, is in many data analysis procedures extremely sensitive to small differences in estimates of $|g_1(\theta,\tau)|$, any unknown bias should be avoided. Therefore it is preferable to invert directly the intensity autocorrelation data for the particle size distribution. This last goal can be achieved by the SVR method.

Singular Value analysis and Reconstruction (SVR)

Experimental intensity autocorrelation data are very often sampled at equidistant time intervals $\{\tau_j = jT \mid j = 1, 2, 3, \dots 2m\}$

The basis of the SVR method is such that for equidistant time delays, eq. (6) can be written as

$$G_2(\theta_k, jT_k) = y_{kj} = A_k(jT_k) + B_k\left[\sum_{i=1}^{n} c_{ki}(\theta_k)\,\lambda_{ki}^j(\theta_k)\right]^2 + \varepsilon_{kj} \qquad (7)$$

where the exponential factors λ_{ki} are defined as

$$\lambda_{ki} = \exp\left[-\Gamma_{ki}(\theta_k,d_i)\,T_k\right] \tag{8}$$

The slowly decaying baseline can be considered as an additional exponential factor :

$$A_k(jT_k) = b_{k1}\,\mu_{k1}^j$$

with $\mu_{k1} \approx 1$ so that eq. (7) can be written as

$$y_{kj} = \sum_{q=1}^{r} b_{kq}\,\mu_{kq}^j + \varepsilon_{kj} \tag{9}$$

where

$$\mu_{kq} = \lambda_{ki}\lambda_{kj} \qquad\qquad (i,\,j = 1 \dots n\,;\,q = 2, 3 \dots r)$$

and

$$b_{kq} = c_{ki}c_{kj} \qquad\qquad (i,\,j = 1 \dots n\,;\,q = 2, 3 \dots r)$$

<u>Single-angle data anlysis</u>.   For data sampled at a single scattering angle θ eq. (9) can be written as

$$y_j = \sum_{q=1}^{r} b_q\,\mu_q^j + \varepsilon_j \tag{10}$$

In order to separate the information part of a set of 2m autocorrelation data, i.e. the sum of exponentials from the noise part, a singular value decomposition of the Hankel matrix H, with elements $H_{ij} = y_{i+j-1}$ $(i,\,j = 1, 2 \dots m)$ is computed. For typical experimental data a few large singular values emerge above a mass of almost equal small ones (see e.g. figures 1-3). The latter ones are ascribed to noise, while the former ones constitute the signal content of the data. Further details of the procedure are given elsewhere [7]. To resume the algorithm goes as follows :

1. Construct the Hankel matrices H and H^s, with elements

    $$H_{ij} = y_{i+j-1}, \quad H_{ij}^s = y_{i+j} \qquad (i,\,j = 1, 2, 3 \dots m)$$

    from the data set $\{y_j\,|\,j = 1, 2, 3 \dots 2m\}$.

2. Compute the singular value decomposition of

    $$H = U\Sigma V^T$$

    (the superscript T denotes transposition)

    $$\Sigma = \text{diag}(\sigma_1, \dots, \sigma_m) \qquad \sigma_i \geq \sigma_{i+1} \qquad\qquad \forall\,i$$

3.   Determine the apparent noise level and choose the rank r of the reconstruction. This decision can be made using the conditions

$$(\sigma_r - \sigma_{r+1})/\sigma_{r+1} > 2$$

and

$$(\sigma_{r+1} - \sigma_{r+2})/\sigma_{r+2} < 2$$

Define the submatrices U_r and V_r consisting of the first r columns of U and V and define the r x r matrix

$$\Sigma_r = \text{diag} (\sigma_1, \ldots , \sigma_r)$$

4.   Compute the eigenvalue decomposition of

$$M = \Sigma_r^{-1/2} U_r^T H^s V_r \Sigma_r^{-1/2} = XAX^{-1}$$

whereby

$$A = \text{diag} (\mu_1, \mu_2 \ldots \mu_r)$$

This step yields the set of exponential factors, from which the particle diameters can be computed.

5.   The intensity weights can be computed by a linear least-squares procedure with the exponential factors obtained in step 4.

Multi-angle data analysis. In a collective analysis p data sets $\{y_{kj} \mid j = 1 \ldots 2m, k = 1 \ldots p\}$ collected at equidistant time delay's at p different scattering angles are available. By a suitable tuning of the sample times T_k to the scattering angles θ_k or vice versa the exponential factors in eq. (9) are independent of scattering angle so that for this particular case eq. (9) can be written as

$$y_{kj} = \sum_{q=1}^{r} b_{kq} \mu_q^j + \varepsilon_{kj} \qquad (k = 1 \ldots p) \qquad (11)$$

The SVR algorithm can be generalized for the collective analysis of p such sets of intensity correlation functions measured at different scattering angles. The details of the generalized SVR procedure will be described elsewhere [10].

3 EXPERIMENTAL

Latex dispersions were obtained from Dow Chemical (220 and 482 nm in diameter) and from Duke Scientific (155 nm diameter). These dispersions were diluted with purified and filtered water (Millipore, Milli-Q system and 0.02 μm filters) to concentrations by weight of the order of 10^{-5} gr/cm^3.

Light scattering experiments were performed with ALV equipment (ALV model SP-86 goniometer, EMI 9863 QB photomultiplier, ALV 3000 structurator-correlator ; ALV, Langen, Federal Republic Germany). The light source was a

Spectra Physics model 2020 Argon-ion laser operating at 514.5 nm wavelength in the light stabilization mode at about 80 mW vertically polarised with respect to the scattering plane. The measurements were carried out at 18.6° C.

Data analysis was performed on several PC's with 80286/80287 and 80386/80387 processors and a T800 transputer board with 2 Mb memory (Archipel, Annecy, France).

In the computation of Mie scattering functions of spheres we have used the expressions as given by Van der Hulst [11].

4 SOME RESULTS AND DISCUSSION

<u>Single angle measurements</u>

Intensity autocorrelation functions sampled at $2m = 128$ equidistant time delays were recorded at a scattering angle of 90° for several monodisperse latex dispersions. For a monodisperse sample the information content can be modelled by a sum of two exponentials so that in this case eq. (10) can be written as

$$y_j = b_1 \, \mu_1^j + b_2 \, \mu_2^j + \varepsilon_j \tag{12}$$

Eq. (12) implies that of the $m = 64$ singular values of the Hankel matrix H only two large ones will emerge above 62 small and almost equal ones. This fact can be illustrated with a typical plot of the singular values obtained from a data set for a monodisperse sample (figure 1).

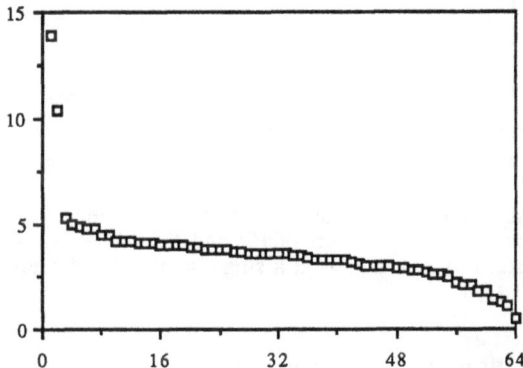

<u>Figure 1</u>   Singular values of the Hankel matrix of the intensity autocorrelation function of a 220 nm diameter monodisperse latex, sampled at 128 equidistant time windows at a single angle of 90° (semi-logarithmic scale).

Also typical is that the first exponential factor μ_1 is almost equal to 1 (actually slightly smaller indicating a slowly decaying baseline) and that the value of the particle diameter can be recovered within a few percent from the second exponential factor μ_2. It was also illustrated that the gap ($\sigma_2 - \sigma_3$) between the smallest "signal" singular value σ_2 and the largest "noise" singular value σ_3 is proportional to the signal-to-noise ratio [7]. Another interesting feature of SVR is that the analysis of a 128 points data set takes typically no more 10 to 20 s depending whether or not a transputer board is used. With a method such as Contin or Maximum Entropy typical computing time is 5 to 10 min.

In a second application measurements were performed on a binary mixture of monodisperse latices with diameters of 155 and 482 nm at a scattering angle of 50°. The mixture was prepared in such a way that the intensity weights of both components were equal. The intensity autocorrelation function was sampled at 128 equidistant time windows. For such a mixture eq. (10) can be written as

$$y_j = \sum_{q=1}^{4} b_q \, \mu_q^j + \varepsilon_j \tag{13}$$

Hence it is expected that four large singular values will emerge above 60 small ones. However as can be seen in figure 2 only three singular values emerge implying that only three of the four exponentials in eq. (13) can be recovered from the noisy data.

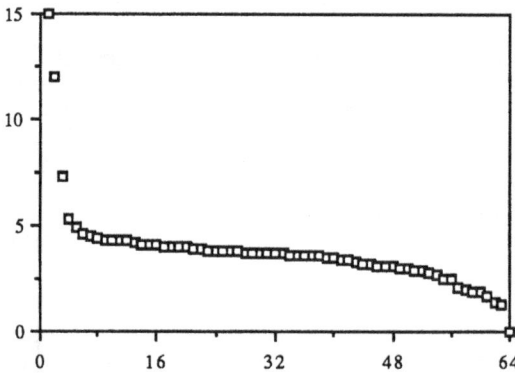

Figure 2 Singular values of the Hankel matrix of the intensity autocorrelation function of a binary mixture of 155 and 482 nm latices, sampled at 128 equidistant time windows at a single angle of 50° (semi-logarithmic scale).

In order to investigate for which noise levels the four exponentials can be recovered from a single angle measurement simulated data for a 1 by 1 intensity weighed binary mixture of 155 and 482 nm latices with different noise levels were analyzed. It appears that the four exponentials can, according to the conditions given in point 3 of the SVR algorithm, be recovered only for noise with relative variance not exceeding 3.10^{-6}, which is not achievable for experimental data. For noise levels comparable to the noise levels in real experiments it appeared that only three or even only two of the four exponentials could be recovered from the noisy simulated data.

In an attempt to recover the four exponentials several data sets collected at different scattering angles were analyzed. In order to assess the feasibility sets of intensity autocorrelation functions at five scattering angles in the range 35° to 61° were simulated for the previously investigated binary mixture. Provided that the basic time windows T_k in eq. (8) are tuned to the corresponding scattering angles θ_k the exponential factors in the five data sets are the same for all five sets so that in this case eq. (11) can now be written as

$$y_{kj} = \sum_{q=1}^{4} b_{kq} \, \mu_q^j + \varepsilon_{kj} \qquad (k = 1 \dots 5) \qquad (14)$$

Note that although the factors μ_q are independent of scattering angle, the coefficients b_{kq} vary with scattering angle.

In the reported angular range the larger 482 nm particles scatter about 6 times stronger at the smallest angle, whereas at the largest angle the smaller particles scatter about 10 times more. From the multi-angle SVR analysis of simulated noisy data sets it appeared that the four exponentials could now be recovered for noise levels with relative variance up to 10^{-4}. Since such noise levels can be achieved in PCS experiments, measurements of the intensity autocorrelation functions at five scattering angles in the range 35° to 61° were performed on the binary mixture of 155 nm and 482 nm diameter latex dispersion. The five sets of 128 data were analyzed with the multi-angle SVR algorithm. The 64 singular values are shown in figure 3.

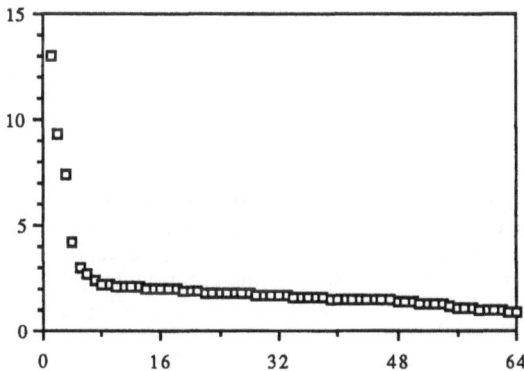

Figure 3 Singular values of the Hankel matrix of a set of 5 intensity autocorrelation functions of a binary mixture of 155 and 482 nm latices, sampled at 128 equidistant time windows at five scattering angles in the range 35° to 61° (semi logarithmic scale).

Now, according to the criterion for separating the information content from the noise, four exponential factors could be recovered from the multi-angle data set. Again the first exponential factor corresponds to a slowly decaying background. From the three other exponential factors both particle sizes d_1, d_2 and their harmonic average

$$\bar{d} = 2 \left(d_1^{-1} + d_2^{-1} \right)^{-1}$$

can be computed. The values recovered by the multi-angle SVR analysis i.e. $d_1 = 477$ nm, $d_2 = 161$ nm and $\bar{d} = 255$ nm, compare well with the expected values which are respectively 482, 155 and 235 nm.

In a second step in the SVR analysis the set coefficients $\{ b_{kj} \mid j = 1 \ldots 4,$ $k = 1 \ldots 5 \}$ can be recovered from the five autocorrelation functions. From this set the ratio $I_{12} = i_1/i_2$ of the scattering power i_1 of the 482 nm particles to the scattering power i_2 of the 155 nm particles can be computed at each scattering angle. The results are reported in table 1. Also reported for comparison is the intensity ratio computed from the Mie scattering coefficients for homogeneous spheres with relative refractive index of 1.195, i.e. for polystyrene latex in water, assuming the scattering power of both components to be equal at a scattering angle of 50°. This last assumption corresponds to a number ratio n_2/n_1 of smaller particles to larger particles of about 42.

Table 1 Intensity ratio I_{12} in a binary mixture of 155 and 482 nm latex particles as a function of scattering angle. Experimental values as obtained without prior knowledge by the SVR multi-angle analysis and calculated values for homogeneous spheres assuming an equal scattering power of both components at a scattering angle of 50°.

Scattering angle	Intensity ratio experimental SVR analysis	Intensity ratio calculated
35.9°	6.3	5.61
45.9°	2.1	1.88
50.0°	0.98	1
55.5°	0.44	0.32
60.4°	0.15	0.10

From these results it is clear that the expected intensity weighed ratio can be recovered without any prior knowledge with the SVR multi-angle analysis in very short computation times : typical computing time for a complete analysis of a set of 5 x 128 data points takes about 20 sec using a transputer board.

5 CONCLUSIONS

The results presented illustrate some of the possibilities of the SVR analysis of single-angle and multi-angle PCS data. In particular the singular value analysis of the Hankel matrix of the data appears to be a practical tool for the determination of the number of exponential decays that can be recovered from noisy data. The method offers stable reconstructions and reliable estimates of the decay times without any prior knowledge and data analysis can be performed with a PC in short times. Finally the resolution can be significantly improved by the SVR analysis of sets of multi-angle data, without using prior knowledge about the variation of particle scattering power with scattering angle.

ACKNOWLEDGEMENTS

The assistance of M. Weckx in typing the manuscript and of M. Van Laethem in carrying out part of the experimental work was greatly appreciated. The research was supported in part by DSM-Research (Geleen, The Netherlands) and by the Belgian "Nationaal Fonds voor Wetenschappelijk Onderzoek (NFWO)" and the Belgian "Fonds voor Kollectief Fundamenteel Onderzoek (FKFO)".

REFERENCES

1. See e.g. various contributions in H. Barth, 'Modern methods of particle size analysis', Wiley, New York, 1984.
2. N. De Jaeger, H. Demeyere, R. Finsy, R. Sneyers, J. Vanderdeelen, P. Van Der Meeren and M. Van Laethem, 'Particle Sizing by photon correlation spectroscopy. Part I', Part. Part. Syst. Charact., 1991, 8, 179.
3. R. Finsy, M. Van Laethem, N. De Jaeger, R. Sneyers, J. Vanderdeelen, P. Van Der Meeren, H. Demeyere and E. Geladé, 'Particle size distributions of bimodals : a comparative study of submicron sizing instrumentation' in N. De Jaeger and R. Williams 'Advances in measurements and control of colloidal processes', Butterworth-Heinemann, Oxford, 1991, Chapter 29, p. 388.
4. S. Provencher, Comput. Phys., 1982, 27, 229.
5. A. Livesey, P. Licinio, M. Delaye, J. Chem. Phys., 1986, 84 , 5102.
6. P.G. Cummins and J.E. Staples, Langmuir, 1987, 3, 1109.
7. R. Finsy, P. De Groen, L. Deriemaeker, M. Van Laethem, J. Chem. Phys., 1989, 91, 7374.
8. See e.g. B. Berne and R. Pecora, 'Dynamic light scattering', Wiley, New York, 1976, p. 62.
9. R. Finsy, Part. Part. Syst. Charact., 1990, 7, 74.
10. R. Finsy, P. De Groen, L. Deriemaeker, E. Geladé and J. Joosten, 'Data analysis of multi-angle photon correlation measurements without and with prior knowledge', in preparation.
11. H.C. Van der Hulst, 'Light scattering by small particles', Wiley, New York, 1957.

Evaluation and Optimization of Photon Correlation Spectroscopy Data Analysis Software

P. Van der Meeren*, J. Vanderdeelen and L. Baert

STATE UNIVERSITY OF GHENT, FACULTY OF AGRICULTURAL SCIENCES, DEPARTMENT OF
PHYSICAL AND RADIOBIOLOGICAL CHEMISTRY, COUPURE LINKS 653, B-9000 GENT, BELGIUM

* Senior Research Assistant of the Belgian National
Fund for Scientific Research (N.F.W.O.)

ABSTRACT

Using experimental as well as simulated PCS data the influence of the data-analysis software on the calculated
particle size distributions was investigated. Hereby,
both the commercial Automeasure software (Malvern) and
the freely available CONTIN program were compared. Simu-
lating data corresponding to increasingly wider monomodal distributions, it was shown that the width estimated by the Automeasure software was rather determined
by the analysis range and by the noise level, whereas
CONTIN yielded much more accurate results. Besides, simulated bimodal distributions revealed that the Automeasure software only yielded accurate bimodal particle size
distributions if the ratio of the larger to the smaller
particle size was between 1.5 and 2.0. The CONTIN software, on the other hand, yielded bimodal distributions
when the actual modi were at least double. Besides, the
Automeasure software overestimated the size of the
smaller particles, whereas CONTIN underestimated the
smallest modus. Finally, it was shown that additional
information could be obtained by comparing the estimated
size distributions obtained at different angles.

1 INTRODUCTION

Photon correlation spectroscopy is a very fast and reproducible sub-micron particle sizing technique.[1-4] It
was developed during the seventies and originally provided only some mean diameter as well as an estimation
of the width of the particle size distribution. Since
then, data-analysis software has been continuously
upgraded.[5,6] Nowadays, detailed particle size distributions are provided either on a number-, a weight- or an
intensity-basis. Besides, the operations and calculations are completely computer-driven, making the apparatuses not only very user-friendly but at the same time
also rather black box systems. Therefore, a thorough
investigation of the main variables seems necessary to

enable a profound evaluation of the large amount of information obtained. In this respect, it is known that the particle concentration, the wavelength and the scattering angle may influence the estimated particle size.[7,8] In the present paper it is demonstrated that besides these experimental conditions, the data-analysis software used can highly affect the results. Hereby, both experimental and simulated autocorrelation functions were analysed using the Automeasure software and the CONTIN program.

2 MATERIALS AND METHODS

PCS Apparatus

A PCS 100 Spectrometer (Malvern) including a 15 mW He-Ne laser (NEC) was used. The samples were contained in small glass cuvettes which were kept at 25 °C during the data-acquisition. The data were processed by a multi 8-bit 7032 CN digital correlator having 4 times 16 channels with a dilation factor of 4.

Polystyrene latices

Both a 109 and a 261 nm polystyrene latex were bought from Duke Scientific Europe (Leusden, The Netherlands). They contained 10% (w/v) monodisperse, spherical polystyrene particles, as well as 0.02% of sodium azide to avoid microbial contamination.

PCS Software

The data-acquisition and on-line data-analysis was performed by the Automeasure software, version 3.1 (Malvern). Using a cumulant analysis method the z-average diffusion coefficient D_z is estimated,[9] from which the inverse z-average diameter $d_{1/z}$ is calculated.

$$D_z = \frac{\sum N_i \cdot MW_i^2 \cdot D_i}{\sum N_i \cdot MW_i^2} \qquad [1]$$

$$<d> = \frac{k \cdot T}{3 \cdot \pi \cdot \mu \cdot D_z} = \frac{k \cdot T}{3 \cdot \pi \cdot \mu} \cdot \frac{\sum N_i \cdot MW_i^2}{\sum N_i \cdot MW_i^2 \cdot D_i} \qquad [2]$$

with : N = particle concentration (m^{-3})
MW = molecular weight (kg/mol)
D = diffusion coefficient (m^2/s)
k = Boltzmann constant $(1.38 \cdot 10^{-23}$ J/K)
T = temperature (K)
μ = viscosity (Pa·s)

As $MW_i = \pi \cdot d_i^3 \cdot \Gamma \cdot N_A / 6$, equation [2] may be written as :

$$<d> = \frac{\sum N_i \cdot d_i^6}{\sum N_i \cdot d_i^6 \cdot (1/d_i)} = \frac{\sum N_i \cdot d_i^6}{\sum N_i \cdot d_i^5} = d_{1/z} \qquad [3]$$

with : Γ = density (kg/m^3)
N_A = Avogadro number $(6.02 \cdot 10^{23}$ mol$^{-1})$

Besides, the Automeasure software includes a non-negatively constrained least squares technique, which estimates the intensity-weighed particle size distribution.

Using a GWBASIC program the experimental data were converted into ASCII-format, enabling their off-line analysis by the CONTIN program. The latter is a very

Table 1. Influence of the analysis range on the intensity-average
diameter, d_I, the width of the calculated distribution and
the fit quality according to the Automeasure 3.1 software,
and on the weight-average diameter, d_w, and the width of
the distribution estimated by the CONTIN program.

Analysis range	d_I (nm)	fit quality	d_w (nm)
1.1	259 ± 3	0.0177	257 ± 4
2.0	262 ± 17	0.0173	255 ± 18
5.0	267 ± 43	0.0166	255 ± 24
10.0	273 ± 54	0.0167	255 ± 23
50.0	281 ± 72	0.0173	256 ± 31
100.0	288 ± 81	0.0186	257 ± 33
500.0	308 ± 110	0.0248	255 ± 33

powerful FORTRAN program developed by Provencher[10] for
inverting noisy linear algebraic and integral equations.
Besides, a FORTRAN program was developed to simulate the
noisy autocorrelation function corresponding to a given
particle size distribution.

3 RESULTS AND DISCUSSION
Analysis Range
The analysis range is defined as the ratio of the
upper to the lower limit in the particle size distribution. As a matter of fact, this parameter has to be
large enough in the case of unknown samples in order to
prevent species with extreme diameters being lost.
The influence of this parameter was demonstrated
using data originating from a 261 nm polystyrene latex
($3.0 \cdot 10^{11}$ particles/ml), observed at a scattering angle
of 90 degrees. Cumulant analysis yielded an inverse
z-average diameter of 260.3 nm. The results of the
intensity-weighted particle diameter distribution
according to the Automeasure software and the weight-
distribution estimated by CONTIN are summarised in table
1. From this table it becomes obvious that the analysis
range parameter influenced the results to a large
extent : the estimated width of the distribution -ex-
pressed as the standard deviation of an assumed Gaussian
distribution- largely increased with increasing analysis
range. This phenomenon was due to the fact that the
classes within the distribution became more widely
spaced, so that the mean diameter of the best fitting
class differed more from the actual value as the analysis range increased ; this deviation was compensated by
assigning a contribution to the higher and lower classes
so that the resulting fitted autocorrelation function
resembled the experimental one as closely as possible.
As was already observed by Taylor et al.[11], the adequate
conversion of the intensity- to a weight- or a number-
distribution is strongly limited by this broadening
effect, so that erroneous bimodal number- or weight-
distributions were sometimes observed.[12]

Table 2. Data-analysis of the theoretical autocorrelation function
of a population of spherical particles with a diameter of
50.0 ± 0.1 nm, onto which an increasing noise level was
imposed.

noise level	cumulants		Automeasure 3.1		CONTIN
	$d_{1/2}$ (nm)	PDI	d_I (nm)	fit quality	d_w (nm)
0.00000	50	0.000	50 ± 2	0.0006	50 ± 1
0.00010	50	0.001	50 ± 4	0.0037	50 ± 2
0.00020	50	0.001	51 ± 4	0.0055	50 ± 2
0.00050	50	0.003	51 ± 5	0.0093	50 ± 2
0.00100	50	0.005	52 ± 6	0.0135	49 ± 4
0.00200	50	0.011	52 ± 7	0.0184	48 ± 6
0.00500	50	0.028	54 ± 8	0.0285	45 ± 10
0.01000	50	0.059	55 ± 9	0.0401	44 ± 12

The increase of the intensity-averaged diameter
also followed from the peak broadening effect. Only for
perfectly monomodal distributions the average values on
number-, weight- or intensity-basis are identical. As
the polydispersity of a distribution increases, the num-
ber-average diameter decreases, whereas the intensity-
average value increases.

From the calculated fit quality (table 1) it is
also concluded that one data-set may give rise to many
different solutions, each with a similar probability.
Only when the standard deviation exceeded 25% of the
mean value, did the fit quality significantly increase.

From table 1 it follows that the results of the
CONTIN program were much more reliable ; except for
analysis range 1.1 -which is not useful for real sam-
ples- the mean diameter and the estimated width remained
remarkably constant upon increasing analysis range.

Noise Level

In order to obtain as realistic simulated data as
possible, noise was superimposed according to a subroutine of the CONTIN program. The influence of the noise
level was investigated using simulated data corresponding to a Gaussian particle size distribution (50 ± 0.1
nm). The results mentioned in table 2 indicate that the
mean diameter estimated by cumulant analysis was rather
independent of the noise level. On the other hand, the
polydispersity index (PDI) increased approximately proportional to the fit quality. Hence, it is concluded
that the former index provides information about the
quality of the data, rather than about the width of the
distribution. As more noise was imposed onto the autocorrelation function, the average diameter and the width
of the intensity-distribution, as well as the fit
quality parameter, estimated by the Automeasure software
at a constant analysis range, largely increased.

On the other hand, the results of the   CONTIN program seemed to be less sensitive to the noise level provided this was below 0.005 ; higher values   resulted in
an   underestimated   weight-average diameter as well as a
largely overestimated width of the distribution.

Keeping account of the fact that a noise   level   of
0.0002 matches the   analytical   data, it was concluded
that the CONTIN program yielded the most   accurate   distributions.

Broad Monomodal Distributions

Autocorrelation   functions were   simulated   which
corresponded to monomodal   particle   size   distributions
with   an   intensity-mean   diameter of   50   nm   and with
increasing distribution width. A noise level   of   0.0002
was selected, and the analysis range was fixed at 5. The
results of the data-analysis (table 2) reveal   that   the
inverse   z-average diameter as estimated by the cumulant
analysis technique decreased at increasing   width.   This
observation   may   be explained by the fact that equation
[3] indicates that the   inverse   z-average   diameter   is
proportional to the fifth power of the diameter, whereas
the scattering ability of small spherical   particles   is
proportional   to   the   sixth   power. The polydispersity
index increased approximately proportional to the   width
squared,   in   accordance to its definition by Koppel ;[13]
however, the   calculated   width   was   almost   twice   its
actual   value.   The estimated intensity distributions of
the Automeasure software seemed to have only   a   limited
reliability. As opposed to the mean diameters, the standard deviations of the   estimated   distributions   didn't
agree   at all with the reality : the calculated width of
the distribution was independent of the effective   standard   deviation   when   the   latter   was   less   than 10%,
whereas bimodal distributions were obtained   for   larger
values   ;   hereby,   large values were observed for the fit
quality, indicating   the   limited   reliability.   Keeping
account of the previous paragraph, it was concluded that
the width of the estimated   particle   size   distribution
was   almost   independent   from the actual width, but was
rather determined by the noise level of the experimental
data and by the analysis range.

On   the   other   hand,   the CONTIN program yielded a
very reasonable agreement between the estimated and   the
actual distribution, although the standard deviation was
always overestimated. The decrease of the weight-average
diameter   was   still   more pronounced as compared to the
evolution of the inverse z-average diameter,   since   the
former   is proportional to the third power of the diameter. Taking the ratio of the fourth to the third   moment
of   the estimated weight-distribution, an average diameter was obtained that was weighed according to the sixth
power   of the diameter ; these values all ranged between
50 and 54 nm.

From table 3 it is   derived   that   all   PCS   data-
analysis   software   has   in   fact   a   broadening effect.
Besides,   it becomes obvious that different   mean   values

Table 3. Results of the data-analysis of the simulated autocorrelation functions, corresponding to an intensity-weighted particle size distribution with an average diameter of 50 nm and with a variable standard deviation (SD). The noise level was fixed at 0.0002, and the analysis range was 5.

SD (nm)	cumulants		Automeasure 3.1		CONTIN
	$d_{1/z}$ (nm)	PDI	d_I (nm)	fit quality	d_w (nm)
0.1	50	0.001	51 ± 4	0.0055	50 ± 2
1.0	50	0.002	51 ± 4	0.0055	50 ± 2
2.0	50	0.004	51 ± 4	0.0055	49 ± 4
5.0	50	0.021	51 ± 5	0.0055	47 ± 7
10.0	48	0.086	52 ± 8*	0.0069	41 ± 12
15.0	45	0.210	57 ± 13*	0.0176	22 ± 16*

* bimodal distribution
* analysis range 25

may arise for the same sample, thus complicating the comparison of results from different software programs. Overall, it is concluded that the particle size distributions estimated by the CONTIN program were more reliable as compared to the Automeasure results.

Bimodal distributions

The resolving ability of the software was investigated using the simulated autocorrelation functions of a 50/50 mixture, on intensity basis, of two monodisperse samples. The first species was characterised by a diameter of 50 nm, whereas the second contained particles of respectively 60, 75, 100, 150 and 200 nm, the standard deviation being 0.1 nm in all cases.

Supposing the scattering ability of the particles to be proportional to the sixth power of their diameter, equation [3] may be rearranged :

$$d_{1/z} = \frac{\Sigma \, I_i}{\Sigma \, I_i/d_i} \qquad [4]$$

with : I_i = relative amount of light scattered
by particles of diameter d_i (%)

A similar expression exists for the weight average diameter :

$$d_w = \frac{\Sigma \, I_i/d_i^2}{\Sigma \, I_i/d_i^3} \qquad [5]$$

From [4] it was calculated that the inverse z-average diameters of the different mixtures amounted to 55, 60, 67, 75 and 80 nm, whereas the weight-average diameters were according to [5] respectively 54, 56, 56, 54 and 52 nm. Table 4 reveals that the cumulant analysis technique yielded very accurate estimations of the mean particle size. The polydispersity index is not included, since this characteristic actually is only meaningful for monomodal distributions.

Only under certain circumstances was the Automeasure able to reconstruct the imposed distribution. When

Table 4. Data-analysis of the simulated autocorrelation functions
of bimodal mixtures of spherical particles with a diameter
of respectively 50 ± 0.1 and X ± 0.1 nm.

X (nm)	AR	cumulants $d_{1/z}$ (nm)	Automeasure 3.1 d_I (nm)	Automeasure 3.1 fit quality	CONTIN d_w (nm)	CONTIN d_w (nm)
60	5	54.6	56 ± 5	0.0055		53 ± 7
75	5	60.1	66 ± 10*	0.0088		51 ± 15
100	5	66.9	94 ± 21*	0.0346	40 ± 5	105 ± 7
150	10	76.5	153 ± 20	0.0785	35 ± 3	154 ± 8
200	15	84.4	208 ± 28	0.0950	33 ± 3	205 ± 10

* bimodal distribution

the diameter of both species differed very little, one
broad peak was found, even at the smallest analysis
range. On the other hand, only the diameter of the
larger species was found in the case of very different
species. Hereby, very large values of the fit quality
were obtained. As a rule of thumb, it may be stated that
estimated particle size distributions yielding fit
quality values above 0.05 generally have to be rejected.
For the sake of completeness, it has to be mentioned
that at analysis range (AR) 5 bimodal distributions were
obtained for the 50/150 and the 50/200 mixture ; as however these distributions contained two cut peaks in both
extreme classes of the distribution, these results were
rather forced and had to be rejected. Thus, accurate
results were only obtained when the ratio of the diameters of both species ranged from 1.5 to 2.

When the size of both species was sufficiently different, CONTIN yielded fairly good results. Besides,
CONTIN is better designed to handle multimodal distributions since it mentions the average diameters of all
subpopulations, as well as of the complete distribution.
Keeping account of the fact that the CONTIN program
contains a penalty function that reflects the irregularity of the distribution, its better resolving ability
as compared to the Automeasure software seemed remarkable. In table 4 it is seen that the estimated diameters
were mutually more different than the actual ones : the
size of the smaller species was underestimated and the
diameter of the larger particles was slightly overestimated. Stock and Ray[6] observed that this behaviour was
due to oversmoothing of the chosen solution.

Subsequently, the simulated autocorrelation functions of different mixtures of 50 and 100 nm spheres
were analysed. The ratio mentioned in table 5 indicates
the relative contribution to the total amount of scattered light. Table 5 indicates that cumulant analysis
was very appropriate to characterise a population by
only two parameters, albeit that the information provided didn't allow to discern between a multimodal or a
broad monomodal distribution.

Table 5. Data-analysis of the simulated autocorrelation functions corresponding to bimodal distributions containing 50 ± 0.1 nm and 100 ± 0.1 nm particles in different proportions.

50 nm / 100nm	cumulants		Automeasure 3.1		CONTIN	
	$d_{1/2}$ (nm)	PDI	d_I (nm)	I (%)	d_w (nm)	W (%)
$\frac{99}{1}$	50.3	0.008	49 ± 7 67 ± 2	99.3 0.7	49.6 ± 3.9	100.0
$\frac{98}{2}$	50.6	0.015	49 ± 7 67 ± 2	97.9 2.1	49.4 ± 4.4 1209.8	99.7 0.3
$\frac{95}{5}$	51.3	0.035	48 ± 7 68 ± 4	91.2 8.8	49.2 ± 4.9 467.3 ± 39.3	99.4 0.6
$\frac{90}{10}$	52.6	0.067	45 ± 5 72 ± 5	83.3 16.7	48.5 ± 5.3 230.3 ± 7.4	99.8 0.2
$\frac{75}{25}$	57.1	0.144	45 ± 6 88 ± 10	65.2 34.8	45.3 ± 6.3 128.7 ± 5.5	97.8 2.2
$\frac{50}{50}$	66.9	0.208	50 ± 5 100 ± 14	16.1 83.9	39.9 ± 5.6 105.5 ± 4.6	90.1 9.9
$\frac{25}{75}$	80.7	0.179	62 ± 4 104 ± 12	3.9 96.1	36.7 ± 4.8 100.0 ± 7.4	75.6 24.4
$\frac{10}{90}$	91.5	0.096	102 ± 8	100.0	36.0 ± 4.3 98.7 ± 8.1	50.6 49.4
$\frac{5}{95}$	95.7	0.054	101 ± 7	100.0	35.1 ± 3.9 98.2 ± 8.4	30.7 69.3
$\frac{2}{98}$	98.3	0.023	100 ± 7	100.0	33.4 ± 3.0 98.0 ± 8.5	10.1 89.9
1/99	99.2	0.012	100 ± 7	100.0	97.8 ± 9.1	100.0

In the intensity distributions of the Automeasure software the diameter of the larger species was underestimated ; this phenomenon was more pronounced as the relative importance of these particles diminished. Moreover, the importance of the smaller particles was strongly underestimated, so that they were even no longer detected when they scattered less than 10% of the total amount of light.

CONTIN had a very good resolving ability : except when a species contributed only 1% of the total amount of scattered light, a bimodal distribution was always observed. As opposed to the Automeasure software, CONTIN overestimated the percentage of the smaller particles. This is due to the light scattering phenomenon : as the scattered intensity of small particles is proportional to the sixth power of the radius, the weight percentage of 50 nm spheres has to be 8 times larger than the weight-percentage of 100 nm particles to obtain a 50/50 ratio on intensity basis. Besides, table 5 indicates that CONTIN largely overestimated the larger particles and underestimated the smaller particles when they represented only a minor fraction whereas the diameter of the main fraction was always accurately estimated.

Table 6. Determination of the particle size distribution of a 95/5 (w/w) mixture of two polystyrene latices with a diameter of respectively 109 en 261 nm by the cumulant analysis technique ($d_{1/z}$, PDI), the Automeasure Laplace inversion (d_I) and the CONTIN software (d_w).

	scattering angle (degrees)						
	60	90		120		150	
d_I (nm)	258 ± 51	15%	194 ± 17	29%	165 ± 30	7%	135 ± 10
		85%	285 ± 40	71%	258 ± 54	93%	302 ± 72
d_w (nm)	216 ± 28	59%	56 ± 20	50%	57 ± 22	26%	86 ± 15
		41%	265 ± 34	50%	241 ± 51	74%	319 ± 54
$d_{1/z}$ (nm)	231.9	228.0		188.5		185.8	
PDI	0.093	0.147		0.178		0.253	

Mixtures of polystyrene latices

In order to control the above findings, a mixture of a 109 and a 261 nm polystyrene latex was prepared. Due to the large difference in scattered intensity, these latices were mixed in a 95/5 (w/w) basis, so that their contribution to the experimental autocorrelation function was of the same order of magnitude. The sample was measured at a scattering angle of 60, 90 and 120 degrees. The Automeasure and the CONTIN program yielded bimodal distributions, except for the 60 degrees scattering angle. This was due to the fact that the light scattering by larger particles is especially pronounced at smaller angles, so that the importance of the 261 nm particles increases as the scattering angle decreases. This behaviour is also responsible for the decrease of the inverse z-average diameter and the increase of the polydispersity index upon increasing scattering angle.

From table 6 it follows that the Automeasure software highly overestimated the diameter of the smaller particles ; this effect was especially pronounced at small scattering angle whereby the contribution of these particles became minimal. As was already observed from the analyses of the simulated data, the CONTIN software displayed an opposite behaviour : this program underestimated the size of the 109 nm latex. Since the error of the estimation also decreased at larger scattering angles, the solutions of both data-analysis software programs converged. Hence, the information available from both programs enables a more reliable conclusion concerning the actual particle size distribution of an unknown sample.

4   CONCLUSIONS

Despite of the fact that PCS is known as a very fast and extremely user-friendly technique, some precautions are needed to obtain reliable information. Even when data-acquisition is performed under optimal experimental conditions, problems may arise due to data-analysis. First of all, it is important to recognise the broadening effect of all software ; due to this effect false bimodal distributions may arise. Comparing the Automeasure and the CONTIN data-analysis software some remarkable differences were observed : using simulated autocorrelation functions it was shown that the CONTIN program provided a much more realistic estimation of the width of monodisperse distributions, whereas moreover its resolving power for bimodal distributions was superior to the Automeasure software. However, despite of its distinct advantages, the CONTIN program displayed also some shortcomings. Therefore, the most reliable information could be extracted from the results of both data-analysis programs, keeping account of their specific properties. Besides, measurements at different angles seemed advisable for bimodal samples since in this way autocorrelation functions with a different weighing of the contributing species were obtained, thus maximising the information content of the sample.

5 ACKNOWLEDGEMENTS

Thanks are indebted to Dr. S. Provencher for providing us with the CONTIN software. Dr. M. Van Laethem is kindly acknowledged for the many helpful hints and fruitful discussions.

6 REFERENCES

1. H.Z. Cummins and E.R. Pike, 'Photon Correlation and Light Beating Spectroscopy', Plenum Press, New York, 1974.
2. B.J. Berne and R. Pecora, 'Dynamic Light Scattering with Applications to Chemistry, Biology and Physics', Wiley-Interscience, New York, 1976.
3. B.B. Weiner, 'Modern Methods of Particle Size Analysis', H.G. Barth (Ed.), Wiley & Sons, New York, 1984, p. 93.
4. R. Pecora, 'Dynamic Light Scattering : Applications of Photon correlation Spectroscopy', Plenum Press, New York, 1985.
5. E.F. Grabowski and I.D. Morrison, ' Measurements of Suspended Particles by QELS', B. Dahneke (Ed.), John Wiley, New York, 1983, p. 199.
6. R.S. Stock and W.H. Ray, J. Polym. Sci.: Polym. Phys. Ed., 1985, 23, 1393.
7. P. Van der Meeren, J. Vanderdeelen and L. Baert, 'Particle Size Analysis 1988', P.J. Lloyd (Ed.), Wiley & Sons, Chichester, 1988, p. 101.
8. N. De Jaeger, H. Demeyere, R. Finsy, R. Sneyers, J. Vanderdeelen, P. Van der Meeren and M. Van Laethem, Part. Part. Syst. Charact. (in press).
9. J.C. Brown, P.N. Pusey and R. Dietz, J. Chem. Phys., 1975, 62, 1136.
10. S. Provencher, Comp. Phys. Comm., 1982, 27, 229.
11. T.W. Taylor, S.M. Scrivner, C.M. Sorensen and J.F. Merklin, Appl. Opt., 1985, 24, 3713.
12. P. Van der Meeren, J. Vanderdeelen and L. Baert, Part. Part. Syst. Charact. (submitted for publication).
13. D. Koppel, J. Chem. Phys., 1972, 57, 4814.

Using the Phase Doppler Particle Analyzer for *In-situ* Sizing of Fine Spherical Particles

R. C. Rudoff, S. V. Sankar and W. D. Bachalo

AEROMETRICS, INC., 550 DEL RAY AVE., UNIT A, SUNNYVALE, CA 94086, USA

1 INTRODUCTION

The phase Doppler Technique is a laser-based interferometric approach for simultaneously measuring particle size and velocity.[1,2] This laser-based method is similar to the well-known laser Doppler velocimeter (LDV) and can make these measurements non-intrusively and with high accuracy. The Phase Doppler Particle Analyzer (PDPA) has been successfully applied to many difficult two-phase flow environments with spherical particles including gas-turbine and rocket fuel injection, cavitation and bubble formation, and similar two-phase flows.[3,4]

There is presently much interest in the use of the PDPA for the sizing of aerosols and other fine particles in the less than 10 micron size range. This is due to current emphasis on the development of medical and similar nebulizers which deliver aerosols in this size range. Accurate delivery of a dosage demands a well-controlled size range for proper deposition. There is also much activity in the area of LDV seed particle sizing.[5]

Sizing of such particles demands fine and accurate resolution from the PDPA. Some recent works have attempted to point out limitations in the instrument response function (PDPA calibration) for the small particle size range.[6-8] However, it can be shown both theoretically and experimentally that with the correct PDPA configuration, even very small particles, down to 0.5 um can be sized with adequate resolution.

In this paper, the above theoretical results of an Aerometrics developed model are presented and experimental comparisons shown. In addition, a miniature probe for in-situ use as an individual or multiplexed probe is detailed. This probe has many applications for process control and monitoring.

2 THEORETICAL STUDY

The sizing of spherical particles much larger than the laser wavelength has been previously described via geometrical optics.[9] This work established the response of the PDPA for particles large than 10 um in both the forward and backscatter orientations. However, for particles smaller than 10 um, the geometrical optics approach is not as reliable as the Lorenz-Mie theory in accurately predicting the details of the light scattering phenomena. Thus, the Lorenz-Mie theory was applied in this study.

Figure 1 shows the coordinate system for theoretical analysis. Two incident laser beams, *beam 1* and *beam 2* lying in the x-y plane, intersect at an angle, γ, forming a measurement volume. Spherical particles passing through the probe volume in the x direction scatter the incident light beams. The scattered light interacts to form spatial fringe patterns. The PDPA receiving lens is placed at a distance R from the measurement volume, at an angle θ from the x-y plane and at an angle ϕ from the y-z plane. The coordinate system (x',y',z') of the receiving lens may be obtained via a coordinate transformation of the $(x,y,z,)$ system.

The receiving aperture of the PDPA is partitioned into three areas, Figure 2. The light from each of these areas is collected and delivered to three separate photodetectors located within the receiver behind a single aperture. Each photodetector outputs a Doppler burst similar to the output of an LDV system. However, the phase differences, η_{12} and η_{13} of the detector outputs may be utilized to determine the size of a particle moving thorough the measurement volume.

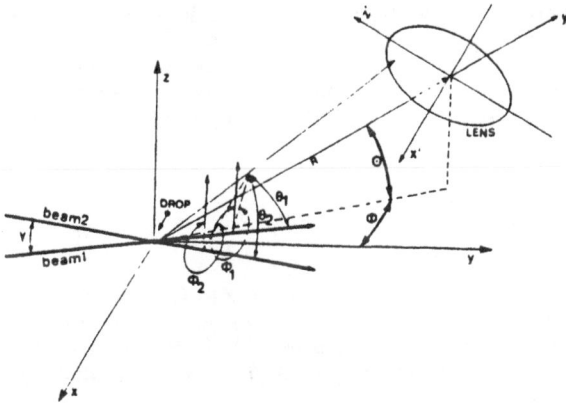

Figure 1   Coordinate system for theoretical analyses

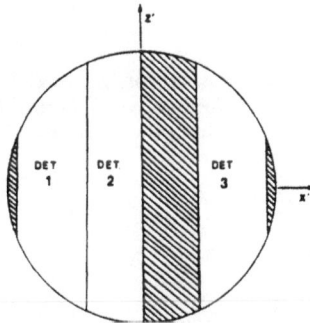

Figure 2   Schematic of PDPA receiving aperture

To generate the calibration curves of detector phase difference versus particle diameter, a spatial integration of the scattered light over each of the three collection areas must be performed. To achieve this, the first step is the calculation of the scattering amplitude functions, $S_{mn}(\theta_n)$ at several points on a fine rectangular mesh placed on the receiving lens. The Lorenz-Mie theory with the numerical algorithm of Wiscombe[10] was used to determine the scattering amplitude functions. The subscript m stands for the direction of the electric field polarization, where $m = 1$ and $m = 2$ are polarizations perpendicular and parallel to the scattering plane, respectively. The second subscript n identifies the two incident laser beams and may be 1 or 2. The amplitude function is complex, and related to the complex electric field E_{mn} by

$$E_{1n} = \frac{i}{k_n r} e^{-ik_n r + i\omega_n t} \cos \phi_n S_{1n}(\theta_n) \tag{1}$$

and

$$E_{2n} = -\frac{i}{k_n r} e^{-k_n r + i\omega_n t} \sin \phi_n S_{2n}(\theta_n) \tag{2}$$

In equations 1 and 2, κ_n and ω_n are the wavelength and frequency, respectively, of the scattered light from the nth beam. For each point (x', z') on the receiving aperture, it is possible to calculate $\theta_1, \theta_2, \phi, \phi$, the scattering beam angles for *beam 1* and *beam 2*. At each of these points on the receiver surface, the scattered light from each beam interfere to yield a light intensity which varies temporally at the Doppler difference frequency, $\omega_D = \omega_1 - \omega_2$. The resulting intensity for any polarization can be expressed as

$$\begin{aligned} I_m(x', z') &= \left(|E_{m1}|^2 + |E_{m2}|^2 \right)/2 \\ &\quad + |E_{m1}||E_{m2}| \cos(\omega_D t + \beta_{m2} - \beta_{m1}) \\ &\quad \text{for } m = 1,2 \end{aligned} \tag{3}$$

The terms β_{m1} and β_{m2} appearing in the above equation are the phases of the scattered electric fields and may be obtained from equations (1) and (2). By integrating the scattered light intensity over the collection areas of each photodetector, Figure 2, the resulting intensity can be expressed as

$$\begin{aligned} T_j &= \frac{1}{A_j} \sum_{z'} \sum_{z'} \sum_{m=1}^{2} I_m(z', z') \Delta x' \Delta z' \\ &= D_j + C_j \cos(\omega_D t + \eta_j) \\ &\quad \text{for } j = 1,2,3 \end{aligned} \tag{4}$$

where A_j is the area of the jth detector, $\Delta x'$ *and* $\Delta z'$ are the integration mesh sizes, m is the polarization direction, and D_j and C_j are the DC and AC levels of the output of the jth detector. From equation (4) the phase difference between the detectors

$$\eta_{12} = \eta_1 - \eta_2 \tag{5}$$

and

$$\eta_{13} = \eta_1 - \eta_3 \tag{6}$$

can be easily determined.

When these calculations are performed for a range of particle sizes, calibration curves of phase difference versus particle size can be generated for any desired optical configuration and particle refractive index. It should be noted that for large particles relative to the Gaussian beam diameter, the phase differences may vary somewhat as the illumination is non-uniform. However, for the particles considered in this study, which are much smaller than the beam waist this effect is negligible.

3 EXPERIMENTAL APPROACH

In order to determine the validity of the just described model, as well as allowing determination of the applicability of the PDPA for fine particles an experimental study was conducted. In this study, the PDPA was used to size polystyrene latex (PSL) particles dispersed in water and air. Only the water results are presented in this paper. The Duke Scientific PSL particles, ranging in size between 0.705 and 15.0 um, are NIST (formerly NBS) traceable and are claimed to be spherical and uniform in diameter to within 2%. The refractive indices are 1.60 and 1.20 in air and water, respectively.

The first test was with PSL dispersed in distilled water. The PSL was measured in triangular test cells at the various desired light collection angles between 20 and 70 degrees. The PSL was dispersed and kept in motion via a magnetic stirrer.

A standard PDPA system was used for sizing the PSL. The transmitter used a He-Ne laser (632.8 nm) and a 200 mm focal length transmitter lens. The rotating phase diffraction grating of the transmitter allowed the choice of beam intersection angle of 3.47 and 13.74 degrees. (track 1 and track 3, respectively). The receiver used a 238 mm focal length, f 2.25 lens for collecting the scattered light. As will be discussed at the end, this large system can be replaced by a much smaller fiber optic probe.

4 RESULTS AND DISCUSSION

The results of the theoretical calculations of the PDPA's response were determined for a variety of optical configurations. As will be seen, some oscillations were seen to be present, but through the proper choice of optics, the severity of the oscillations could be substantially mitigated and adequate resolution obtained.

Figures 3 and 4 show typical PDPA calibration curves for water droplets in air. In Figure 3, a 500 mm, f 4.7 receiver focal length, 1.39° beam crossing angle and a 30° collection angle yield fairly severe oscillations in the less than 10 um region. Accuracy is certainly no better than +/- 1.0 um, inadequate for many measurement needs such as LDV seeding.

However, as shown in Figure 4, if the correct optical configuration for the PDPA is chosen, resolution can be substantially enhanced. For a 238 mm, f 2.25 receiver lens, 13.74° beam crossing angle, and a 74° crossing angle the resolution is improved to +/- 0.3 um, a considerable improvement. The model shows that both the faster lens and larger crossing angle serve to improve resolution by providing better phase sensitivity.

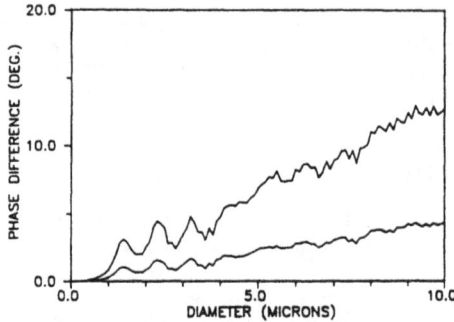

Figure 3   Predicted PDPA response curve for fine water droplets
at a mean collection angle of 30°. 500 mm focal
length, f 4.70 receiver, beam crossing angle 1.39°

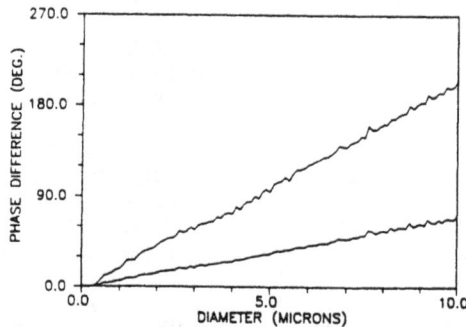

Figure 4   Predicted PDPA response curve for fine water droplets
at a mean collection angle of 74°. 238mm focal length,
f 2.25 receiver, beam crossing angle 13.74°

The receiver location also plays a key role in the resolution limits of
the system. The 74° angle corresponds to the Brewster's angle for water
droplets. This optimizes the contribution of the refracted light from the
drop, while minimizing the reflected light for the chosen beam polarization.
Since the interference of the reflected and refracted light can result in
oscillations of the calibration curve[9] the correct choice of collection angle
is needed for optimal results. Also, the contribution of any diffracted light
at an angle of 74° is nil, even for the smallest particles. All these factors
work in concert to give the excellent resolution shown in Figure 4. Similar
results have been calculated for glass beads, but at a different optimal
collection angle due to the different index of refraction.

The model results were then validated experimentally. The model
was used for determining the calibration curves for PSL in water at
collection angles of 20, 30, 50 and 70 degrees. Histograms of phase
difference for each size particle were generated with the PDPA at each
collection angle. For each histogram, 5000 particles were measured. The
phase corresponding to the peak of the histogram was chosen as the
measured phase. The results of this study are shown in Figures 5 and 6
for the 50 and 70 degree cases.

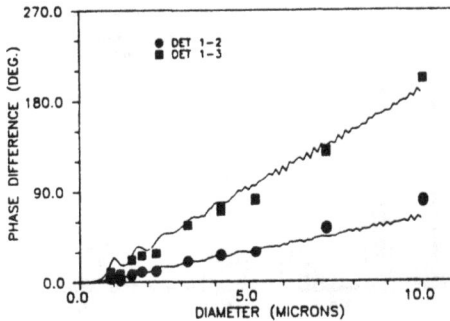

Figure 5   Predicted and measured phase differences for PSL in water at a mean collection angle of 50°. 238mm focal length, f 2.25 receiver, beam crossing angle 13.74°

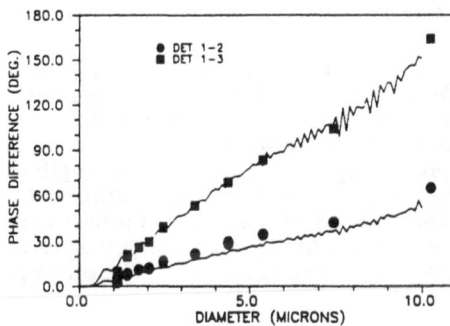

Figure 6   Predicted and measured phase differences for PSL in water at a mean collection angle of 70°. 238mm focal length, f 2.25 receiver, beam crossing angle 13.74°

The data presented in Figures 5 and 6 show excellent agreement between the modelled and experimental data. This verifies the models predictive capability. The data clearly shows the relationship of collection angle and resolution. The 50° data has an uncertainty of about +/- 0.6 um, while the 70° data uncertainty is approximately +/- 0.4 um. The 20° and 30° data not shown have even more uncertainty, due in part to the diffracted light also present at these angles for small particles. Similar results were obtained for PSL aerosols in air.

To more fully demonstrate the resolution capability of the PDPA, the size distribution of a PSL mixture consisting of 0.7, 3.0, 7.0, 9.9, and 15.0 um particles suspended in water was measured. As Figure 7 shows, all the data agrees within the predicted uncertainly of +/- 0.4 um.

5 PROBE DEVELOPMENT

The standard PDPA probe is rather large for many experimental applications, particularly the monitoring of processes where spherical particles or aerosols are being produced and changing as they progress along a pipe or duct. The standard PDPA probe is not submersible in working fluids, which is also of importance for some measurement situations.

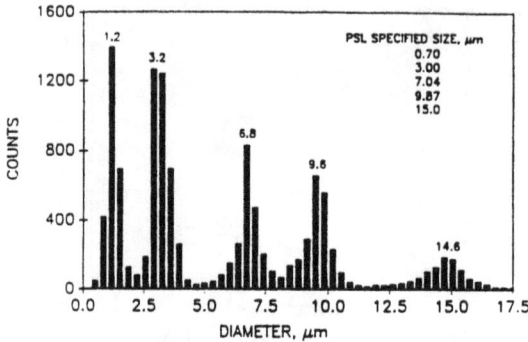

<u>**Figure 7**</u>   Measured size distribution for a mixture of PSL
                  particles dispersed in water.

         For these circumstances, fiber optic based PDPA probes have been
developed. While these may range from 100 mm diameter down to 12 mm
diameter, the smallest "miniprobes" are of particular interest for process
monitoring. The miniprobe is shown is Figure 8. It is linked via a single
mode polarization preserving fiber of any desired length to a remote He-Ne
laser. The probe may be used singly or in a multiplexed configuration
allowing sequential monitoring of a number of process points. The probe
is very rugged and will maintain alignment on a long term basis. It may be
inserted in the flow at any region of interest. The probe performance is
similar to a standard PDPA.

         An example of the measurement of a commercial nebulizer aerosol
is shown in Figure 9. The histogram of size and velocity is shown in Figure
9a with the associated mean sizes and velocities. Figure 9b shows the
histogram and cumulative distribution of number and volume, while
Figure 9c portrays the size-velocity correlation of the flow.

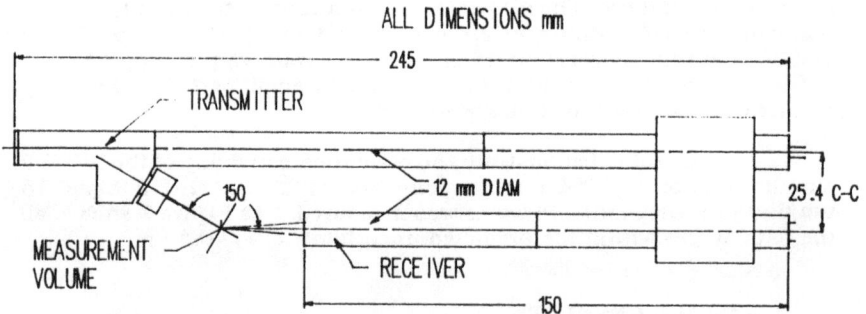

<u>**Figure 8**</u>   Schematic of PDPA miniprobe

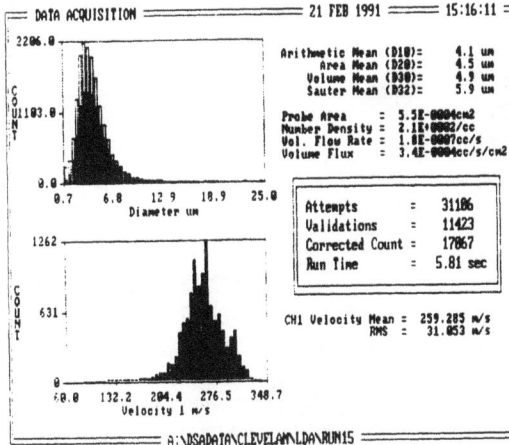

a) Size and velocity distribution

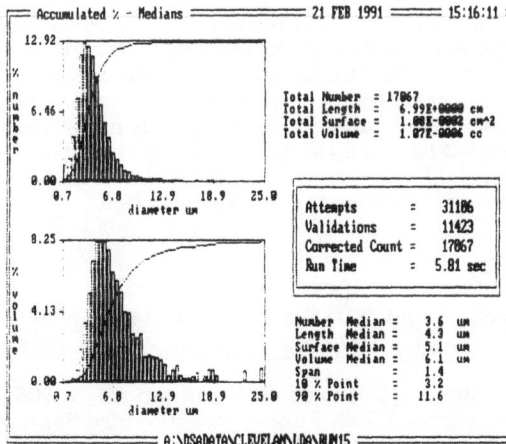

b) Number and volume distribution

<u>Figure 9</u>   Sample PDPA results for nebulizer output

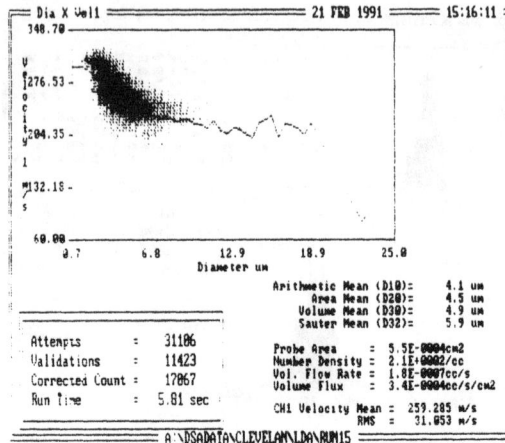

c) Size-velocity correlation

Figure 9   Sample PDPA results for nebulizer output

6 SUMMARY AND CONCLUSIONS

The theoretical and experimental results of this study clearly point out
that the magnitude of oscillations of the calibration curve of the PDPA can
be minimized via the correct choice of optics and collection angle. For fine
particles (<10 um), resolutions of up to +/- 0.3 um may be realized. This
resolution should be adequate for many aerosol measurements The
excellent agreement between the experiment and model point out the
validity of the model for the prediction of the PDPA response. A small in-
situ probe for process monitoring is also described.

REFERENCES

1.   W.D. Bachalo and M.J. Houser, Optical Engineering, 1984, 23, 583.
2.   W.D. Bachalo, Optical Particle Sizing, Plenum Publishing, 283,
      1988.
3.   W.D. Bachalo, R.C. Rudoff, S.V. Sankar, ASTM 1083, 209. 1990.
4.   C.F. Edwards and R.C. Rudoff, Twenty-Third Symp. (Int.) on Comb.,
      The Combustion Inst., 1990.
5.   G.S. Jones and D.Y. Kamemoto, L.R. Gartrell, AIAA-90-0502, Reno,
      Nevada, 1990.
6.   S.A.M. Al-Chalabi, Y. Hardalupas, A.R. Jones, and A.M.K.P Taylor,
      Proc. Int. Symp. on Optical Particle Sizing: Theory and Practice,
      1987.
7.   Y. Hardalupas and A.M.K.P. Taylor, Experiments in Fluids, 1988, 6,
      137.
8.   S.R. Martin, L.E. Drain, M.L. Yeoman, and D.M. Livesley, Proc.
      Fourth Int. Symp. on the Applications of Laser Anemometry, 1988.
9.   W.D. Bachalo and S.V. Sankar, Proc. Fourth Int. Symp. on the
      Applications of Laser Anemometry, 1988.
10.   W.J. Wiscombe, Report No. NCAR/TN-140 STR, 1979.

Application of the Phase Doppler Technique to Optically Absorbent Liquids

U. Manasse, Z. Jiang, Th. Wriedt and K. Bauckhage

UNIVERSITY OF BREMEN, FB4/VERFAHRENSTECHNIK, D-2800 BREMEN 33, GERMANY

1 INTRODUCTION

The phase-Doppler technique for simultaneous measurements of diameters and velocities of spherical particles has been successful in the analysis of spray atomization of both water[1-3] and molten steel[4,5] as well as in the diagnostic of model experiments with glass spheres or bubbles in water flows[6-8].

However difficulties arise in applying the phase-Doppler anemometry to process fluids with an optical absorption that lies between zero and very high. In addition these fluids are often inhomogeneous as in the case of metal flakes in paints or dissolved instant coffee in spray drying processes.

2 THEORY

A number of authors have described the phase-Doppler technique for simultaneous measurements of diameter and velocity of spherical particles[1,7,9-11]. Therefore we confine ourselves to a short description of the PDA and focus our attention on Mie scattering theory and geometrical optics.

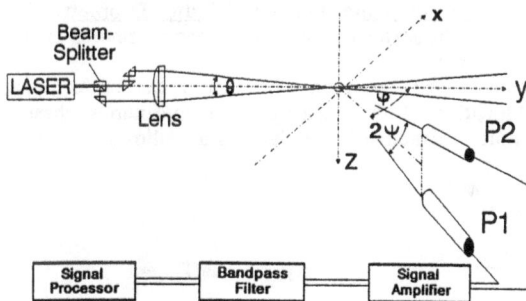

Figure 1   Phase-Doppler anemometer (schematic)

Phase-Doppler Anemometry

In Figure 1 a typical phase-Doppler-anemometer (PDA) is sketched. Based on a Laser-Doppler-anemometer and supplemented by a second photo-detector, the two detectors are usually arranged symmetrically to the interference planes with equal elevation angles ψ.

The time difference, measurable as a phase difference Φ between the Doppler bursts at the first and the second photodetector, is simply related to the diameter d of the particle that crosses the measuring volume as long as one light scattering component - either refraction or reflection - dominates. Equation (1) shows this relationship, where b is a function of the geometrical arrangement of the set-up.

$$d = \frac{1}{2b} \left(\frac{\lambda}{\pi n_c} \right) \Phi \tag{1}$$

n_c refractive index of the continuous phase

$$b_{Reflection} = \sqrt{2} \; [(1 + \sin\frac{\theta}{2}\sin\Psi - \cos\frac{\theta}{2}\cos\Psi\cos\varphi)^{\frac{1}{2}}$$

$$- (1 - \sin\frac{\theta}{2}\sin\Psi - \cos\frac{\theta}{2}\cos\Psi\cos\varphi)^{\frac{1}{2}}] \tag{2}$$

$$b_{Refraction} = 2 \; \{ \; [1 + n'^2 - \sqrt{2} \; n'(1 + \sin\frac{\theta}{2}\sin\Psi + \cos\frac{\theta}{2}\cos\Psi\cos\varphi)^{\frac{1}{2}}]^{\frac{1}{2}}$$

$$- [1 + n'^2 - \sqrt{2} \; n'(1 - \sin\frac{\theta}{2}\sin\Psi + \cos\frac{\theta}{2}\cos\Psi\cos\varphi)^{\frac{1}{2}}]^{\frac{1}{2}} \} \tag{3}$$

n_d refractive index of the dispersed phase

$$n' = \frac{n_d}{n_c} \tag{4}$$

For an actual application of the phase-Doppler method it is necessary to find off-axis-angles φ at which the measured value of the phase shift Φ shows a linear or continuous and unambiguous dependence on the particle diameter d. This is directly related to the scattering behaviour of the liquid droplets that can be exactly described only by the Mie scattering theory[12].

Light Scattering

Light Scattering of Homogeneous Liquid Droplets. The Mie theory describes exactly the light scattering by homogeneous spheres of arbitrary size that are uniformly illuminated.

For the calculations based on Mie theory the optical properties such as refraction and absorption of the dispersed and continuous phase are included in the form of a complex refractive index defined as follows:

$$n = \quad n' - i \, \kappa. \tag{5}$$

$$\kappa = \frac{K \cdot \lambda}{4 \pi} \qquad \text{absorption coefficient} \tag{6}$$

n' \qquad refractive index

Van de Hulst[13] has shown that for spheres larger than the light wavelength and with refractive indexes sufficiently different from the surroundings, the amplitude functions derived from the geometrical optics approach were, in the asymptotic limit, equal to the Mie amplitude functions. For Mie parameters $\alpha > 10$, the scattering of light can be separated into the simplified theories of diffraction, refraction and reflection.

Figure 2 shows the phase difference Φ versus droplet diameter d for water with a real refractive index n' = 1.3333 at an off-axis angle φ of 30° (The calculation are based on the programmes of Adrian and Earley[14] in combination with a subroutine of Wiscombe[15]).

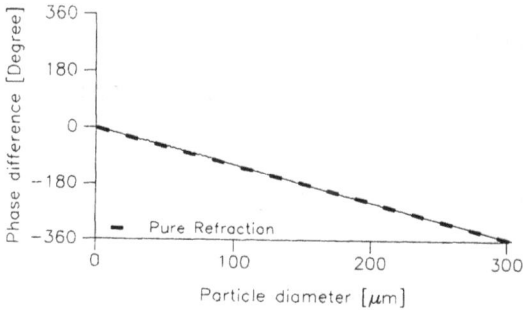

Figure 2   Mie calculations for water (n = 1.3333), φ = 30°, parallel polarized light

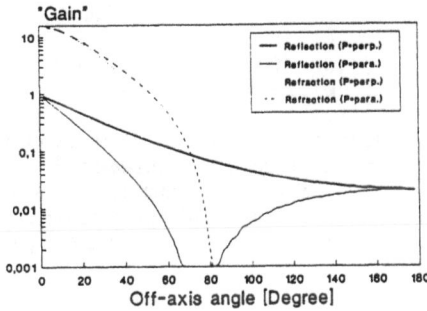

Figure 3   Geometrical optics for water (n = 1.3333), P = para - parallel polarized light/P = perp - perpendicular polarized light

By using the lineargeometrical optics Φ-d-dependancy can be easily understood. In Figure 3 the gain, as defined in[13], versus the off-axis angle φ is shown. It can be seen that for parallel polarized light the refracted light dominates at an off-axis angle φ of 30°. (Note: Negative phase means refraction, positive phase means reflection.)

Light Scattering of Inhomogeneous Liquid Droplets. Many process fluids such as dissolved instant coffee, milk, paint etc. are not only optical absorbent but also inhomogeneous. Light scattering theories concerning inhomogeneous liquid droplets are still in their infancy [16-20].

All of these theories have the disadvantage that they are only capable of calculating the scattering behaviour of drop sizes in the order of the light wave length λ - partly because of the large storage capacity and the computation time that is required, partly because of the validity range of the approximations.

For our purposes the "effective-medium"-theory[21] seems to be the most promising. According to that it is possible - under certain assumptions - to assign to a microscopic inhomogeneous fluid an effective macroscopic refractive index and absorption.

By using an effective complex refractive index the liquid is treated as macroscopic homogeneous.

Then the scattering behaviour of the microscopic inhomogeneous liquid can as well be calculated by means of the Mie theory.

The validity of this theory strongly depends on the size and shape of the inhomogeneities and their density. It has to be checked by measurements for every liquid.

3 EXPERIMENTAL SET-UP

The exact geometrical parameters are given in Table 2.

Table 2 PDA optical geometries
Transmission optics

Laser wavelength	λ	488 nm		
Lens focal length		1200 mm		
Beam crossing angle	θ	1.88°		
Measuring volume	d_z	565 μm		
	d_y	564 μm		
	d_x	34394 μm		

Receiving optics

Off-axis-angle	φ	30°	60°	90°
Collection aperture dia.		52	52	52 mm
Distance from measuring volume		1000	800	800 mm
Elevation angle	ψ	1.83°	2.47°	4.0°

The frequency information and phase difference of the bandpass-filtered Doppler-bursts were obtained by using a FFT-processor[22].

Photography served as a reference method for size measurement. The droplets were illuminated by a Xenon-flashlamp stroboscope (illumination time = 150 ns) and were enlarged 63 times by means of a microscope.

Droplet Generation

The impulse-jet technique was comprehensively discussed by Heinzel and Hertz[23] for computer ink jet printers. With the application of a mechanical disturbance a pressurized liquid jet is broken into a stream of uniformly sized droplets. Based on the work of[24] this task was fulfilled by a piezo-electric tube.

4 EXPERIMENTS

Phase-Doppler technique applied to an optical absorbent homogeneous liquid

By diluting strongly absorbent black ink in distilled water it is possible to adjust the imaginary part of the refractive index for being able to investigate the influence of absorption on the Φ-d-correlation.

An analysis of the particles inside the black ink led to the result that their size is smaller than 0.5 μm. Therefore we treat it as a homogeneous liquid. Nevertheless due to its pigments it is still a microscopic inhomogeneous liquid.

Table 3 contains the complex refractive indexes of the chosen ink solutions. The refraction was determined by means of an Abbe-refractometer whereas the absorption with the help of a spectrophotometer. These quantities are treated as the effective refractive indexes.

Figure 4 to 8 show the Φ-d-dependencies predicted by Mie theory for some ink solutions at an off-axis angle of 30°.

According to that it should be possible to apply the phase-Doppler technique for particle sizing by using parallel polarized laser light and the Φ-d-relation for dominant refracted light up to an imaginary part of the refractive index in the order of 10^{-3} - by adding that the particle size range goes down to 150

μm. However the experimental set-up (i.e. perpendicular polarization and signal order) and Φ-d-relation for dominant reflection has to be used for an imaginary part of the refractive index higher than 10^{-3}. But the lower end of the sizing range is cut, because due to the contribution of the refracted light in this size range the Φ-d-relation is ambiguous.

For an imaginary part of the refractive index in the order of $1 \bullet 10^{-2}$ all of the refracted light is absorbed.

By the aid of three different orifice diameters (25, 50 and 100 μm) monodisperse droplets of three diameters were produced for every ink concentration.

<u>Table 3</u>   Refractive indexes of ink solutions

concentration c (ink : water) weight in per cent	refractive index $\lambda = 488$ nm, $T = 20°C$
0	1.3333 – i 0.0
2	1.3366 – i 0.78E-4
4	1.3372 – i 6.43E-4
7	1.3378 – i 9.03E-4
12	1.3396 – i 1.28E-3
25	1.3404 – i 2.49E-3
50	1.3441 – i 3.96E-3
75	1.3472 – i 6.37E-3
100	1.3514 – i 0.89E-2

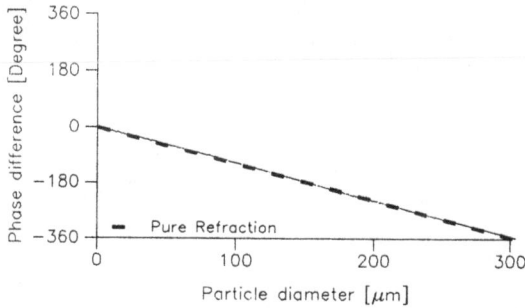

Figure 4   Mie calculations for ink solution, c = 2 % φ = 30°, parallel polarized light

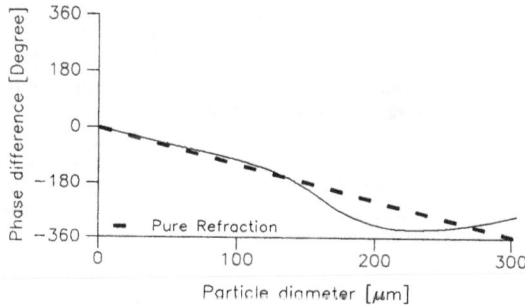

Figure 5   Mie calculations for ink solution, c = 7 % φ = 30°, parallel polarized light

Particle Size Analysis

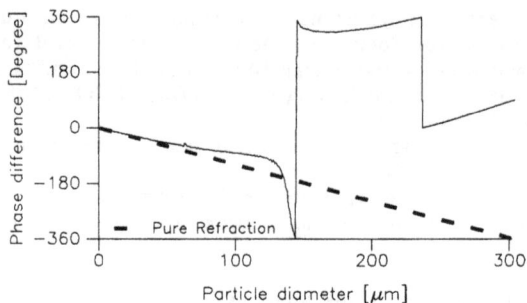

Figure 6   Mie calculations for ink solution, c = 12 % φ = 30°, parallel polarized
            light

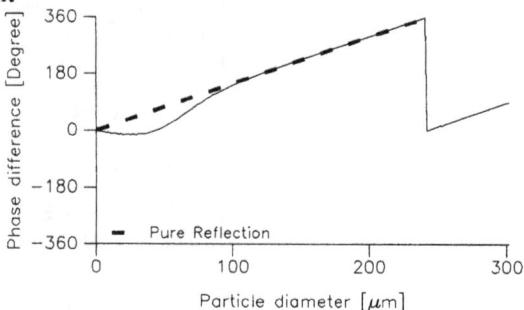

Figure 7   Mie calculations for ink solution, c = 25 % φ = 30°, perpendicular
            polarized light

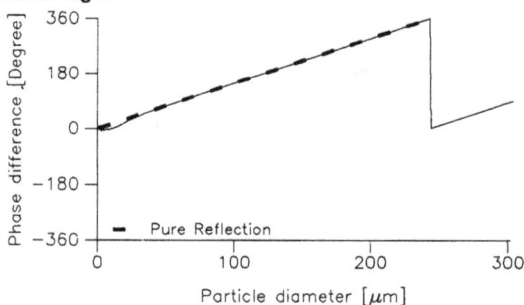

Figure 8   Mie calculations for ink solution, c = 100 % φ = 30°, perpendicular
            polarized light

In Table 4 the results of the phase-Doppler measurement applied to 2000
particles and of the photography are summarized.

It was impossible to get results by phase-Doppler measurements for a
droplet 150 μm in diameter with an ink concentration of 12 per cent, because the
visibility of the signals was very low as can also be seen by theory (Figure 9).

For all ink concentrations the results of the phase-Doppler technique and
photography agree very well.

<u>Table 4</u>   Comparison of phase-Doppler results and photography, $\varphi = 30°$

| Ink concentration | Photography | | Result by PDA | | after Eq. |
| | d | Δ d | d_{max} | Δ d | (1) and ( ) |
[%]	[μm]		[μm]		
	40		40	3.2	
0	76	6	77	6.2	3
	192		199	15.9	
	50		53	4.2	
2	76	6	75	6.0	3
	151		159	12.7	
	51		57	4.6	
4	76	6	81	6.5	3
	139		147	11.8	
	63		60	4.8	
7	101	6	99	7.9	3
	142		156	12.5	
	55		50	4.0	
12	90	6	83	6.6	3
	-		-	-	
	60		50	4.0	
25	88	6	97	7.8	2
	152		146	11.7	
	51		41	3.3	
50	88	6	82	6.6	2
	164		168	13.4	
	58		63	5.0	
75	81	6	89	7.1	2
	154		141	11.3	
	51		61	4.9	
100	88	6	97	7.8	2
	150		159	12.7	

Note: The maximum error in adjustment of the optical set-up leads to an uncertainty in determining the droplet diameter by phase-Doppler technique of
about 8 %. The vagueness of photography is in the order of about ± 6 μm.

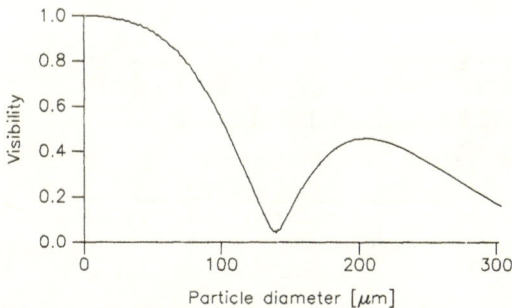

Figure 9   Visibility by Mie calculations for ink solution,
c = 12 %, φ = 30°, parallel polarized light

According to Figure 3 the refraction/reflection ratio for parallel polarized
light at an off-axis angle of 30° is about 30 : 1, whereas at an off-axis angle of
60° it is 100 : 1. Therefore it should be possible to enlarge the sizing range for ab-
sorptions cancelling out the dominance of refracted light.

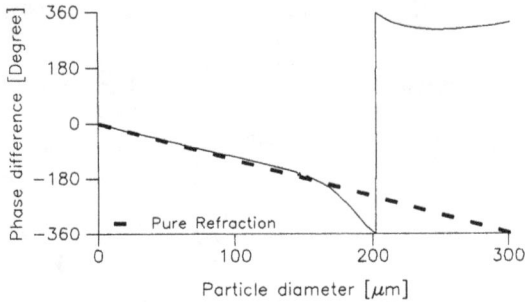

Figure 10   Mie calculations for ink solution, c = 12 % φ = 60°, parallel polarized light

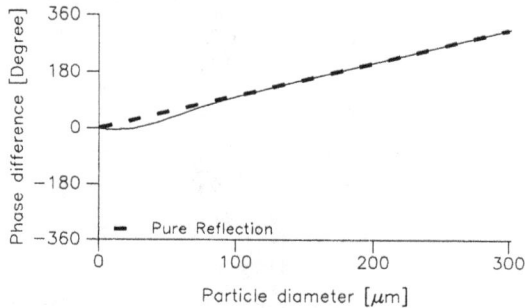

Figure 11   Mie calculations for ink solution, c = 25 % φ = 60°, perpendicular polarized light

Figure 10 to 11 show the Φ-d-relation predicted by Mie theory (φ = 60°) for ink solutions with a concentration of 12 and 25 per cent, respectively. Although the sizing range for an ink solution of 12 % is enlarged the sizing range for an ink solution of 25 % is still the same. The results of the particle sizing are compared in Table 5.

Table 5   Comparison of phase-Doppler results and photography, φ = 60°

Ink concentration	Photography		Result by PDA		after Eq.
[%]	d [μm]	Δ d	d_max [μm]	Δ d	(1) and ( )
12	46		51	4.1	
	77	6	71	5.7	3
	157		154	12.3	
25	51		25 [56]	4.5	
	101	6	98	7.8	
	139		152	12.2	

Discrepancies between these results arise for droplets with a diameter of 51 μm and an ink concentration of 25 per cent. It ensues from the application of the Φ-d-relation according to the geometrical optics approach, i.e. the assumption that only one light scattering component exists. However Φ-d-relation predicted by Mie theory is not linear any more.

The errors in sizing can be reduced or even avoided by the use of a modified Φ-d-relation (see Eq. (1)):

As long as the Φ-d-relation is a monotonous and continuous function, the particle diameter d is unequally related to the phase difference Φ by

$$d = c(\Phi) \ \Phi \qquad (1')$$

$$0 \leq \Phi < 2\pi$$
$$V \quad -2\pi \leq \Phi < 0$$

The function c (Φ) can be determined by the Φ-d-relation predicted by Mie theory. These proceedings yield the values in brackets (see Table 5). The difference between the results by PDA and photography is reduced. By this modified Φ-d-relation a wider range of PDA-applications should be possible.

Phase-Doppler technique applied to an optical absorbent inhomogeneous liquid

As described above the complex refractive index of a powdered milk solution (8 per cent, n = 1.3400 - i 2.57E-4) and an instant coffee solution (8 per cent, n = 1.3422 - i 4.47E-4) were determined. Treating these quantities as the macroscopic refractive indexes it should be possible to apply the PDA for sizing at an off-axis angle φ of 30° and 60° - as predicted by Mie theory. However in both cases the size distributions of monodisperse droplets were too broad. This may be due to the scattering of the refracted light caused by the inhomogeneities inside the droplets.

A repetition of the experiments at an off-axis angle φ of 90°, where reflection is the only light scattering component (see Figure 12 and 13), provided smaller size distributions and agreement between PDA and photography as summarized in Table 6. The width of the size distributions may follow from the surface roughness caused by the inhomogeneities.

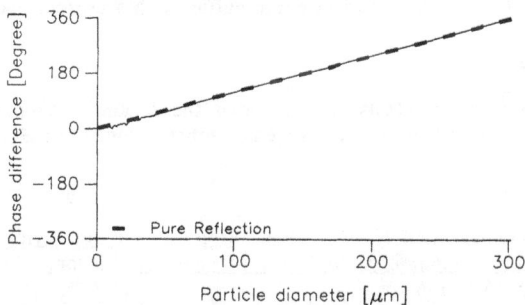

Figure 12   Mie calculations for powdered milk solution c = 8 % (n = 1.34 - i 2.57E-4), φ = 90°, perpendicular polarized light

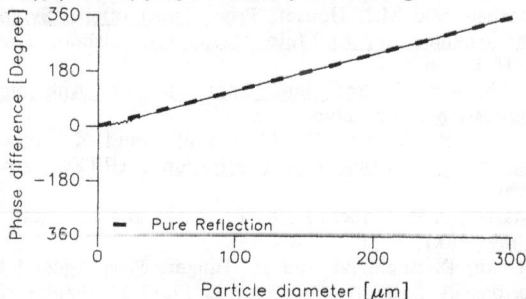

Figure 13   Mie calculations for instant coffee solution c = 8 % (n = 1.3522 - i 4.47E-4), φ = 90°, perpendicular polarized light

<u>Table 6</u>   Comparison of phase-Doppler results and photography, $\varphi = 90°$

Fluid	Photography d [μm]	Δ d	PDA d_{max} [μm]	Δ d	after Eq. (1) and ( )
powdered milk	63		57	4.6	
solution	182	6	183	86.6	3
(8 per cent)	177		167	13.3	
Instant coffee	51		51	4.0	
solution	95	6	99	7.9	3
(8 per cent)	152		141	11.3	

5   CONCLUSIONS

It was shown that the phase-Doppler technique is a powerful tool for simultaneous measurements of diameters and velocities of spherical particles in application to optical absorbent homogeneous *and* inhomogeneous liquids.

As long as the inhomogeneities inside the particles are small compared to the laser wavelength it should be possible to treat the fluids as macroscopic homogeneous and to apply the PDA also at off-axis angles, where refracted light dominates.

If the inhomogeneities are much larger than the laser wavelength, it is recommended to employ the PDA at off-axis angles, where the detected light does not pass through the particle, that means where reflected light is the only scattering component.

So far the Mie theory seems to be a powerful tool for describing the scattering behaviour of optical absorbent liquid droplets and to select the right parameters of the optical set-up in combination with a correct Φ-d-relation.

<u>Acknowledgement</u>

The authors gratefully acknowledge the financial support for this work provided by the Deutsche Forschungsgemeinschaft, Bonn/Bad Godesberg.

REFERENCES

1.   K. Bauckhage and H.-H. Flögel, Proc. Second Int. Symp. on Applications of Laser Anemometry to Fluid Mechanics, Lisbon, Portugal, July 2-4, 1984, p-18.1, 1-6

2.   M.L. Yeoman, L.E. Drain, D.M. Livesley and S.R. Martin, Proc. Fourth Intern. Symp. on Applications of Laser Anemometry to Fluid Mechanics, Lisbon, Portugal, July 11-14, 1988, p 2.8, 1-9

3.   W.D. Bachalo and M.J. Houser, Proc. Third Intern. Symp. on Application of Laser Anemometry to Fluid Mechanics, Lisbon, Portugal, July 7-9, 1988, p.18.3, 1-6

4.   K. Bauckhage et al., 2nd Inter. Conf. on Laser Anemometry - Advances and Applications, Strathclyde, 1987

5.   H.-M. Liu, B. Seuren, V. Uhlenwinkel and K. Bauckhage, Proc. 4. European Symp. Particle Characterization PARTEC, April 19-21, 1989, Nuremberg

6.   W.W. Martin, A.H. Adbelmessih, J.J. Liska and F. Durst, Intern. J. Multiphase Flow, 1981, 7, pp.433-460

7.   M. Saffman, P. Buchhave and H. Tanger, Proc. Second Intern. Symp. on Applications of Laser Anemometry to Fluid Mechanics, Lisbon, Portugal, July 2-7, 1984, p.81, 1-8

8. R.W. Sellens, PhD Thesis, Waterloo, Ontario, 1987
9. F. Durst and M. Zaré, <u>Proc. of LDA Symp.</u>, Copenhagen, Denmark, 1975, pp.403
10. W.D. Bachalo and M.J. Houser, <u>Optical Engineering</u>, 1984, <u>23</u>, pp.583-590
11. K. Bauckhage, <u>Part. Part. Syst. Charact.</u>, 1988, <u>5</u>, pp.16-22
12. G. Mie, <u>Annalen der Physik</u>, 1908, <u>25/4</u>, pp.25
13. H.C. van de Hulst, 'Light scattering by small particles', Wiley & Sons, New York, 1981
14. R.J. Adrian and WL.Earley, Technical report, Department of Theoretical and Applied Mechanics, University of Illinois, 1974
15. W.J. Wiscombe, NCR-report TN-140, STR, June 1979
16. G.H. Goedecke and S.G. O'Brien, <u>Applied Optics</u>, 1988, <u>27</u>, pp.2431-2438
17. R. Holland, L. Simpson and K.S. Kunz, <u>IEEE Transactions on Electromagnetic compatibility</u>, 1980, <u>EMC-22</u>, pp.203-209
18. W.M. McClain and W.A. Ghoul, <u>J. Chem. Phys.</u>, 1986, <u>84</u>, pp.6609-6622
19. M.A. Morgan and K.K. Mai, <u>IEEE Transactions on Antennas and Propagation</u>, 1979, <u>AP-27</u>, pp.202-214
20. D.H. Schaubert, D.R. Wilton and A.W. Glisson, <u>IEEE Transactions on Antennas and Propagation</u>, 1984, <u>AP-32</u>, pp.77-84
21. C.F. Bohren, <u>J. of the atmosph. Sc.</u>, 1986, <u>43</u>, pp.468-475
22. Th. Wriedt, J. Heuermann, K. Bauckhage and A. Schöne, <u>3rd Int. Conf. Laser Anemometry - Advances and Applications</u>, Swansea, GB, 1989
23. J. Heinzel and C.H. Hertz, 'Ink-jet printing advances in electronics and electron physics', Academic, New York, 1985, <u>65</u>, pp.91-171
24. Th. Döking, Konstruktiver Entwurf, Ruhr-Universität Bochum, Institut für Thermo- und Fluiddynamik, Jan. 1990

Extension of Phase Doppler Technique to Sizing and Material Recognition of Sub-micron Particles

Amir Naqwi and Franz Durst

FRIEDRICH-ALEXANDER UNIVERSITÄT, ERLANGEN-NÜRNBERG, LEHRSTUHL FÜR STRÖMUNGSMECHANIK, CAUERSTR. 4, 8520 ERLANGEN, GERMANY

ABSTRACT

Based on the analysis of the response of a phase/Doppler system to a Rayleigh scatterer, basic requirements for the application of this technique in the Rayleigh and 'near-Rayleigh' range are established. Response curves of the corresponding phase/Doppler systems are computed for various materials, using Mie scattering theory. Possibilities of particle sizing, as well as particle material recognition, in the submicron range, are highlighted. Measurements, with submicron latex spheres in water, have demonstrated the feasibility of the application of the phase/Doppler technique in the near-Rayleigh range.

1 INTRODUCTION

Principles of the phase/Doppler system have been traditionally described in terms of geometrical optics[1-4]; i.e., the technique has been studied in detail for the limiting case of very large particles. Calculations based on Mie scattering theory have been performed, usually, to examine if the behavior of a given particle agrees with that of an idealized reflecting or refracting particle. This approach has restricted the applications of the phase/Doppler method to rather large particles. Unlike other particle sizing techniques, the limiting behavior of phase/Doppler anemometry (PDA) for very small particles has not been explored in detail.

It is shown herein that the heterodyne signals, collected at two different locations in a particular optical arrangement, are completely out-of-phase for Rayleigh scatterers. Such an optical layout may be used for the recognition of Rayleigh scatterers and sizing of slightly larger particles, which are referred to as 'near-Rayleigh' scatterers.

Since different materials approach the Rayleigh limit at different values of the Mie parameter, it is possible to recognize the particle material, in a mixture of particles with substantially different refractive indices. The present technique is particularly suitable for sizing of very small absorbing particles, which reach the Rayleigh limit at a very small particle diameter.

In the subsequent text, particular attention is given to the case of Rayleigh scatterers. Signal characteristics for near-Rayleigh scatterers are examined using Mie calculations. Some experimental results are given to illustrate the feasibility of the proposed concepts for submicron particle sizing.

2 SCATTERING BY SMALL PARTICLES IN A PDA SYSTEM

Characteristics of the scattered light produced by a spherical particle, exposed to two plane waves, have been discussed by several investigators; e.g., Hong & Jones[5], and Pendleton[6]. In a recent study, Naqwi & Durst[7] have generalized the previous formulations, so that, polarization of the individual beams can be specified independently, and the cases of arbitrarily oriented circular and rectangular apertures may be considered. Such a formulation is necessary for evaluating innovative applications of the dual-beam system.

The optical arrangement is illustrated in Fig. 1. The scattering particle is illuminated with two laser beams propagating in xz-plane. The center of the receiving aperture is defined in terms of two angles: off-axis angle ϕ and elevation angle ψ. These angles are so defined that if the coordinate system is first rotated through ϕ about x-axis and subsequently through ψ about y-axis, then the z-axis points towards the center of the aperture.

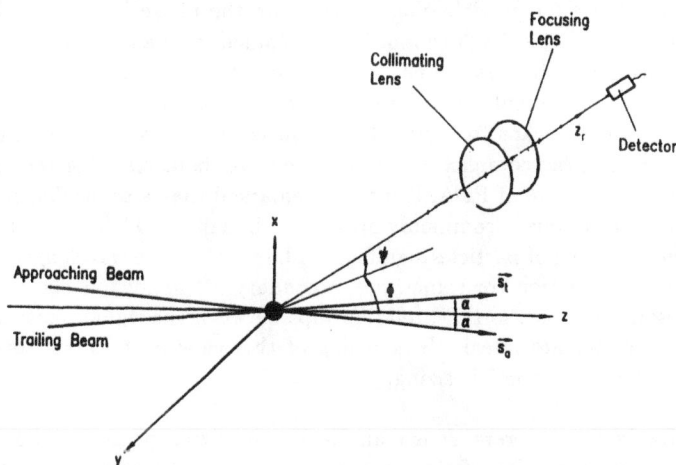

Fig. 1: Optical arrangement for PDA.

The scattered power collected by the detector is given as

$$P_s = \frac{I_o}{k^2} \left\{ \overline{G} + 2\Re \left[\overline{H} \exp\left(-i\omega_D t\right) \right] \right\},\tag{1}$$

where $\overline{G}$, and $\overline{H}$ are the values of signal pedestal and fluctuation respectively, which have been integrated over the receiving aperture. The symbols I_o, k, ω_D, and t denote the incident light intensity, the wavenumber of light, the circular frequency of heterodyne Doppler shift, and time respectively. The parameter $\overline{H} = \overline{H}_r + i\overline{H}_i$ is a complex quantity, whose amplitude and phase represent the amplitude and the phase angle of signal oscillations respectively. In order to extract the signal properties, Eq. (1) may be written in the following form:

$$P_s = \overline{P}_s \left[1 + \mathcal{V} \cos\left(\omega_D t + \Delta\right) \right];\tag{2}$$

where mean scattered power

$$\overline{P}_s = \frac{I_o \overline{G}}{k^2},\tag{3}$$

signal visibility,

$$\mathcal{V} = \frac{2\sqrt{\overline{H}_r^2 + \overline{H}_i^2}}{\overline{G}},\tag{4}$$

and the signal phase,

$$\Delta = \tan^{-1}\left(\frac{-\overline{H}_i}{\overline{H}_r}\right).\tag{5}$$

The scattered light signals, expressed by Eq. (2), have a phase shift Δ with respect to the incident light intensity pattern traced by the center of the particle during its passage through the fringe volume. This phase shift — which depends upon the particle size, material and the receiving geometry — constitutes the basis for particle diagnostics using the phase/Doppler technique. Obviously, the phase shift Δ cannot be ascertained from a single measured signal. At least two detectors are needed to record the difference of phase shifts, observed at two different receiving locations. The task of the PDA designer is to find ways and means to invert the measured phase data to determine the particle size and the complex refractive index. Without claiming the originality of this perspective of PDA, it may be remarked that a somewhat narrower view of this technique is commonly presented. Usually, PDA is introduced as a technique for sizing of particles significantly larger than the wavelength, having reflection or refraction (or sometimes, secondary refraction) as the dominant scattering mechanism, so that, relationships between the signal phase and the particle diameter are linear. Broadening of this view enables one to employ PDA for submicron particle sizing.

For Rayleigh scatterers, $\overline{H}$ is real. Hence, the signal phase is either 0 or π, depending upon the sign of this parameter. An arrangement is considered in Fig. 2, where the signal phase changes from zero to 180° within the scattered field. As shown in Fig. 2(a), for a beam intersection angle of 90° and electric vectors lying parallel to the plane of the beams, large portions of the scattered field above and below the yz-plane have a negative value of $\overline{H}$. The signals from these regions of the scattered field are out-of-phase with respect to the fringes.

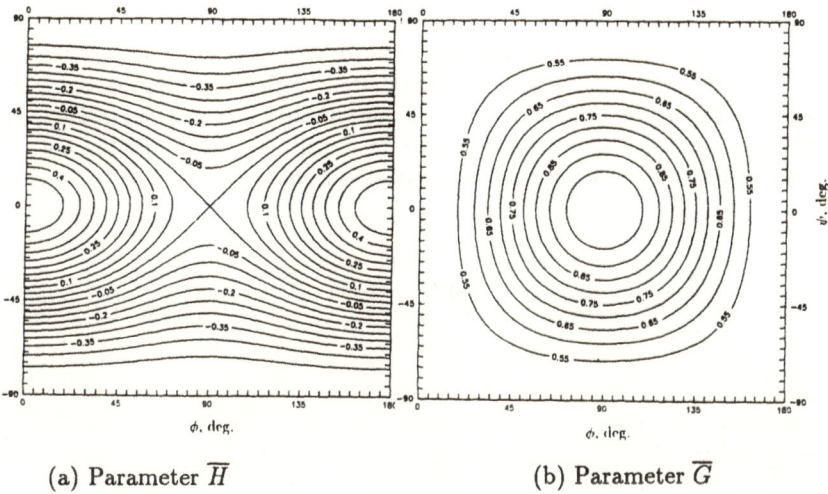

(a) Parameter $\overline{H}$ (b) Parameter $\overline{G}$

Fig. 2: Scattered light properties in the Rayleigh limit.

Comparing Fig. 2(a) with Fig. 2(b), it is obvious that strong and highly visible signals are collected along the x-axis (ϕ: arbitrary and $\psi = \pm 90°$) and z-axis ($\phi = 0$ or 180° and $\psi = 0$). The signals along the x-axis are in-phase, whereas, those along z-axis are out-of-phase. This phase relationship is valid only in the limit of Rayleigh scattering. It is found to disappear with the increasing size of the particle, providing a criterion for recognition of the Rayleigh scatterers. As shown later, the line of demarcation between the Rayleigh scatterers and the larger particles may be quite distinct for certain particle materials. Hence, a phase jump is encountered in the phase-diameter relation at a certain particle diameter. For other materials, the departure from the limiting behavior may be rather gradual with an increase in the particle size. These characteristics of the phase/Doppler signals may be utilized for particle material recognition in the submicron range.

With the arrangement considered in Fig. 2, the most visible in-phase and out-of-phase signals are collected along z-axis and x-axis respectively. From a consideration of Rayleigh scattering, it can be shown that the signals collected along x-axis and z-axis are proportional in strength to $\sin^2 \alpha$ and $\cos^2 \alpha$ respectively. In practice, it is desirable to focus both beams by a single lens. The largest half-angle between the two beams, achievable with a single lens, is usually limited to 15°. Hence, out-of-phase signal is about 7% in strength as compared to the in-phase signal.

With the approach of Rayleigh scattering, the phase-diameter relationship follows a power law, which has been given by Raszillier[9]. It is shown that the phase varies with the fifth power of the particle diameter for dielectric particles and third power of diameter for perfectly conducting particles. The limiting behavior of phase for both very large and very small particles is shown in Fig. 3, q denotes the normalized particle diameter $kd_p/2$. This figure corresponds to a PDA system designed for sizing of large particles. Hence, the phases for the small particles are too small to be measurable. Nonetheless, it clearly shows that the phase-diameter relations with decreasing particle size do become monotonic. Hence, there is a possibility of using phase/Doppler technique for very small particles. The next task is to work out the optical arrangements which would yield large values of signal phase for small particles.

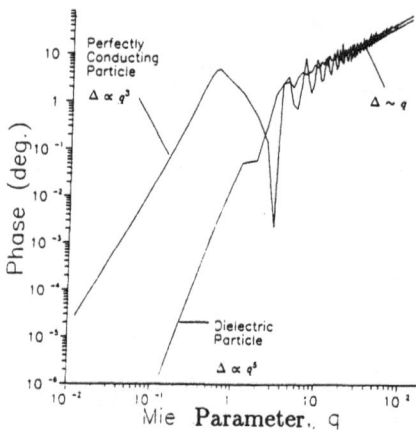

Fig. 3: The limiting behavior of phase in PDA.

3 SUBMICRON PARTICLE SIZING AND MATERIAL RECOGNITION

As shown above, the nature of scattering from a particle approaching the wavelength of laser is substantially different from that of large particles. However, smaller particles also yield phase signals which contain information about their size as well as material. Some aspects of the application of phase/Doppler systems to small particles are considered here through a few examples.

Iron particles are considered as an example of small absorbing particles. The real and imaginary parts of the refractive index of these particles are 1.51 and 1.63; see Born and Wolf[10]. The response of the iron particles is compared with that of small glass beads, which are taken as dielectric particles with refractive index 1.51. A comparison between these two cases demonstrates how the material of a small particle may be recognized in a mixture of particles with different optical properties.

The signal properties of metallic particles and glass beads are computed for $\phi = 0$ and the receiving-cone angles of 20°. Electric vectors of the two beams lie in the plane of beam intersection. Several elevation angles ψ have been used, which are mentioned on the diagrams in Fig. 4. The wavelength of the light and the angle of intersection between the beams are taken as 514.5 nm and 30° respectively. The phase shifts plotted in Fig. 4 are relative to an in-phase signal, which may be collected in the forward scattering direction ($\phi = \psi = 0$). As shown in Fig. 4, smooth phase-diameter relations are obtained over certain particle size ranges. The smallest measurable diameter of an iron particle is estimated to be about 40 nm.

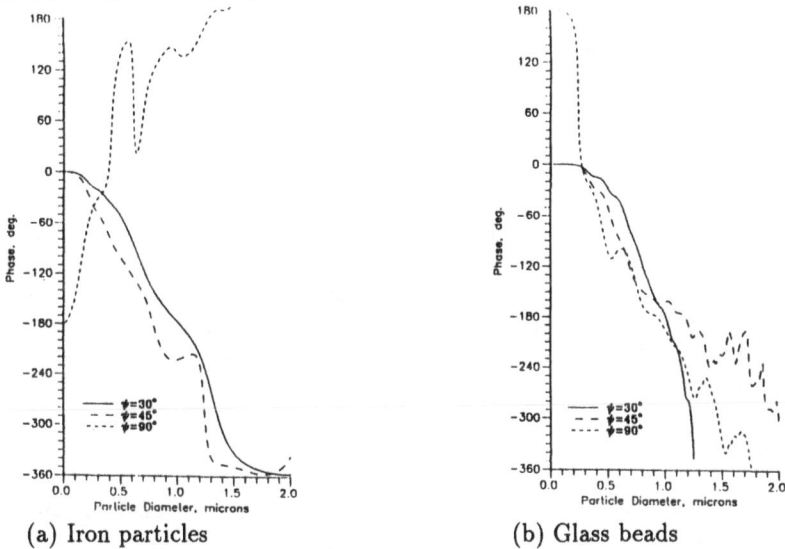

(a) Iron particles   (b) Glass beads

Fig. 4: Phase-diameter relations for submicron particles.

Considering a mixture of glass beads and the metallic particles, it may be seen that the two types of particles may be easily distinguished from each other due to very different response curves. For particle diameters smaller than 0.25 μm, the glass beads have an out-of-phase signal at $\psi = 90°$, whereas the metallic particles exhibit a smaller absolute value of the phase shift. For the diameter range 0.5–1 μm, the phase shift of the signals at $\psi = 90°$ relative to the reference signal have opposite signs for the two types of particles, provided that the phase signals are sampled in a range $(-180°, 180°)$. Hence, the sign and the magnitude of the signal phases may be used for recognizing the particle material and obtaining a measure of its size.

The existence of a monotonic PDA response for small particles has been demonstrated above. However, the measurable size range appears to be limited to 1–2 μm if only the monotonic portions of the phase-diameter curves are used. An increase in the size range is desirable for many applications in the field of aerosol mechanics. This goal may be achieved by utilizing several simultaneous phase measurements, taken at different elevation angles, to overcome the ambiguities associated with the nonmonotonic behavior. Such a procedure is common in the solution of inverse problems. In the field of particle sizing, the work of Card & Jones[11] could be quoted as an example, where simultaneous measurements at different locations are used to overcome the ambiguities related to the nonmonotonic response of the visibility technique. Whereas Ref. 11 deals with a sample of particles, instead of individual particles, a relatively simple method, applicable to multiple phase measurements on single particles is presented by Naqwi et al.[12] Such methods will help extend the measurable size range as well as enhance the ability to recognize the particle material.

It is clear that any processing method used to overcome the ambiguities due to nonmonotonic response, would become ineffective if the fluctuations in the response curve are too large and too frequent. Hence, smooth response curves, even if nonmonotonic, are desirable. Fortunately, the phase-diameter relations for small particles, as shown in Fig. 4, are fairly smooth. As an example, the response curve of iron particles at $\psi = 90°$ may be considered, where a measured phase of 60° corresponds to three discrete values of diameter. Out of this set of three values, the correct diameter will be easily identified when this measurement will be compared with one more phase measurement at a different elevation angle. In the situation considered in Fig. 4, two phase measurements are likely to be sufficient to resolve the particle material and to yield the correct value of the particle diameter.

A peculiar feature of small particle sizing is the large variation in scattered power with the particle size. An optical system designed for the size range 0.1–5 μm, would typically span five decades in scattered power. Such a large variation in signal strength requires one to use a dual-detector arrangement in which the scattered light is split up into two parts and focused onto two detectors with different sensitivities. In this way, a large range of signal strengths may be covered.

3 EXPERIMENTAL VERIFICATION

The principle of phase/Doppler system for sizing of submicron particles was demonstrated experimentally, using the optical layout similar to that shown in Fig. 4. An Argon-Ion laser with wavelength 0.5145 μm and output power 3 W was used for the measurements. Two quarter wave-plates were employed to rotate the polarization of the light arbitrarily. Two parallel laser beams of almost equal intensities were produced with the help of a beam splitter and a mirror. A dielectric non-polarizing beam splitter was used so that the splitting ratio was insensitive to the polarization of the incoming beam. The two parallel beams were about 50 mm apart. They were focused by a large numerical aperture lens with a focal length of 100 mm.

The measuring volume produced by intersection of the focused beams was placed inside a glass tube containing a flow of distilled water seeded with submicron latex spheres. A square section glass tube was chosen as the flow channel, in order to minimize distortion of the laser beams. The seeded water was de-aerated by heating, prior to measurements. In order to avoid deposition or agglomeration of the latex particles, an appropriate amount of sodium lauryl sulphate was added as a surfactant.

Scattered light was collected by two detectors in the plane of the beams. The signal strength was always satisfactory in the forward direction. However, measurable signals were collected by the elevated receiver, only after tilting the glass tube within the plane of the beams, so that the angle of incidence of the scattered light relative to the tube surface, became small.

Avalanche photodiodes were used for collecting the signals, which were transferred to a transient recorder. The digitized signals were processed by a microcomputer, using the cross-spectral density method, described by Domnick et al.[13]

The measured phase-difference signals collected by two detectors at elevation angles of −3.06° and 33.13° are shown in Fig. 5(a). These measurements correspond to a beam half-angle of 11.15° and a receiving-cone angle of 5.1°. The modes of the probability distribution of the measured signal phases from monosize particles are plotted in Fig. 5(a). The solid line represents the theory. There is a remarkably good agreement between the theory and the measurements. The vertical bars at the data points represent the uncertainty in determining the mode of the distribution.

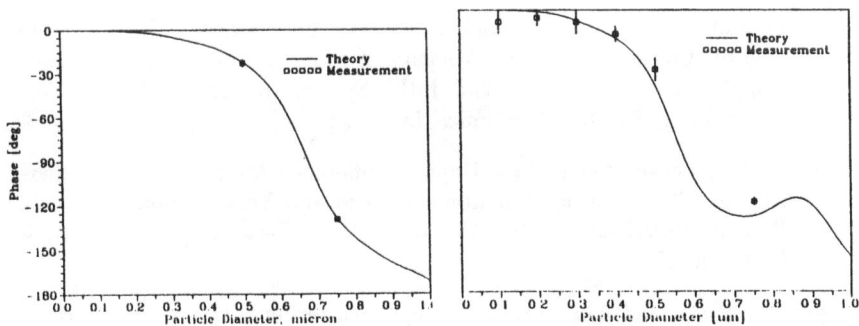

(a) $\psi = 33°$ & -3°   (b) $\psi = 15°$ & 0°

Fig. 5: Phase versus diameter for latex particles.

Another set of measurements, taken at elevation angles of 0 and 45°, is plotted in Fig. 5(b). The half-angle of the beams was 10.46° for these measurements. The receiving cone angle of the elevated receiver was reduced to 3.9°, so that, the phase-diameter relation becomes non-monotonic beyond a particle diameter of 0.75 μm. The measured modes of distribution for six different particle sizes are shown in Fig. 5(b). As before, the vertical bars indicate the uncertainties in determining the modes. Due to relatively weaker signals at the elevated receiver, the agreement between the theory and the experiment is not as good as that in Fig. 5(a). For particle diameters of 0.1 and 0.2 μm, the results are obviously biased due to the presence of undesirable particles, which could not be completely eliminated from water by distillation.

4 CONCLUSIONS

Properties of the scattered fields, relevant to phase/Doppler system design for small particles, are described. An examination of the phase/Doppler response to Rayleigh scatterers, has been helpful in designing the optical system for sampling of both Rayleigh and near-Rayleigh scatterers. The computed response of an optical system, designed for sizing of small particles, is verified experimentally, using submicron latex spheres.

REFERENCES

[1] F. Durst, and M. Zaré, "Laser Doppler Measurements in Two-Phase Flows," Proceedings of LDA-Symposium, Copenhagen, 403–429 (1975).

[2] W.D. Bachalo, and M.J. Houser, "Phase/Doppler Spray Analyzer for Simultaneous Measurements of Drop Size and Velocity Distributions," Optical Engineering, **23** (1984) 583–590.

[3] M. Saffman, P. Buchhave, and H. Tanger, "Simultaneous Measurements of Size, Concentration, and Velocity of Spherical Particles by a Laser Doppler Method," Proc. 2nd Intl. Symposium on Appl. of Laser Anemometry to Fluid Mechanics, Lisbon, (1984).

[4] K. Bauckhage, "The Phase-Doppler-Difference Method, a New Laser-Doppler Technique for Simultaneous Size and Velocity Measurements, Part 1: Description of the Method," Part. Part. Syst. Charact., **5** (1988) 16–22.

[5] N.S. Hong, and A.R. Jones, "A Light Scattering Technique for Particle Sizing Based on Laser Fringe Anemometry," J. Phys. D: Appl. Phys., **9** (1976) 1839–1848.

[6] J.D. Pendleton, "Mie and Refraction Theory Comparison for Particle Sizing with the Laser Velocimetry," Applied Optics, **21** (1982) 684–688.

[7] A. Naqwi, and F. Durst, "Light Scattering Applied to LDA and PDA Measurements. Part 1: Theory and Numerical Treatments; Part 2: Computational Results and their Discussion," submitted to Part. Part. Syst. Charact. (1991).

[8] C.F. Bohren, and D. Huffman, *Absorption and Scattering of Light by Small Particles* (Wiley, New York, 1983).

[9] H. Raszillier, "Das Elektromagnetische Streuproblem der Laser-Doppler-Anemometrie," Report LSTM 252/T, University of Erlangen (1989).

[10] M. Born, and E. Wolf, *Principles of Optics* (Pergamon, 1986), p. 621.

[11] J.B.A. Card, A.R. Jones, "Measurement of the Refractive Index of Atomized Liquid Drops by Light Scattering," Proc. 2nd Intl. Congress on Optical Particle Sizing, Arizona (1990) 316–324.

[12] A. Naqwi, F. Durst, and X.Z. Liu, "Particle Material Recognition using a Phase Doppler System," Intl. Conf. on Multiphase Flows, Tsukuba (1991).

[13] J. Domnick, H. Ertel, and C.Tropea, "Processing of Phase-Doppler Signals Using the Cross Spectral Density Function," Proc. 4th Int. Symp. on Application of Laser Anemometry to Fluid Mechanics, Lisbon (1988).

A New Device for Particle Characterization by Single Particle Light Extinction Measurements

H. Umhauer

INSTITUT FÜR MECHANISCHE VERFAHRENSTECHNIK, UND MECHANIK DER UNIVERSITÄT
KARLSRUHE (TH), 7500 KARLSRUHE, GERMANY

1 INTRODUCTION

Light extinction measured from individual particles has long since been the basis for single particle size analyses. The individual particles successively pass through a specifically delimited and homogeneously illuminated measuring volume and block part of the light. Hence, such devices are sometimes referred to as "light blockage instruments". The change in the measured luminous flux is proportional to the projected area of the particle concerned, for in the case of particles which are extremely large with respect to the wavelength of the light used, the extinction coefficient k is independent of the particle's diameter and refractive index. (This presentation deals with particles in the size range of around 1 mm).

In the case of non-spherical particles, however, their *shape influence* becomes a problem since the measured projected area of a non-spherical particle depends on its orientation within the measuring volume. Repeated measurements of the same particle will then yield a range of different projected areas according to the statistical orientation of the particle. An unequivocal measurement from just one single passage through the measuring volume is therefore restricted. The consequence is a loss of resolution. As a result of the shape influence the primarily measured number distributions of the equivalent diameter (defined as the diameter of a sphere possessing the same projected area) become broader. When converting to volume distributions, for example, greater or lesser deviations emerge. Particularly in the case of a single particle sizing technique this is unfortunate, since such techniques generally yield a high degree of resolution; a characteristic which is of value for many applications.

One possibility of counteracting this loss of resolution when analyzing non-spherical particles is to conduct the extinction measurements in a *number* of different directions instead of in just *one single* direction. The paper introduces a device which utilizes *three orthogonal* optical sub-systems to simultaneously derive three projected areas from each individual particle. From the three values, an *average* particle projected area and standard deviation is derived. The average projected area and standard deviation yield an enhanced unequivocality and more representative information than may be derived from unidirectional measuring systems. In the case of an orthogonal alignment of the three individual projected areas, the possible differences should, with the exception of special orientation instances, be generally most comprehensively accounted for.

The following describes the basic concept and operation of the device. Furthermore, the results of measurements are presented which demonstrate the improvement attained with respect to the unambiguity or resolution in the case of measurements concerning different types of non-spherical particles.

2   BASIC CONCEPT AND OPERATION OF THE DEVICE

To begin with, the basic concept of a single particle extinction size analysis is to be explained in more detail. A suitable configuration is demonstrated in Fig. 1. Other arrangements are feasible.

A condenser, diffuser, lenses and masks convert the source-light to a quasi parallel beam through which the particles to be measured must individually pass. The cross-sectional area of this beam (measuring area) is A_b. The light intensity I should remain the same across the whole beam. When no particles are present, the sensor measures the luminous flux ϕ_0, given by

$$\phi_0 = I \cdot A_b \tag{1}$$

When a particle passes through the beam, the signal trend depicted in Fig. 1 (bottom left) emerges. As long as the particle is completely within the measuring cross-section the luminous flux remains reduced to ϕ. For this, eq. (2) is valid:

$$\phi = I (A_b - k \cdot A_p) \tag{2}$$

A_p is the particle's projected area (in the case of a sphere this is identical to its cross-section). k is the extinction coefficient. According to definition, the product between both parameters is equal to the extinction cross-section: $A_{ext} = k \cdot A_p$. The measured values ϕ_0 and ϕ are converted by a computer to the relative measuring parameter U:

$$U = \frac{\phi_0 - \phi}{\phi_0} = k \cdot \frac{A_p}{A_b} \tag{3}$$

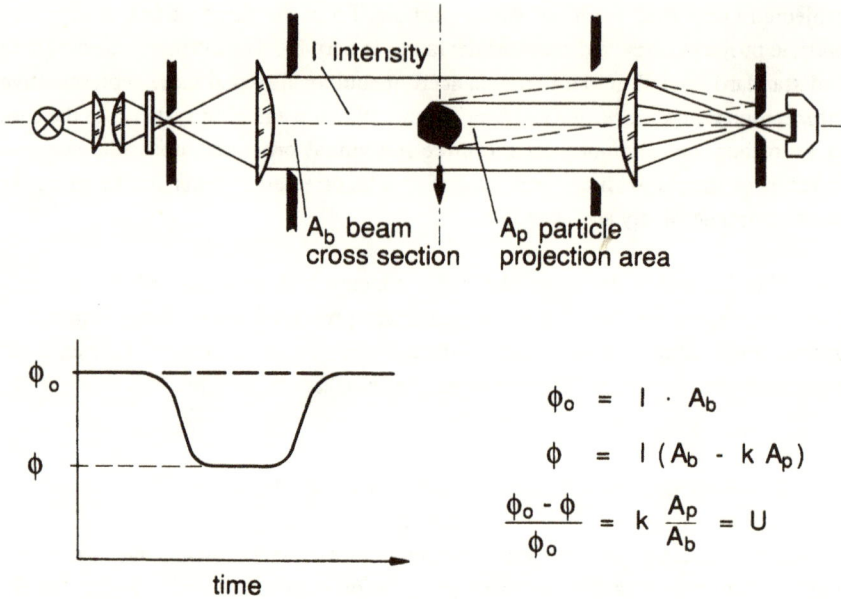

Fig. 1:   Principle of an optical system for the measurement of the extinction
            cross-section or projected area of individual particles.

            Time-dependent trend of the measured signals $\phi = \phi(t)$;
            correlation between the various parameters (see text).

U is therefore dimensionless. Its value is between 0 and 1, whereby A_p must
always be smaller than A_b.

For particles which are extremely large in comparison to the wavelength of the
light source (and only such particles are considered here), the extinction coefficient
k remains independent of the particle size and the refractive index n of the particle
substance. In this case, the theoretical value k approaches "2" [1]. In metrological
practice, however, k possesses modified values nearer to "1", since all sensors
operate with a finite aperture [2]. The reasons are well known and do not require
any further discussion. The aim is not to determine the extinction coefficient but the
size of particles. For this purpose it must only be ensured that k and hence k/A_b
remain constant. The measuring area A_b is to be observed as an apparatus-specific
constant which governs the measuring range. Therefore a strict proportionality
exists between U and A_p. As such, the primary dispersity parameter to be measured
is A_p, which in the case of non-spherical particles can be allocated to the diameter of
the sphere of identical projected area.

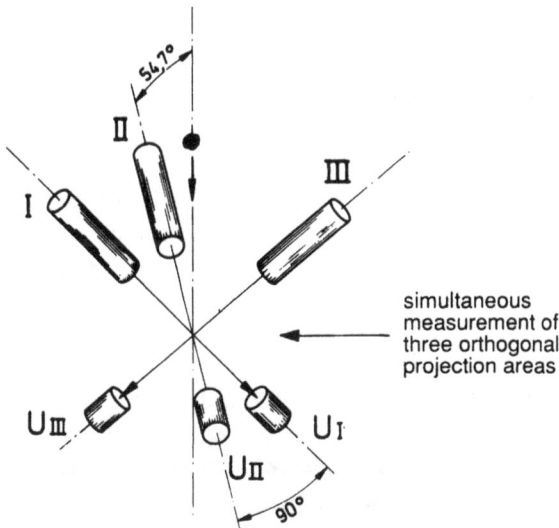

$$U^* = \frac{1}{3}\left(U_I + U_{II} + U_{III}\right) \qquad (4)$$

$$\delta^* = \sqrt{\sum_{i=I}^{i=III} \frac{1}{2}\left(U_i - U^*\right)^2} \qquad (5)$$

Fig. 2: Schematic of the alignment of the three optical sub-systems (I, II and III) for the determination of the average projected areas of the particles.
The axes of the optical sub-systems are mutually orthogonal. The angle between the vertical body-diagonals (tube axes) and the axis of each respective optical sub-system is hence 54,7°.
U^* is the arithmetic mean value of the individual values U_I, U_{II} and U_{III}, measured by the three sub-systems,

δ^* is the respective standard deviation.

From Fig. 1, one can easily recognize the difficulties which emerge in the case of non-spherical particles. The individually measured projected areas then depend on the particle's orientation within the measuring cross-section of the light beam. A single measurement cannot deliver an unequivocal specification of the particle size. The consequences have already been described in the introduction.

If one was to repeat the measurements for a certain particle any number of times for random orientations within the measuring cross-section, then one would not obtain a single reproducible value (such as in the case of a sphere), but a wider or narrower value spectrum. The *mean* value of this spectrum would then be an unequivocal measure of the particle size. Whilst such a process is principally feasible, it is nevertheless extremely impractical. For further aspects of the problematics see [3, 4, 5].

A somewhat less effective but realizable method is to determine at least *three* different mutually orthogonal projected areas of the particle, and to calculate the respective mean value. A suitable measuring equipment set-up is schematically shown in Fig. 2.

The equivalent optical systems I, II and III illustrated in Fig. 1 are positioned such, that their axes describe an orthogonal optical tripod. The particles pass along the vertical body diagonal through the optical intersection zone of the three systems which create the measuring volume. Hence, from each particle, three values (U_I, U_{II} and U_{III}) are measured *simultaneously*, which correspond to the three orthogonal projected areas of the particle. From these measurements, the mean value U^* and standard deviation δ^* can be evaluated in the manner described in Fig. 2 (eq. 4 and 5). In comparison to a single value which would be obtained from just one system, the mean value and standard deviation incorporate more information. The mean value U^* must principally lead to an enhanced definiteness and restore the resolution. The standard deviation δ^* gives an insight into the degree of deviation from the true spherical shape. The results presented in section 3 will demonstrate the extent to which the improvement in resolution justifies the additional apparatus expense and effort involved.

A simplified cross-sectional view of the apparatus is illustrated in Fig. 3. The three optical sub-systems and the particle feed are mounted to a spherical frame, fixed to a base-plate via a cylindrical shaft. The light sources of the optical sub-systems are respectively mounted in the head of the upper protrusions. The sensors, electrical transducers and filters are located in the arms below. The light sources are fed from a stabilized power supply.

In the simplest case, the particles are conveyed to the inlet of the vertical tube by a vibrating feed so that they may fall freely without any flow assistance. The maximal counting rate for the device presented is 50 particles / s. The measuring range extends from 0.5 mm to 6 mm. A modification of the measuring system for the size analysis of larger or smaller particles does not present any difficulty.

The result of an analysis is usually the number distribution density $q_o(d_A)$ and the cumulative number distribution $Q_o(d_A)$. d_A is the diameter of the sphere with a cross-section identical to the projected area of the particle (see eq. 7).

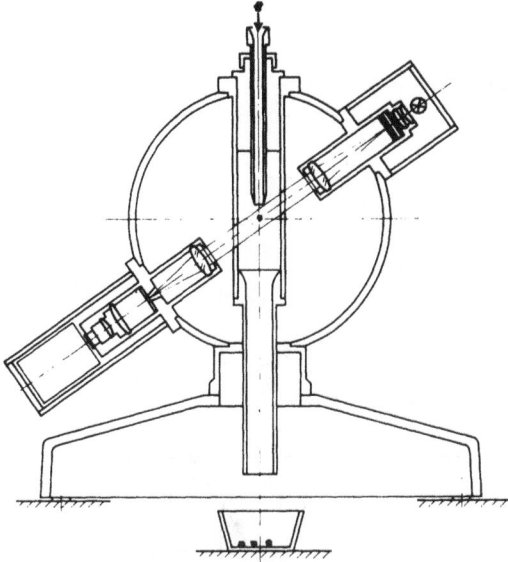

Fig. 3:   Meridional cross-section through the device in a plane
            cutting one of the optical sub-systems (simplified).
            The light source is at the top-right, the detector at the
            bottom left.

3   MEASUREMENTS AND RESULTS

In order to test the measuring system and verify the measuring principle,
measurements were initially conducted using steel balls and then with non-
spherical, but regularly-shaped particles. The measurements with the steel balls
allow an assessment of whether the three optical sub-systems operate to an
absolutely equivalent standard. From the results, one can also assess the highest
attainable measuring reproducibility, the accuracy and resolution of such a system
(i.e. in the absence of any shape influence).

The regularly-shaped particles (cube, cuboid, cylinder, cone etc.) were
individually made in the size range of between 1 mm and 6 mm. Regular bodies
allow the average projected area and volume to be calculated from their linear
dimensions. In the case of the average projected area, this can either be calculated
numerically by a computer or with the aid of the Cauchy theorem:

$$\bar{A}_p = \frac{1}{4} S \tag{6}$$

S is the geometrical surface area of the particle; a prerequisite for the validity
of eq. 6 is that no concave surface regions exist.

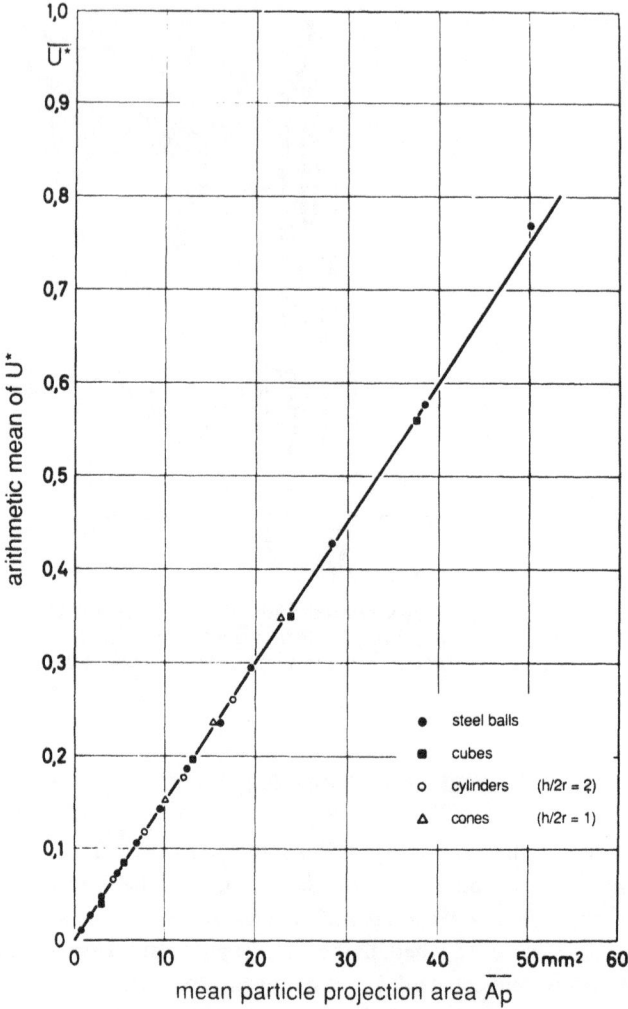

Fig. 4:   Dependence between the mean $\overline{U^*}$ of the relative parameter U* and
the mean particle projected area $\overline{A_p}$ for variously shaped particles.

For each of the above particles, (at least) 100 measurements were conducted
with random particle orientation within the measuring volume. From the results, the
frequency distribution of U* (or the respective d_A) was determined, together with
the mean and accompanying standard deviations. Upon plotting the mean values $\overline{U^*}$
as a function of the mean projected area $\overline{A_p}$, then in this size range, (when $d_A \gg \lambda$)
the values for all particles should comply with the requirements of eq. (3), i.e. all
values should sit exactly on a straight line which passes through the zero point. The
results in Fig. 4 verify that a universal and unequivocal correlation exists between
the mean value $\overline{U^*}$ of the relative parameter U* and the mean projected area $\overline{A_p}$ for

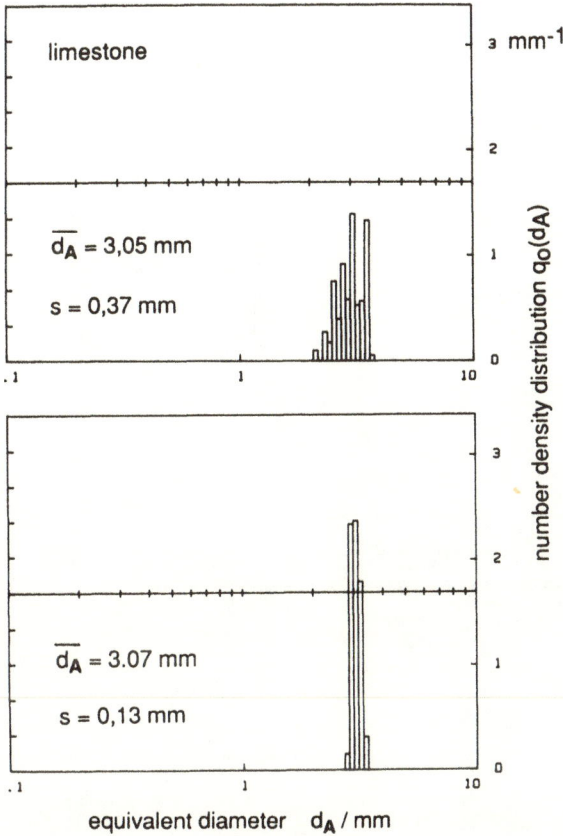

Fig. 5:   Measurements of a single limestone particle ($\overline{d_A}$ = 3,1 mm).
The figure illustrates the normalized number density distributions
of the equivalent diameter d_A.
Top: Result using the values U_I measured with a single optical system
Bottom: Result using the values U^* derived with the tri-optical system

all particles. One can assume that the values for completely irregularly shaped
particles comply accordingly. For the equivalent diameter d_A, the following holds:

$$d_{\overline{A}} = \sqrt{4\,\overline{A}_p / \pi} \quad = \quad \text{const} \cdot \sqrt{\overline{U}^*} \tag{7}$$

(Please note the difference between $d_{\overline{A}}$ and $\overline{d_A}$, see also Fig. 5). Eq. (7) is
accordingly also valid for single measurements. For a single measurement of U^*
(which is usual for common particle size analyses), the correlation depicted in Fig. 4
remains fulfilled, although A_p becomes scattered. The significance of the reduction of
this scattering by the use of U^* instead of U_I, for example, is to be discussed with
the aid of the results demonstrated in Figs. 5 and 6.

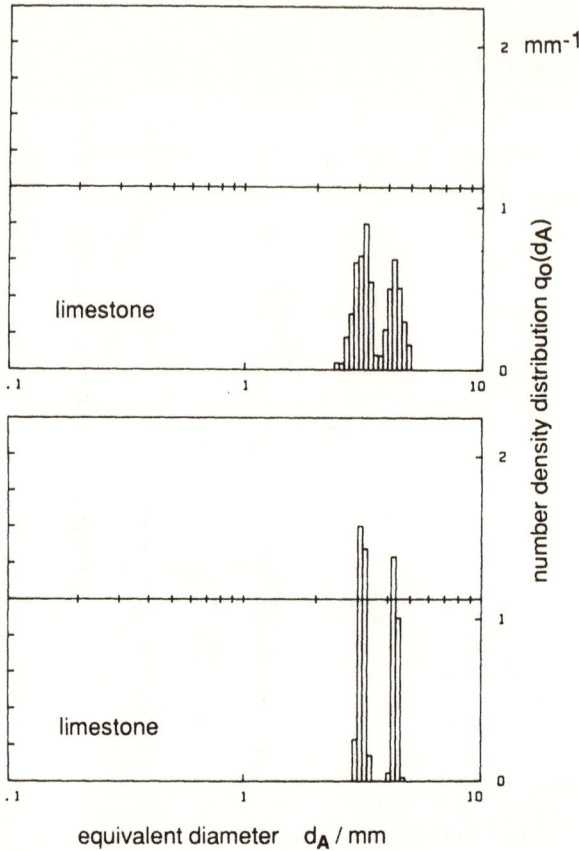

Fig. 6:   Results of the measurements from a collective comprising two
          narrow and closely adjoining limestone particle fractions.

          Top: Result of the measurements using just one optical system (U_I).
          Bottom: Result of the simultaneous measurements using the
          tri-optical system (U^*)

A series of 100 measurements was also conducted using a limestone particle
with random orientation within the measuring volume (as with the cube, cylinder
and cone etc.). The upper diagram in Fig. 5 demonstrates the frequency distribution
of the equivalent diameter using the values measured from just one single optical
system (e.g. U_I), the lower diagram depicts the frequency distribution when the
equivalent diameter is derived from the mean U^* value. One can clearly see that the
lower distribution is significantly narrower. The standard deviations $\delta*$ of both
distributions differ by nearly a factor of 3. This difference can be observed as a
measure of the resolution improvement. Such comparisons have also been
conducted in the same manner for the various regularly-shaped particles. Similar
differences emerged to various degrees, depending on the particle shape.

A more impressive resolution gain is demonstrated by the following example: Two different narrow limestone fractions of equal quantity were prepared with closely adjoining size distributions. An analysis was conducted for both fractions as a total collective in the usual manner i.e. each particle passed through the measuring volume just once. The resulting frequency distributions $q_0(d_A)$ were again compared with d_A derived from U_I and U^* respectively (see Fig. 6). The lower diagram demonstrates how the different fractions more clearly emerge when measured using the tri-optical system. This stems from the fact that the reduction in the shape influence allows a more concise registration of each individual fraction. With this enhanced resolution, the two fractions could be even closer together, and still be "separately" registered.

4 FINAL REMARKS

The presentation of further results and a more comprehensive discussion together with subsequent aspects cannot unfortunately be conducted here due to the restrictions imposed on the publication length. The few examples, however, should demonstrate that for a size analysis, the shape influence of non-spherical particles is indeed effectively reduced by the presented technique and apparatus which yields a significantly enhanced resolution. The measures involved are not too laborious and are justified by the improvements achieved.

The device allows a rapid, contact-free analysis especially for small sample quantities without necessitating any preparation. The key feature is the enhanced resolution concerning non-spherical particles. Special applications arise above all where relatively narrow particle fractions are at hand, and where a correspondingly high resolution is consequently desired. A whole range of examples for this may not only be extracted from chemical engineering products, but also from the food-processing sectors e.g. cereal grains.

REFERENCES

1. M. Kerker, "The Scattering of Light",
    Academic Press, New York, San Francisco, London, 1969

2. G. Tonna, Aerosol Science, 1974, 5, 579.

3. M. Bottlinger and H. Umhauer,
    Part. Part. Syst. Charact., 1989, 6, 100.

4. H. Umhauer and M. Bottlinger,

5. M. Bottlinger, PhD Thesis, Universität Karlsruhe (TH), 1989

6. M.A. Cauchy, CR Hebd. Seances Acad. Sci., Paris, 1841, 13.

The Particle Size Analysis of Flocs Using the Light Obscuration Principle

R. J. Akers, A. G. Rushton, I. Sinclair and
J. I. T. Stenhouse

CHEMICAL ENGINEERING DEPARTMENT, LOUGHBOROUGH UNIVERSITY OF TECHNOLOGY,
LOUGHBOROUGH, LEICS. LE11 3TU, UK

1   INTRODUCTION

In the context of this paper flocs are aggregates of particles in suspension. These aggregates, which may be induced by the application of specific treatments or occur naturally, are commonly used to enhance the performance of solid liquid separation operations such as filtration and sedimentation when the size of the primary particles is so small that the operation would not be practical on them in an untreated state. This is likely to occur when the primary particle size is <10μm. In many systems of processing interest the primary particles may have a size in the sub-micron range, eg pigments.

For the study and control of processes involving flocs and for research into flocculation phenomena there is often a need to know the size and size distribution of the flocs present.

Such a size analysis is difficult because:

(a) the floc size may be very large with respect to the primary particle size, eg 2mm flocs of 0.6 μm primary particles so that wide dynamic ranges are needed,

(b) the flocs may be very irregular in shape,

(c) flocs have a low (20%) to very low (<2%) volume fraction of solids in them,

(d) their structure and porosity distribution is usually very irregular,

(e) they are generally extremely fragile and liable to disruption in sampling and measurement systems,

(f) the concentration of flocs may range from an extremely dilute to a concentrated suspension depending on the context , and

(g) the process of flocculation, ie aggregate growth may occur whilst the sample is waiting for analysis and during the analysis itself.

Because of these constraints many particle size methods are not applicable to flocs. For example the uncertainty and variability of floc density within the outer envelope volume, the possible irregularity of shape and their fragile nature means, that all types of sedimentation method are inapplicable.

The Coulter principle has been applied to flocs but is severely constrained by the very severe shear stress that occurs within the vortex of fluid passing through a Coulter orifice. Indeed the Coulter Counter has been used as an on-line method for studying the disruption of flocs[1]. If a floc were entirely disrupted just before or in the orifice it is reasonable to assume that the additive volume of the fragments would be detected as one coincident group of particles. This would however assume that the fragments were of the same particle volume concentration as the parent, an unsafe assumption for flocs. However breakup in an orifice is not normally complete and the observed size may also depend on the amount of distortion the floc experiences in the orifice. From the behaviour of other types of porous particle in the Coulter Counter it is often assumed that the instrument measures the external envelope volume if the flocs are not disrupted[2]. Attempts have been made to measure individual well characterised flocs under realistic orifice conditions but the experiments are very difficult to do and the results remain ambiguous[3].

An attempt was made to apply the Malvern light scattering method to the present problem and some work was done this way. However the primary particles being used were smaller than the lower end of the particular light scattering instrument used and as has been reported[4] the method gives anomalous results under these conditions. Additionally the present work was particularly concerned with the size distribution of the relatively very small number of large flocs which would contribute very little to the scattering pattern.

This paper describes a method of floc size determination that was developed during a study of the disruption of flocs in turbulent flows[5].

2   THE LIGHT OBSCURATION PRINCIPLE

The possibility of using a HIAC™ light obscuration sensor was examined as it was thought that having a relatively smooth fluid passage this particular device might not prove so disruptive to flocs passing through it as is the Coulter orifice. Figure 1 is a schematic diagram of the HIAC sensor. Initially experiments were carried out using a 6 channel HIAC counter, model 4100 RT, fitted with a type HR-120 HC sensor having a maximum sensing size of 120 μm. To prevent blockage this size is smaller than 150 μm, the smallest dimension of the flow passage through the sensor head. The light path has an area corresponding to the square of the maximum flow channel width, defined by square stops in the light path. The HIAC sensor gives a pulse output where the pulse height is proportional to the projected area of the shadow on the photo-detector of the object being sized. The lower sensing size of the HIAC sensor is determined by the signal-to-noise ratio of the whole system of light source, sensor and electronics and for small particles by the assumption that the shadow is an accurate representation of the projected particle area, ie that forward scattering does not occur. It was claimed that this particular sensor could be used to reliably detect and size particles down to 2 μm diameter. At this diameter the particle obscures 1.4×10^{-4} of the projected flow passage area, ie that fraction of the incident light is obscured, and as the signal to noise ratio at this size based on the calibration data supplied is less than 1 this claim should be regarded with caution.

The basic problem of this type of device is the need to distinguish small fluctuations in a steady maximum light level, where the ultimate noise level will be due to the detecting device, the measuring circuits and any noise present in the illuminating light. Furthermore as the instrument operates by measuring fluctuations from the maximum bright light level means must be provided to allow it to compensate for changes due to aging of the bulb and other causes of long term drift. In the standard instrument the output consists of +ve going voltage pulses of up to 20V on a baseline of -10.0V Before processing the pulses are shifted so that

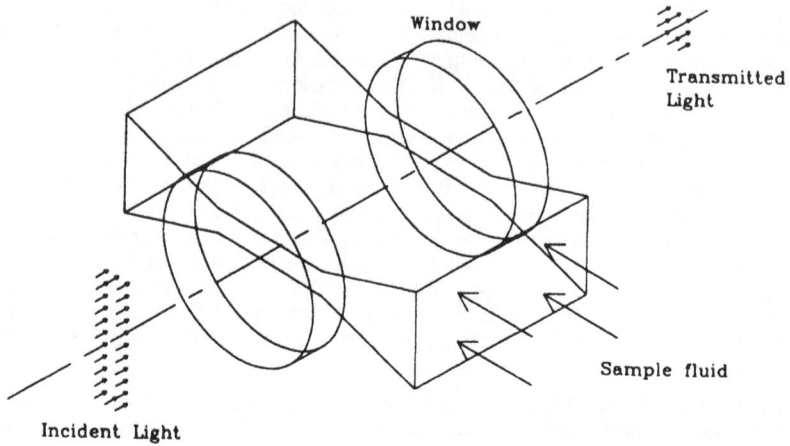

Figure 1 Schematic of the HIAC Light Obscuration Sensor.

the response to individual particles is a positive pulse of size V_h, (0 - 10V) on a baseline of 0.000V. This is achieved using a special type of lamp power supply that is controlled by feedback from the detector.

For studying particle size changes during floc break-up the 6 channels available on the standard instrument were considered to give far too coarse a size distribution. It was decided to attempt to use the available HIAC sensor assembly with a pulse height analyser (PHA) and a square root signal conditioning amplifier to give a linear pulse height/particle diameter response so that more points would be available on the distribution curve. Using a 1024 channel analyser means that any analysis would be restricted to an equivalent diameter ratio of $\sqrt{1024} = 32$, assuming that the system was linear, had no zero offset and was not noise limited.

The instrument is calibrated on the assumption that

$$V_h = A d^k \ + \ N \tag{1}$$

where A,k and N are calibration constants, N corresponding to the experimentally determined noise level and k and A to the gradient and intercept of a log V_h vs. log d calibration plot, determined using standard latices. Each sensor is supplied with a calibration curve of this type. In the curve supplied the calibration is extrapolated from 7 μm down and the signal/noise ratio cited above is determined from the constants for this particular calibration. It was found that the sensor over-estimates particle counts by about 10% when not sufficiently isolated from vibration generated by the vacuum pump used to draw the sample through it. This was attributed to vibration of the incandescent lamp filament.

3 THE MODIFIED HIAC SENSOR/PULSE HEIGHT ANALYSER SYSTEM

The Lamp Control System

In the standard HIAC instrument the average light level, which for normal sampling is very close to the background light level, is used in a feedback circuit to adjust the intensity of the illuminating lamp and maintain the stable baseline signal of $V_h = 0$.

This feedback mechanism compensates for non-transient changes in light transmission such as those caused by long-term fouling of the sensor channel wall, absorbance by the fluid medium due to colour and long term drift. The time taken for a particle to transit the cell is of the order of 50 μs and the time constants of the feedback system are much longer than this.

The lamp control circuit shown in Figure 2 was designed to simulate those necessary parts of the HIAC six channel analyser system.

There is a need to transform the raw signal from the detector/preamplifier system from pulses of 0 - 20 V on a baseline of -10 V to pulses from 0 - 10 V on a baseline of 0 V.

The input signal at i/p is first halved and added to the output of IC1 which is set to 10 V by RV1. A copy of this signal, isolated by IC2 running at unity gain, is fed to the square rooting amplifier from o/p2. The signal is then passed through IC3, which serves to isolate the voltage divider and lamp control parts of the circuit, to an integrator based on IC4. This strips the approximately 50 μs pulses caused by 'genuine' particles from the signal by allowing them to pass through the filter, C6 and R7. IC4 amplifies and inverts the long term drift to provide a controlling signal for the lamp. This is subtracted from a preset voltage through an audio amplifier, IC5, used to drive the lamp. This amplifier is driven lop-sidedly so that the carefully aged (and expensive) HIAC lamp is protected from over-voltage. By using a preset voltage of ≈ 4.0 V, which is the time averaged voltage of the lamp supply, the feedback loop through the lamp/photo-diode sensor is only required to compensate for drift and not provide the entire lamp voltage through an open gain op. amp. The original HIAC circuit did not have this feature. It is essential in making equipment of this type to choose components having the lowest possible noise levels and to design the layout and enclosures to eliminate sources of internal and external noise.

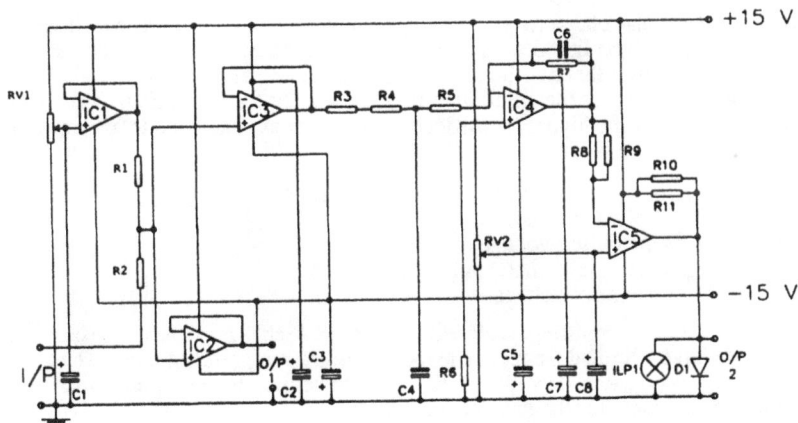

Figure 2 Lamp Control Circuit

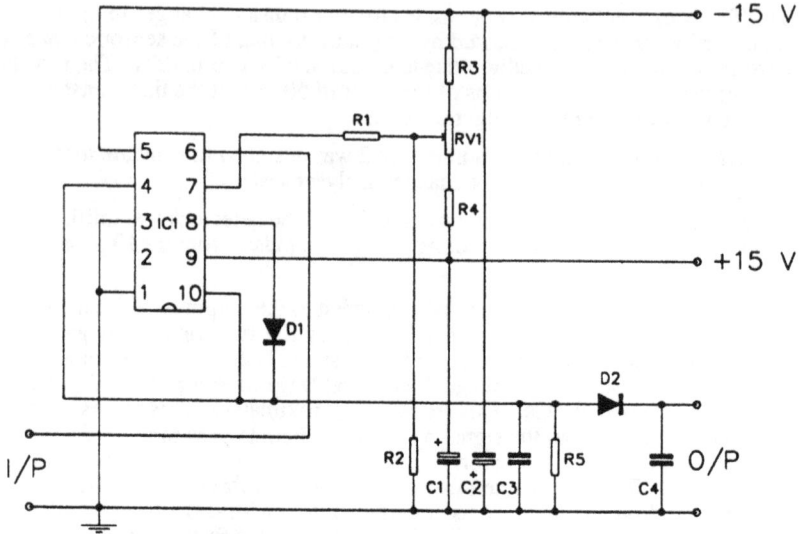

Figure 3 Square Root Conditioning Amplifier

Conditioning Amplifier and Pulse Height Analyser

To expand the dynamic range it was decided to investigate the use of a square rooting conditioning amplifier between the sensor assembly and the pulse height analyser. A detailed study was made of suitable operational amplifiers available at the time and it was decided to use a hardware programmable precision function operational amplifier, type AD534 (Analog Devices). This is a precision amplifier where the response characteristics can be selected by external circuitry. Because of the data compression that occurs in the square rooting operation it is necessary to choose an amplifier that has the requisite precision and also has the speed of response necessary for the amplification of pulses. In the configuration shown in the figure and for the highest specification amplifier available the total error in the transfer function is given by:

$$V_o = (10(V_h - V_c))^{1/2} \quad = \quad 0.25\% \quad F.S.D \tag{2}$$

where V_o is the output voltage
                 V_h is the input (*HIAC*) pulse voltage and
                 V_c is a control voltage.

Many operational amplifiers designed to generate highly non-linear transfer functions have a limited frequency response and/or long settling time. The amplifier chosen had a slew rate of 20 V/μs and settling time of 2 μs, which are more than adequate for the present application.

As the transfer function is only valid for real outputs it is essential that the input voltage to the amplifier always be positive. This condition is maintained by applying a control voltage, V_c, which is set to be slightly more than half the peak-to-peak noise voltage of the input, V_h, when no particles are in the sensor. The source of noise at this stage is mainly in the photo-diode and its pre-amplifier.

The input to the amplifier was obtained from the buffered output of the lamp control circuit, which was in the range 0-10 V. The measured input noise level was ± 5 mV. The output voltage was also in the range 0-10 V and the output noise in the absence of pulses had an amplitude of 0.6 V. This condition corresponds to the amplifier having maximum gain.

Figure 3 shows the circuit designed for the square root amplifier.

Pulse height analysers normally operate by detecting a change in the input voltage level which starts the measurement process. From then on the input voltage is applied to a low leakage capacitor. After a preset interval of time the input pulse is assumed to have passed its maximum and the charge on the capacitor, which is proportional to the maximum voltage reached, is allowed to discharge through a special circuit which maintains a linear fall in voltage with time, ie the time taken to discharge is proportional to the peak height. During this fall the system scans over its memory channels so that when the capacitor is deemed by a comparator circuit to have reached zero volts an increment is added to the appropriate channel. During the measurement phase the instrument does not respond to further pulses. These analysers are normally designed for counting nuclear and other rapid events and are intended for far shorter pulses (« 20 µs) than would correspond to those from a HIAC type detector which are typically 50 µs in duration.

The design of pulse height analyser used, made by Nuclear Medical Equipment Ltd., is an example of another type of instrument that can measure pulses at frequencies from DC up. In this design the pulse is not considered to have ended until the voltage has returned to a selected threshold which may be set as near to zero as noise levels permit hence this type of circuit is independent of the pulse rise time. The peak height measuring is similar to that described above. One characteristic of this design is that a multiple peak due to coincident events will appear as one pulse corresponding to the highest level reached. A dead time could be set which enabled the total sampling time to be held constant irrespective of the cumulative run down of all the pulses detected. The analyser used was designed to fit into two expansion slots of an Apple][™ microcomputer and software was provided to control it and to enable the data retrieval to be programmed.

4   CALIBRATION OF MODIFIED HIAC DETECTOR

The detector was calibrated using Coulter standard PVDB latices. Figure 4 shows the plot obtained when latices of 5.78, 9.7, 12.9, 18.2, 18.6, 37.7 and 91.4 µm were used. This line was fitted to an expression of the type

$$ch \ = \ m.D \ + \ c \tag{3}$$

where ch is the channel number
          D is the diameter of a spherical particle of
                   equivalent projected area, and
          m,c are calibration constants.

Linear regression of this line gives the calibration factors:

Channel Number = $10.7278 \times$ Diameter $- 70.17097$

with a correlation coefficient of 0.9991.

The intercept on the line is due to the offset that was entered into the setting of the sensor output to prevent the locking which would occur if noise presented a momentary negative voltage to the amplifier. This intercept corresponds to a lower sizing limit of 6.5 µm.

Particle Size Analysis

The size distribution given on the Coulter calibration certificate was compared with the distribution available from the modified HIAC system for a 37.7 μm diameter latex. Figure 5 shows the comparison plot.

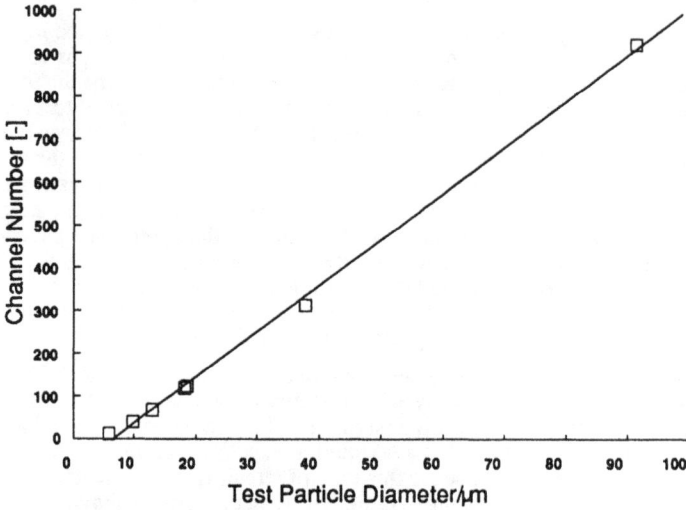

Figure 4 Calibration Curve with Standard Latices

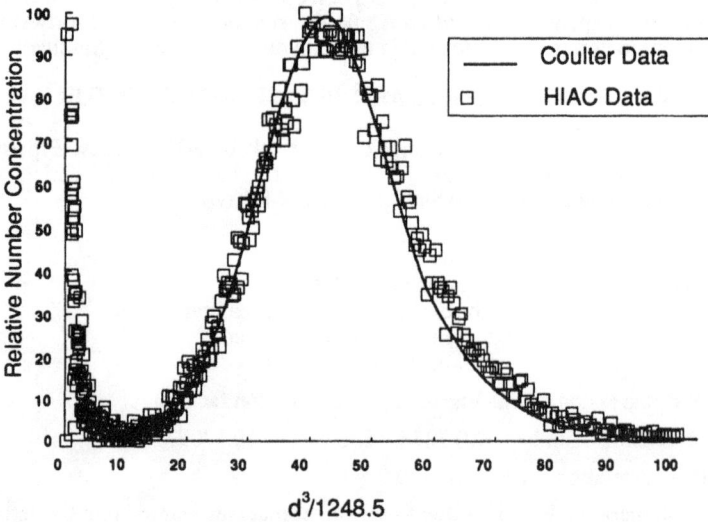

Figure 5 Comparison of Modified HIAC and Coulter Calibration Latex Size Distribution

The Coulter Calibration is presented as a Multisizer® chart record with the modal diameter given and the 28.2 and 46.0 μm points marked on the arbitrary 0-100 x-axis scale. The plot is cut off at $x=5$ corresponding to $d=18.4$ μm. Examination of the data shows this to be a scale in d^3 with an x-offset of 3, so that:

$$d^3 = [x + 3].1248.5 \ (\mu m^3) \tag{4}$$

The Coulter data was read from the chart, corrected for the offset of 3 and plotted. A spline curve is shown fitted to these points in the figure. The HIAC data was transformed to the same scale for comparison and is shown in the figure as individual data points. Agreement of the distributions around the maximum are excellent. The HIAC data shows a significant number of particles in the <18μm region where data is not given for the Coulter. This is above the noise level of the system and is believed to be due to smaller contaminant particles in the calibration standard as used after extensive dilution.

5   FLOC DEGRADATION IN THE HIAC SENSOR

A sample of flocculated suspension was circulated through a system consisting of a peristaltic pump, a beaker fitted with a low speed (15 rpm) stirrer and the sensor head. The mean fluid detention time was $\approx$ 5 min at a flow rate through the sensor of 40 ml/min. The size distribution was determined 1, 6, 21, 51 and 151 minutes after starting the circulation. It was necessary to pump for about 1 minute before any reading was made to allow for the dead volume between the beaker and sensor.

Figures 6a - 6e are histograms of the relative particle volume per channel obtained for each of the sampling times. Towards the right hand side of the plots, eg Figure 6b, at 6 minutes, the heights of the individual columns of the histogram are seen to follow a family of envelope lines sloping upward to the right. Each of the envelope lines corresponds to a small integral number of particles in that channel. This is a consequence of the use of so many counting channels and the small number of particles per channel at large sizes. Examination of the lines will show that they fit a cubic function passing through the origin.

At 1 minute, when there is very little opportunity for particle degradation the distribution by volume is seen to comprise mainly large particles with some smaller debris. At 6 minutes, when about 3/4 of the sample will have passed the detector once, there is a mixture of particles in the original size band and many more fines than before. This trend continues until at 151 minutes (30 passes) the sample has almost completely degraded.

From the main body of the work[5] it is known that flocs are rapidly disrupted, ie in milliseconds, if they are exposed to a region of high enough energy dissipation and that all susceptible flocs passing through that region will be disrupted. Hence it may be concluded by comparison of Figures 6a and 6b that whilst a significant number of flocs are destroyed in the first pass they are still detected and sized before separation of the fragments occurs. It is reasonable to conclude from the evidence of these plots that there are regions within the HIAC sensor where there are rates of energy dissipation sufficient to degrade flocs but that not all particles passing through the sensor encounter one of these regions.

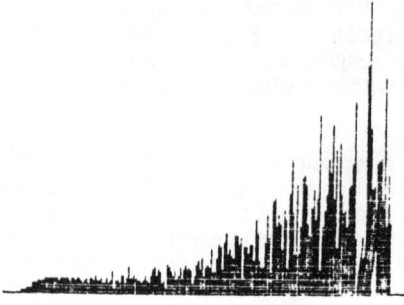

Figure 6a 1 minute

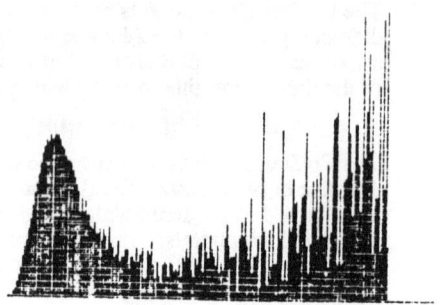

Figure 6b 6 minutes

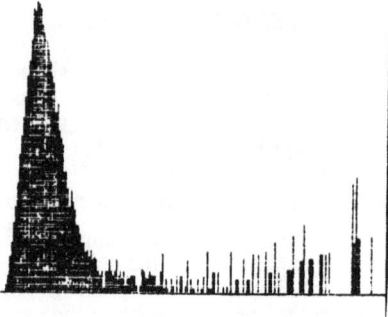

Figure 6c 21 minutes

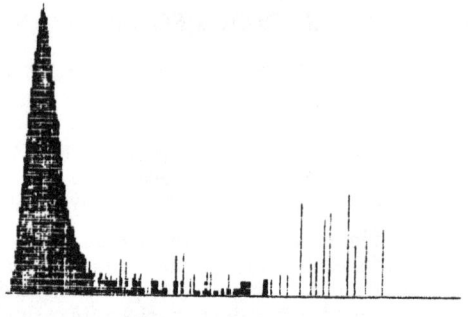

Figure 6d 51 minutes

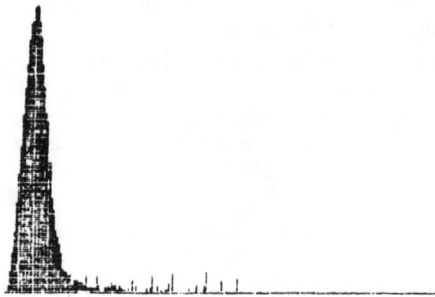

Figure 6e 151 minutes

<u>Figure 6</u>   Floc Degradation with Passage through HIAC Sensor

6 CONCLUSIONS

(1) The HIAC sensor is shown to exhibit a linear projected area/pulse height response over a dynamic range of approx 15:1 in diameter, ie between the range 6.5 to 94 μm. It is likely that this linear range would extend to the upper size limit of the detector used (120 μm) but calibration latices of that size were not available to test.

(2) It is possible to use a high precision operational function amplifier to produce pulses corresponding to the square root in amplitude of the input signal at the operational speed of the HIAC detector.

(3) At the lower end the measuring range is limited by electrical noise in both the measuring/detection system and the light source. It is necessary to apply an offset to the input of the square root amplifier to prevent it from being locked by noise taking it below 0V. This is the region where a rooting amplifier has maximum gain (and hence error and noise) and will have the effect of precluding the method from being used to count and size primary particles in many systems.

(4) Using the correct choice of pulse height analyser, conditioning amplifier and lamp control circuit, it is possible to convert a HIAC sensor into a 1024 channel instrument.

(5) When tested using Coulter calibration latices the modified HIAC sensor gives a very similar response to a size distribution measured by the Coulter Multisizer when the data is correctly transformed.

(6) Whilst the system does cause some degradation of very weak floc structures, this degradation is not so much as to make the method unusable.

(7) Comparison of the size of the flocs as shown by optical microscopy[5] with those measured by the sensor suggests that the sensor system is measuring the external floc envelope size, irrespective of the internal porosity, and that when used as described is a practical floc sizing technique.

ACKNOWLEDGEMENT This work was carried out as part of a project sponsored by the Specially Promoted Programme in Particle Technology of the Science and Engineering Research Council

REFERENCES

1. A. A. Hanna, J. M. Cohen and G. G. Roebuck, Measurement of floc strength by particle sizing, Jour.AWWA.,1967, 59, 843-858

2. D. Horák, J. Peska, F. Svec and J. Stamberg, The influence Effect of the porosity of discrete particles upon their apparent dimensions as measured by the Coulter principle, Powder Technology, 1982, 31(2), 263-268.

3. R. J. Akers, unpublished data

4. G. Butters and A. L. Wheatley., In 'Particle Size Analysis 1981', Heydon, London, 1981, pp 164-175

5. A. G. Rushton, PhD Thesis, Loughborough University of Technology, 1990

New Techniques in Sub-micron Particle Size Analysis: The Controlled Reference Method

Paul Cloake

LEEDS AND NORTHRUP EUROPE, WHARFDALE ROAD, TYSLEY, BIRMINGHAM B11 2DJ, UK

INTRODUCTION

Under normal circumstances, Particles suspended in a fluid will be subject to random collisions between the particles and the thermally excited molecules of the fluid. When the particles approach a size of up to three microns in diameter these collisions result in observable random motion, referred to as Brownian Motion. While the direction and velocity of this motion is random, the velocity distribution of a large number of monosized particles averaged over a period will approach a known functional form. Particle velocity can vary from between 5 microns per second, for 3 micron particles, to 6000 microns per second for particles of the order of .003 microns.

Two major factors affect the nature of Brownian Motion, those being Temperature and Viscosity. Obviously if the molecules within the suspending fluid have a higher thermal energy due to higher temperature they will impart more energy during the collisions and subsequently increase the velocity distributions within the suspension. It can be said that the velocity median will be directly proportional to the temperature of the fluid. Conversely the more viscous the fluid the greater the resistance to movement of particles within that fluid. Viscosity is usually the greater effect. With the combination of both these factors, particle velocities can be affected by up to 2% per degree C.

Any measurement system for particle size distribution utilizing Brownian Motion as it's basis must include compensation for both temperature and viscosity effects.

Velocity distribution of Particles suspended due to Brownian Motion can be represented by fig. 1.

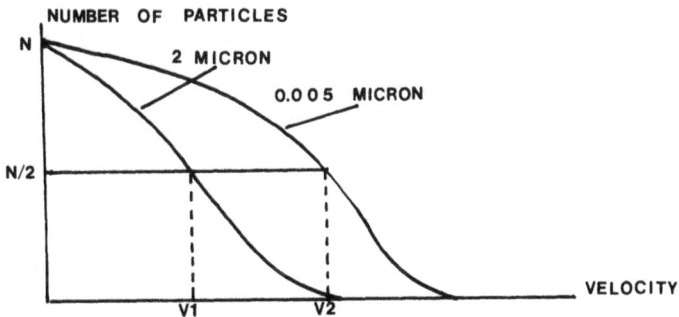

NUMBER OF PARTICLES

2 MICRON

0.005 MICRON

VELOCITY

Figure 1   Velocity distribution

DOPPLER EFFECT.

The use of light scattering as a means of determining particle size information is commonplace. As light is incident on a particle it will scatter that light in all directions regardless of the optical properties of the particle. If the particle is stationary the frequency of the incident light will be the same as the scattered light. If however the particle is moving the scattered light will be shifted in frequency due to the Doppler Effect. The frequency shift will be proportional to the velocity of the particle. As particle velocity in Brownian Motion is related to the size of the particle these shifted frequencies will yield information on the size distribution within this suspension.

HOMODYNE DETECTION.

In order to determine meaningful data on particle size distribution within a suspension, it is necessary to examine the Doppler shifted light with respect to an un-shifted sample or reference beam.

As the particle velocities are so small compared to the velocity of light, the Doppler frequency shifts can only be detected using frequency beating techniques.

To achieve this a specialized wave guide was fabric-ated . The wave guide consists of Y shaped fiber optic arrangement terminating on a glass plate. Light from a diode laser source is transmitted down one leg of wave guide and transmitted through the glass fluid interface. The laser has a frequency of 780 nm. At the glass fluid interface a component of the laser beam is reflected back down the opposite leg of the waveguide to a solid state photodetector. This is referred to as the reference beam. This will consist of a large amplitude component FT. As this is a direct reflection of the laser output this component will be relatively large and constant.

As the transmitted laser beam passes through the glass fluid interface, it is incident on the particles in

the fluid. The scattered light is shifted in frequency
in all directions. The 180° scattered light is transmitted to the photodetector. There will be a large number
of these shifted components, as the particles of different
sizes will be moving at different velocities and directions.
These signals will be smaller in amplitude than the
reflective beam. Refer to fig. 2.

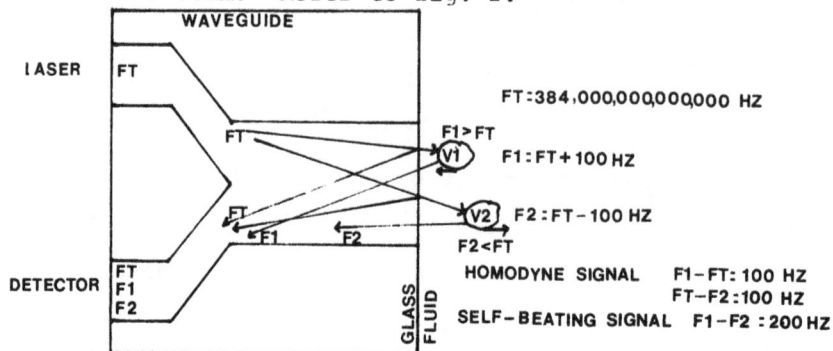

Figure 2   Homodyne detection

Self beating can be described as the result of comparing
the signals from pairs of particles within the sample
against each other. The homodyned signals are represented
by F1 and F2. The self beating frequencies can be considered as interference.

     Because the controlled reference method utilizes a
constant reflected beam it has two major advantages over
other methods using self beating.
(a) In order to detect the scattered light the incident
beam must first penetrate the cell and the scattered light
must get out. This will require very low concentrations
of material in order to avoid multiple scattering. Because
the measurement takes place at or close to the glass fluid
interface, the controlled reference method is much less
susceptable to high concentrations of material.
(b) In conventional methods, because particles of the
order of 3 microns or below scatter light inefficently, a
photo-multiplier tube is usually required to amplify the
signals. This is not the case in the controlled reference
method and as such the signal to noise ratio is much higher
resulting in better repeatibility.

FREQUENCY SPECTRUM.

     In a real situation during sample measurement, large
numbers of particles will be moving at random on or close
to the glass/fluid interface, generating large numbers of
frequency shifted signals due to the laser light that is
incident upon them. These signals when combined with the
reference signal FT produce a wide spectrum of Homodyne
Frequencies. This is represented at the output of the
photodetector as a random signal. This signal is sampled
by an analog to digital converter and is digitized. The
digitized signal is treated by a digital signal processor.

The digital signal processor, using fast fourier transform techniques generates a power spectrum of the original signal. The power spectrum is what is used to determine the particle size distribution of the material being analyzed.

For monodisperse particles the power spectrum will take the form of a Lorentzian Function. Fig. 3.

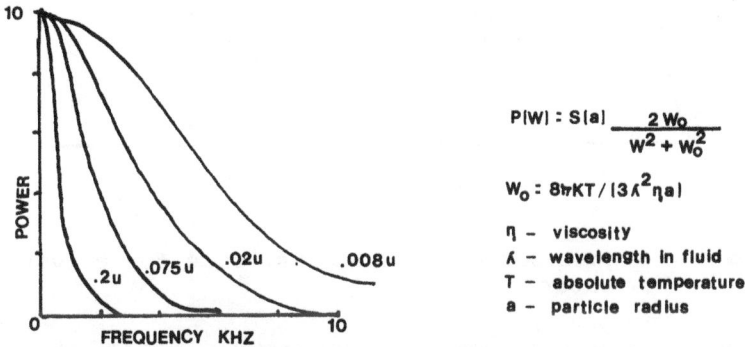

$$P(W) = S(a) \frac{2 W_0}{W^2 + W_0^2}$$

$$W_0 = 8\pi KT / (3\Lambda^2 \eta a)$$

η – viscosity
Λ – wavelength in fluid
T – absolute temperature
a – particle radius

Figure 3 Power spectrum

As size decreases the power spectrum plot will shift to higher frequencies. A normal sample will consist of distribution of various particle sizes which will be reflected by a power spectrum that is a combination of Lorentzian functions weighted by the volume concentration of each particle size. Another factor that will affect this spectrum will be the scattering efficiency of the particle. This will be discussed later. The software routine of the analyzer will deconvolve this information to produce a volume distribution of the sample.

SCATTERING EFFICIENCY

Further parameters must be taken into account when trying to analyze materials of the size range of .005 to 3 microns. Most particles in this range are transparent and give rise to optical interference effects. As light impinges on a transparent particle a component of the light is scattered normally and a component enters the particle and travels within the particle at a different velocity dependent on the refractive index of the material. Some of this light will hit the back wall and 'bounce' backwards and exit at the front wall and travel towards the probe on the same path as the normally scattered light.

The light frequency will be the same but the phases may be different. The mis-matching of phases can give rise to either constructive or destructive interference. Signal amplitude will be affected by the nature of the interference. The size of the particle and the refractive index will determine what is referred to as the scattering efficiency of the material.

This is the paramter S in the Lorentzian function.

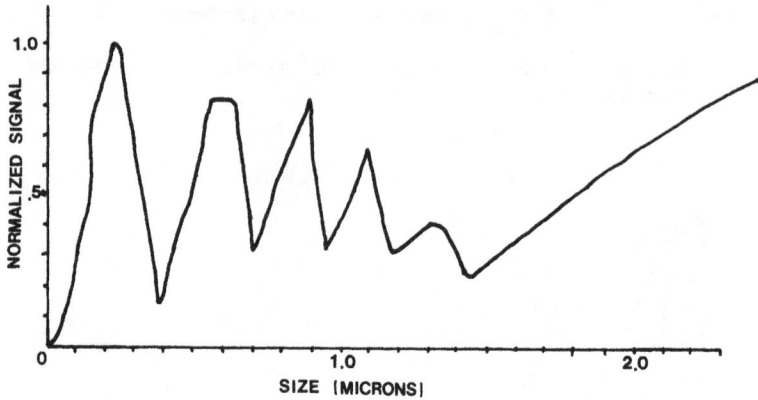

Figure 4   Signal per unit volume *vs.* size for polystyrene
in water

This plot is essentially the scattering efficiency of the
material for a combination of fluid and particle refractive
indices.   Other fluid and particle combinations will yield
similar plots but with the peaks at different positions.   On
expanding the graph in the region of 0 to 0.10 microns i.e. the
Rayleigh range it is seen that the scattering efficiency changes
rapidly for changes in particle size.

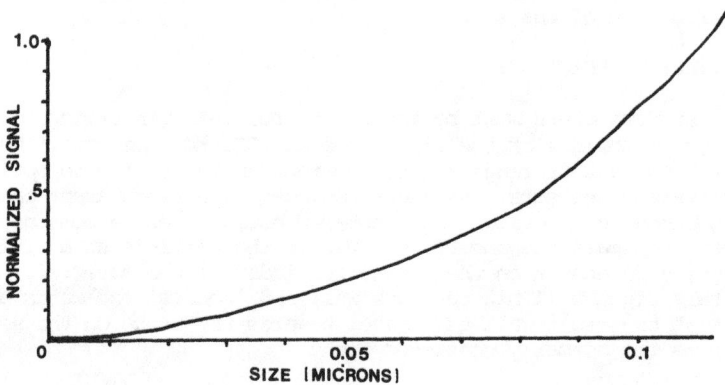

Figure 5   Scattering Efficiency - Rayleigh Range

For monodisperse materials the interference effects are minimal and the power spectrum of the material is well defined. If however there is a range of materials of different sizes within the sample these effects will become significant. Generally the more efficient scatterers will bias any analysis in their favour. For true volume distribution the calculation will require compensation for scattering efficiency as an intrinsic part of that calculation. Information on the optical properties of the material being analyzed is required by the analyzer to correctly define the scattering efficiency of the material.

In conclusion to accurately determine volume weighted particle size distribution in materials of the order of .005 to 3 microns, an instrument must compensate for :

  a) Temperature.
  b) Viscosity of suspending fluid.
  c) Refractive index of material and suspending fluid.
  d) Scattering efficiency of the material.

The software routine presents the distribution using a process of iterative deconvolution without an 'a priori' assumption of the distribution. This allows multi-model materials to be analysed.

References :

1. P. J. Freud, M. N. Trainer
    Leeds and Northrup,
    North Wales, Pennsylvania.
    Research Department.

2. Albert Tennay,
    Leeds and Northrup,
    North Wales, Pennsylvania.

Particle Characterization for Health and Safety Auditing

A. P. Rood

THE OCCUPATIONAL MEDICINE & HYGIENE LABORATORIES, 403 EDGWARE ROAD, LONDON NW2 6LN, UK

INTRODUCTION

The new Control of Substances Hazardous to Health (COSHH) regulations require employers to make an assessment of the risks to health which arise from exposure to hazardous substances during work activities.

As part of this assessment employers must establish what measures are necessary to prevent or adequately control exposure to hazardous substances, and what further precautions need to be taken to protect peoples' health. These involve the use and maintenance of control measures, information and training for employees and, in some circumstances, routine exposure monitoring and health surveillance.

Exposure to substances hazardous to health can result in the intake into the body by inhalation, ingestion or absorption through the skin or by a combination of these. However, inhalation is the main route of entry for most substances into the body, and I shall review the physical characteristics of dusts in this context.

Dust, in the popular and generally accepted sense, means minute solid particles of matter that may form clouds in the air, having been released as a result of various kinds of activity or disturbance. When solid material is broken, a wide range of product is formed and energy acquired in the process by the dust. By virtue of this kinetic energy, spread of the dust in the air occurs until sedimentation by gravity precipitates the material. Transport of fine powders similarly liberates dust into the air, as will the release of aerosol from plant and process.

It is the inhalation of such material which needs to be assessed under COSHH. The maximum concentration of an airborne substance to which an employee may be exposed by inhalation under any circumstance is specified by regulation, as is the method of measurement. These maximum exposure limits are published in EH40 by the Health & Safety Executive. A list of occupational exposure standards are also published according to which there is no evidence of injurious effect if not exceeded.

The occupational exposure standard should be the target for a given substance in the workplace, and it will be prudent for an employer to reduce exposure below this value. An assessment of the material and process involved requires not only the characterisation and measurement of the substance in the air, but the potential for such releases.

The general approach necessary to control occupation exposure to dusts is outlined below. Total inhalable dust is the fraction of airborne material which enters the nose and mouth during breathing and is therefore available for deposition in the respiratory tract. Respirable dust is intended to simulate the fraction which penetrates to the gas exchange region of the lung. Where dusts contain components which have their own assigned occupational exposure limits, all the relevant limits should be complied with.

Sampling for Total Inhalable and Respirable dusts

The personal sampler for total inhalable dust has a multiple entry (see Fig. 1) and a pump unit capable of maintaining a smooth flowrate of 2.0 $\pm$ 0.1 l/min throughout the sampling period. High volume samplers up to 50l/min can be used to obtain short term exposure data under certain conditions.

The personal sampler for respirable dust involves a cyclone preselector (see Fig. 2) operating at 1.9 $\pm$ 0.1 l/min, and a parallel plate elutriator serves as a high volume area sampler (Fig. 3).

Personal samplers are usually clipped on the workers lapel, and high volume area samplers are ideally suspended at head height away from obstructions and potential fresh air inlets. For accurate determinations, filters should be conditioned by leaving them in the balance room overnight before each weighing. Membrane filters should be passed over a static eliminator to dissipate excess charge.

The volume of air passed through the filter is calculated by multiplying the flowrate by the sampling time (usually a full shift), and the weight gain of the filter is divided by the is volume to give the exposure.

Examples of Materials Assessed in this way

	Total Inhalable	Respirable
Asiprin	5 mg/m^3	-
Barium Sulphate	-	2 mg/m^3
Beryllium	1.5	-
Calcium Carbonate	10	5
Carbon Black	3.5	-
Cristobalite	0.15	0.05
Limestone	10	5
Mica	10	1
Quartz	0.3	0.1
Titanium Dioxide	10	5

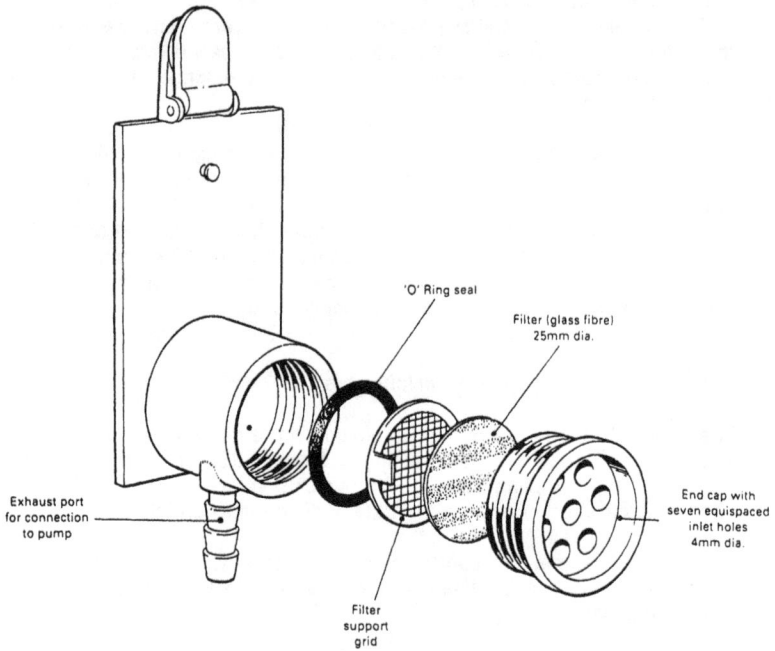

'O' Ring seal

Filter (glass fibre)
25mm dia.

Exhaust port
for connection
to pump

End cap with
seven equispaced
inlet holes
4mm dia.

Filter
support
grid

**Figure 1   Modified UKAEA personal sampling head used for
             gravimetric dust sampling**

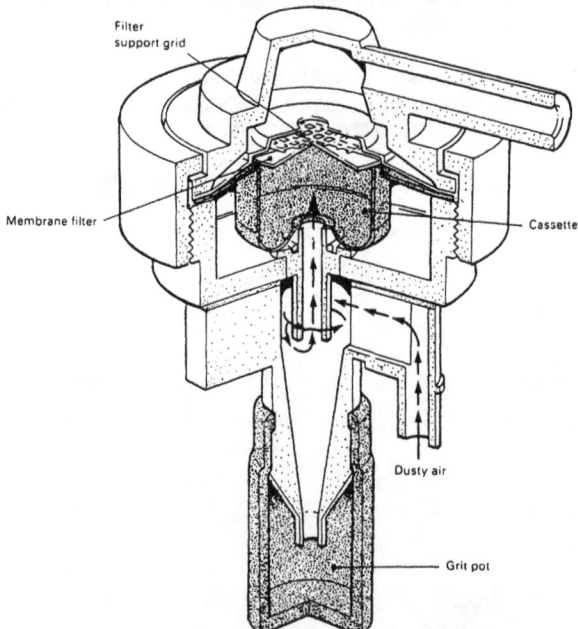

Filter
support grid

Membrane filter

Cassette

Dusty air

Grit pot

Figure 2   Sampler with cyclone elutriator

Examples of Material Assessed by their own Specific Methods

The occupational exposure standard for cotton dust is 0.5 mg/m^3 total dust less fly, using area samplers, although personal sampling may be adopted in the near future.

Asbestos control limits are determined by fibre counting with the phase contrast optical microscope, and are:

(a) For asbestos consisting or containing any crocidolite or amosite :

(i) 0.5 f/ml over 4 hours
(ii) 1.5 f/ml over 10 mins.

In addition, exposure levels apply over the longer terms:

(a) Where exposure to crocidolite and amosite occur, 48 fibre - hours/ml in 12 weeks, or
(b) 120 fibre - hours/ml for other forms of asbestos

Man-made mineral fibres also have a special method, not only 5mg/m^3 gravimetric, but 2 f/ml counted in the microscope. In this case a fibre means a particle with length >5µm, average diameter <3µm, and the ratio length to diameter >3 to 1. Fibres shall be counted with a phase contrast microscope of such a quality and maintained in such a condition at all times during use that Block 5 on the HSE/NPL Test.

Slide Mark II would be visible when used in accordance with the manufacturer's instructions. The microscope magnification shall be between 400 x and 600 x during counting, the difference in refractive index between the fibres and the medium in which they are immersed shall be between 0.05 and 0.3. The results shall be regularly tested by quality assurance procedures to ensure that the results are in satisfactory agreement with a national quality assurance scheme.

From these examples it can be seen that HSE takes seriously the risk of lung disease and intends to reduce exposure using the COSHH regulations as the main thrust of its policy, time will tell how effective this proves.

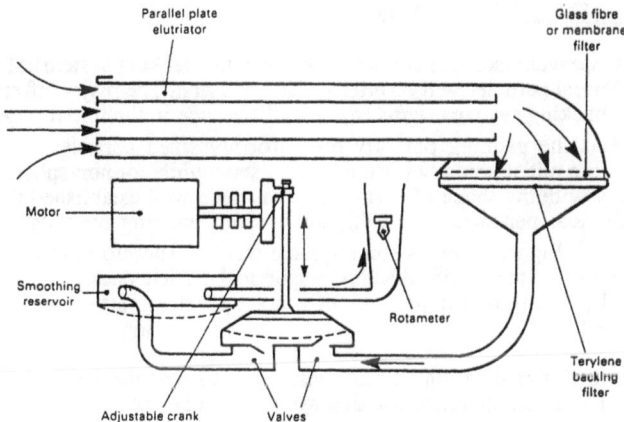

Figure 3   Sampler with parallel plate elutriator

Determination of the Packing Size of a Non-spherical Particle: An Emperical Approach

A. B. Yu[1] and N. Standish[2]

[1] CSIRO DIVISION OF MINERAL AND PROCESS ENGINEERING, P.O. BOX 312, CLAYTON, VIC. 3168, AUSTRALIA
[2] DEPARTMENT OF MATERIALS ENGINEERING, THE UNIVERSITY OF WOLLONGONG, P.O. BOX 1144, WOLLONGONG, NSW 2500, AUSTRALIA

1 INTRODUCTION

Recent development in the study of the packing of solid particles has shown that the dependence of the porosity of an assemblage of spherical particles on the particle size distribution involved can be generally described by a packing model[1-5]. On the other hand, experimental results have indicated that the packing concepts of spheres can be extended to the packing of non-spherical particles[6-10]. Therefore, it is likely that a packing model for spherical particles may be extended to non-spherical particles. However, in order to predict the porosity of non-spherical particle mixtures in a more precise way, it is felt that the packing size of a non-spherical particle should be introduced and quantified first.

The purpose of this paper is to propose an empirical approach to the characterization of the packing size of a non-spherical particle with our present knowledge of the packing of solid particles.

2 THEORETICAL CONSIDERATION

Packing of Non-Spherical Particles

For convenience, the similarity between the packing systems of spherical and non-spherical particles will be briefly discussed first. The major difference between the two packing systems stems from the difference in the initial porosity which is defined as the packing porosity of uniformly sized particles[6,7]. For spherical particles the initial porosity is usually a constant while for non-spherical particles it may vary with the shape of particles. It has been well established that monosized non-spherical particles give a higher random packing porosity than spherical particles[8,9]. Figure 1 shows some typical results. Therefore, in the discussion of the similarity of the spherical and non-spherical particle packing systems, the effect of initial porosity on the porosity of a non-spherical particle mixture should be considered.

With this realization, it has been reported that the similarity between the spherical and non-spherical packing systems can be more conveniently discussed by means of the term specific volume variation which is defined as the difference between the volumetrically weighted mean of the initial specific volumes and specific volume of a particle mixture[12]. Thus, the specific volume variation, ΔV, of

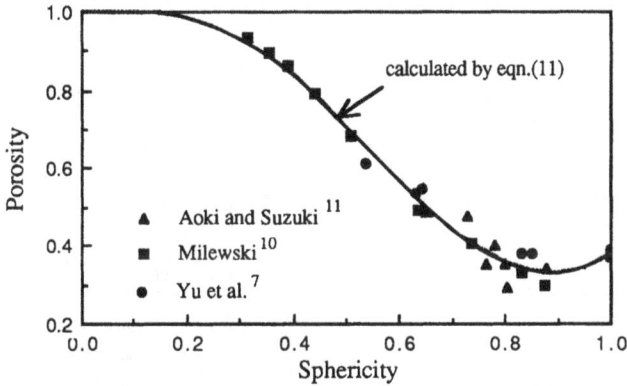

Figure 1   Variation of initial porosity with sphericity

a n-component mixture of particles is given by

$$\Delta V = \sum_{i=1}^{n} . V_i X_i - V \tag{1}$$

where V is the specific volume of the mixture, and V_i and X_i are, respectively, the initial specific volume and fractional solid volume of ith (i=1,2,...,n) component in the mixture. The initial specific volume can be by definition calculated from the initial porosity of a component.

As shown in Figure 2, the results in Figure 2(a) can be transformed to those in Figure 2(b) according to eqn.(1). Noting the constancy of $\sum . V_i X_i$ for spherical particles, it is obvious that the results in Figure 2(b) are more comparable with the well-established results of binary mixtures of spherical particles as shown in Figure 3. This compatibility may be extendible to the packing of multi-component mixtures[7,12]. Therefore, the packing behaviour of non-spherical particles is similar to that of spherical particles. As a consequence, the packing concepts of spherical particles can be used in the study of the packing of non-spherical particles. Perhaps

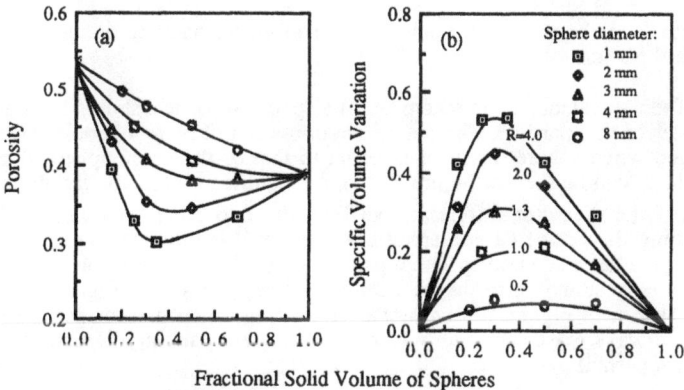

Figure 2   Packing of binary mixtures of fibers (L:D=30.0:4.0 mm) and spheres of different sizes.

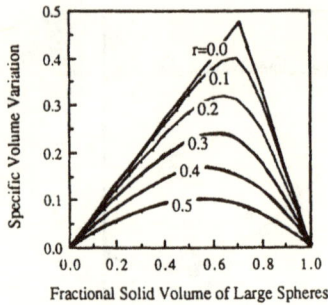

Figure 3
The packing of binary mixtures of spheres.[13]

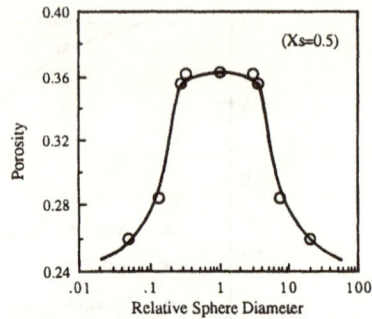

Figure 4 Porosity of binary mixtures of spheres of different sizes for constant fractional solid volume.[16]

one of the most important benefit from this conclusion is that a packing model for spherical particles may be extended to non-spherical particles. To do so, however, it is necessary to quantify the packing size of a non-spherical particle first as discussed below.

Equivalent Packing Diameter

Equivalent spherical diameters are usually determined by measuring a size dependent property of the particle and relating it to the diameter of a sphere[14,15]. Accordingly, the equivalent packing diameter of a non-spherical particle can be defined as the diameter of a sphere having the same packing behaviour as the particle. However, 'the same packing behaviour' referred to here must be clarified further. This can be done by examining the packing of the binary mixtures of given uniformly sized spherical or non-spherical particles and spheres of different sizes as discussed in what follows.

As shown in Figure 4, for a given fractional solid volume of spheres, the increase of sphere diameter will first increase the packing porosity of the binary mixtures to a maximum, then further increase of sphere diameter will decrease the packing porosity. In fact, for spherical particles this kind of plot can be also extracted from the results shown in Figure 3. On the other hand, for given uniformly sized non-spherical particles packed with spheres of different sizes, similar relationship to that shown in Figure 4 has been observed as illustrated in Figure 5. This observation, as it is expected to be found for any non-spherical particles[12], is here used to quantify the equivalent packing diameter of a non-spherical particle.

The equivalent packing diameter of a spherical particle should be equal to the diameter of that particle. The maximum porosity in Figure 4 should therefore be obtained when sphere diameter is equal to that of the given spherical particles though, as can be seen in Figure 4 also, no significant porosity variation may be found if the diameter difference between the two components is small[4,16,17]. Extending this idea to a non-spherical particle, it is obvious that the equivalent packing diameter of a non-spherical particle should be defined as the diameter of the spheres corresponding to the maximum packing porosity in Figure 5. Since the curve shown in Figure 5 is unimodal, according to the present definition, there should always exist one, and only one, equivalent packing diameter for a non-spherical particle.

Naturally, it is of interest to find out what this maximum packing porosity implies. For convenience, let us consider the binary packing mixtures of particles

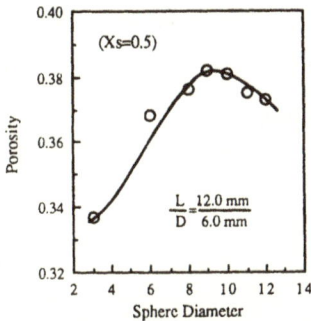

Figure 5 Porosity of binary mixtures of fibers and spheres of different sizes.[12]

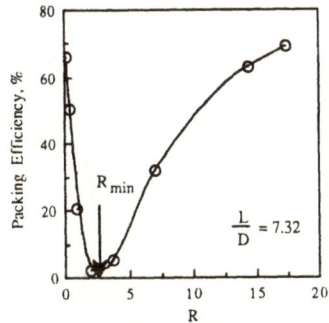

Figure 6 Packing efficiency vs. diameter ratio of sphere to fiber.[10]

and spheres of an equivalent packing diameter. As one may expect, since the two components have the same packing sizes, there is no interaction (joint action) between them, i.e. there is no porosity change resulting from the difference in sizes[4]. In this case, the packing of these two components is just equivalent to the system in which each size is placed as separate layers. Therefore, the specific volume of this particular binary system, V, can be written as

$$V = X_s V_s + X_{ns} V_{ns} \qquad (2)$$

where X_s, V_s and X_{ns}, V_{ns} are the fractional solid volume and initial specific volumes of the spheres and the non-spherical particles, respectively. Therefore, the maximum packing porosity shown in Figure 5 should be determined by eqn.(2). In general, the specific volume variation of a mixture of particles of same equivalent packing diameter should be always equal to zero, irrespective of the fractional solid volumes and initial specific volumes involved. This result explains why it is more convenient to use the specific volume variation in the discussion of the similarity between the packing systems of spherical and non-spherical particles.

The equivalent packing diameter of a non-spherical particle should be a function of its size and shape. Probably the so-called concept 'packing size' was first used by Meloy[18] who also felt that it should be possible to quantify the packing size of a non-spherical particle from its shape analysis. However, at least to our knowledge, to date there have been no publications dealing with this problem in the literature. In fact, it is very difficult, if not impossible, to develop such a method. Therefore, we have here proposed an alternative approach to this problem. As discussed below, the resultant equivalent packing diameter of a non-spherical particle can be readily related to its 'geometrical size' and shape.

3 CORRELATION OF PACKING SIZE WITH SPHERICITY

It is highly desirable to relate the packing size of a non-spherical particle to its size and shape analysis. For this purpose, the packing of fiber-sphere (cylinder-sphere) binary mixtures[6,7,9,10], as it has been extensively investigated, is worthwhile being examined further.

Figure 2(b) shows the specific volume variation, ΔV, of the binary mixtures of fibers, which can be viewed as a given non-spherical particle, and spheres of different sizes. It can be observed that for a given ratio of sphere diameter to cylinder diameter R, ΔV increases to a maximum and then decreases with the increase of the fractional solid volume of spheres. This maximum variation of specific volume, $\Delta V^{max}(R)$, changes with the value of R. When R tends to be equal to zero or infinity, the maximum variations of specific volume can be calculated according to the Furnas theory[13]. They are respectively

$$\Delta V^{max}(0) = \frac{V_s(V_{ns}-1)}{V_{ns}+V_s-1} \tag{3}$$

and

$$\Delta V^{max}(+\infty) = \frac{V_{ns}(V_s-1)}{V_{ns}+V_s-1} \tag{4}$$

Milewski[10] has found that the packing efficiency, which is defined as the ratio of $\Delta V^{max}(R)$ to $\Delta V^{max}(+\infty)$, varies with R and the general relation can be illustrated in Figure 6. It is evident that there is a minimum packing efficiency and this minimum packing efficiency is equal to zero. Thus, corresponding to this minimum packing efficiency should be the spheres having the same diameter as the equivalent packing diameter of the particular fiber. Therefore, the relationship between the equivalent packing diameter and the size and shape of a non-spherical particle can be established by using his experiment results.

In practice, the shape of a particle is usually represented by its sphericity which is defined as the ratio of the surface area of a sphere having the same volume as the particle to the actual surface area of the particle[14,15]. Since the equivalent geometrical diameters, such as equivalent volume, surface and surface volume diameters, of a particle can be measured[14,15,18], it would be advantageous to relate the equivalent packing diameter to these equivalent spherical diameters.

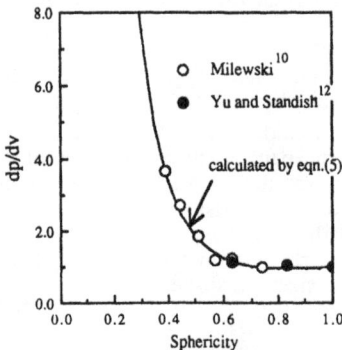

Figure 7 Relation between dp/dv and ψ.[12]

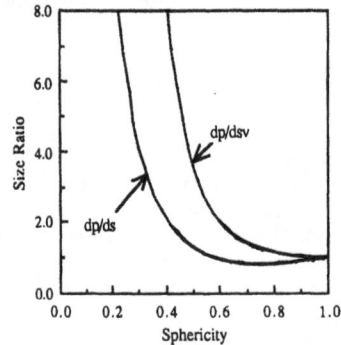

Figure 8 Packing size as a function of sphericity.

Figure 7 shows the experimentally determined relation between the ratio of the equivalent packing diameter to the equivalent volume diameter and the sphericity of a particle.

These results can be fitted by the following equation[12]:

$$\frac{d_p}{d_v} = 3.1781 - 3.6821\frac{1}{\psi} + 1.5040\frac{1}{\psi^2} \tag{5}$$

where d_p and d_v are respectively the equivalent packing and volume diameters while ψ is the sphericity of a particle. If the equivalent surface and surface volume diameters of the particle are represented by d_s and d_{sv}, respectively, then noting that[14]

$$\psi = \frac{d_v^{\,2}}{d_s^{\,2}} \tag{6}$$

and

$$d_{sv} = \frac{d_v^{\,3}}{d_s^{\,2}} \tag{7}$$

the dependence of $\frac{d_p}{d_s}$ or $\frac{d_p}{d_{sv}}$ on sphericity can be readily obtained as given in Figure 8. It should be noted that most of the other equivalent spherical diameters can be related to the above three diameters[14,15], so that the results in Figures 7 and 8 have actually provided a useful bridge to connect the equivalent packing diameter with other equivalent spherical diameters of a particle.

However, as can be observed in Figures 7 and 8, the ratios of the equivalent packing diameter to these three equivalent diameters of a particle varies with the particle shape, especially for particles of low sphericity (less than 0.6). Therefore, it is likely that none of these equivalent spherical diameters can be properly used in characterizing the packing behaviour of particle mixtures.

The usefulness of the approach proposed above will be demonstrated by the examples in the porosity estimation of non-spherical particle mixtures below.

4 APPLICATION

From the foregoing discussion, it is clear that the packing concepts of spheres can be extended to the packing of non-spherical particles. It seems that a packing model for spherical particles can be extended to non-spherical particles without any problem once the equivalent packing diameter of a non-spherical particle is determined. However, there may be a technical problem in doing so as discussed below.

For the reasons of generality, we will first consider the system composed of n components of equal-density particles. Component i (i=1,2,...,n) has equivalent volume diameter d_{vi}, sphericity ψ_i and fractional solid volume X_i. Under a given packing condition which may be reflected by the initial porosity of spheres ε_s, the porosity of the system considered ε should be expressed as

$$\varepsilon = f(\varepsilon_s, X_1, X_2, ..., X_n, d_{v1}, d_{v2}, ..., d_{vn}, \psi_1, \psi_2, ..., \psi_n) \tag{8}$$

On the other hand, from the earlier discussion, component i should have an equivalent packing diameter d_{pi}, the values of which can be determined by eqn.(5). Therefore, if the sizes of all the components in the considered system are represented by their equivalent packing diameters respectively, then we will have an equivalent packing system composed of spherical particles only. This concept may be schematically illustrated by Figure 9. The porosity of this equivalent spherical packing system should be expressed as

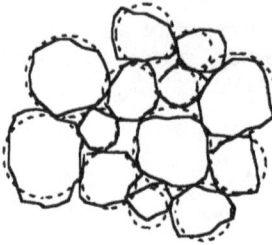

Figure 9 Two-dimensional illustration of a non-spherical particle packing system (solid line) and its equivalent spherical particle packing system (broken line).

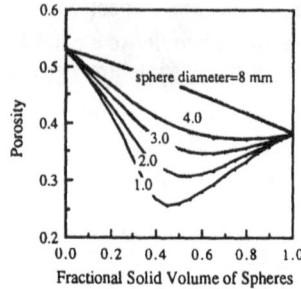

Figure 10 Predicted porosities of binary mixtures of fibers (L=4.0, D=30.0 mm) and spheres of different sizes (refer to Figure 2).

$$\varepsilon_e = f(\varepsilon_s, X_{p1}, X_{p2}, ..., X_{pn}, d_{p1}, d_{p2}, ..., d_{pn}) \qquad (9)$$

where $d_{pi} > d_{pj}$ ($1 \leq i < j \leq n$). Once the fractional solid volume X_{pi} ($i=1,2,...,n$) is determined, the porosity of this equivalent spherical packing system can be estimated by one of the existing models without any difficulty[1-5]. A simple theory has been recently proposed to relate the porosity of the non-spherical packing system ε to the porosity of its equivalent spherical packing system ε_e[12]. However, the theory fails to account for the packing of particles of low sphericities (less than 0.6). Therefore, at this stage of development, other simplified methods may have to be employed.

In fact, as soon as the equivalent packing diameters are known, the porosity of a packing system can always be estimated by the simplified approach to the real packing problem encountered in practice as, for example, suggested in the literature[3,5]. In this simplified approach, the initial porosities of the components involved are used as prescribed information in a model calculation. Therefore, the porosity of the system considered can be given as[5]

$$\varepsilon = f(X_1, X_2, ..., X_n, d_{p1}, d_{p2}, ..., d_{pn}, \varepsilon_1, \varepsilon_2, ..., \varepsilon_n) \qquad (10)$$

The initial porosity ε_i can be determined by the results shown in Figure 1 which has been formulated as[12]

$$\varepsilon(\psi) = \varepsilon_s^{15.521 \cdot \psi^{3.853} - 14.521 \cdot \psi^{4.342}} \qquad (\varepsilon_s = 0.38) \qquad (11)$$

In this case, the porosity of a non-spherical particle mixture can be estimated as has been reported in the literature[3,5]. However, for simplicity, the following discussion on the validity of eqn.(10) will be limited to the packing of binary mixtures of non-spherical particles only. Furthermore, since the packing of the fiber-sphere binary systems has been investigated by a number of investigators[6,7,9,10], the predictability of the present approach may be checked by comparing the theoretical predictions with some typical reported experimental results[7,10].

It has been verified that the packing of binary mixtures of spheres can be accurately described by the Westman conic equation[4,19]. Therefore, the specific volume of any binary mixture of particles can be written as

$$V = \frac{-B + \sqrt{B^2 - 4AC}}{2A} \tag{12}$$

where

$$A = \left(\frac{1}{V_2}\right)^2 + \frac{2G}{V_2(V_1-1)} + \frac{1}{(V_1-1)^2}$$

$$B = -\frac{2V_1X_1}{V_2^2} + \frac{2G}{V_2(V_1-1)}(V_2X_1-V_2\cdot X_1-V_1X_1) + \frac{2(V_2X_1-V_2\cdot X_1)}{(V_1-1)^2}$$

$$C = \left(\frac{V_1X_1}{V_2}\right)^2 - \frac{2G}{V_2(V_1-1)}V_1X_1(V_2X_1-V_2\cdot X_1) + \frac{(V_2X_1-V_2\cdot X_1)^2}{(V_1-1)^2} - 1$$

where the coefficient, G, can be approximated by the following equation[4,12]:

$$\frac{1}{G} = \begin{cases} -0.0641r + 1.9077r^2 & (r \le 0.0741) \\ 1 & (r > 0.741) \end{cases} \tag{13}$$

where r is the ratio of d_{p2} to d_{p1}.

Figure 10 shows the estimated porosity for the conditions given in Figure 2. By inspection of the results in Figure 2(a) and Figure 10, it is evident that the estimated porosities are in reasonably good agreement with the measurements. The validity of the proposed approach to the packing of non-spherical particle mixture can be further confirmed by the results shown in Figure 11. Bearing in mind the relatively large error in the measurement of the porosity of a non-spherical particle mixture, it is therefore considered that the porosity of a non-spherical particle mixture can be estimated theoretically though further studies are necessary in order to develop a more perfect theory. In fact, the present results clearly indicate that the resultant information by the present approach can at the least provide a useful guide in practice.

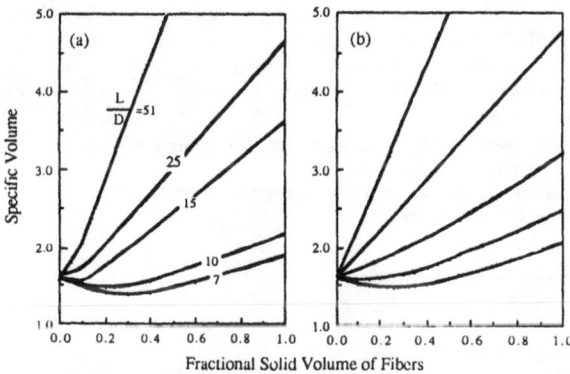

Figure 11 Packing of binary mixtures of spheres (R=10) and fibers of different sizes: (a), measurements,[10] (b), predictions.

5   CONCLUSIONS

The results presented in this paper clearly indicate that the equivalent packing diameter of a non-spherical particle can be readily defined and determined from its packing behaviour. For the purpose of application, the equivalent packing diameter of a particle has been empirically related to its equivalent spherical diameters such as volume, surface and surface volume diameters, respectively. The proposed approach is verified by the reasonably good agreement between the estimated and the measured porosities of fiber-sphere binary mixtures.

However, it should be noted that the present work is based on the limited quantitative data available in the literature. For example, eqn.(5) was only partially verified (see Figure 7). Furthermore, the proposed approach is to a great degree established based on the results of fiber-sphere binary mixtures. The effect of other shapes on the porosities of non-spherical particle mixtures should be investigated. Therefore, it is considered that in order to characterize the packing size of a particle and hence the structural properties of particle mixtures properly, it seems that much more experimental work needs to be done in order to obtain the required store of quantitative data, especially for the packing of particles with low sphericities.

ACKNOWLEDGEMENT

One of the authors (A. B. Yu) is grateful to CSIRO for providing a postdoctoral fellowship. Special thanks are due to Dr. J.K. Wright, Assistant Chief of Div. of Min. and Proc. Eng., for his support.

NOMENCLATURE

d = particle size
D = cylinder diameter
d_p = equivalent packing diameter
d_s = equivalent surface diameter
d_{sv} = equivalent surface volume diameter
d_v = equivalent volume diameter
L = cylinder length
n = component number of a considered system
r = size ratio, d_{p2}/d_{p1}
R = diameter ratio of sphere to fiber
R_{min} = diameter ratio of sphere to fiber corresponding to the minimum point in Figure 6
V = specific volume of a considered system
V_i = initial specific volume of ith component, $V_i=1/(1-\varepsilon_i)$
V_s = initial specific volume of spheres, $V_s=1/(1-\varepsilon_s)$
ΔV = specific volume variation
$\Delta V^{max}(R)$ = maximum specific volume variation of a fiber-sphere binary system
X = fractional solid volume of ith component
X_p = fractional solid volume of a component in an equivalent spherical packing system
X_s = fractional solid volume of spheres

Greek Letters
ε = porosity of a considered packing system
ε_e = porosity of an equivalent spherical packing system
ε_i = initial porosity of ith component
ε_s = initial porosity of spheres
ψ = particle sphericity

subscript
i = ith component

REFERENCES

1. J.A. Dodds, <u>J. Colloid and Interface Sci.</u>, 1980, <u>77</u>, 317.
2. T. Stovall, F. De Larrard and M. Buil, <u>Powder Technol.</u>, 1986, <u>48</u>, 1.
3. N. Ouchiyama and T. Tanaka, <u>Ind. Eng. Chem. Fundam.</u>, 1989, <u>28</u>, 1530.
4. A.B. Yu and N. Standish, <u>Powder Technol.</u>, 1988, <u>55</u>, 171.
5. A.B. Yu and N. Standish, Estimation of the porosity of particle mixtures by a linear-mixture packing model, <u>Ind. Eng. Chem. Res.</u>, in print.
6. T.L. Star, <u>Am. Ceram. Soc. Bull.</u>, 1986, <u>65</u>, 1293.
7. A.B. Yu, R.P. Zou and N. Standish, Packing of non-spherical particle mixtures, submitted for publication.
8. G.C. Brown and others, 'Unit Operations', John Wiley & Sons, New York, 1950, p. 210-16.
9. R.M. German, 'Particle Packing Characteristics', Metal Powder Industries Federation, Princeton, New Jersey, 1989.
10. J.V. Milewski, Packing concepts in the utilization of filler and reinforcement combinations, Handbook of Fillers and Reinforcements for Plastics, Katz, H. S. and Milewski, J. V. (eds.), Van Nostrand Reinhold, New York, 1978, p. 66-78.
11. R. Aoki and M. Suzuki, <u>Powder Technol.</u>, 1970/71, <u>4</u>, 102.
12. A.B. Yu and N. Standish, Characterization of non-spherical particles from their packing behaviour, submitted for publication.
13. C.C. Furnas, <u>Ind. Eng. Chem.</u>, 1931, <u>23</u>, 1052.
14. T. Allen, 'Particle Size Measurement', Chapman & Hall, 3rd ed, London, 1981.
15. N. G. Stanley-Wood, Enlargement and Compaction of Particulate Solids, Butterworths, London, 1983.
16. H.J. Fraser, <u>J. Geol.</u>, 1935, <u>43</u>, 910.
17. K. Ridgway and K.J. Tarbuck, <u>Chem. Proc. Eng.</u>, 1968, <u>49</u>, 103.
18. T. P. Meloy, Particulate Characterization: Future Approaches, Handbook of Powder Science and Technology, in M. E. Fayed and L. Otten (eds.), Van Nostrand Reihold Company Inc., New York, 1984, p. 69-98.
19. A.E.R. Westman, <u>J. Am. Ceram. Soc.</u>, 1936, <u>19</u>, 127.

Design and Performance of the 'HSE' Fibre Monitor

A. P. Rood

THE OCCUPATIONAL MEDICINE & HYGIENE LABORATORIES, 403 EDGWARE ROAD, LONDON NW2 6LN, UK

INTRODUCTION

Worldwide concern over the risk of developing cancer from the inhalation of asbestos and other fibres has led to the monitoring of workers exposed to the material, and a good estimate of the dose response curve has now been arrived at in the case of asbestos (Doll & Peto, 1985). Epidemiological studies have also indicated a thirty year latency for the onset of lung cancer and mesothelioma from the time of initial exposure. This observation has given rise to concern in the general public, and a call to eliminate exposure to asbestos, especially in schools where the latency might give rise to disease in early adulthood (Dalton, 1979). A dose response curve has not been established for other fibres, but a statistically significant excess of lung cancer has been observed in workers exposed to man-made mineral fibres (MMMF) of rock and slag wool (Saracci et al, 1984). Although the exact cause is controversial, the International Agency for Cancer Research has catagorised these materials as possible human carcinogens.

Workplace exposure is usually assessed by personal sampling in which a known volume of air is sucked through a membrane filter which is subsequently prepared for examination by phase-contrast optical microscopy, and respirable fibres counted using predetermined rules (HSE, 1990). 0.5 & 0.2 f/ml control limits for chrysotile and amphibole asbestos exposure respectively has been introduced in the UK, and a 2f/ml maximum exposure to MMMF has been agreed, running in parallel with a 5mg/m^3 gravimetric limit.

Monitoring of air quality after asbestos stripping is another important procedure requiring fibre measurement. In the UK it is necessary to determine if an area is suitable for re-occupation by taking a series of clearance membrane filter samples to establish if the air concentration, measured by PCOM, is below 0.01 f/ml.

In the USA, the electron microscope has been used to assess filters taken after asbestos removal. The preparation of filters for either the light microscope or the electron microscope by necessity requires a delay before fibre counting can be undertaken, and the assessment itself further adds to the time before results can be given. An hour is typical for sampling, followed by another hour for optical assessment, and up to twenty four hours for electron microscope assessment. Real time monitoring of worker exposure or building contamination is thus not possible using microscope techniques.

Lilienfeld (1979) described an instrument that monitors fibre concentration in sampled air by laser light scattering from fibres oscillating in phase with an electric field, and the instrument has been employed for workplace and environmental monitoring in the USA. Work by Shenton-Taylor and Iles (1986) has suggested that silicon carbide dust responds in a fibre like fashion in this instrument. More recently (Al-Chalabi et al 1990) have described a light scattering instrument able to detect fibres in air down to a fibre diameter of 1μm. Light scattering from aligned fibres has been reported by Timbrell (1980), but only for a magnetically aligned sample on a membrane filter.

DESIGN AND CONSTRUCTION DETAILS

A portable instrument needs to be battery operated, simple in operation and robust in use. We have stripped down a standard 41/min sampling pump (Rotheroe and Mitchell L2SF) and fitted it with associated electronics for fibre precipitation and sensing into a hand-held unit running from rechargeable nickel-cadmium cells. The instrument top block (see Fig. 1) is cast from epoxy resin in which the corona needles, the precipitator electrode, and the detectors are set. The baseplate on which the block seals is made from aluminium.

Fibre alignment in the instrument is based on the aerosol spectrometer of Prodi et al (1982), but a simplified design of inlet was used to accelerate the incoming air; two razor blades set with their edges parallel and 1mm apart are angled at 20° to the incoming air (Fig. 1), and cause the fibres to align in the direction of flow. The sheath air used by Prodi to surround the aerosol was not found necessary in our design, but mechanical flatness and accurate positioning of the blades were found important. The air containing the aligned fibres then passes through a corona discharge generated by three needles operating at 3.5KV with respect to the baseplate (Whitby, 1961). Positive ions collide with the incoming dust and fibres giving them an overall charge. A precipitator plate downstream from the charging region precipitates the dust and fibres (Liu et al 1967) onto a removable glass slide under the sensors. Both corona and precipitator voltages are generated by battery driven generators (Brandenberg 512AA).

The fibres precipitated onto the glass slide retain their alignment and the difference in light scattering parallel and at right angles to the flow (Timbrell & Gale, 1980) is used to determine the number of fibres. A calibration curve can be derived by counting, in the PCOM, fibres sampled from a range of concentrations. Four sensors (RS BPW21) are mounted below the slide to detect the scattered light emanating from a green light emitting diode collimated to a spot in the middle of the slide, the optimum angle of these detectors has been determined from a separate experiment looking at the angular dependence of the scattered light. The collimation produces a 3mm spot of uniform intensity within the area of dust deposit. A matt black surface under the microscope slide serves to absorb light transmitted through the glass. Signal differences can be observed by phase-locked amplification of the output from detectors at right angles to each other, holding the signals electronically, and displaying the difference on a voltmeter. The signal sum from all four detectors can also be acquired and displayed.

A backup filter (5μm Millipore SM) between the instrument chamber and the pump prevents any dust that has not been precipitated from reaching the pump. Transmission electron microscopy can be used to identify fibres that escape precipitation by preparing the filter by the direct transfer technique (Burdett and Rood, 1983).

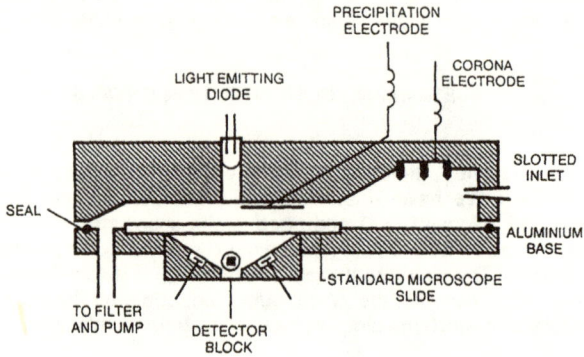

Fig.1 - Section through instrument top block

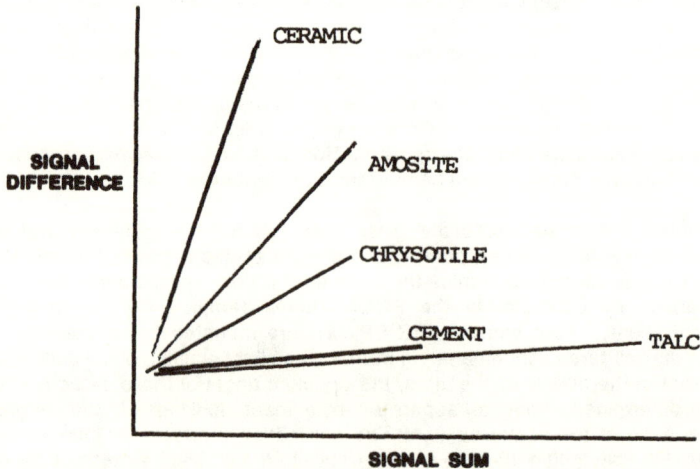

SIGNAL SUM

RESULTS

The performance of the instrument has been checked on two groups of fibres, asbestos and ceramic glass, and on a range of test dusts including talc. In the figures below, the sum and difference signals are plotted, two distinct forms of response can be seen, the fibres show a strong positive slope, and the dusts (including talc) a flat response.

The signal strength for fibres reduces with diameter, and the curly nature of chrysotile has a similar effect. Exposure standards can be measured in just a few minutes using the instrument even in the presence of background dust. The figure below shows how such as cement contributes to the sum signal, but not to the difference to any significant extent when sampled with fibres.

REFERENCES

Al-Chalabi, S A M; Jones, A R; Savaloni, H; and Wood, R (1990) Meas. Sci. Technol. 1, 29-35

Burdett, G J; and Rood, A P; (1983) Envir. Sci. Tech. 17.11, 643-8

Dalton, A J P; (1979) "Asbestos Killer Dust", BSSRS Publications London, ISBN Pb 0950254134

Doll, R; and Peto, J; "Effects on Health of exposure to Asbestos", HMSO London, ISBN 011883700

Health and Safety Executive (1990) "Asbestos fibres in Air" MDHS 39/3, HMSO London

Lilienfeld, P; Elterman, P B; and Baron, P (1979) Am. Ind. Hyg. Assoc. J. 40, 270-82

Liu, B Y H; Whitby K T; and Yu, H S (1967) Rev. Sci. Ins. 38.1, 100-2

Prodi, V; De Zaiacomo, T; Hochrainer, D; and Spurny, K; (1982) J. Aerosol Sci. 13.1, 49-58

Saracci, R; (1984) Biological Effects of Man-made Mineral Fibres Vol 1, WHO Copenhagen ISBN 92 890 1026 6

Shenton-Taylor, T; and Iles, P; (1986) Ann. occup. Hyg. 30.1, 77-87

Timbrell, V; and Gale, R W; (1980) "Biological Effects of Mineral Fibres I" IARC Scientific Pub. No 30, 53-60

Whitby, K T; (1961) REv. Sci. Inst. 32.12, 13251-55

Fast Separation and Characterization of Micron Size Particles by Sedimentation/Steric Field-flow Fractionation: Role of Lift Forces

P. S. Williams, M. H. Moon and J. C. Giddings

FIELD-FLOW FRACTIONATION RESEARCH CENTER, DEPARTMENT OF CHEMISTRY,
UNIVERSITY OF UTAH, SALT LAKE CITY, UTAH 84112, USA

1 INTRODUCTION

In its early stages of development, *field-flow fractionation (FFF)* was limited to the analysis of macromolecules and submicron size particles with diameters ranging from a few nm to about 1 μm[1-3]. Since the discovery of *steric FFF*[4], this form of FFF has become an increasingly powerful method for measuring the particle size distribution (PSD) of 1-100 μm particles[5]. For both submicron and micron size particles, FFF has the advantage of high resolution and relatively high speed[6]. These characteristics are illustrated in Figure 1, which shows the detector signal *versus* time curve (or fractogram) generated in the separation of polystyrene (PS) latex beads in both size ranges. The separation of the larger (>1 μm) latex is particularly fast, requiring only three minutes. The PSD is obtained directly from the fractograms using appropriate software and calibration procedures (see later).

Steric FFF, used for micron size particles, is not only fast and accurate but it has the overriding advantage of allowing the physical collection of narrow size fractions. With such fractions, one can use microscopy and other methods to confirm the accuracy of calibration and examine variations in particle shape, density, chemical composition, etc., as a function of size. These fractions provide an optional "window" for cross checking results and extending the characterization of particulate materials beyond that commonly available.

The mechanism of FFF entails the use of a driving force that impels particles transversely across a thin (<1 mm) ribbonlike channel. As the particles approach one of the channel walls (termed the accumulation wall), their motion is brought to a halt by opposing forces or influences. For submicron particles, the opposing transport is driven by Brownian motion. Micron size particles, on the other hand, may approach contact with the wall before their motion is halted by steric exclusion. Particles of different size are thus held at different mean

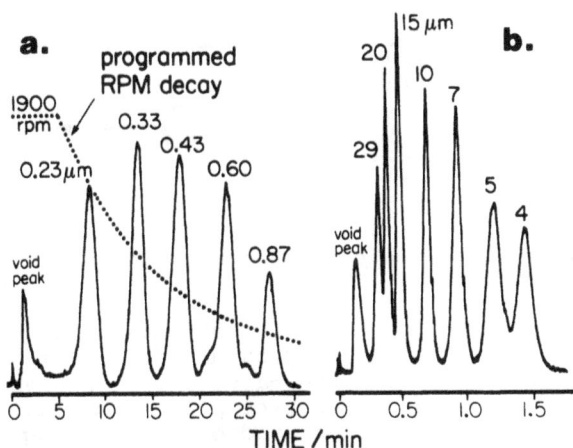

Figure 1 (a) Fractionation by normal mode
sedimentation FFF of submicron polystyrene (PS)
particles using Channel I (see Experimental section)
at a flowrate of 1.5 mL/min and a field strength power
programmed[7] (with p=8) from 609 gravities (1900 rpm)
after hold period of 5 min; (b) sedimentation/steric
FFF fractionation of supramicron PS microspheres using
Channel I at a flowrate of 15.0 mL/min and a field
strength of 635 gravities (1940 rpm).

elevations above the wall.

Separation begins with the onset of flow down the
channel. The flowing fluid displaces particles along the
channel length. However, since the flow profile is parabolic (see Figure 2), particles held at different elevations will be displaced at different velocities[2,3]. For
steric FFF, larger particles, with centers held further
from the wall, are displaced more rapidly toward the exit
than smaller particles. The particles are separated and
eluted through a detector at different times. The
differential elution is recorded and converted to a PSD.

During flow displacement, the migrating particles,
rather than remaining in contact with the wall, are driven
a short distance δ above the accumulation wall by
hydrodynamic lift forces. The slightly elevated position
of the particles is illustrated in Figure 2. The speed of
elution increases with δ. However, δ is variable,
increasing with flowrate and decreasing with the applied
force. Thus the influence of the lift forces can be
flexibly modulated by changes in flowrate and in the
primary field strength.

Several kinds of driving forces can be used in FFF
including those of electrical, magnetic, and thermal (i.e.,
temperature gradient) origin. However, the most important
driving forces for particles are sedimentation and
crossflow. The latter, utilized in *flow FFF*, is based on a
crossflow of fluid entering and exiting the channel through

Particle Size Analysis

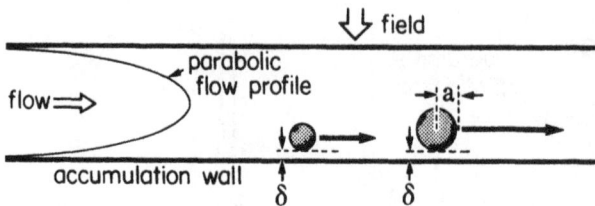

Figure 2 Schematic diagram of particle positions and differential particle displacement in sedimentation/steric FFF.

permeable walls. The crossflow drives particles toward the accumulation wall. This method is highly versatile for both submicron and micron size particles[8]. However, in this paper we focus on sedimentation FFF, in which the primary driving force is generated by a centrifuge. The channel encircles the centrifuge axis so that flow and field direction are perpendicular. When operated in the steric mode, the resulting FFF subtechnique is termed *sedimentation/steric FFF* or *Sd/St FFF*. More complete descriptions of various forms of FFF are found in the references.

2   THEORY FOR STERIC MIGRATION

Early experiments in steric FFF[9,10] revealed a dependence of the particle retention time on both field strength and channel flowrate. It was concluded that hydrodynamic lift forces must influence the position of the particles relative to the accumulation wall. A balance between the primary driving force and the lift forces would establish the position of each particle. This equilibrium position then determines the particle velocity and hence its time for elution through the system.

The presence of a hydrodynamic lift force due to the effect of fluid inertia has been predicted[11-15]. This force, if unopposed, tends to drive particles toward either of two stable equilibrium positions located at 0.19 and 0.81 of the distance across the channel thickness. Its magnitude and direction is described approximately by

$$F_L' = 13.5 \ \pi \ \frac{\rho <v>^2 a^4}{w^2} \ g\left(\frac{x}{w}\right) \tag{1}$$

where

$$g\left(\frac{x}{w}\right) = 19.86 \left(0.19 - \frac{x}{w}\right)\left(0.5 - \frac{x}{w}\right)\left(0.81 - \frac{x}{w}\right) \tag{2}$$

and where ρ is the fluid density, $<v>$ is the mean fluid velocity, a is the particle radius, w is the channel thickness and x is the distance from the accumulation wall to the particle center. By convention, the lift force is positive when it acts away from the accumulation wall. The coefficient of g(x/w) was chosen for consistency with the limiting magnitude of the lift force predicted by Cox and Hsu[14] as the particle approaches the wall. The theory requires certain restrictions on flow velocity and also requires that a<<w. A more serious restriction from our point of view requires a/x<<1, that is, particles must be relatively far from the bounding walls. The theory therefore breaks down just in the region of interest for steric FFF. Even if still valid, the predicted force would be insufficient to counter the sedimentation forces typical of steric FFF.

Systematic studies in our laboratory[16] have indicated the presence of a second lift force that acts on the particles. These studies involved the measurement of retention times for a set of PS particle standards under wide ranging conditions of flowrate and field strength. The results yielded an empirical lift force dependence described by

$$F_L" = C \frac{a^3 s_0}{\delta} \tag{3}$$

where C is an empirical constant for a given system, s_0 is the shear rate of undisturbed fluid at the wall, and δ is the shortest distance between the wall and the particle surface (see Figure 2). For conditions typical of steric FFF, $F_L"$ generally dominates F_L'. The equilibrium position is therefore given when this force counterbalances the sedimentation force

$$\frac{4}{3}\pi a^3 \Delta \rho G = C \frac{a^3 s_0}{\delta} \tag{4}$$

where $\Delta \rho$ is the density difference between particle and fluid and G is the field strength (acceleration). Hence

$$\delta = \frac{3C s_0}{4\pi \Delta \rho G} \tag{5}$$

Thus δ is independent of particle size provided particle density is constant. This property is specific to sedimentation/steric FFF due to the cancellation of the powers of a. For flow FFF, by contrast the driving force is proportional to a, and δ is therefore expected to increase with a^2.

Goldman, Cox, and Brenner[17] have shown that a particle entrained in a sheared flow bounded by a plane wall will migrate with a velocity somewhat smaller than that of an undisturbed fluid streamline at the position of the particle center. The degree of retardation is a function of the ratio of δ/a

$$v_p = v \cdot f(\delta/a) \qquad\qquad\qquad (6)$$

where v_p is the particle velocity and v is the undisturbed fluid velocity at the particle center. The retardation function $f(\delta/a)$ approaches unity for $\delta \gg a$ but approaches zero as δ approaches zero.

Within the thin parallel-walled FFF channel, viscous drag gives rise to a parabolic velocity profile

$$v = 6\langle v\rangle \frac{x}{w}\left(1 - \frac{x}{w}\right) \qquad\qquad (7)$$

where v is the local carrier fluid velocity at distance x from the accumulation wall. Assuming a particle is held at some constant distance from the accumulation wall throughout its elution by the balance of sedimentation and lift forces, the retention ratio, defined as the ratio of particle velocity to mean fluid velocity, is given by

$$R = \frac{v_p}{\langle v\rangle} = 6\ f(\delta/a)\ \frac{x_{eq}}{w}\left(1 - \frac{x_{eq}}{w}\right) \qquad (8)$$

where x_{eq} is here the equilibrium distance from the particle center to the wall, equal to $(a+\delta)$. For $x_{eq} \ll w$ this reduces to

$$R = 6\ f(\delta/a) \cdot (a+\delta)/w \qquad\qquad (9)$$

The retention time t_r for a given particle size is then

$$t_r = \frac{t^0}{R} = \frac{wt^0}{6\ f(\delta/a) \cdot (a+\delta)} \qquad\qquad (10)$$

where t^0 is the channel void time, equal to the time for elution of a nonretained material.

3   CALIBRATION PLOTS

Expressions were derived by Goldman, Cox, and Brenner[17] for the limiting behavior (as $\delta/a \to \infty$ and 0) of the function $f(\delta/a)$. Neither limit applies to most cases of steric FFF. However, a cubic spline approach can be used to interpolate between the values of $f(\delta/a)$ that they numerically calculated and tabulated. Since δ is expected to be independent of a (equation (5)), the size dependence of R and t_r is governed by the product of $f(\delta/a)$ and $(a+\delta)$. The logarithm of this product is plotted in Figure 3 against $\log(a)$ (over a 1 to 30 μm range in radius a) for δ values of 0.02, 0.05, 0.1, 0.2 and 0.5 μm. The plots are seen to be fairly linear over this size range.

From equation (10) we get

$$\log t_r = \log\left(wt^0/6\right) - \log\left(f\left(\delta/a\right).\left(a+\delta\right)\right) \tag{11}$$

It follows that plots of $\log(t_r)$ versus $\log(a)$, or alternatively versus $\log(d)$, where d is the particle diameter, are expected to be very nearly linear. Such linear calibration plots have routinely been obtained[7]. The slope of such plots (by convention always made positive) is defined as the diameter-based selectivity

$$S_d = \left| \frac{d \log t_r}{d \log d} \right| \tag{12}$$

The selectivity is expected to equal the slopes of the plots shown in Figure 3. Although these show slight curvature, mean slopes range from 0.87 for $\delta=0.02$ μm to 0.75 for $\delta=0.5$ μm. The same exercise carried out for flow FFF (where δ increases with the square of a) results in $S_d>1$, in this case increasing as δ-values increase. The prediction that $Sd>1$ is in agreement with experimental observation[18].

A density compensation procedure has recently been developed[5] whereby a calibration plot obtained using, for example, polystyrene bead standards can be used to obtain a particle size distribution for a material of different density. Equation (5) shows that δ is inversely dependent on the product $\Delta\rho G$. If a change in $\Delta\rho$ is compensated by a change in G so that the product remains constant, then δ will be independent of $\Delta\rho$ and particles of a given diameter, even if different in density, will elute in the same time. The procedure for obtaining the PSD of particles of arbitrary density from the recorded fractogram has also been described[5].

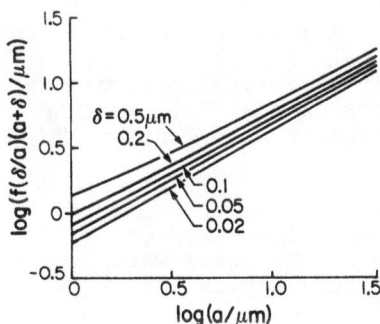

Figure 3 Plots of $\log(f(\delta/a)\cdot(a+\delta))$ *versus* $\log(a)$ for indicated δ.

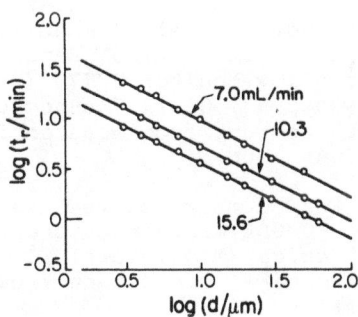

Figure 4 Experimental calibration plots for PS standards from 3 to 60 μm using Channel II at 800 rpm and flowrates of 7.0, 10.3, and 15.6 mL/min.

4   EXPERIMENTAL

The sedimentation FFF system used in this study is similar
to the model S101 Colloid/Particle Fractionator from
FFFractionation, Inc. (Salt Lake City, UT, USA). Two
different channels were used in this work. Channel I has a
nominal thickness w of 127 μm, a breadth b of 1.0 cm, and a
tip-to-tip length L of 90 cm. The void volume V^0, measured
as the elution volume of a nonretained sodium benzoate
peak, is 1.18 mL. The radius r_0 of rotation is 15.1 cm.
For Channel II w = 254 μm, b = 2.0 cm, L = 90 cm, r_0 = 15.7
cm, and V^0 = 4.20 mL.

The carrier was doubly distilled and deionized water
containing 0.1% FL-70 detergent (Fisher Scientific,
Fairlawn, NJ, USA) and 0.02% sodium azide as a bactericide.
The particle standards were polystyrene latex beads of
density 1.050 g/mL and nominal diameters 59.7, 47.9, 29.4,
19.58, 15.00, 9.87, 7.04, 5.002, 3.983, 3.002, 0.868,
0.596, 0.426 μm from Duke Scientific (Palo Alto, CA, USA),
and 0.330, 0.232 μm from Seragen Diagnostics (Indianapolis,
IN, USA). (Hereafter these standards are designated as 60,
48, 29, 20, 15, 10, 7, 5, 4, 3, 0.87, 0.60, 0.43, 0.33, and
0.23 μm, respectively.) To test our method, a nominal 8-58
μm polydisperse glass bead sample (SRM 1003a) with a
density of 2.41 g/mL from the National Institute of
Standards and Technology (NIST) was used.

Sample volumes of 50 μL of 40 mg/mL glass bead suspensions were injected directly into the channel using a
microsyringe. Following injection, flow was stopped and
the centrifuge set spinning to reach the desired rpm. The
flow was resumed for the separation.

An HPLC pump, Kontron Model 410 (Kontron Electrolab,
London, UK) was used for most runs. An FMI Lab Pump Model
QD-2 (Fluid Metering, Inc., Oysterbay, NY, USA) was used
for the fast steric analysis of Figure 1b and for flushing
the channel to keep it clean. The eluted samples were
monitored by an Altex model 153 (Beckman Instrument,
Fullerton, CA, USA) UV detector for Channel I and by a
Spectroflow Monitor SF770 (Kratos Analytical Instruments,
Westwood, NJ, USA) UV detector for Channel II. A strip
chart recorder from Houston Instrument Corp.(Austin, TX,
USA) was used.

Fractions of eluted glass beads were collected by a
model FC-80K microfractionator from Gilson Medical
Electronics (Middletown, WI, USA). The beads were examined
by an optical microscope from Olympus Optical Co. (Tokyo,
Japan).

5   RESULTS

Calibration plots are shown in Figure 4 for the series of PS standards from 60 to 3 μm eluted from Channel II at 800 rpm (112 gravities) at three different channel flowrates: 7.0, 10.3, and 15.6 mL/min. The plots yield selectivities of 0.77, 0.74, and 0.74, respectively, values in reasonable agreement with our theoretical model.

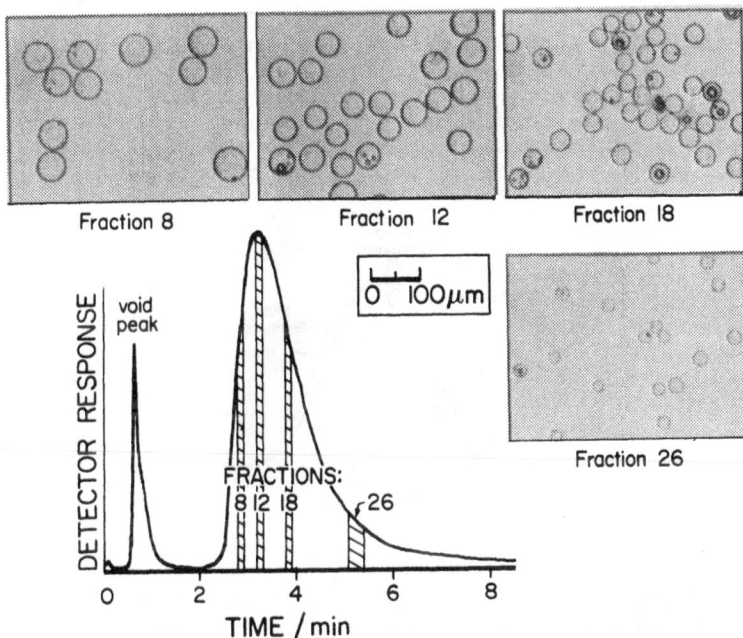

Figure 5 Fractogram of glass bead sample using Channel II at 150 rpm and a flowrate of 7.0 mL/min. Optical micrographs correspond to indicated fractions.

Compensation for the change in Δρ on going from PS standards to glass beads required altering the rotation rate from 800 to 150 rpm. The glass bead fractogram obtained at 7.0 mL/min is shown in Figure 5. (This run, since it is carried out at the lowest flowrate in the largest volume channel, is considerably slower than runs optimized for high speed.) Micrographs of fractions 8, 12, 18 and 25 are also shown. The uniformity of the particle sizes within each fraction is apparent.

In Figure 6 we show the number, surface area, and mass distributions calculated for the glass bead sample. Since the detector responds to the scattering of light at the particle surfaces, it produces a signal proportional to a fraction's particle surface area. Thus the area distribution is obtained directly from the detector response curve via a change of scales; the number and mass distributions are determined from this result[5].

The cumulative mass distribution is compared with the NIST specification in Figure 7. The agreement is reasonably good, the FFF result indicating a slightly narrower distribution and slightly higher mean particle size.

Finally, to show the self-consistency of the FFF technique, mass distributions obtained at flowrates of 7.0 and 10.3 mL/min are compared in Figure 8. The agreement is excellent. We note that the calibration plot and glass bead analysis carried out at 10.3 mL/min were obtained several days after the other runs. Elution times and selectivities may vary slightly with the passage of time depending on the condition of the channel surfaces, but provided a calibration run is made in the same time frame as an unknown then the variation is accounted for.

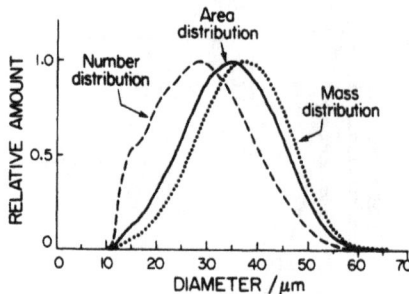

Figure 6 Number, area, and mass distributions obtained by sedimentation/steric FFF for glass bead sample.

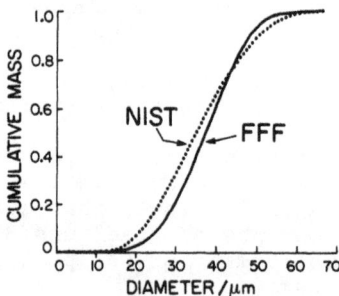

Figure 7 Cumulative mass distributions by FFF compared with NIST specification.

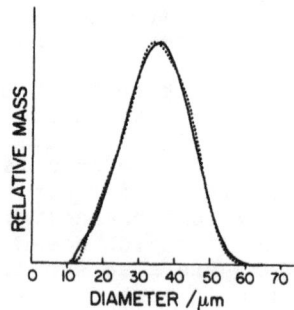

Figure 8 Mass distributions for glass bead sample from FFF at 7.0 mL/min (———) and 10.3 mL/min (....).

6   CONCLUSIONS

The apparent lift experienced by micron size particles migrating close to a wall has not been satisfactorily explained. Our empirically derived lift force expression is consistent with the observation that calibration plots of log(retention time) *versus* log(diameter) are invariably quite linear over a wide range of particle sizes. Work is in progress to more fully elucidate the nature and mechanism of the lift forces. In the meantime, rapid and accurate particle size analysis, with the advantage of collectable size fractions, can presently be achieved based on empirical calibration and density compensation.

ACKNOWLEDGMENT

This work was supported by Grant CHE-8800675 from the National Science Foundation.

REFERENCES

1.   J.C. Giddings, Sep.Sci., 1966, 1, 123.
2.   J.C. Giddings, J.Chromatogr., 1976, 125, 3.
3.   J.C. Giddings, M.N. Myers and J.F. Moellmer, J.Chromatogr., 1978, 149, 501.
4.   J.C. Giddings and M.N. Myers, Sep.Sci.Technol., 1978, 13, 637.
5.   J.C. Giddings, M.H. Moon, P.S. Williams and M.N. Myers, Anal.Chem., accepted.
6.   J.C. Giddings, M.N. Myers, M.H. Moon and B.N. Barman, in 'Particle Size: Assessment and Characterization', T. Provder, Ed., ACS Symp.Series, American Chemical Society, Washington, D.C., in press.
7.   P.S. Williams and J.C. Giddings, Anal.Chem., 1987, 59, 2038.
8.   S.K. Ratanathanawongs and J.C. Giddings, in 'Particle Size: Assessment and Characterization', T. Provder, Ed., ACS Symp.Series, American Chemical Society, Washington, D.C., in press.
9.   J.C. Giddings, M.N. Myers, K.D. Caldwell and J.W. Pav, J.Chromatogr., 1979, 185, 261.
10.   K.D. Caldwell, T.T. Nguyen, M.N. Myers and J.C. Giddings, Sep.Sci.Technol., 1979, 14, 935.
11.   R.G. Cox and H. Brenner, Chem.Eng.Sci., 1968, 23, 147.
12.   B.P. Ho and L.G. Leal, J.Fluid Mech., 1974, 65, 365.
13.   P. Vasseur and R.G. Cox, J.Fluid Mech., 1976, 78, 385.
14.   R.G. Cox and S.K. Hsu, Int.J.Multiphase Flow, 1977, 3, 201.
15.   J.A. Schonberg and E.J. Hinch, J.Fluid Mech., 1989, 203, 517.
16.   P.S. Williams, T. Koch and J.C. Giddings, in preparation.
17.   A.J. Goldman, R.G. Cox and H. Brenner, Chem.Eng.Sci., 1967, 22, 653.
18.   J.C. Giddings, X. Chen, K.-G. Wahlund and M. N. Myers, Anal.Chem., 1987, 59, 1957.

Influence of Zone Broadening on Particle Size Analysis by Sedimentation Field-flow Fractionation

Y. Mori, K. Kimura and M. Tanigaki

DEPARTMENT OF CHEMICAL ENGINEERING, KYOTO UNIVERSITY, KYOTO 606 JAPAN

1 INTRODUCTION

Field-flow fractionation (FFF) is a relatively new analytical technique applicable to the separation of fine particles, polymers and macromolecules in solutions. Recent efforts concerned with Sedimentation field-flow fractionation (SdFFF) is to separate a wide variety of particulate species and to apply it to the particle size measurement. That is because SdFFF has advantages that it employs the fractional collection sorted by the particle mass, and has a high resolution over a wide range of particle size compared to other methods of sub-micrometer particle size determination.

However, SdFFF has inherent disadvantages of the chromatographical technique, that is the effects of the carrier solution and the zone broadening. In other elution techniques such as size-exclusion chromatography or hydrodynamic chromatography, many researchers have reported the effects of the carrier solution, especially the ionic strength of the eluent, which are explained by the colloidal interaction forces between particles and the packing material or the channel walls[11,13,14]. These effects also exist between the particles in a FFF channel and the walls of the channel. Recently, there are some papers[6,8] in which these effects and the role of the surfactant in the carrier solution were discussed. From the results of these papers[5,10], the particle size can be calculated from the retention time of the peak signal of the fractogram by the simplified theory, if the used carrier solution is of certain ionic strength with an ionic surfactant, such as 1 mol m^{-3} ionic strength and 0.1 % surfactant concentration.

On the other hand, it is more difficult to obtain the particle size distribution from the fractogram because of the existence of distortions in the fractogram due to the zone broadening by the nonequilibrium effect. Theoretical treatment in this phenomenon is well known as the deconvolution of the zone broadening[12]. Since this treatment is the inverse problem in the field of mathematics, the deconvoluted results, that is the particle size distribution, depends on the employed method of the deconvolution[12].

As a first step of the evaluation of the effect of zone

broadening on the size distribution, this paper describes the comparison between the zone broadening observed in experiments with mono dispersed latex particles and that from computer simulation by using the method of finite differences. The operating conditions of SdFFF are also discussed to lessen this zone broadening.

2 THEORY

General Principles

Figure 1 shows the outline of the principle of SdFFF and gives an overview of the experimental equipment. The separation occurs inside a narrow channel, in which smooth and parallel walls confine the carrier so that the flow is laminar. When an external field, the centrifugal force field in the case of SdFFF, is applied in the direction perpendicular to the flow, i.e. x axis, the particles are concentrated against the wall. The formed concentration gradient induces a diffusion flux in the reverse direction. After a certain time, a steady state concentration profile is reached which depends on the particle mass. Light particles are distributed near the center of the channel compared with heavy particles. The average transport speed of light particles, therefore, is higher than that of heavy particles, because of the parabolic flow profile in the channel.

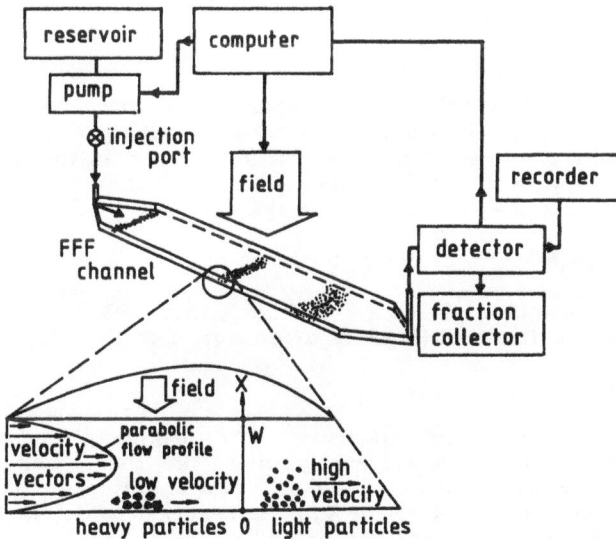

Figure 1   Schematic experimental arrangement and the principle of separation.

Particle Movement in Channel

The motion of particles in the channel of FFF is determined by the combined action of the flow, the applied field, and the Brownian migration[1]. The flux of the particles, J, can be written as a vector sum of the x axis component, J_x, in the direction of the induced external field, and the z axis component, J_z, in the direction of the fluid flow.

$$\partial c / \partial t = -\partial J_x / \partial x - \partial J_z / \partial z \tag{1}$$

$$J_x = -D \left(\partial c / \partial x \right) + Uc \tag{2}$$

$$J_z = -D \left(\partial c / \partial z \right) + v_p c \tag{3}$$

Here, D is the diffusion coefficient of the particle whose diameter is D_p, and is expressed as follows by the use of the Stokes-Einstein relationship,

$$D = kT / (3\pi \mu D_p) \tag{4}$$

where k is the Boltzmann constant and T is the absolute temperature. The expression of U is derived by the balance of the viscous resistance force of Stokes' law and the centrifugal force from the external field.

$$U = -\pi D_p^2 \Delta \rho\, G \,/(18\pi \mu) \tag{5}$$

Here G is the field strength and $\Delta \rho$ is the density difference between particle and carrier. The negative sign in Eq. 5 means that the sign of the x axis is taken to be in the opposite direction of the field.

In Eq. 3, $v_p(x)$ can be replaced by $v(x)$, because small particles could safely be assumed to move with the same velocity as the fluid at the center of the particles. $v(x)$ is the velocity of the laminar flow in the channel whose thickness is w, and is expressed as follows by using the average flow velocity, $\langle v \rangle$.

$$v(x) = 6\langle v \rangle [(x/w) - (x/w)^2] \tag{6}$$

The computer simulation in this paper was carried out by the method of finite differences using Eqs. 1 to 6.

Retention

In the "standard" theory of FFF originally presented by Giddings[2,3], the concentration gradient in the x direction is assumed to be in a steady state or quasi-equilibrium condition in Eq. 2. Also, the first term in Eq. 3 is assumed to be much smaller than the second one. We then obtain

$$-D \left(\partial c / \partial x \right) + Uc = 0 \tag{7}$$

$$J_z = vc \tag{8}$$

The retention ratio R of particles is defined as the ratio of the elution time for the particles, t_R, to that for a nonretained peak, t_o, as in usual chromatography, and its expression was obtained by Hovingh et al[7].

$$R = t_o/t_R = 6\lambda \{ \coth[1/(2\lambda)] - 2\lambda \} \tag{9}$$

$$\lambda = D/(Uw) \tag{10}$$

For highly retained sample components, the following second-order approximation is valid in place of Eq.9.

$$R = 6\lambda(1 - 2\lambda) \qquad \text{for } R < 0.5 \tag{11}$$

Field Programing Technique

The commercial Du Pont SdFFF instrument uses the time-delayed exponential decay force-field programing mentioned by Yau et. al[15]. The centrifugal force field is changed according to

$$G(t) = G_0 \qquad \text{for } t \leq \tau \tag{12}$$

$$G(t) = G_0 \exp\{ -(t -\tau)/\tau \} \qquad \text{for } t > \tau \tag{13}$$

Here G_0 is the initial field strength, τ is the exponential decay time constant. In this field programed SdFFF, the retention factor R becomes a function of time, depending on the field strength at that time, and then the average retention factor can be expressed as follows:

$$\langle R \rangle = t_0/t_R = t_R^{-1} \int_0^{t_R} R(t) \ dt \tag{14}$$

The time-dependent retention $R(t)$ is still expressed by Eq. 9, if the changing rate of the field is slow so that each particle can move to the equilibrium position which is determined by the force of the field and the Brownian motion of the particle, that is the "secondary relaxation effect" can be neglected.

Zone Broadening

As in other chromatography techniques, zone broadening in FFF operated at constant field is discussed in terms of the plate height[1,4]. For a uniform channel length L, the observed plate height H is obtained from the experimental fractogram, using the variance of a peak in time units, σ_t^2.

$$H = (\sigma_t/t_R)^2 L = H_l + H_n + H_p + \Sigma H_i \tag{15}$$

In Eq. 15, H_l accounts for the zone broadening caused by the longitudinal diffusion and is expressed by $2D/(R\langle v \rangle)$. This effect can be neglected at usual FFF operating conditions. The second term, H_n, is the nonequilibrium effect on the plate height.

$$H_n = \chi w^2 \langle v \rangle / D \tag{16}$$

χ is the complex function of λ, but in the limit of high retention it reduces to

$$\chi = 24\lambda^3 \qquad (17)$$

The third term in Eq. 15, H_p, is the contribution of the particle size polydispersity to the plate height

$$H_p = L\{(1/R)(dR/d\lambda)(d\lambda/dD_p)\}^2\sigma_d^2 \qquad (18)$$

where σ_d is the standard deviation of the particle size distribution. Equation 18 reduces to a simpler form in the limit of small λ as

$$H_p = 9L(\sigma_d/D_p)^2 \qquad (19)$$

The fourth term is Eq. 15 is the sum of the instrumental contributions from injection, detection, dead volume of the system, and flow irregularities. This term is not considered in the present work.

3 EXPERIMENTAL

The commercial instrument produced by Du Pont (SF^3 ™ Analyzer) was used, and was usually operated at the field programming mode (henceforth called TDE-SdFFF). When the experiment was operated at constant centrifugal speed in this instrument (henceforth called CF-SdFFF), a long decay time constant, τ, was chosen. The retention time and the particle diameter were calculated from raw data from the detector using the measured channel dimensions and the modified software, which were reported previously[9]. The channel is 204 μm in thickness, 51.01 cm in length and has 3.065 cm^3 channel volume. The radius of the rotor is 9.52 cm. The carrier used in all experiments was the degased distilled water containing 0.1 % di-2-ethylhexyl sodium suifosuccinate (Aerosol OT). The flow rate was 2 cm^3 min^{-1}.

Polystyrene latex (PSL) beads of two different sizes were used as 0.1 wt.% sample solution in this study. The nominal diameters of them are 269 ± 7 nm (NBS #1691) and 305 nm (8.4 nm standard deviation, Seragen Diagnostics Inc.). The density of the both latexes were taken to be 1050 g m^{-3}.

4 RESULTS AND DISCUSSION

Simulation of Zone Broadening

The nonequilibrium effect on the zone broadening should be evaluated by the computer simulation using Eqs. 1 to 3. In the present study, the simplification by using Eq. 8 instead of Eq. 3 was employed, because the zone broadening due to the longitudinal diffusion is of minor effect. For the simulation without the nonequilibrium and the secondary relaxation effects, Eq. 7 was used instead of Eq. 2.

Figure 2 shows the influence of the field strength on the experimental fractogram, when 269 nm PSL was measured at CF–SdFFF. The abscissa in Figure 2 is converted to the particle size scale from time scale of the detector signal for easy comparison, because the retention time depends on the field strength. The result at 1000 rpm rotor speed shows a wide distribution and a larger diameter at the peak signal, compared with other cases. This indicates the existence of the nonequilibrium effect. Figure 3 shows the simulated fractograms, which agree well with the measured ones in Figure 2. The difference of the fractograms on the operated rotor speed in Figure 2 can be concluded to be due to the nonequilibrium effect. When the nonequilibrium effect is not considered in the simulation, the results at any rotor speed are almost same as the one at 2900 rpm in Figure 3.

The secondary relaxation effect in the case of TDE–SdFFF is shown in Figure 4. The solid lines show the experimental results obtained by using 305 nm PSL and 5000 rpm initial centrifugal speed condition. The broken lines are the simulated curves considering the secondary relaxation effect. The constant field result is also indicated in Figure 4, which was obtained at the condition of 2000 rpm rotor speed. At decreased decay time constant, the increased peak width and the apparent particle size at the peak absorbance are observed which indicates the zone broadening due to the secondary relaxation effect. The simulated curves fit well with the experimental ones except at the larger particle size. These additional shoulder and peaks in the experimental results may be due to the agglomeration of the sample particles.

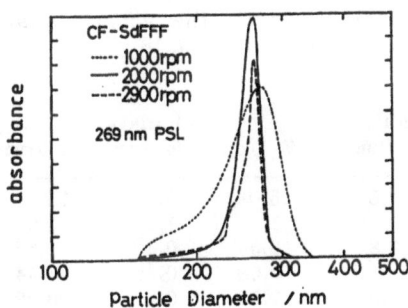

Figure 2 Effects of the rotor speed of the centrifuge on the zone broadening in the experiments of CF–SdFFF.

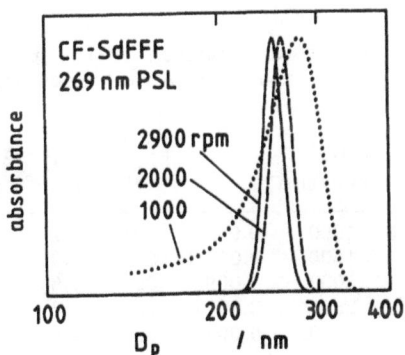

Figure 3 Simulation results of nonequilibrium effects at various centrifugal speed of CF–SdFFF.

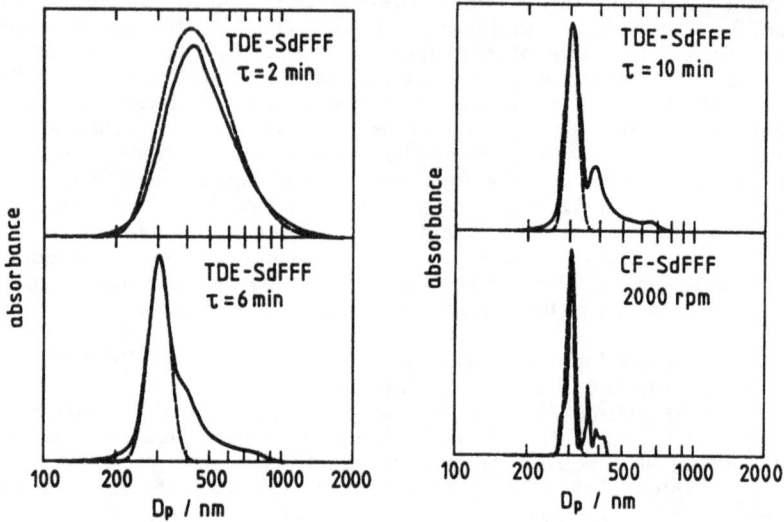

<u>Figure 4</u> Influence of the decay time of TDE-SdFFF on the zone
           broadening. Solid and broken lines are the experiments
           and the simulations at 5000 rpm initial centrifugal speed,
           respectively. The results of CF-SdFFF were obtained at
           2000 rpm centrifugal speed.

<u>Table 1</u> Comparison of the plate height among the experimental,
          simulation and calculation with Eqs. 16 and 18

rotor speed/min^{-1}	λ	Experiment H /cm	Simulation H /cm	Calculation[‡]		
				H_n /cm	H_3 /cm	H_7 /cm
1000	0.038	2.0 – 3.1	4.6	15.0	15.1	15.3
1500	0.017	2.3 – 2.9	–	1.6	1.7	2.0
2000	0.0095	0.7 – 1.6	0.8	0.3	0.4	0.66
2500	0.0061	0.3 – 0.7	–	0.08	0.15	0.44
2900	0.0045	0.2 – 0.4	0.6	0.03	0.1	0.39

[‡] H_3 and H_7 are the sums of H_n and H_p, which are obtained by
   using 3 nm and 7 nm as the standard deviation of the particles,
   respectively.

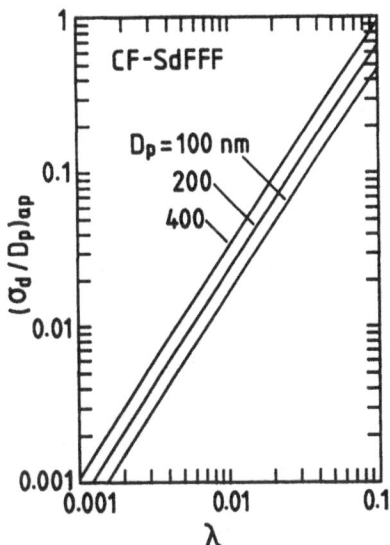

Figure 5 Apparent polydispersity as a function of particle average diameter and λ. The condition of the calculation is the same as the experimental one.

Nonequilibrium Effect

Table 1 shows the observed plate heights, which were obtained from the CF-SdFFF experiments of 269 nm PSL and the simulations at the same conditions, together with the calculations using Eqs. 16 and 18. The standard deviation of the particle size was taken to be the reported minimum and maximum values of 269 nm PSL, i.e. 3 and 7 nm, in the calculation of the plate height due to the particle polydispersity. Since the total plate height, H, in the case of the simulation depends on the number of the division in the flow direction (z axis), it is difficult to compare with the calculated results. However, there exists clearly a discrepancy in the influence of the field strength to the zone broadening between the simulation and the calculation. Comparing them with the experimental results, the simulation might be more similar than the calculation, especially at low centrifugal speed conditions. However, when 7 nm standard deviation of particle size is used, the plate height of the calculation agrees well with the observed plate height of experiment at higher field strength. The calculated plate height due to the nonequilibrium effect may overestimate at the weak field strength, even when the complex expression of χ is used instead of Eq. 17.

One of the methods to evaluate the nonequilibrium effect is the comparison of the "apparent" polydispersity converted from the plate height due to that effect. For this purpose, the ratio $(\sigma_d/D_p)_{ap}$ was calculated as the apparent polydispersity, by equating Eq. 18 to Eq. 16. Figure 5 shows the apparent polydispersity as a function of D_p and λ. If the ratio of the standard deviation and the average diameter of the actual particulate sample, σ_d/D_p, is so larger than $(\sigma_d/D_p)_{ap}$, the zone broadening due to the nonequilibrium effect could be neglected, and the particle size distribution converted from the experimental fractogram would be almost real. In the measurement with broadly distributed particles, therefore, the operating condition can be selected wider than that in the case of the sample having narrow standard deviation, that is allowing larger λ.

Secondary Relaxation Effect

Figure 6 shows the apparent average diameter converted from the experimentally obtained fractogram, ignoring the secondary relaxation effect. The apparent diameter increases as τ decreases, when the secondary relaxation effect would become larger. From the results of the apparent diameter at maximum absorbance compared with the nominal ones, both conditions of CF-SdFFF at 2000 rpm and TDE-SdFFF at 10 minutes' decay time are reasonable for reducing the nonequilibrium and the second relaxation effects. The average diameter by weight is still higher even if at those good operating condition. One of the reasons of this phenomena is the agglomerates of particles.

Figure 7 shows the apparent particle size distribution by weight on the log-normal distribution chart. The influence of decay time on the size distribution is more clearly shown, that is the difference in the bigger diameter side between CF-SdFFF and TDE-SdFFF operated at a large decay time constant is larger compared with that in the average diameter. As imagined from the curve at CF-SdFFF and TDE-SdFFF of large decay time constants, those results indicate the existence of agglomeration in the sample.

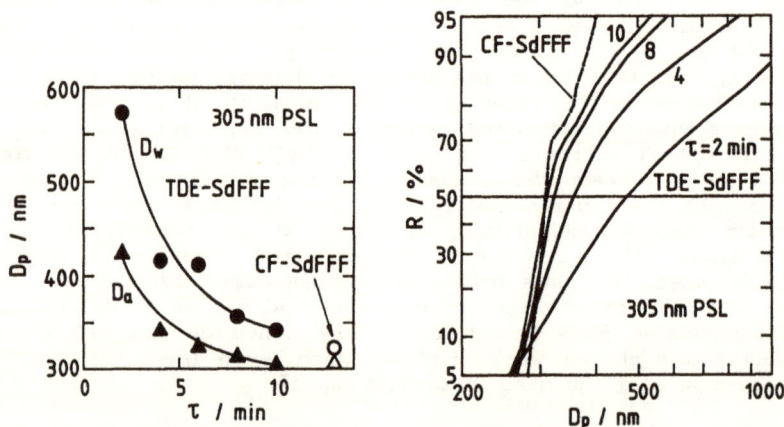

Figure 6 Influence of decay time constant on particle size at maximum absorbance, D_a, and average diameter by weight distribution, D_w.

Figure 7 Log-normal size distributions of 305 nm PSL at various decay time constants at TDE-SdFFF and at 2000 rpm CF-SdFFF.

5 CONCLUSION

Sedimentation field-flow fractionation is one of the most superior submicron size measuring methods, but has the essential disadvantage of the zone broadening. Although the field programming technique has short analysis time compared with the constant field operation, the former fractogram would be affected more by the zone broadening due to the secondary relaxation effect. However, the distortions in the size distribution due to the zone broadening might not be so serious, if the operating field strength is strong enough so that the nonequilibrium effect can be neglected.

REFERENCE

1. K. D. Caldwell, 'Modern Methods of Particle Size Analysis', H. G. Barth, Ed., John Wiley and Sons, New York, 1984, Chapter 7, p.211.
2. J. C. Giddings, *Sep. Sci.*, 1966, *1*, 123.
3. J. C. Giddings, *J. Chem. Phys.*, 1968, *49*, 81.
4. J. C. Giddings, P. S. Williams and R. Beckett, *Anal. Chem.*, 1987, *59*, 28.
5. M. E. Hansen and J. C. Giddings, *Anal. Chem.*, 1989, *61*, 811.
6. T. Hoshino, M. Suzuki, K. Ysukawa and M. Takeuchi, *J. Chromatogr.*, 1987, *400*, 361.
7. M. E. Hovingh, G. H. Thompson and J. C. Giddings, *Anal. Chem.*, 1970, *42*, 195.
8. Y. Mori, B. Scarlett and H. G. Merkus, *J. Chromatogr.*, 1990, *515*, 27.
9. Y. Mori, H. G. Merkus and B. Scarlett, *Part. Part. Syst. Charact.*, 1990, *7*, 99.
10. Y. Mori, K. Kimura and M. Tanigaki, *Anal. Chem.*, 1990, *62*, 2668.
11. D. C. Prieve and P. M. Hoysan, *J. Colloid Interface Sci.*, 1978, *64*, 201.
12. M. R. Schure, B. N. Barman and J. C. Giddings, *Anal. Chem.*, 1989, *61*, 2735.
13. C. A. Silebi A. J. McHugh, 1978, *AIChE J.*, *24*, 204.
14. H. Small, *J. Colloid Interface Sci.*, 1974, *48*, 147.
15. W. W. Yau and J. J. Kirkland, *Anal. Chem.*, 1984, *56*, 1461.

The Impact of Fractal Geometry on Fineparticle Characterization

B. H. Kaye

DEPARTMENT OF PHYSICS, LAURENTIAN UNIVERSITY, SUDBURY, ONTARIO, CANADA P3E 2C6

INTRODUCTION   What Is Fractal Geometry

Unlike most invited speakers I cannot review 25 years of progress in my presentation. Fractal geometry only burst upon the English speaking scene in 1977 when Benoit Mandelbrot's book "Fractals: Form, Chance and Dimension" was published[1]. When I inspected a copy of that book in July, 1977, I found it full of what appeared to be pictures of fineparticles and things such as fibrous filters. When I received my copy of Mandelbrot's book I was preparing a lecture for presentation at the Particle Size Analysis conference to be held at Salford in the fall of 1977. Stimulated by Mandelbrot's fractal theories I was able to prepare a presentation for the Salford Conference entitled "Characterization of the Surface Area of a Fineparticle Profile by its Fractal Dimension"[2]. The audience response to this presentation endorsed my personal evaluation of fractal geometry that it was a valuable tool for describing the structure of many fineparticles, fineparticle systems and ancillary systems important to the fineparticle technologists such as filter systems. As applications of fractal geometry in powder science and technology have evolved many different fractal based descriptive parameters for describing fineparticle systems have been developed. Thus, in the scientific literature today one reads about, amongst others, structural and textural boundary fractal dimensions, Korcak fractals and Sierpinski fractals[3]. The popular scientific literature has proliferated fantastically complex and beautiful pictures of what are known as "Fractal Systems". Such systems are of tremendous interest to the mathematician but only of passing interest to the applied material scientist[4]. It is perhaps useful to define what is known as "Applied Fractal Geometry". In applied fractal geometry, fractal dimensions are used to describe the structure of systems whereas, in Graphic or Mathematical fractals the infinitely complex structure of some mathematical systems are explored. The basic concept of applied fractal geometry is that, one can add a fractional number to the topological dimension of a system to describe the ruggedness of that system in the topological space that it occupies. Thus, lines are topologically one-dimensional and the structure of a rugged line can be described by the addition of a fractional number to the topological dimension as illustrated by the data shown in Figure 1. In this review, we will discuss the various fractal dimensions that have been used to describe fineparticle systems and the physical significance of such fractal

Topological
Dimension

Fractal
Dimension

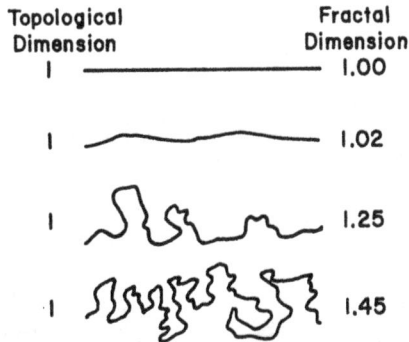

Figure 1   The basic concept of the fractal description of rugged systems is that one can add a fractional number to the topological dimension of systems to describe its space filling properties[5].

dimensions.   Techniques for measuring the structure of the fractal dimensions will be briefly mentioned but for detailed studies of the measurement techniques the reader is referred to technical publications on this topic[5,6].

2   AEROSOL AGGLOMERATES AND AGGREGATES

Fineparticle specialists have always appreciated that a complete description of a fineparticle system would have to include some measure of shape and texture.   Pioneers in the field such as Heywood and Hausner tackled the problem of shape by suggesting the use of geometric shape factors[7-9].   The use of such shape factors as aspect ratio (length of the profile divided by the width of the profile) were relatively successful for simple systems such as rock fragments. However, the complete description of rugged convoluted profiles such as those exhibited by many industrially important pigments such as carbonblack and fumed titanium dioxide eluded the best efforts of specialists to characterize their structure.   Over a period of several years up until 1977, I had been involved in attempts to characterize the structure of carbonblacks focusing on the task of describing some carbonblack profiles first described and published by Medalia[10].   On reading Mandelbrot's book on fractal geometry, I was immediately tantalized by the fact that many of his systems looked like carbonblack profiles and that Mandelbrot was able to calculate fractal dimensions for ideal systems, however, Mandelbrot did not tell the analyst how to characterize real profiles such as those of the carbonblack shown in Figure 2.   Mandelbrot had described how a giant might discover fractal geometry for himself by striding around the coastline of Great Britain and finding that as he reduced his strides his estimates of the perimeter increased.   In a famous publication on "How long is the coastline of Great Britain?"   Mandelbrot pointed out that if one plotted the perimeter estimate of Great Britain estimated by striding around the coastline with a given step on log-log graph paper then the estimates of the perimeter versus step size would generate a straight line heading for infinity[11]. Furthermore, Mandelbrot showed that the slope of such a line could be used to deduce the fractal dimension descriptive of the ruggedness of the coastline.   In Figure 2(a) a modification of the giant stepping

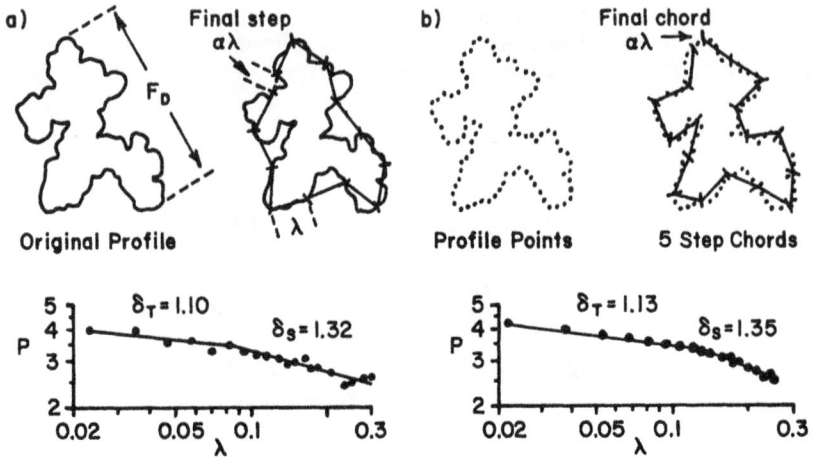

Figure 2   Describing the structure of carbonblack profiles originally studied by Medalia constituted the first successful use of applied fractal geometry.
(a) Structured walk exploration of the carbonblack
(b) The equipaced method of exploration of the profile

λ = resolution of inspection.
$\alpha\lambda$ = fractional final step
P = perimeter estimate
F_D = the Feret diameter of the profile
δ_S = structural fractal dimension
δ_T = textural fractal dimension

routine, using compasses to step around the boundary of the carbonblack profile is shown. The summary of the perimeter estimates against step size plotted on log-log graph paper is described as a Richardson Plot in honour of Lewis Fry Richardson who pioneered the study of problems associated with the indeterminacy of real coastlines[11,12]. The Richardson Plot data for an exploration of the carbonblack profile is shown in Figure 2(a). Immediately, one sees that an interpretive problem arises with real profiles such as the carbonblack became the Richardson Plot of the carbonblack profile data manifests more than one straight line.

It is useful to differentiate between ideal fractal systems and natural fractal systems. An ideal rugged coastline, as typified by the mathematical curve known as the Koch Triadic Island shown in Figure 3, has a Richardson Plot which tends to infinity with a single slope. Carbonblack profiles, such as that shown in Figure 2(a), are natural fractal systems and natural systems are said to be statistically self-similar. They can manifest more than one fractal dimension as in the Richardson Plot of Figure 2(a). At first this multiplicity of datalines, and hence of deduced fractal dimensions for a natural profile, was a source of concern to scientists attempting to explore the fractal geometry of real systems. It is now apparent however, that natural systems can manifest several fractal dimensions at various levels of

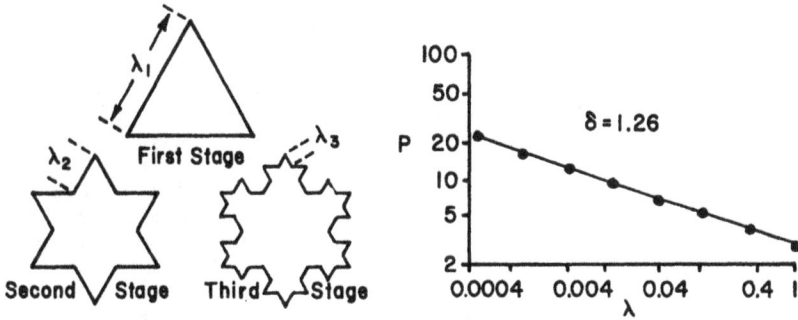

Figure 3 The "coastline" of a Kock Triadic Island is infinite and has a single boundary fractal dimension of 1.26.

λ = resolution of inspection
P = perimeter estimate
δ = fractal dimension

scrutiny and that the fractal dimension over a given range of resolution gives information on the formation dynamics operative to produce the aspect of the systems structure being inspected[13]. Thus, recent experimental studies have shown that a fractal dimension of the order of 1.40 at a coarse range of inspection is typical of an agglomerate that has been formed by the collision of primary agglomerates to produce relatively open and rugged agglomerates. On the other hand, the value of the order of 1.15 is typical of primary dense agglomerates formed in the first stage of turbulent consolidation of a population of monosized spheres generated in a fuming and/or precipitation process[5]. High resolution data of agglomerates produced by fuming often manifest the fractal dimension of 1.08 which appears to be indicative of the close packing of spheres forming the texture of the agglomerate[3,5,6,14]. Although the striding around a profile with compasses is physically easy to visualize as a method for estimating the boundary ruggedness of a profile a simpler procedure which generates data in a relatively simple manner is the equipaced method originally developed by Schwarz and Exner[15]. Their procedure is illustrated in Figure 2(b). The boundary is first digitized and stored in the memory of a computer. Next a polygon is constructed to the rugged profile by drawing a chord between a specified number of steps all around the profile as illustrated. A series of perimeter estimates are then generated by varying the number of steps between the two ends of the chord. Data generated in this manner is shown in the Richardson Plot of Figure 2(b).

In the early days of the applications of fractal geometry to a study of aerosol agglomerates, it was thought that the fractal dimension was simply a ruggedness number. It is now becoming apparent that the number is similar to a dimensionless constant used to look at complex phenomena. The magnitude of the fractal dimension gives the scientist some indication of the relative significance of forces generating a structure. Thus, just as the Reynolds number tells the physicist about the ratio of frictional to inerterial forces, the fractal dimension of a system indicates to the specialist the way in which the system has

formed.   Currently, most of the data available on the study of the
structure of aerosol agglomerates is empirical data generated by
inspecting two-dimensional projections of complex structures.
However, studies are under way to study three-dimensional growth
agglomeration of aerosol systems and the aim of such studies is to
compare real structures with those modelled on computers[16-18].   When
the structures of the model and a real system coincide, it can be assumed
that the kinetic formation rules built into the modelling algorithm
probably match possible formation dynamics for a real system.   Much of
this work is devoted to the study of commercially important pigments
and/or respirable dusts formed by a fuming process[16-19].

   The basic principle used in modelling the growth of agglomerates
can be appreciated from the two-dimensional systems illustrated in
Figure 4.   The space in which turbulent aggregation is modelled is
divided into a set of squares known as pixels.   At the centre of this space
a nucleating centre is located.   Small quantities of material, equal in size
to a pixel, are then allowed to undertake a randomwalk in the modelling
space.   If and when this wandering pixel meets the nucleating centre
orthogonally, it is considered to join the nucleating centre.   A further
pixel is then released into the randomwalk space until it either vanishes
from the space or joins the growing cluster.   If it is assumed that the
incoming pixel has 100% chance of joining the growing agglomerate if
it meets an element of the agglomerate orthogonally a well known
spider type agglomerate, known as a Witten-Sander agglomerate, is
generated[20].   Many scientists studying the structure of fumes and
pigments use a parameter known as the density fractal dimension of the
system to describe the effective density of such structures.   The density,
or mass fractal dimension as it is called by some workers, is a measure of
how efficiently the agglomerate occupies space.   A simple technique for
measuring the density fractal dimension of a two-dimensional
agglomerate is based on the interception of the arms of the agglomerate
by a series of search rings as illustrated in Figure 4(b).   It can be shown
that the Witten and Sander fractal dimension has a mass density fractal
of 1.74.   The density fractal dimension of many different systems such as
silica gels has been measured by Schaeffer and co-workers using x-ray
scattering and neutron scattering[19].   This interception technique can
be extended to a study of three-dimensional agglomerates modelled on a
computer[5].

   One of the basic parameters that can be varied in the growth
algorithm of the agglomerates is the probability of sticking on
encounter.   A model built with lower probability of sticking probably
corresponds to lower temperature in the region of the flame in which
collisions are taking place and/or changes in the kinetic energy of the
approaching pixels.   In Figure 4(c) a series of DLA agglomerates
generated using different sticking probabilities are shown.   It can be
seen that as the probability of sticking goes down the density of the
agglomerate goes up.   The boundary fractal for the agglomerates
discussed earlier in this review is an enveloping curve wrapped around
the gross overall structure of the agglomerates as illustrated by the set
of boundary profiles shown in the figure.   In the study of the physical
properties of agglomerates, a high boundary fractal dimension will
probably create a high reflectivity and/or chemical activity of a
pigment.   The higher the fractal dimension of a respirable dust the
smaller its effective aerodynamic diameter and the higher the burden

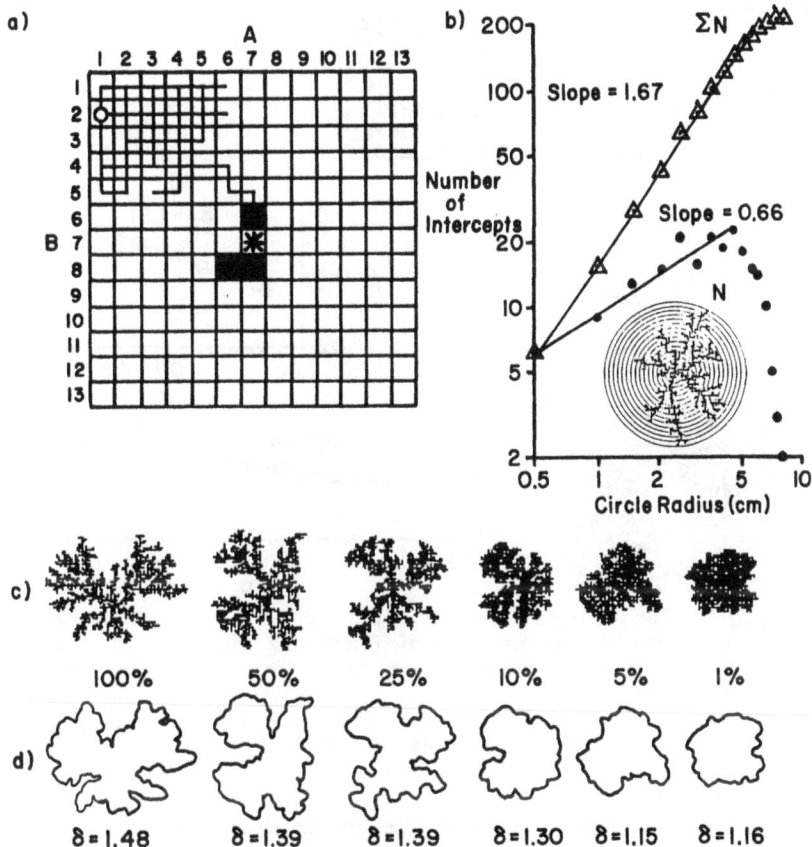

Figure 4   The growth processes of agglomerates can be modelled on computers using a technique known as diffusion limited aggregation (D.L.A.)[8].
(a) Randomwalk kinetics of the D.L.A. technique
(b) Density fractal dimension by intercept procedure.
(c) Appearance of a series of (DLA) agglomerates using various sticking probabilities in the generative algorithm
(d) Boundary profiles for the agglomerates of (c).

of toxic chemicals that it is capable of carrying into the lung.   Overall it seems that a combination of experimental modelling, direct empirical studies, and the use of fractal dimensions plus shape factors are creating a whole new technology for studying aerosol fineparticles and commercially important systems such as drugs, pigments and respirable dust hazard.

3   SIERPINSKI FRACTALS AND THE STRUCTURE OF POROUS BODIES

In Figure 5 the construction algorithm for a fractal curve known as the "Sierpinski Carpet" is shown.   By varying the construction algorithm various carpets can be constructed and they are characterized by

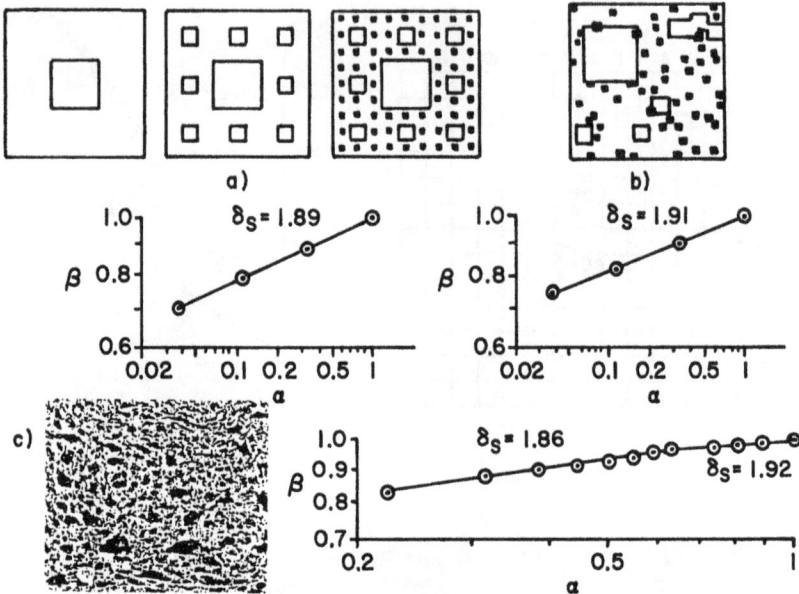

Figure 5   The Sierpinski fractal is proving to be a useful model for characterizing the fractal structure of filters and other porous bodies.
(a) Steps in the construction of an ideal Sierpinski carpet and the related Sierpinski fractal.
(b) Self-similar carpet and its related Sierpinski fractal.
(c) A section of a slice of bread and its Sierpinski fractal.

α = size of the hole just resolved
β = residual area of the carpet after holes to size α are removed
δ_S = Sierpinski fractal dimension

looking at the rate at which the carpet disappears as one varies the level of resolution used to inspect the holes in the carpet[1,3,21]. Thus, in the graph of Figure 5(a) the rate at which the carpet disappears as one sees holes of various sizes is shown.   The Sierpinski fractal dimension of such a system is defined as (2 - m) where m is the value of the slope of the curve linking the data for the disappearing surfaces as the hole size inspected decrease in size.   Note in all Richardson Plots, both for the boundary fractal data reported earlier and for the disappearing carpets, the physical slope of the line on the graph is only the appropriate magnitude for use when calculating fractal dimensions if the two logarithmic scales of the graph paper are the same.   If different magnitude log scales are used to display the data, as is often the case for Sierpinski fractal studies, then one must use the mathematical slope of the line when calculating the fractal dimensions.   Again when using Sierpinski fractal dimensions, it is useful to draw distinctions between an ideal and a natural fractal.   Natural bodies such as slices through a porous ceramic or bone will not look like the idealized carpet of Figure 5(a).   If however, one randomizes the various contributory holes in the carpet, as shown in Figure 5(b), such a system is defined as a statistically self-similar fractal system.   The statistical version of the ideal Sierpinski carpet has a lower fractal dimension than its ideal counter part because holes fall within holes in the stochastically self-similar structured system of Figure 5(b). In Figure 5(c), a slice of bread which is a good model of a sponge or depth filter, is shown along with its fractal dimension.   It is an interesting fact that again many natural fractals exhibit more than one straight line on their Richardson data plots.   In the case of the bread, it appears that the data for coarse resolution inspection represents a few big holes that represent the appearance of the bread.   The fractal dimension at the high resolution inspection is representative of the texture of the bread experienced by the eater of the bread.   If one increases the resolution to very high levels, one obviously starts to encounter the flesh of the bread. Therefore, the data from the tail end of the Sierpinski fractal information will become asymptotic to the solid content of the bread.

The Sierpinski fractal does not contain any other information than that embodied in a size distribution of holes presented in the classical manner.   However, the fractal dimension is often a convenient method of summarizing the "hole structure" of the system and as such, can be a useful descriptive parameter that can be linked to physical performance and/or physical properties.   Kaye has recently reviewed the use of fractal dimensions to discuss the structure of fibrous filters and has suggested that the structure of deposits on filter fibres may be useful in deducing information on the capture forces operative within the filter[21].   Other studies of the role of fractal dimensions in the study of filter performance have been described by Trottier et al[22].

In Figure 6 the basic type of information which could prove to be very useful to the fineparticle characterization specialist when looking at systems such as porous material is illustrated by some recent studies of the structure of healthy and diseased bone[6].   In an illness known as osteoporosis there is a decrease in bone mass with decreased density and an enlargement of bone spaces producing porosity and fragility, a condition that results from disturbance of nutrition and mineral metabolism.   In Figure 6 healthy bone and bone from a patient suffering from osteoporosis are shown.   It can be seen that the Sierpinski fractals of the two systems are very different and apparently characteristic of the two structures.   It is well known that the acceptance by the body of synthetic bone and the growth of bone tissue into a prosthetic device made of synthetic bone depends upon the pore structure of the synthetic bone.   Therefore, it would appear that the matching of the Sierpinski fractal for a synthetic bone with a corresponding natural bone may be an important step in predicting the acceptability of an implant of synthetic bone by the body.   Studies of the pore structure of synthetic bone are in progress.   A full report will be prepared in the not too distant future.   Characterization of the roughness of a surface is another area of applied fractal geometry of growing interest to the fineparticle specialist.   Kasper and co-workers have characterized the fractal structure of surfaces used in clean room technology and have been able to link particle shedding characteristic of surfaces with the fractal dimension of the substance[23].   Klinzing and co-workers have been able to link the fractal dimension of a surface to the erosion forces operating on those surfaces during pneumatic conveying[24].

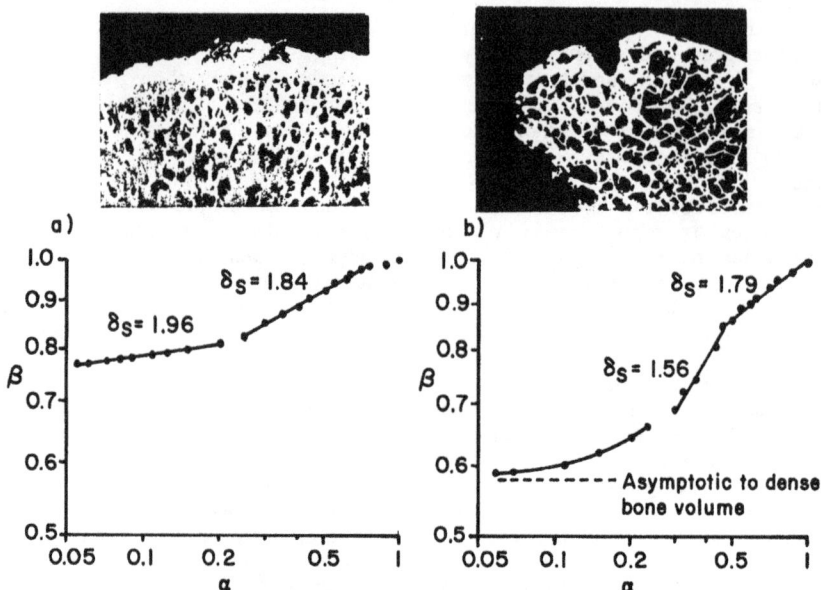

a)

b)

Figure 6   The Sierpinski fractal is a useful parameter for describing the overall structure of porous bodies such as healthy and diseased bone.
(a) Section through a healthy bone and its Sierpinski carpet data.
(b) Structure studies of a bone with osteoporosis.

α = size of the hole just resolved

β = residual area of the carpet after holes to size α are removed

δ_S = Sierpinski fractal dimension

4   IMPACT OF FRACTAL GEOMETRY ON GAS ADSORPTION STUDIES AND MERCURY INTRUSION STUDIES OF POROUS BODIES

In classical discussions of the measurement of surface areas by gas adsorption, it has always been mentioned that uncertainty in the knowledge of the size of the molecule being adsorbed in the studies is a source of uncertainty[8].   Workers in this area of study have usually assigned the reason for discrepancies between surface area estimates based on the use of different gas molecules to this source of uncertainty. It is now becoming apparent that in fact studying the surface area of a solid using different sized gas adsorbant molecules is a method of estimating the fractal roughness of the surface.   Thus, Avnir has reviewed the gas adsorption literature and reported that in every case where a surface was studied with more than one type of gas a Richardson type plot of surface area against molecular size of adsorbant resulted in the generation of a dataline which could be interpreted as a fractal dimension of the rough surface[25].   From a fractal perspective discrepancies between surface area estimates based on different adsorption gases is not uncertainty but information on the fractal roughness of the surface.   This fact is revolutionizing the surface area measurements of solids by gas adsorption.   The fractal roughness of reactive surfaces is of particular interest to chemists studying catalytic reactions.   The overall flow through the porous body supporting a catalyst is also proving to be an active branch of applied science.

Other specialists making use of the fractal description of porous bodies are petroleum engineers interested in secondary oil recovery, and pharmaceutical specialists describing the structure of transdermal continuous released drug systems.   Such studies are making use of another area of fractal geometry which is known as percolation theory[3,26].

The reinterpretation of mercury porosimetry data for certain types of porous bodies is proceeding.   If a porous body has been assembled in a way similar to that of an Apollonian gasket shown in two-dimensions in Figure 7, then the Washburn equation, which interprets increased penetration of a porous body at higher pressures as being due to the invasion of progressively smaller throats, is a valid model for the structure of the body[27-29].   Thus, one would anticipate that sedimentary rocks should have pore structures which are interpreted as fractal dimensions using mercury porosimetry data to deduce the value of the fractal dimension.   The Menger sponge, which is a three-dimension analog of the Sierpinski fractal, has also been used to interpret the pore structure of bodies[30].

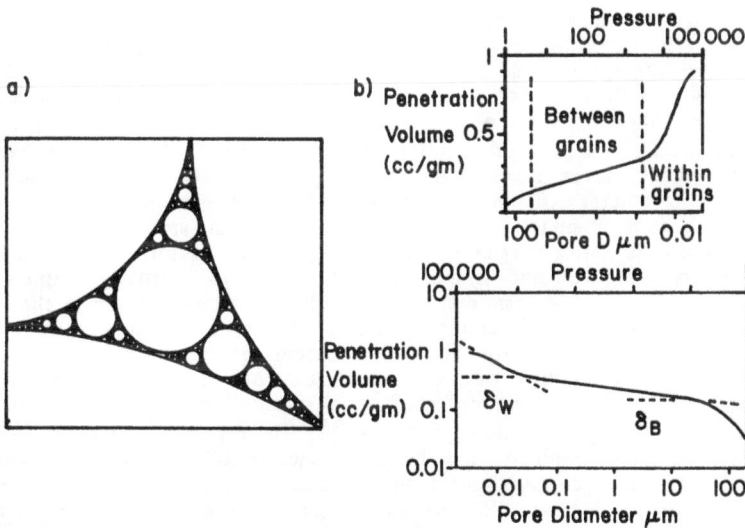

Figure 7   The pore structure of some porous bodies, those similar in structure as a mathematical system known as an Apollonian gasket, should be describable by fractal dimensions deduced from mercury intrusion data
(a) The appearance of an Apollonian gasket
(b) Mercury intrusion data revised to a form compatible with Fractal dimensions.

δ_B = fractal dimension related to the packing of the grains
δ_W = fractal dimension of the pores within a grain

5   FRACTAL GEOMETRY AND PULVERIZATION EFFICIENCY STUDIES

Not only is fractal geometry revolutionizing the significance of surface area measurements by gas adsorption but a fractal perspective on the significance of surface area data is causing reassessment of one of the major goals of surface area measurements in the mineral processing industry and powder production research.   At a superficial level, it is obvious that the surface area of a powder represents part of the energy put into a pulverizing process.   Since however, the theories of fractal geometry demonstrate that one can never measure the surface area of a powder in the absolute sense, then all magnitudes of surface area reported in the literature are operative estimates dependent on the operation used to estimate the surface area.   It is now becoming clear that the measured surface area can only be a relative measure of pulverizing efficiency.   Thus, from a simple perspective, the statement that the surface area of all rugged fractures is in essence infinity would indicate that the single breakage of a rock will always create an infinite surface which would violate the first law of thermodynamics[3].   The resolution of this paradox is to realize that all real surface areas are not infinite but indeterminate and that one can never have absolute knowledge of the magnitude of the surface area of a powder.   It has been pointed out that fractal geometry also offers the possibility of a new approach of crushing and grinding research.   In the past there have been many extensive studies of the fragmentation produced by slow crushing of objects such as glass spheres.   It is now evident that the utility of such studies is extremely limited since, in fact, most crushing and grinding procedures to produce very fine powders are not slow crushing processes but involve a phenomena known as ballistic fracturing.   Ballistic fracture is the interaction of reflected shock waves to create zones of tension in which the body fails and flies apart by tension and not crack propagation by compression[31,32].   The reflection of such shock waves is dependent both upon the point of impact and the irregular geometry of the fragment and these parameters will vary from impact to impact.   Therefore, the classical study of interacting mechanisms of slow crushing to the dynamic fracturing of irregular profiles is irregular and makes predictive studies virtually impossible. Modern studies in deterministic chaos then indicate strongly that the scientist could return to empirical studies of direct fracture of complex systems if he would gain an understanding of ballistic fracture.   In ballistic fracture studies one can make use of what is known as the Korcak fractal.   The Korcak fractal is a measure of the size distribution of the elements of a fragmented body[33].   In Figure 8 a tracing from a high speed photograph of the ballistic fracture of a piece of resin is shown.   The size distribution of such fragments is plotted in Figure 8(b). The fact that a log-log plot of such fragments yields a straight dataline indicates that the survival of a large fragment involves a rare combination of interacting cracks.   In the language of the stochastic expert, such a relationship represents what is known as a $1/f$ relationship and the slope of the line f can be interpreted as a fractal dimension[34].   It has been suggested that the fractal dimension of the fragments is related to the rugged boundary of individual fragments and in an ideal case the relationship is

$$1/f = \frac{\delta}{2}$$

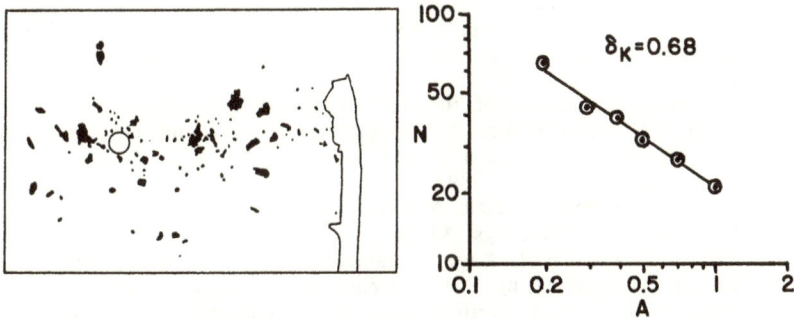

Figure 8   The fractal boundary of the fragments produced by ballistic disintegration of a body may be related to the size distribution of the fragments produced by such a disintegration procedure.

A = area of the fragment
N = number of fragments of the stated size or larger
δ_K = Korcak fractal dimension

where δ is the structure boundary fractal dimension of the fragments

Even if the relationship involved turns out to be more complex than this simple relationship it is obvious that such a suggestion is worth exploring as a means of predicting and improving fragmentation studies.   It is hoped that this brief survey of the different aspects of fractal geometry and their impact on characterization studies will stimulate others to carry out more experiments in this area.

ACKNOWLEDGEMENTS

The author wishes to acknowledge the help of G. Clark in the preparation of the diagrams for this report.

REFERENCES

1.   The basic concepts of fractal geometry were introduced to the English speaking world in the book by B. B. Mandelbrot entitled, "Fractals, Form, Chance and Dimension".   The book was published by W. H. Freeman and Company, San Francisco, 1977.   This book is a translation and extension of a book published in french two years earlier.   In 1983 Dr. Mandelbrot published an extended and revised version of his 1977 book entitled, "The Fractal Geometry of Nature", W.H. Freeman and Company 1983.   Dr. Mandelbrot considers this book to be the definitive text on fractal geometry. He has indicated that he prefers that the second edition of the book should always be quoted as the basic reference for fractal geometry.

2. B. H. Kaye, "Characterization of the Surface Area of a Fineparticle by its Fractal Dimension", Proceedings of the Salford Conference on Particle Size Analysis, M. J. Groves, Editor. Hayden and Sons Ltd., 1977, pp. 250-259

3. A convenient review of the various fractal dimensions used in fineparticle science and technology are to be found in B.H. Kaye, "A Randomwalk Through Fractal Dimensions", VCH Publishers, Weinheim Germany, 1989.

4. B.B. Mandelbrot, "Fractals A Geometry Of Nature", New Scientist, September 15, 1990, Pgs. 37-43.

5. The publication B.H. Kaye, "Characterizing The Structure Of Fumed Pigments Using The Concepts Of Fractal Geometry", in press, Particle and Particle Systems Characterization contains many references leading into the specialist literature on Size Characterization.

6. Detailed techniques for carrying out the measurement of the fractal dimension of fineparticle systems are presented in the book, B.H. Kaye, "Discovering The Surprising Patterns Of Chaos And Complexity", to be published by VCH Publishers, Weinheim, Germany, publication date early 1992.

7. The pioneer work of Heywood and Hausner is reviewed in the standard textbooks on fineparticle characterization see for examples References 8 and 9.

8. T. Allen, "Particle Size Analysis", Chapman and Hall, 4th Edition, 1990.

9. B.H. Kaye, "Direct Characterization of Fineparticles", J. Wiley, New York, 1981.

10. A. I. Medalia and G. J. Hornik, "Pattern Recognition Problems in the Study of Carbonblack", Pattern Recognition, 1975 4, 155.

11. B.B. Mandelbrot, "How Long is the Coast of Britain, Statistical Self Similarity and Fractional Dimensions", Science, 1967, 155, pp. 636-638

12. The most accessible reference to, and discussion of, the work of Lewis Fry Richardson on the Indeterminacy of Natural Coastlines is to be found in the books of Reference 1.

13. B.H. Kaye, "Fractalicious Structures And Probable Events", an address given to the Maths Educators of Canada at their annual conference, Vancouver, 1990, published as part of the proceedings of the conference. The same material is presented in extended form in Reference 6.

14. B.H. Kaye, "Characterizing The Health Hazard of Respirable Dusts", (In preparation).

15. H. Schwarz, H.E. Exner, "The Implementation of the Concept of Fractal Dimensions on a Semi-Automatic Image Analyzer", Powder Technol., 1980, 27, 207-213.

16. A.J. Hurd, W.L. Flower, "In Situ Growth and Structure of Fractal Silica Aggregates in a Flame", Journal of Colloid and Interface Science, March, 1988, 122, No. 1.

17. R. Richter, L.M. Sander and Z. Cheng, "Computer Simulations of Soot Aggregation", Journal of Colloid and Interface Science, July 1984, 100, No. 1, , 203-209.

18. P. Meakin, Section 3.12 in the "Fractal Approach to Heterogeneous Chemistry", Ed. D. Avnir , John Wiley & Sons, 1989.

19. D.W. Schaeffer, "Polymers, Fractals and Ceramic Materials", Science, February 24, 1989, 243, 1023-1027.

20. T.A. Witten, L.M. Sander, Phys. Rev. Lett, 1981, 47, 1400.

21.   B.H. Kaye, "Describing Filtration Dynamics From The Perspective Of Fractal Geometry", accepted for publication in KONA, an international Powder Science and Technology Journal, published by the Hosokawa Micron International Inc., Americas Block, 10 Chatham Road, Summit, New Jersey, USA, 07901

22.   R.A. Trottier, R.C. Brown, "Production of Neutral Submicrometre Aerosols and Their Use In Testing Filters" R.A. Trottier, I. Stenhouse, "Possible Links Between The Fractal Structure Of Dust Capture Tree Deposits In A Fibrous Filter And Loading Effects", extended abstracts of these papers are in the proceedings of the 5th Annual Conference, of The Aerosol Society, Loughborough University of Technology, Loughborough, England, March 26-27, 1991.

23.   G. Kasper, S. Chesters, H.Y.. Wen, M. Lundin, "Fractal-Based Characterization of Surface Texture", <u>Applied Surface Science</u>, 1989, <u>40</u>, 185-192.

24.   A. Zaltash, C.A. Myler, S. Dhodapkar, G.E. Klinzing, "Application of Thermodynamic Approach to Pneumatic Transport At Various Pipe Orientations", <u>Powder Technol</u>, 1989, <u>59</u>, Pgs. 199-207.

25.   D. Avnir, editor, "The Fractal Approach to Heterogeneous Chemistry", John Wiley & Sons, 1989.

26.   L.E. Holman, H. Leuenberger, "The Effect of Varying the Composition of Binary Powder Mixtures and Compacts on Their Properties: A Percolation Phenomenon", <u>Powder Technology</u>, 1990, <u>60</u>, Pgs. 249-258.

27.   C. Orr, "Application of Mercury Penetration in Material Analysis", <u>Powder Technol.</u>, 1969-70, <u>3</u>, Pgs. 117-123.

28.   S.H. Ng, C. Fairbridge, B.H. Kaye, "Fractal Description of the Surface Structure of Coke Particles", <u>Langmuir</u>, (3), May-June, 1987, <u>3</u>, Pgs. 340-345.

29.   B.H. Kaye, "The Apollonian Gasket as a Model For Evaluating Mercury Intrusion Porosimetry Data", paper presented at the Rosemont Powder Technology Conference, Chicago, Illinois, 1988.

30.   W. Freisen, R.J. Mikula, Canmet Divisional Report, ERP-CRL 86-128.   Available from Energy Mines and Resources Canada, Canmet Technology Information Division, Technical Enquiries, Ottawa, Ontario, K1A OG1, Canada.

31.   B.H. Kaye, "Describing The Structure of Fineparticle Populations Using The Korcak Fractal Dimension", Particle Size Analysis, Conference Guildford, April 19-20,1988, proceedings published by the Royal Institute of Chemistry, Great Britain.

32.   For a discussion of the process of ballistic failure of pieces of material see J.E. Gordon, "The New Science of Strong Materials, or Why You Don't Fall Through The Floor", 2nd Ed., Penguin Books, 1976.   In the United States this book (1984) is available as a Princeton University Paperback , Princeton University Press, 41 William Street, Princeton, NJ, 08540.

33.   J. Korcak, "Deux Types Fondamentaux, de Distribution Stastitique", <u>Bull. Inst. Int. Stat.</u>, 1938, <u>3</u>, Pgs. 294-299.

34.   B.J. West, M. Shlesinger, "The Noise in Natural Phenomena", <u>American Scientist</u>, January-February, 1990, <u>78</u>, Pgs. 40-45.

Using Mathematical Morphology to Perform Particle Sizing

Andrew Morris

LEICA CAMBRIDGE LTD., CLIFTON ROAD, CAMBRIDGE CB1 3QH, UK

1 INTRODUCTION

The technique of image analysis allows quantitative results to be obtained from images. The technique used to obtain the image and the size of the features which it contains are relatively unimportant to the analysis process, since a calibration coefficient relating to the size of the image to the size of the "real world", can always be set. The analysis technique involves a controlled data reduction. The original image contains a high number of separate points, each set to a specific grey level. The result of analysis of an image can be a single piece of information; for example does a sample confer with the specification; a single number such as the average particle diameter; a size distribution of the features present or a list of each individual feature's attributes and characteristics.

As the image analysis technique has developed so the number of measurements possible has increased. The number of techniques which are available to segment or separate the objects for measurement from the background have also increased. These techniques have enabled more and more applications to be addressed using image analysis with ever more complex images being analysed. The hardware used to perform the analysis has also been improved, meaning more and more particles are being analysed in shorter and shorter timescales.

This paper begins with an overview of the analysis method and continues to describe a method called mathematical morphology. This technique is used to simplify grey images prior to their measurement and therefore has many uses in image analysis.

The paper finally discusses some real life applications where image analysis has proved invaluable and lastly a technique called granulometry. This method relies on mathematical morphology to construct size distribution from the most complex of images which could not be analysed by any classical method.

2 HISTORICAL OVERVIEW

Image analysis has been providing scientists with quantitative results from images for many years. The classical technique is to start with a digitised grey level image containing the objects to be measured. This image can contain objects of any size (e.g. lakes in a satellite image or pollen grains in a SEM image). Providing that correct calibration is employed then quantitative measurements will be produced in the units of calibration. In order to extract the extent of the features of interest the grey level image is thresholded between a pair of grey levels into a corresponding binary image. For example all the dark features can be extracted and measured against a lighter background, alternatively all the light features on a dark background or those within a slice of grey levels. Examples of features detected might include inclusions in a steel sample or the nuclei of cells dispersed on a microscope slide.

The problems associated with segmenting light particles against a non–uniform background are shown in Figure 1. At the higher threshold setting the features on the right–hand edge are not segmented, whilst at the lower threshold those on the left are merged together.

Measurements on the features usually comprise a list of parameters for each feature. Each object in the binary image is therefore classified according to its area, width, height, shape, number of holes and so on. Imagine the original grey level image consists of part of a page of printed text. A threshold could easily be set extracting the letters into a binary plane. A feature analysis would create a list of measurements, or feature vector, on each feature or object within the image. Examination of the data will yield, for example, the difference in width between the letters 's' and 'm'.

Image analysis systems of this type work well for high contrast images where all the features are similar in grey level, and can therefore be segmented easily. As image contrast or quality falls, so the results produced from such a system increasingly become inaccurate and open to criticism.

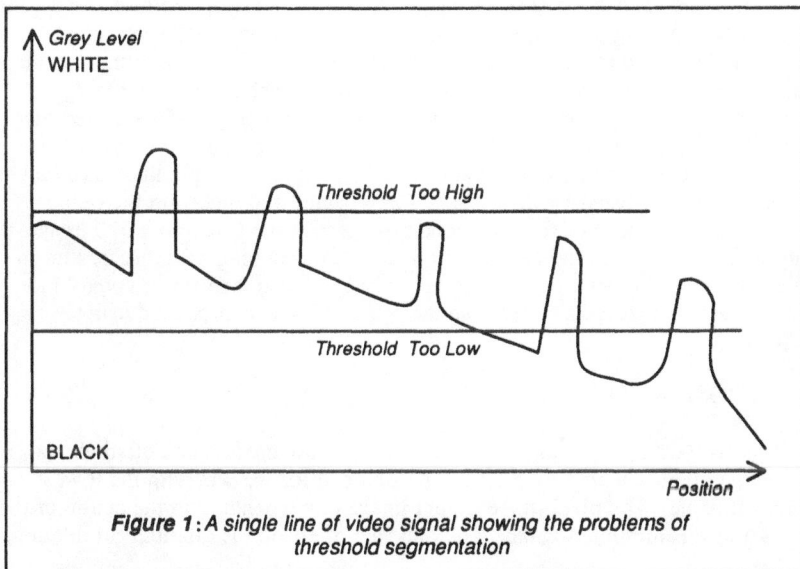

Figure 1: *A single line of video signal showing the problems of threshold segmentation*

The image analysis technique is effectively a method of controlled data loss, firstly at the thresholding stage and secondly when a list of numbers or results is created from a binary image. If we assume that the original grey image contains 255 grey levels (that is 8 bits) then we are effectively throwing away seven eighths of the data when performing the thresholding operation. This is an extremely unforgiving operation to perform, and some method of cushioning this data loss is required.

3 IMAGE MORPHOLOGY

Morphological processing provides a way of transforming grey level images so that unwanted detail is selectively discarded. When sufficient unwanted detail has been removed from the grey image then the process of segmenting and measuring the features of interest is made easier. In this paper I will describe the basic Morphological processes and also the way in which they are used to solve demanding applications.

The word morphology sounds rather grand, however it really just means "shape", thus morphological grey level processing is concerned with extracting the maximum information from the shape (or arrangement of the pixels) within a grey level image.

4 BASIC CONCEPTS

The basic morphological operations are performed in a similar way to linear convolutions. Simple convolution operations (such as Sobel edge detection) take an input image and run a mask or kernel over every pixel in turn, at each point calculating a new pixel value, which is placed into an output image.

In morphology the kernel is renamed a structuring element. As with convolution the SE is centred on every pixel in the input image in turn and a value for the corresponding pixel in the output image is calculated. The important difference lies in the method of calculation for the pixels placed into the output image. In convolution ALL pixels covered by the kernel are used in the calculation of an output pixel's value, whereas in morphology only selected pixels in the SE are used. The choice of which pixels to use and how to combine them depends on the morphological operation in question; for example the maximum or minimum pixels covered by the SE could be selected and placed in the output image. The high speed of the grey–tone processor in the Quantimet 570 image analyser combined with the relative simplicity of morphological processes mean far larger structuring elements can be sensibly employed than linear convolution kernels. Large processing effects can therefore quickly be produced using morphology.

5 EROSION AND DILATION

The two best known morphological processes are erosion and dilation. In erosion the pixels covered by the SE are examined for the one with the lowest numerical value. The pixel in the output image corresponding to the centre of the SE is set to this minimum value. The opposite of erosion is dilation. In dilation,

Original Image

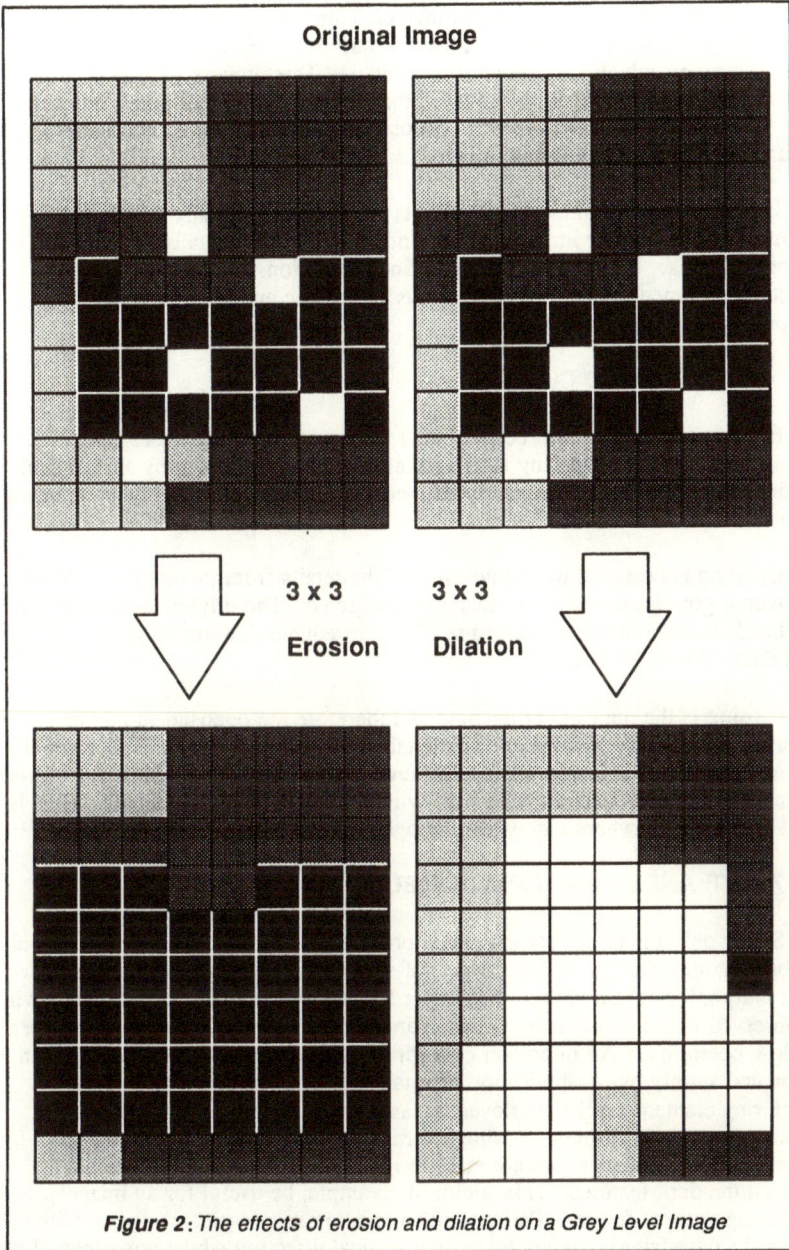

3 x 3   3 x 3

Erosion   Dilation

Figure 2: *The effects of erosion and dilation on a Grey Level Image*

the maximum value covered by the SE is chosen and copied to the output image.

Erosion produces two effects when the input and output images are compared. Firstly the output image is darker than the input image. This is not surprising since the minimum pixel values within the SE have been selectively chosen. The second effect is that bright detail within the input image which are smaller than the SE will be removed in the output image. This is because as the SE passes over the light detail it selectively chooses a dark pixel from the background thus removing pixels which correspond to the details. This process will occur at the edges of large light features, however in their centres , where all pixels are set to high values, the SE will be forced to choose a higher value pixel, and therefore the feature will appear in the output image.

In dilation exactly the opposite effects are observed. Firstly the general brightness of the output image is raised and secondly dark details are removed in the output image, (again "details" are defined as regions smaller than the structuring element used). Figure 2 shows an erosion and dilation when performed on a grey level image.

6 OPENING AND CLOSING

Erosion and dilation can be combined into two further processes called opening and closing. Opening is defined as an erosion followed by an identical dilation whilst closing is conversely defined as a dilation followed by an identical erosion.

Opening is designed to remove small light details from images but to leave the overall grey level and larger features unaffected. The original erosion removes the details but also darkens the image. The subsequent dilation however re–lightens the entire image.

Closing is the opposite of opening and therefore the opposite should be observed, namely the removal of dark details in the output image. The original dilation brightens the image whilst also removing the dark details; this is followed by an erosion which darkens the image without reintroducing the details. This is shown in Figure 3 where a grey level image is shown both opened and closed.

7 SIZE AND DIRECTIONAL EFFECTS

So far only the basic morphological processes have been described. It should not be forgotten that the size and shape of structuring elements can be altered for most morphological processes. If a larger SE is employed then larger effects will be observed in the output image – for example larger details removed using open or close operations. An important concept of morphology is that large SE's can be created simply by re–application of smaller SE's. The orientation of the structuring element can be employed in cases where directional effects are required in the output image. Using a vertical opening operation will merge darker features vertically, but not effect bright regions which lie horizontally between the dark features. This could, for example, be useful for an image containing vertical fringes. Performing an opening in a vertical direction will ensure the dark fringes are unbroken in a vertical direction whilst not affecting the bright regions separating the fringes. It is very important that the structuring element should be matched to the processing which is required.

8 ADVANCED TRANSFORMS

Morphological operations are not limited to the four already described; there

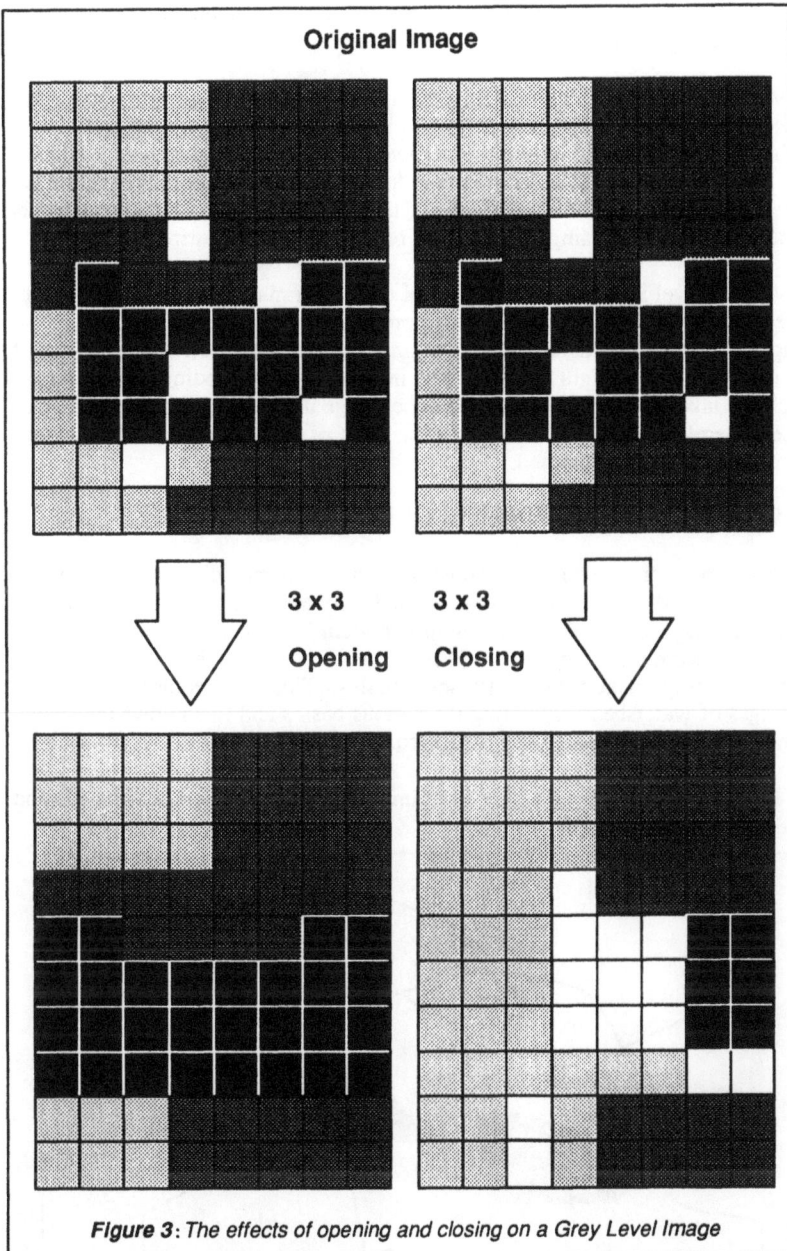

Figure 3: *The effects of opening and closing on a Grey Level Image*

is in fact a cornucopia of choice. Some of the more common functions are described here in order to provide a flavour of the processing effects which can be achieved. By subtracting the result of an open or close operation from the original image the details within the image can be highlighted, (light details in the case of opening and dark details in the case of closing). This is known as a top–hat transform. Again the size and shape of the structuring element can be altered in order to highlight different size and shaped details.

Another technique, called the morphological gradient, is used to determine edges within images irrespective of their orientation. The maximum and minimum points covered by the SE are determined and their differences taken. This difference is the value sent to the output image. Although this function will extract edges from images irrespective of their gradient, a directional bias can be introduced simply by using an elongated or directional structuring element.

A grey level image can be thought of as a relief map where the bright points correspond to the mountain tops. Often processing is required which will completely remove all dark features within an image, irrespective of their size, but will not process the bright features. The morphological 'flooding' operation will fill all the lakes or dark features within the image until they are brimming. This process therefore removes detail about the relief of the lake beds whilst leaving the mountain peaks untouched.

9 THE WATERSHED OPERATION

A complex grey level image contains a lot of information, however the zones of influence of the depressions is not immediately apparent. Extracting this information is often required for example in grain boundary reconstruction or analysis of two dimensional electropheretic gels. The watershed operation segments a grey image into its catchment basins. This process therefore highlights the ridges of watersheds separating the various basins and this allows the basins themselves to be segmented and measured.

Figure 4 shows part of a grey level image with the different watersheds and dividing lines highlighted

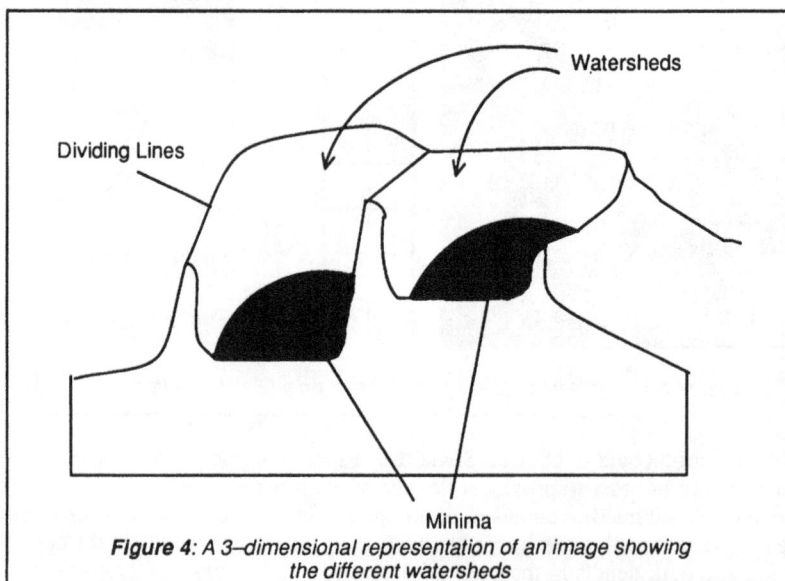

Figure 4: A 3–dimensional representation of an image showing the different watersheds

Many morphological transforms exist, only a few of which are described in this article. They are useful because they selectively extract these features within an image which are of interest whilst removing the unwanted characteristics.

Having described some of the techniques of morphological processing, the next stage is to examine some application examples and the way in which these techniques can be used to extract the required data.

10 SIZING OF ALUMINIUM GRAINS

The first application is to detect grain boundaries in an aluminium sample. The metal is highly polished and imaged using incident light in an optical microscope. The requirement is to identify the boundaries of the grains as closely as possible so that the size of the grains can be related to the physical properties of the metal. In order to segment the boundaries, extensive grey level morphological processing needs to be performed.

The first stage in the processing removes the small detail from the image, but does not affect the boundaries. This is achieved using erosion and dilation commands. A morphological gradient is used to highlight the edges of the grains and a watershed employed to remove unwanted noise in the centres of the grains and identify the grain boundaries.

The boundaries can be detected and thinned using a binary skeleton, their areas measured and a size distribution constructed of the grains. This can be used to construct a relationship between grain size and the physical properties of the metal.

11 GRANULOMETRY

Some images are completely unsuitable for image analysis using the "threshold then measure" approach. This unsuitability can be caused by factors such as shading across the image, particles touching, particles lying in heaps and so on. Amazingly, using mathematical morphology, size distribution can still be constructed from highly complex images, using a technique call granulometry.

Imagine a dark background overlaid with bright particles. The bright particles fall into three classes, small spherical objects, medium sized spherical objects and lastly large jagged flakes. The density of objects means that it is impossible to construct a size distribution using the "threshold then measure" approach. The binary image obtained from a thresholding operation is overly complex and will not respond favourably to binary processing techniques, and thus a different approach needs to be employed to construct the size distribution. This is where granulometry is employed, as follows. Opening operations of increasing size are applied to the grey image. By subtracting the opened image from the original, the detail within the image which is smaller than the structuring element is high–lighted. By measuring the difference between the opened and original images a measure is obtained of the amount of detail.

The granulometric technique therefore starts with the smallest opening possible. If this is smaller than the small spheres then no detail is present in the output image. As the size of the structuring element reaches the size of the small spheres, so the amount of detail in the output image will rapidly rise. When the size of the element is between the small and medium features, so the amount of detail in the output image is again low As the size of the structuring element reaches the size of the medium features so the amount of detail in the output image will again rapidly rise. This pattern is repeated between the medium and large features.

By noting the amount of detail in each output image, and comparing one with another, then it is possible to construct a standard size distribution histogram, where the horizontal axis is size of detail (or structuring element) and the vertical axis is amount of detail in the output image. Obviously this technique will produce histograms with larger errors than by a more precise technique, however the overall size and shape of the distribution will be correct.

The granulometric technique is interesting because it does not rely on the setting of a threshold value or use of a binary image. It therefore performs particle sizing in a way more familiar to the one used by the human eye and brain.

12 MORPHOLOGY AND IMAGE ANALYSIS

This paper has attempted to describe basic morphological image processing on grey level images. The relevance of these techniques to image analysis is highlighted by the use of two application examples which were unsolvable using classical image analysis, convolution or fourier techniques.

The strength of the morphological technique is that it allows controlled data loss prior to the creation of a binary image. Morphological transforms or tools are available which produce a high variety of effects in the output image, thus a transform can be chosen to selectively discard the information which is unwanted. The output image (from a morphological transform) can act as input to a second transform and so on until the grey level image is suitable for the analysis process.

The morpho–processor at the heart of the Quantimet 570 image analyser performs morphological transforms very quickly. The morphological techniques described in the article are only useful if they can be performed in realistic timescales. Using the Q570 a 3 x 3 erosion takes 27 msec, thus all the transforms described can be performed in realistic timescales. The morphological treatment of images is still a developing science, however very impressive results can be created using the transforms and hardware available today.

Three Dimensional Reconstruction of Particles

Michael Bottlinger

DEUTSCHES INSTITUT FÜR LEBENSMITTELTECHNIK, PROF.-V.KLITZING-STR. 7,
D-4570 QUAKENBRÜCK, GERMANY

1. Introduction

The behaviour of complex systems like aerosols and suspensions depends on numerous physical properties. For example, it is influenced by the viscosity and density of the fluid phase and by the size and density of the constituents of the disperse phase (particles), to name only a few. The effect of particle shapes has often been neglected or considered as too difficult to determine and was included only in a few investigations.

In most cases where particle shapes are included in the investigations, the shape of particles is registered only in two dimensions. Up to now it is not possible to state what kind of error is induced by this proceeding. Only for fractal shapes of a special kind Gentry [1] proved by simulations that a systematical error in the calculated fractal dimension arises from analyzing the projected images of particles instead of their three dimensional shape (Gestalt).

There exist different methods for three dimensional registration of particles:

Stereoscopy and Interferometry
In these cases the surface of the particle is viewed only from *one* side. There exists no information about the other half of the particle. Both methods require preparative steps like projecting a grid on the particle surface by optical means or a complex optical setup. The purely computer based evaluation of stereo images demands to find corresponding points of the particle surface on the two images automatically and is therefore very time consuming.

Reconstruction from projected areas (coplanar)
Here the particle is viewed from different angles in the same plane. From the resulting projected areas a three dimensional reconstruction can be calculated [2; 3]. When the projections from many different angles are included, the visible part of the particles can be reconstructed very precisely. However, there are regions of the particle that are not registered while other parts of the surface are covered more than once.

In this paper a method is presented to register and reconstruct a particle from a minimum of different images (three) and gain the maximum information about the surface.

2. Shape reconstruction from three orthogonal direction

2.1. Experimental technique

The results presented in this paper were achieved by a system where every analyzed particle resting on an object slide, was definitely oriented by special holders and viewed under a microscope. Fig. 1 shows in principle the utilized coordinate system in relation to a system aligned to the object slide. The particle can be imagined to rest in the origin of a three dimensional, Cartesian coordinate system, oriented in the depicted manner. The angle α has a value of approximately 27°.

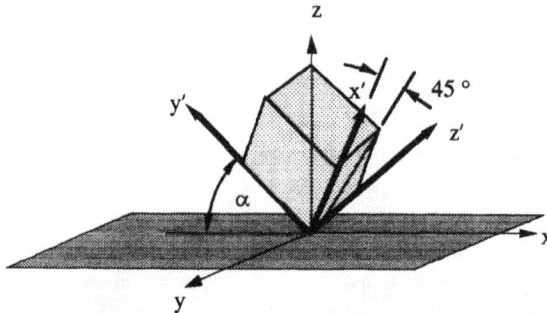

Fig. 1 coordinate system K' (x', y', z') in relation to the coordinate system K (x,y,z) of the object slide (laboratory system)

The particle was illuminated from the background to achieve a good contrast between background and projected image. For image digitizing a Macintosh II computer with an 8-Bit video digitizer (Quickcapture from Data Translation) was used. The gray scale image was transformed into a black and white image, were all subsequent steps of the analysis were performed. Fig. 2 depicts schematically the experimental set-up.

From the particle image the contour line is determined and is approximated by a polygon. This is necessary for the scaling algorithm used later in the three dimensional reconstruction. The program allows to choose polygons with 3 to 20 sides. The exact number of sides depends on the complexity of the particle shape and the requirements of later processing steps, where the three dimensional representation is used for the calculation of physical properties (volume, center of gyration) or modeling of the physical behaviour (e.g. light scattering [4]). The experimentor can place the corners of the approximating polygon by hand (using a special input device (mouse)) or start different algorithms to determine the corners automatically. The programs position the corners either symmetrically - in constant angle steps- or at relative maxima of the radius in given angle segments. The resulting polygon is stored.

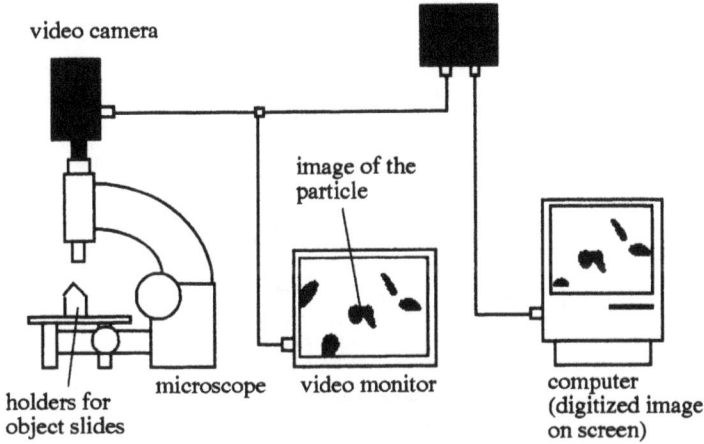

Fig. 2 Experimental set-up for the registration of particles

2.2. Reconstruction

The process explained above is conducted for all three projections. Fig. 3 illustrates the spatial relations of the three contour lines. The arrangement of the outlines shows in which manner the dimensions correspond.

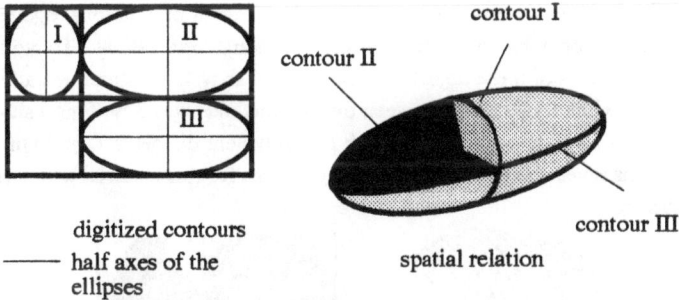

Fig. 3 spatial relation of the contour lines

For the reconstruction of the particle, one of the outlines is chosen and the corresponding polygon is divided into an upper and lower section. The other two contour lines are spatially related, as illustrated at the top of Fig.4. This three-dimensional wire-structure is intersected by 26 planes, parallel to contour I. The intersection planes are equidistantly distributed over the whole object. Every intersection contains four points, defining a rectangle, into which the two sections of the polygon are separately scaled to fit inside. This operation is executed for all 26 intersections. The result of this procedure is 26 equidistant polygons, which approximate the original particle shape.

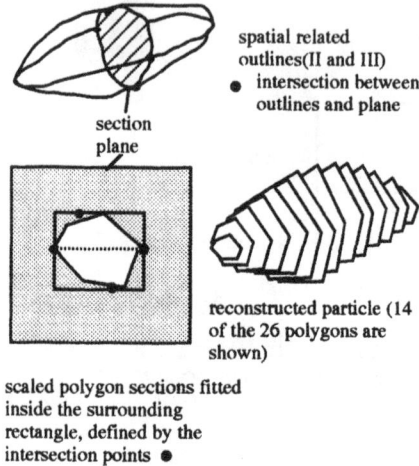

spatial related
outlines(II and III)
● intersection between
outlines and plane

section
plane

reconstructed particle (14
of the 26 polygons are
shown)

scaled polygon sections fitted
inside the surrounding
rectangle, defined by the
intersection points ●

Fig. 4 Reconstruction by scaling the approximating polygon
of contour I (see Fig. 3)

3. Results

Results are presented for different kinds of analyzed objects. Two macroscopic test
objects were manufactured from teflon and PVC with diameters of approximately
10 cm [1]. Fig. 5 shows the macroscopic objects in different orientations together with
the three dimensional reconstruction in corresponding orientations. To verify the
method, the volumes of these objects were calculated as well as experimentally determined. The maximum deviation between the two methods was 3 %. Fig. 6 shows the
reconstruction of a limestone particle with an equivalent diameter x_p = 90 μm [2]. The
reconstructions reveal the typical shape characteristics for this material with a relative
smooth surface.

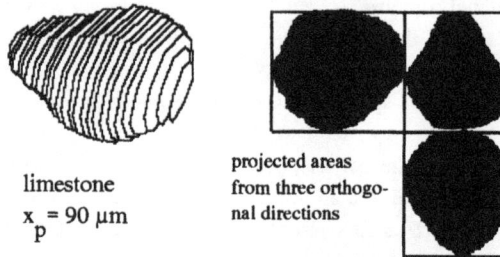

limestone
x_p = 90 μm

projected areas
from three orthogo-
nal directions

Fig. 6 reconstruction of a limestone particle

---

1. The described objects were used in microwave scattering experiments conducted at the University
of Bochum. The experiments and their results are described in [5]

2. x_p is the diameter of the sphere with the same volume

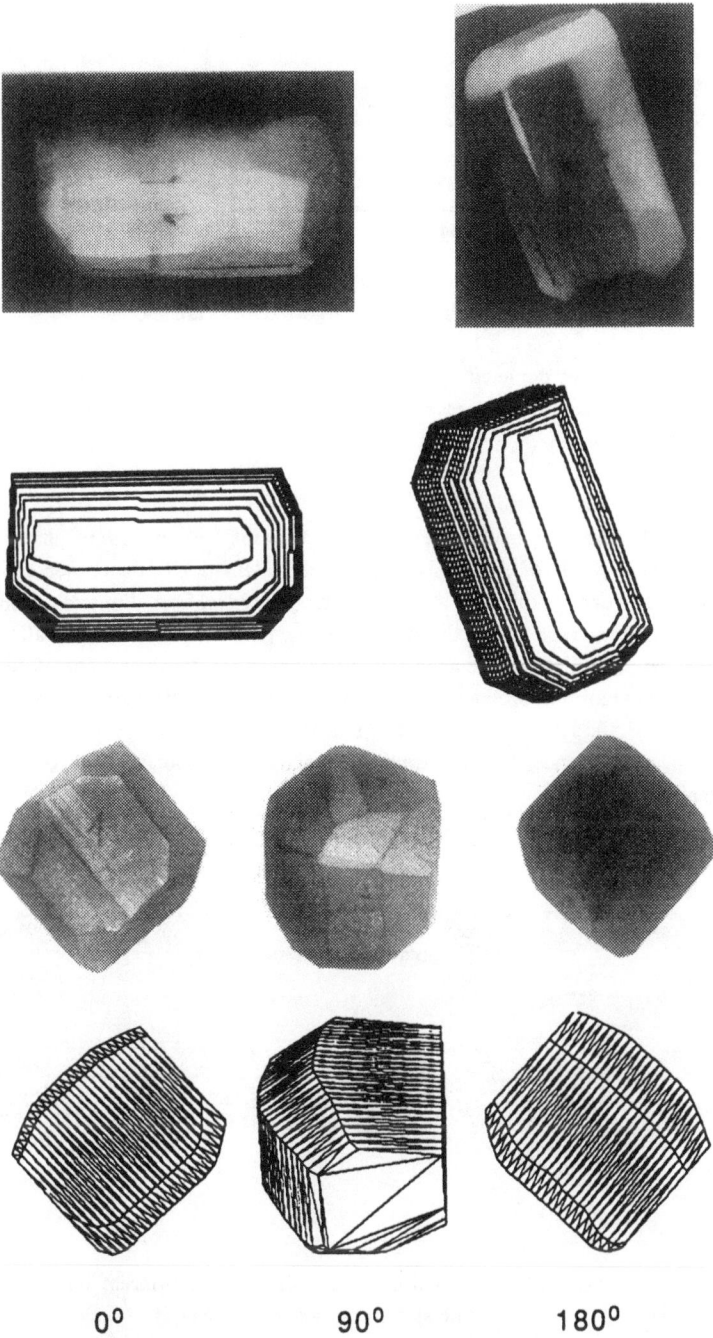

Fig. 5 macroscopic Teflon object (upper part) and PVC object (lower part) compared with their three dimensional reconstructions in different orientations

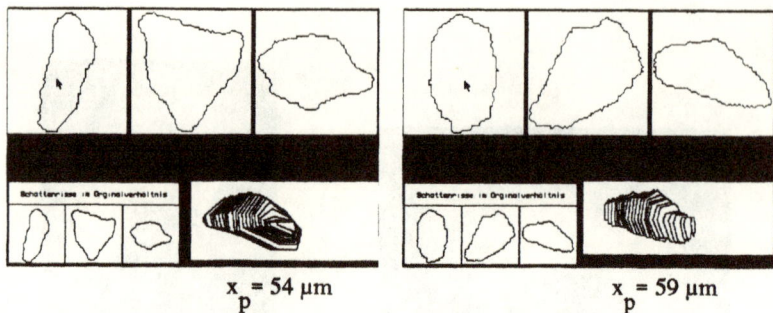

$x_p = 54\ \mu m$ $x_p = 59\ \mu m$

Fig. 7 reconstructions of two different quartz particles.
The contour lines of the projected areas are shown.

The quartz particles depicted in Fig. 7 reflect in contrast the characteristic edges, spikes and the plane regions responsible for certain shape effects in light scattering measurements [6].

4. Discussion of restrictions

The main disadvantage of the presented method is the fact that the results of the reconstruction depend, for certain shapes, from the orientation of the chosen coordinate system. Fig. 8 depicts this fact. A thin circular plate is registered in two different coordinate systems. System A is the one used in the presented results. From all three directions the projections of the analyzed object are of elliptic shape. Therefore the reconstruction algorithm produces an ellipsoid as the three dimensional reconstruction. In example B, the same object is viewed in a special coordinate system adapted to the shape. In consequence, the reconstructed structure corresponds much better to the original.

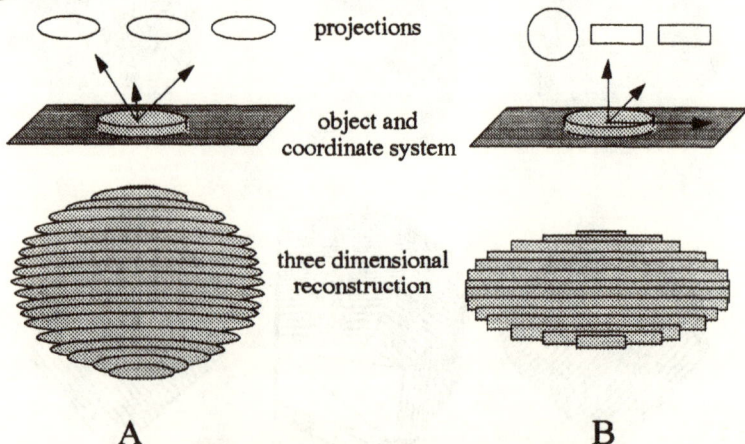

Fig. 8 Reconstruction of a circular plate in different
coordinate systems (K and K' *see Fig 1*)

5. Summary and outlook

The technique presented for the three dimensional reconstruction of particles by analyzing the projected areas from three orthogonal directions proved to be very reliable. The method was used to determine the particle volume and to calculate the three dimensional reconstructions of macroscopic objects and limestone and quartz particles utilized in simulation programs.

It is planned to develop an apparatus to perform three dimensional on-line registration of particles in suspensions. This apparatus will consists of three miniaturized CCD-cameras arranged in the proper angles together with a suitable light source. The technique will be used to investigate particle sizes and shapes in crystallization and agglomeration processes.

References

[1]
Pao, J-R.; Chang, Y-C.; Gentry, J.W.
The Use of Simulated Fractals to Determine the Relation between a Cluster and its Projection
J. Aerosol Sci., Vol 21, Suppl. 1, pp S63-66, 1990

[2]
Weichert, R.; Huller, D.
Volumenbestimmung und Formerkennung unregelmäßig geformter Partikeln mittels dreidimensionaler Bildanalyse
2nd. Europ. Symp. "Partikelmeßtechnik", Nürnberg (1979)

[3]
Huller, D.
Quantitative Formanalyse von Partikeln
Dissertation, Fakultät für Chemieingenieurwesen, Universität Karlsruhe (TH) 1984

[4]
Bottlinger, M.; Umhauer, H.
Modeling of Light Scattering by Irregularly Shaped Particles Using a Ray-Tracing Method
forthcoming publication in "Applied Optics", 1991

[5]
Bottlinger, M.; Umhauer, H.
Simulation of Light Scattering by Irregularly Shaped Particles using a Ray-Tracing Method
Proceeding of the 2nd Int. Cong. on Optical Particle Sizing,
March 5-8, 1990, Tempe, Arizona, USA

[6]
Umhauer, H.; Bottlinger, M.
The effect of particle shape and structure on the results of single particle light scattering analysis
forthcoming publication in "Applied Optics" 1991

Measurement and Modelling of Shape Changes in Developing Biological Systems

D. A. Dunnett[1], A. M. Goodbody[1] and M. Stanisstreet[2]

[1] DEPARTMENT OF APPLIED MATHEMATICS AND THEORETICAL PHYSICS, UNIVERSITY OF LIVERPOOL, LIVERPOOL L69 3BX, UK
[2] SCHOOL OF LIFE SCIENCES, UNIVERSITY OF LIVERPOOL, LIVERPOOL L69 3BX, UK

1 INTRODUCTION: CHANGES IN CELL SHAPE DURING ANIMAL DEVELOPMENT

All animals, including humans, start life as a single cell, the fertilised egg. For the fertilised egg to develop into an adult various processes occur: cells divide to produce a multicellular organism; cells differentiate or specialise to produce the range of tissue types; and tissues change shape to produce the characteristic form of the adult. These processes must occur at the right time, in the right place, and in a coordinated manner, so that the tissues and organs of the adult are correctly arranged.

The shaping of tissues, or morphogenesis, is driven by changes in cell properties, including cell shape. Thus, coordinated changes in the shapes of individual cells comprising a tissue can result in an alteration in the overall form of that tissue. The aims of this paper are: to outline two morphogenetic processes in which changes in cell shape are thought to be important, the development of the central nervous system and the repair of wounds in embryonic tissues; to describe how these processes can be modelled by computer simulations; to demonstrate the importance of quantification of shape changes in the interpretation of such simulations; and to explore the suitability of various shape indices for the quantification of changes in the shapes of cells of real embryos.

2 CELL SHAPE CHANGES IN NEURULATION

A major example of morphogenesis is seen in the early development of the central nervous system; this process is known as neurulation. In the adult the central nervous system comprises an internal fluid-filled tube, the spinal cord, expanded at one end to form the brain. However, in the early embryo the tissue which will develop into the central nervous system originates as an external flat sheet of tissue, the neural ectoderm. Morphogenesis of the central nervous system therefore involves a folding and an invagination of the sheet to form a concave profile. This is followed by the apposition of the lateral folds of the neural ectoderm, the neural folds, and their fusion at the midline to produce an internalised tube, the neural tube, with a continuum of covering tissue (Fig. 1).

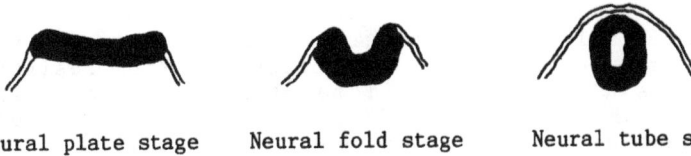

Neural plate stage   Neural fold stage   Neural tube stage

Figure 1   Diagram of neural tube formation in an amphibian embryo, seen as transverse sections of embryos of different stages.

In humans this process of neurulation occurs very early in pregnancy and is vulnerable to failure. If the neural tube fails to close at its anterior end an anencephalic fetus is formed; if the neural tube fails to close at its posterior end then spina bifida abnormalities result [1].

The elevation of the neural tube is accompanied by changes in the shapes of the cells which comprise the neural ectoderm [2-4] and there is evidence that these changes in cell shape make an active contribution to, and are not just a passive response to, the elevation of the neural folds. Hence, electron microscopy shows that the cells have the internal organelles, an apical bundle of microfilaments, needed to change their shape [5-8]. Biochemical analyses show that the proteins which are associated with microfilament contraction and anchorage in adult cells are present in the cells of the neural ectoderm [9,10]. Also, treatment of embryos with agents or metabolic inhibitors known to disrupt microfilament contraction prevents the elevation of the neural folds and causes the collapse of elevated neural folds [11-15]. Thus, active changes in the shapes of individual cells appear to contribute to the forces needed for morphogenesis.

3   COMPUTER MODELLING OF NEURAL TUBE FORMATION

One way to explore the extent to which changes in cell shape offer a sufficient explanation for the tissue shaping seen during neurulation is to use computer models to attempt to simulate the observed changes. The model employed here is a dynamic model which predicts the changes in tissue form which will occur if cells obey certain rules. The model, originated by Odell et al. [16], is complex and requires a mainframe computer. Conceptually, however, the model is quite straightforward. Some cells are passive and have a resistance to externally-imposed compression and extension; these cells behave in an elastic manner. Other cells are active and have the additional property that if their apices are stretched beyond a threshold, the apices contract. This component mimics the contraction of the apical bundle of microfilaments seen in cells of the neural ectoderm.

Initially, cells form a ring around a cavity which also has a resistance to compression. This resistance simulates that which would be offered by the internal tissues of the embryo (Fig. 2). The top 16 cells, representing the neural ectoderm, are active.

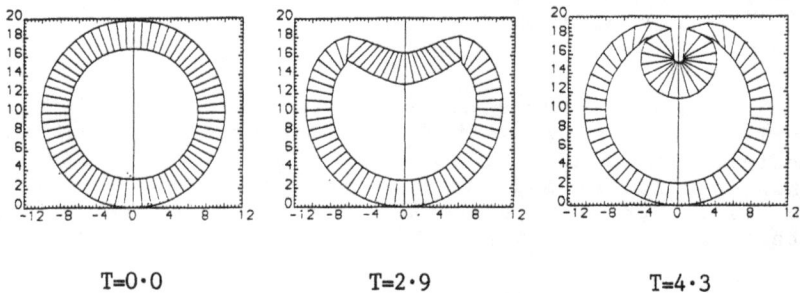

T=0·0   T=2·9   T=4·3

Figure 2 Computer simulation of neural tube formation (after Odell et al. [16]). T is the elapsed time.

The computer programme is designed to solve a series of non-linear, simultaneous, time-dependent differential equations of the form $A.\dot{x} = h$ for the coordinates x. A and h are complicated functions of the vector x. Each step involves the inversion of the matrix A of dimensions $\blacksquare$ x $\blacksquare$, where $\blacksquare$ is of the order of 4 times the number of cells (=128, since the simulation is run as a closed semicircle of 32 cells, and the results are plotted with their mirror image to mimic a transverse section of an embryo). The programme used in the present series of investigations uses a new and more efficient procedure for performing the inversions of the matrix A.

Following 'manual' stimulation of active contraction in the apices of the top two cells to initiate the simulation, the apices of the neighbouring active cells are stretched and so triggered, automatically, to contract. This leads to a wave of contraction propagated, domino-like, from cell to cell. These coordinated changes in cell shape lead to an overall change in tissue form similar to that seen in neural tube formation (Fig. 2).

This model can also be used to simulate abnormalities of neurulation, such as those in which the neural folds fail to elevate normally; this can give insight into how defects of the neural tube might arise. For example, reduction of the ability to contract of a relatively small proportion of the active cells has a profound effect on the overall profile of the neural tube [17], suggesting that the process of neurulation might be vulnerable to a relatively minor perturbation.

4   USE OF SHAPE INDICES IN INTERPRETING COMPUTER SIMULATION OF NEURULATION

In many cases it is necessary to know when the simulations have run to completion. However, it is not possible to tell by observation of the plots whether or not subtle changes are still occurring. For this reason, changes in the shapes of the cells were quantified by calculating and plotting the values of a form factor [18] for the cells throughout the simulation. The form factor used, F, was calculated as:

$$F = 4.\pi.A/p^2$$

where A is the area and p is the perimeter of the cell. This provides a dimensionless ratio which is measure of the efficiency with which a perimeter encloses an area. For a circle, F is unity; for any other shape, F is less than unity. Thus F provides an indication of the degree of deformation from circular of a plane figure.

Examination of the behaviour of the form factor, F, during the simulation of normal neurulation showed that F was sensitive to the changes in cell shape generated by the simulation. Furthermore, the behaviour of F could differentiate between active, contracting cells and passive, non-contracting cells. The value of F for active cells decreased and remained low whereas the value of F for passive cells increased and remained high (Fig. 3). These changes reflect the alteration from the initial cell shape, which is approximately rectangular. Active cells taper as they go undergo apical contraction, and passive cells become less elongated as they are stretched.

Use of the shape index also allows interpretation of the perturbations of the simulation of neurulation caused by altering some of the parameters. For example, reduction of the resistance to compression of the internal cavity disturbs the simulation [17]. Examination of the behaviour of F shows that a reduction causes a delay, and that a greater reduction causes a complete inhibition of the simulation (Fig. 3).

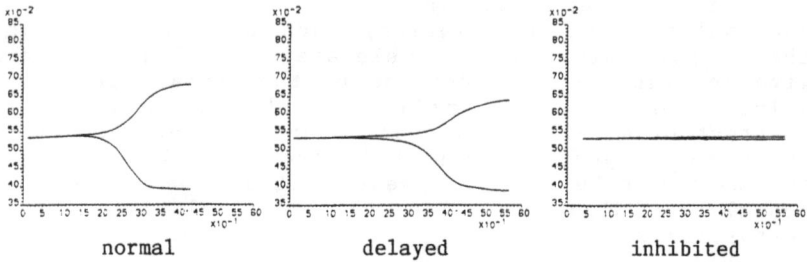

normal   delayed   inhibited

Figure 3 Comparisons of the changes in magnitude of the form
          factor, F, for an active cell and a passive cell during
          normal, delayed and inhibited simulation of neurulation.

5   CELL SHAPE CHANGES IN EMBRYO TISSUE REPAIR

In view of the apparent complexity of the events
of early embryonic development it is perhaps surprising
that embryos also have powers of regeneration or
repair. For example, if a linear wound is made in the
tissue of an amphibian early embryo, much at the same
time as the embryo is undergoing neural tube formation,
the wound will heal rapidly [19](Fig. 4).

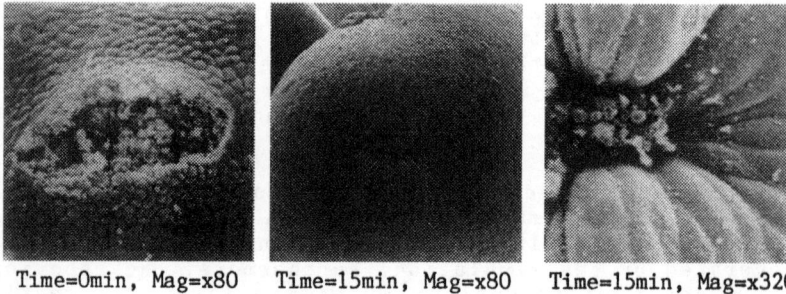

Time=0min, Mag=x80   Time=15min, Mag=x80   Time=15min, Mag=x320

Figure 4 Scanning electron micrographs of repair in tissue from
          amphibian early embryo.

Embryos are normally protected by membranes,
shells or, in the case of the mammals, by the mother's
body. In view of this it is improbable that they would
experience such precise wounding in natural conditions,
and so it is unlikely that embryos would have evolved
special mechanisms for wound healing. Thus, it is
probable that embryo tissue repair is effected by those
cellular mechanisms which are normally employed in
morphogenesis.

This idea is supported by scanning electron
microscopical observations of amphibian embryos
undergoing tissue repair. Such observations show that
this process too is accompanied by changes in the
shapes of cells; cells at the wound margin taper
towards the wound [19](Fig. 4). Furthermore, inhibitors

which prevent wound healing also prevent these changes in cell shape [20-22], supporting the notion that the cell shape changes, like those seen in neural tube formation, play an active role in embryonic tissue repair.

6 COMPUTER MODELLING OF EMBRYO TISSUE REPAIR

We have proposed above that tissue repair in embryos is driven by the same cellular properties that contribute to normal morphogenesis, namely changes in cell shape. So, it is of interest to determine whether the computer model which can simulate normal morphogenesis by mimicking cell shape changes can also simulate the changes seen in tissue repair.

The computer model was modified to simulate changes in the shapes of cells within an array of regular hexagons to represent a surface view of an area of single-layered ectoderm (Fig. 5). The boundary of the reticulum was fixed, and the cells were simulated as being under tension. As previously, cell sides could be made active or passive, and active sides could be made to contract 'manually' or automatically, when the length of the side was stretched beyond a predetermined threshold. In some cases this automatic contraction produced a domino effect; contraction of one cell stretched its neighbour, which in turn contracted. Incisions in the tissue were modelled by removing the sides of some cells to produce a linear opening. A number of simulations were performed, and two examples are given here.

In the first set of conditions contraction of sides was triggered automatically. All of the cells of the reticulum were active and so responded automatically to the stretching of some cell sides which occurs as the wound opens. The results of this simulation suggest that these conditions can not account for the wound closure seen in real embryonic tissues (Fig. 5). The wound starts to close but then re-opens because cells peripheral to the wound become stretched and they contract, pulling the wound open.

The second set of conditions was designed to model the situation in which wound closure is effected by contraction of those sides of cells which are newly exposed by the incision. Here, the cells of the reticulum were passive, with the exception of those sides which bordered the wound. The results of this simulation are reminiscent of those seen in real embryonic tissues. At the end of the wound those the cell sides facing the wound contract, so that the wound narrows. Later, at the lateral border of the wound the cell sides facing the wound contract, reducing the length of the wound (Fig. 6).

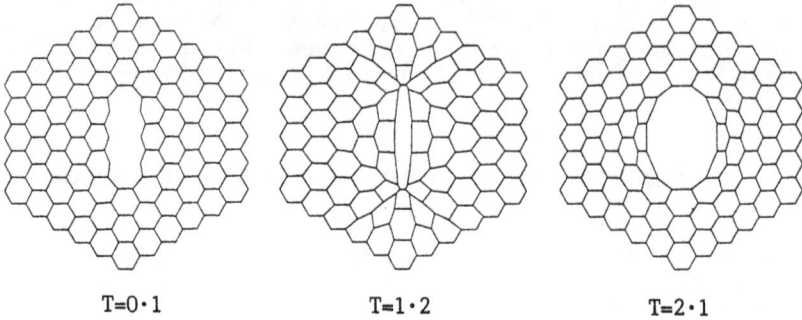

T=0·1   T=1·2   T=2·1

<u>Figure 5</u> Simulation of reaction to incision when all cells are
           active and respond automatically to stretching. T is the
           elapsed time.

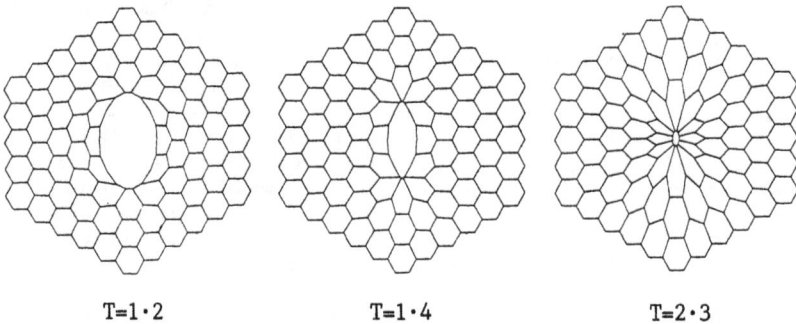

T=1·2   T=1·4   T=2·3

<u>Figure 6</u> Simulation of reaction to incision when cells at the
           wound margin are triggered to contract. T is the elapsed
           time.

7   USE OF SHAPE INDICES IN INTERPRETING COMPUTER SIMULATION OF EMBRYO TISSUE REPAIR

In the simulations of tissue repair, as in the
simulations of neurulation, it is necessary to quantify
changes in shape to determine when the simulation has
reached equilibrium. In this case the form factor, F,
as defined above was calculated for the wound rather
than for a whole series of cells. The behaviour of the
form factor, F, for the outline of the wound during the
first simulation, in which all cells are active, is
shown in Figure 7.

The value of F increases as the wound opens initially,
decreases as the wound starts to close, and then
increases again as the wound is pulled open by
contraction of cells peripheral to the wound. The
constant value of F after this shows that the
simulation has reached equilibrium, and that the wound
will not close again.

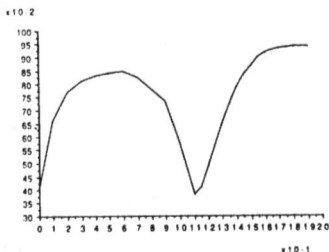

<u>Figure 7</u> Changes in the magnitude of the form factor, F, for the outline of the wound when all cells are active.

8   SUITABILITY OF DIFFERENT SHAPE INDICES FOR QUANTIFYING CHANGES IN CELL SHAPE DURING MORPHOGENESIS

The examples above show that the form factor, F, is a suitable index for providing a quantitative score of the changes in cell shape seen during computer simulations of morphogenetic processes. In general, these changes involve a tapering of a regular figure. As might be expected, however, changes in shape of real cells during the development of real embryos are more complex. Here, for example, cells may become spindle-shaped [23]. A range of shape indices is available [18](Table 1) and the suitability of these for quantifying different types of shape alteration can be tested by calculating and plotting the changes in their values during transformations of geometric figures.

<u>Table 1</u>   Definitions of shape indices (after Schwartz [18])

Index	Variables	
$E1 = \dfrac{Hmin}{Hmax}$	Hmin	= Minimum Feret diameter
	Hmax	= Maximum Feret diameter
$E2 = \dfrac{H(\perp Hmax)}{Hmax}$	$H(\perp Hmax)$	= Feret diameter at right angle to Hmax
	Hmax	= Maximum Feret diameter
$E3 = \dfrac{H1}{H2}$	H1, H2	= Lengths of sides of rectangle of minimum area which circumscribes the figure
$E4 = \dfrac{A^2}{2\pi Ip}$	A	= Area
	Ip	= Polar moment of inertia of figure with respect to it centroid
$E5 = \sqrt{\dfrac{IB}{IA}}$	IB, IA	= Areal principle moments of inertia

Ideally, the shape index should be sensitive (ie the slope of graph should be steep), linear (ie the slope should be linear), independent of other geometric variables (eg size, orientation) and accessible (ie derived from measurable parameters)[24,25]. The aim of determining the behaviour of indices during changes to standard geometrical figures is to see which index most closely matches these criteria. An example of this standardisation is shown in Figure 8. Here, we compare the behaviour of indices during the transformation of a cylinder to become a double truncated cone, to mimic the generation of a spindle-shaped cell. A number of conditions are imposed, to reflect the situation in real embryos. So, there is no alteration in height of the figure, and no change in the volume of its three dimensional projection. The values of the indices are calculated from the profile of the figure, as could be observed by scanning electron microscopy.

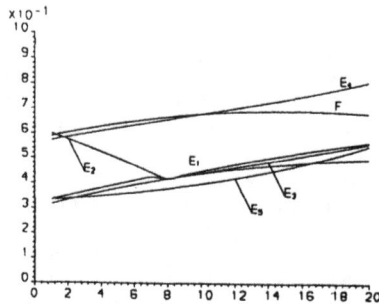

Figure 8 Behaviour of indices F and E1 to E5 during the transformation of a cylinder to a double truncated cone. The abscissa is calculated as 21-20d, where d is the apical diameter. Thus d decreases from 1 to 1/20.

The results show that some indices are clearly unsuitable for this particular transformation. For example, in these conditions F initially increases and then decreases, and E2 shows a sudden transition in its behaviour. Index E4 appears to match the criteria for this particular shape transformation.

9   CONCLUSION

Here we have summarised the evidence that changes in the overall form of tissues, which are of significance to evolutionary as well of embryological changes [26], are partly driven by active changes in the shapes of the cells which constitute the tissues. This idea is supported by the fact that computer models which simulate the effect of such changes in cell shape produce alterations in tissue form reminiscent of those seen in real embryos. Since such computer simulations involve subtle changes in shape, indices of shape are needed to determine the time course of the simulations and whether the simulations have reached a stable equilibrium. Such indices are useful too for quantifying the changes in cell shape seen in real embryos, although it is essential to determine the suitability of indices. This can be achieved by plotting the changes in their values for transformations of regular geometrical figures.

REFERENCES

1. M. Stanisstreet, Med. Sci. Res., 1987, 15, 681.
2. P. Karfunkel, Int. Rev. Cytol., 1974, 38, 245.
3. G. Morriss-Kay, J. Embryol. exp. Morph., 1981, 65 (suppl), 225.
4. G.C. Schoenwolf, Develop. Biol., 1985, 109, 127.
5. T.E. Schroeder, J. Embryol. exp. Morph., 1970, 23, 427.
6. B. Burnside, Develop. Biol., 1971, 26, 416.
7. P. Karfunkel, Develop. Biol., 1971, 25, 30.
8. M.M. Perry, J. Embryol. exp. Morph., 1975, 33, 127.
9. T.W. Sadler, J.L. Lessard, D. Greenberg and P. Coughlin, Science, 1982, 215, 172.
10. T.W. Sadler, K. Burridge and J. Yonker, J. Embryol. exp. Morph., 1986, 94, 73.
11. K.S. O'Shea, Prog. Anat., 1981, 1, 35.
12. G. Morriss-Kay, J. Physiol., 1983, 345, 52P.
13. G. Morriss-Kay and F. Tuckett, J. Embryol. exp. Morph., 1985, 88, 333.
14. M. Smedley and M. Stanisstreet, J. Embryol. exp. Morph., 1985, 89, 1.
15. M. Smedley and M. Stanisstreet, J. Embryol. exp. Morph., 1986, 93, 167.
16. G.M. Odell, G. Oster, P. Alberch and B. Burnside, Develop. Biol., 1981, 85, 446.
17. D. Dunnett, A. Goodbody and M. Stanisstreet, Acta Biothereotica, (in press).
18. H. Schwartz, Microscopie (Wien), 1980, 37 (suppl), 64.
19. M. Stanisstreet, J. Wakely and M.A. England, J. Embryol. exp. Morph., 1980, 59, 341.
20. M. Stanisstreet, J. Embryol. exp. Morph., 1982, 67, 197.
21. M. Smedley and M. Stanisstreet, Cytobios, 1985, 42, 25.
22. M. Stanisstreet, M. Smedley and C.J. Veltkamp, Cytobios, 1986, 46, 25.
23. D.C.P. Moore, M. Stanisstreet and G.E. Evans, J. Anat., 1987, 155, 87.
24. H.E. Exner, M.O.P. News (Kontron), 1978, 6, 1.
25. D.R Johnson, P. O'Higgins, T.J. McAndrew, L.M. Adams and R.M. Flinn, J. Embryol. exp. Morph., 1987, 90, 363.
26. M. Stanisstreet in 'Post-Implantation Mammalian Embryos: A Practical Approach', IRL Press at Oxford University Press, Oxford, 1990.

The Application of Principal Component Analysis to Particle Shape Description

M. Whiteman[1] and K. Ridgway[2]

[1] SMITHKLINE BEECHAM PHARMACEUTICALS, WELWYN GARDEN CITY, HERTS. AL7 1EY, UK
[2] THE SCHOOL OF PHARMACY, UNIVERSITY OF LONDON, LONDON WC1N 1AX, UK

1. INTRODUCTION

The size and shape of particles are tremendously important in the processing of particulate solids because of their effects on bulk and flow properties. Historically much more emphasis has been put on the particle size and size distribution of powders than on particle shape, partly because the effects of particle size are often greater than the shape effects. Perhaps the main reason though is that it is so much easier to define and measure particle size (making broad assumptions about the shape of the particles) than to attempt to quantify shape.

The most common way to quantify shape is to use one of a wide variety of shape factors which may be derived from the physical properties of the particles in question. This practice has certain disadvantages - particles with obviously different shapes may well have similar shape factor values since many shape factors measure deviation from a standard shape and give no indication of the variation in shape within a sample.

Jolicoeur and Mosimann[1] studied the applicability of principal component analysis to size and shape variation in the shells of Midland Painted Turtles, claiming that only multivariate statistical techniques could account for all of the variation jointly and thus provide a unified analytical approach. The work presented here examines the suitability of principal component analysis for particle shape analysis.

The principal component method may be explained quite simply in terms of its application to observations of particle length (x_1), width (x_2) and thickness (x_3).

If we consider a 3-dimensional scatter plot of x_1, x_2 and x_3, (see Figure 1) this would appear as an ellipsoidal swarm of data points. The objective in forming the first principal component is to define the linear combination, y_1, of x_1, x_2 and x_3 which accounts for the largest possible proportion of variance:

$$y_1 = u_{11}x_1 + u_{12}x_2 + u_{13}x_3 \qquad (1)$$

y_1 is in fact the long axis of the scatter ellipsoid and the coefficients u_{ij} are the cosines of the angles between y_1 and x_1, x_2 and x_3.

The second principal component, y_2, is the linear combination of x_1, x_2 and x_3 which accounts for the largest possible proportion of the remaining variance _and_ is orthogonal to (uncorrelated with) y_1.

$$y_2 = u_{21}x_1 + u_{22}x_2 + u_{23}x_3 \qquad (2)$$

y_2 is the intermediate axis of the scatter ellipsoid.

The third principal component, y_3, is the linear combination of x_1, x_2 and x_3 which accounts for the remainder of the variance and is orthogonal to both y_1 and y_2.

$$y_3 = u_{31}x_1 + u_{32}x_2 + u_{33}x_3 \qquad (3)$$

y_3 is the smallest axis of the scatter ellipsoid.

Obviously, larger values of u_{ij} will result in greater variance in y_1, so the magnitude of the variance is standardised by requiring that

$$\sum_{j=1}^{3} u_{ij}^2 = 1 \qquad (4)$$

The determination of the principal components and the proportion of the total variance associated with each component involves the use of matrix algebra and is quite complicated[2]. However, the task may be greatly simplified by using a statistical software package such as SAS (SAS Institute Inc, Cary, NC, USA).

Any number of variates may be used, though interpretation becomes more difficult as the number of variates increases. In general, n variates will give rise to n principal components and as n increases from unity, the proportion of the total variance represented by the nth component decreases.

2. MATERIALS AND METHODS

The materials used, the shape-sorting techniques applied to them and the characterisation methods have been described fully elsewhere[3] but are summarised briefly below:-

Materials

1. Citric acid monhydrate (BDH chemicals Ltd, Poole, UK) –
   MW 210.14;
   Density 1.552g cm^{-3}; white or colourless rhombic crystals.

2. Sodium perborate tetrahydrate (BDM chemicals Ltd, Poole, UK) – WM 153.86;
   Density 1.717g cm^{-3}; white crystalline powder.

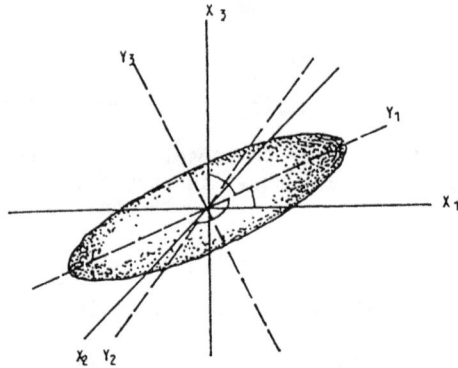

Figure 1 : Principal axes of trivariate observations.

```
DATA ELLIP;
   INPUT LENGTH WIDTH HEIGHT;
   CARDS;
(data)
   ;
PROC PRINT;
PROC PLOT;
   PLOT LENGTH*WIDTH WIDTH*HEIGHT LENGTH*HEIGHT;
PROC MEANS;
PROC CORR;
PROC PRINCOMP COV;
   VAR LENGTH WIDTH HEIGHT:
```

Figure 2 : SAS statements used for statistical analysis of
            particle dimensions.

Sample Preparation

Bulk samples of each material were sieved into closely-sized fractions using square-mesh ASTM Ell sieves arranged in a series. These fractions were divided in two and each of the sub-fractions were classified by slotted sieving or by passage over the Jeffrey-Galion vibratory shape sorter.

Slotted sieving yields samples which have been graded in terms of particle width and thickness whereas sorting on the Jeffrey-Galion apparatus yields samples which have been graded in terms of particle width and approximate projected area (the determining factor is actually area of contact with the deck, which is roughly proportional to projected area).

Sample Characterisation

Each sample was reduced in size by repeatedly passing through an Endecott's ¼" sample splitter and discarding half of the total. When the sample size became too small for the sample splitter, it was further reduced by pouring over a knife-edge until only a few thousand particles remained. 100 particles were taken and allowed to come to rest in their most stable orientation on a glass microscope slide. For each particle, the length, width, projected area and projected perimeter were determined using a microcomputer based image analyser and particle thickness was determined by means of a specially mounted miniature displacement tranducer.

Statistical Treatment of Data

Principal component analysis was performed on the lengths, widths and thicknesses of each size- and shape-sorted fraction using the PRINCOMP procedure of the SAS statistical software package (SAS Institute Inc, Cary, NC, USA). A list of the procedures used is given in Fig 2. The output obtained comprises a statistical analysis for each individual variate, a correlation matrix for the variates, scatter plots for each variate against each of the others in addition to the eigenvectors and eigenvalues.

3. Results and Discussion

Principal component data for 64 size- and shape-sorted fractions of citric acid and sodium perborate are tabulated in Appendix 1.

If we consider the results for Jeffrey-Galion sorted citric acid, we can see that the first component generally has a high length coefficient, a medium width coefficient and a low thickness coefficient. Since all three coefficients are positive, this component is effectively a size descriptor, a weighted mean of the original variates. It accounts for around 60 - 70% of the variance for this group of samples. This is an interesting point, because these samples were very carefully prepared to differ in shape rather than size. The fact that 60 - 70 % of the within-sample variance results from size variation, rather than

shape variation may be seen as a measure of the success of the
shape sorting operation. The second component is usually made up
of a large positive width coefficient with low negative thickness
and length coefficients, while the third component is generally
made up of a large positive thickness coefficient, a small to
medium width coefficient and a medium negative length
coefficient. Occasionally the second and third coefficients are
reversed. The second component accounts for 17 – 31% of the
variance and the third component is responsible for 7 – 16% of the
variance. Because these two components contain negative elements,
they indicate variations in shape – as some dimensions increase,
others decrease.

For slotted-sieved citric acid, the situation is slightly
different. The slotted sieves have minimised the variation in
thickness between particles in a fraction, so neither the first
nor second components has much of a thickness coefficient. The
first component is still basically a size component with the
length being the most important element. The second component is
generally a comparison between width and thickness while the third
component is almost completely a measure of the variation in
thickness.

Considering now the results for sodium perborate sorted on
the Jeffrey-Galion deck, the first component is once again a size
component, with the length coefficient being most important. This
component represents 63 – 78% percent of the variance, slightly
more than for the citric acid samples.   The second component is
made up of a medium to high positive width coefficient, usually a
lower positive thickness coefficient and a medium negative length
coefficient. The third component generally has a high positive
thickness coefficient, a low positive or negative length
coefficient and a low-medium negative thickness coefficient.

The data for slotted sieved sodium perborate are broadly
similar to those for slotted-sieved citric acid.

The proportion of the variance represented by the first
component is usually higher for slotted sieved citric acid than
deck-sorted and the proportion of variance represented by the
third component is usually smaller. the overall rotation of the
principal component axes away from the standard axes is less for
the slotted sieved citric acid.

For sodium perborate there is less difference in the
proportion of variance accounted for by each component between the
two shape sorting methods. The difference in rotation of the axes
between the method is not so apparent for sodium perborate as for
citric acid.

For deck sorted materials on moving from pot 1 to pot 13
(equidimensional to platey, blocky), the length coefficient of
component 1 increases, the width coefficient of component 2
increases and the thickness coefficient of component 3 increases.
It was not possible to pick out such trends for the slotted sieved
samples.

Overall, variances for citric acid were much greater (x5) than those for sodium perborate. The variances for deck sorted materials increase with pot number while those for slotted sieved samples were more uniform.

4. CONCLUSIONS

Principal component analysis is a powerful technique for studying the variance within a sample, but it is important to remember that it can't describe particle shape as such, only the variation in shape. The method can yield different trends and patterns for the same material sorted by different methods or for different materials sorted by the same method; it is capable of reducing a huge volume of data to a few coefficients.

It is a sensitive technique, since the samples studied had been carefully sorted to minimise variations in size and shape. The differences detectable with unsorted samples are likely to be much greater. Although principal component analysis cannot strictly be used for hypothesis testing, its application to two samples should allow the analyst to make a reasonable assumption as to whether the samples come from the same population.

However, despite the many advantages of this way of looking at particle shape data, it remains for the moment a solution in search of a problem.

REFERENCES

1. P. Jolicoeur and J. E. Mosimann, Growth, 1960, 24, 339.

2. T. W. Anderson, "An Introduction to Multivariate Statistical Analysis", Wiley & Sons, New York, 1958.

3. M. Whiteman and K Ridgway, Powder Technology, 1988, 56, 83.

Appendix 1: Results of Principal Component Analysis

CITRIC ACID 850–710μm

Pot 1				Pot 5				Pot 9				Pot 13		
Component	1	2	3		1	2	3		1	2	3		1	2	3
% Variance	69	18	13		63	21	16		59	26	14		67	19	15
l	0.846	0.188	-0.500		0.779	0.455	-0.432		0.806	-0.396	-0.439		0.933	-0.233	-0.276
w	0.514	-0.542	0.665		0.596	-0.753	0.280		0.481	0.872	0.096		0.304	0.093	0.948
t	0.146	0.819	0.555		0.198	0.476	0.857		-0.345	-0.288	0.893		0.195	0.968	-0.158

CITRIC ACID 710–600μm

Pot 1				Pot 5				Pot 9				Pot 13		
Component	1	2	3		1	2	3		1	2	3		1	2	3
% Variance	69	17	14		60	23	16		74	19	7		61	29	11
l	0.763	-0.163	-0.626		0.819	-0.311	-0.482		0.739	0.532	-0.412		0.848	-0.384	-0.364
w	0.609	-0.140	0.780		0.454	0.865	0.214		0.615	-0.782	0.094		0.455	0.881	0.130
t	0.215	0.977	0.007		0.350	-0.395	0.850		0.272	0.323	0.906		0.271	-0.276	0.922

CITRIC ACID 600–500μm

Pot 1				Pot 5				Pot 9				Pot 13		
Component	1	2	3		1	2	3		1	2	3		1	2	3
% Variance	63	22	15		63	24	14		70	20	11		63	29	9
l	0.704	0.596	-0.386		0.948	-0.238	-0.212		0.892	-0.322	-0.318		0.931	-0.309	-0.194
w	0.622	-0.780	-0.069		0.290	0.971	-0.016		0.365	0.927	0.083		0.333	0.937	0.106
t	0.342	0.191	0.920		0.209	-0.036	0.977		0.268	-0.190	0.944		0.149	-0.163	0.975

CITRIC ACID 500–425μm

Pot 1				Pot 5				Pot 9				Pot 13		
Component	1	2	3		1	2	3		1	2	3		1	2	3
% Variance	67	20	13		66	22	13		58	31	12		67	25	8
l	0.793	-0.574	-0.201		0.879	-0.458	-0.130		0.855	-0.465	-0.228		0.898	-0.369	-0.239
w	0.583	0.812	-0.018		0.454	0.889	-0.059		0.465	0.883	-0.054		0.428	0.859	0.282
t	0.173	-0.103	0.980		0.143	-0.007	0.990		0.227	-0.060	0.972		0.101	-0.356	0.929

CITRIC ACID 850-710μm

	t=850-710μm			t=710-600μm			t=600-500μm			t=500-420μm		
Component	1	2	3	1	2	3	1	2	3	1	2	3
% Variance	55	24	22	64	27	9	78	17	6	75	20	5
1	0.859	0.179	-0.480	0.901	-0.429	0.064	0.734	0.676	-0.066	0.833	0.553	-0.011
w	0.480	-0.607	0.633	-0.433	0.897	-0.079	0.678	-0.735	0.023	0.550	0.831	-0.076
t	0.178	0.774	0.607	-0.023	0.099	0.995	0.031	0.063	0.997	0.052	0.057	0.997

CITRIC ACID 710-600μm

	t=710-600μm			t=600-500μm			t=500-420μm			t=420-355μm		
Component	1	2	3	1	2	3	1	2	3	1	2	3
% Variance	73	22	16	72	20	8	75	22	3	73	22	4
1	0.985	-0.169	-0.033	0.824	-0.566	-0.037	0.923	-0.384	0.004	0.792	-0.609	-0.044
w	0.172	0.981	0.093	0.566	0.824	0.000	0.384	0.923	-0.003	0.609	0.793	-0.015
t	0.017	0.097	0.995	0.031	-0.021	0.999	-0.002	0.005	0.999	0.043	-0.015	0.999

CITRIC ACID 600-500μm

	t=600-500μm			t=500-420μm			t=420-355μm			t=355-300μm		
Component	1	2	3	1	2	3	1	2	3	1	2	3
% Variance	79	15	6	72	22	6	69	26	5	81	17	3
1	0.993	-0.092	-0.072	0.923	-0.383	0.023	0.738	-0.675	0.010	0.920	-0.391	0.003
w	0.099	0.989	0.103	0.384	0.921	-0.061	0.674	0.737	-0.037	0.391	0.920	0.034
t	0.062	-0.109	0.992	0.002	0.065	0.998	0.017	0.035	0.999	-0.016	-0.029	0.999

CITRIC ACID 500-425μm

	t=500-420μm			t=420-355μm			t=355-300μm			t=300-250μm		
Component	1	2	3	1	2	3	1	2	3	1	2	3
% Variance	84	13	3	65	30	5	66	30	4	70	26	3
1	0.997	-0.073	-0.021	0.911	-0.412	-0.038	0.770	-0.637	0.021	0.906	-0.423	0.002
w	0.071	0.995	-0.067	0.412	0.911	-0.001	0.638	0.770	-0.019	0.423	0.905	-0.032
t	0.025	0.066	0.998	0.035	-0.015	0.999	-0.004	0.028	0.999	0.012	0.029	0.999

SODIUM PERBORATE 425-355μm

	Pot 1			Pot 5			Pot 9			Pot 13		
Component	1	2	3	1	2	3	1	2	3	1	2	3
% Variance	72	20	8	77	19	4	67	22	11	72	19	9
l	0.647	-0.693	0.318	0.883	-0.458	0.190	0.893	-0.442	-0.087	0.858	-0.511	-0.061
w	0.571	0.164	-0.804	0.379	0.572	-0.727	0.340	0.787	-0.515	0.474	0.830	-0.295
t	0.506	0.702	0.502	0.276	0.680	0.679	0.297	0.430	0.853	0.201	0.224	0.954

SODIUM PERBORATE 355-300μm

	Pot 1			Pot 5			Pot 9			Pot 13		
Component	1	2	3	1	2	3	1	2	3	1	2	3
% Variance	69	22	9	65	25	10	73	17	10	68	26	6
l	0.720	-0.694	-0.013	0.775	-0.626	-0.082	0.750	-0.590	-0.298	0.824	-0.544	-0.155
w	0.503	0.534	-0.680	0.460	0.471	0.752	0.511	0.804	-0.305	0.495	0.826	-0.269
t	0.479	0.483	0.733	0.433	0.621	-0.654	0.419	0.076	0.905	0.275	0.145	0.951

SODIUM PERBORATE 300-250μm

	Pot 1			Pot 5			Pot 9			Pot 13		
Component	1	2	3	1	2	3	1	2	3	1	2	3
% Variance	76	18	6	77	18	5	78	18	5	73	21	6
l	0.818	-0.571	-0.064	0.905	-0.422	0.046	0.887	-0.455	-0.072	0.947	-0.268	-0.175
w	0.408	0.500	0.764	0.340	0.654	-0.676	0.354	0.773	-0.527	0.204	0.928	-0.313
t	0.404	0.651	-0.643	0.255	0.628	0.735	0.295	0.442	0.847	0.246	0.260	0.934

SODIUM PERBORATE 250-212μm

	Pot 1			Pot 5			Pot 9			Pot 13		
Component	1	2	3	1	2	3	1	2	3	1	2	3
% Variance	73	21	6	63	27	10	69	22	9	76	14	10
l	0.664	-0.724	-0.187	0.958	-0.284	-0.028	0.944	-0.314	-0.106	0.909	-0.411	-0.071
w	0.523	0.271	0.808	0.215	0.785	-0.581	0.267	0.910	-0.318	0.333	0.818	-0.469
t	0.535	0.634	-0.559	0.187	0.551	0.813	0.196	0.271	0.942	0.251	0.403	0.880

SODIUM PERBORATE 425-355μm

	t=420-355μm			t=355-300μm			t=300-250μm			t=250-210μm		
Component	1	2	3	1	2	3	1	2	3	1	2	3
% Variance	72	20	9	67	22	12	66	22	12	63	31	6
l	0.976	-0.219	-0.005	0.969	-0.230	0.084	0.856	-0.515	0.035	0.813	-0.581	-0.024
w	0.203	0.910	-0.361	0.245	0.902	-0.356	0.516	0.855	-0.043	0.581	0.814	-0.018
t	0.084	0.351	0.933	0.006	0.366	0.931	-0.008	0.055	0.998	0.030	0.000	0.999

SODIUM PERBORATE 355-300μm

	t=355-300μm			t=300-250μm			t=250-210μm			t=210-180μm		
Component	1	2	3	1	2	3	1	2	3	1	2	3
% Variance	83	11	6	59	31	10	69	26	6	65	31	4
l	0.994	-0.108	-0.003	0.879	-0.463	-0.110	0.904	-0.427	-0.020	0.843	-0.537	-0.037
w	0.102	0.952	-0.290	0.432	0.873	-0.226	0.422	0.898	-0.121	0.537	0.844	-0.021
t	0.035	0.288	0.957	0.200	0.151	0.968	0.069	0.101	0.992	0.042	-0.002	0.999

SODIUM PERBORATE 300-250μm

	t=300-250μm			t=250-210μm			t=210-180μm			t=180-150μm		
Component	1	2	3	1	2	3	1	2	3	1	2	3
% Variance	73	19	9	80	17	3	66	25	9	61	30	9
l	0.981	-0.193	0.007	0.980	-0.199	0.010	0.995	-0.056	-0.085	0.561	0.828	0.009
w	0.189	0.954	-0.232	0.166	0.786	-0.595	0.045	0.991	-0.112	0.820	-0.554	-0.142
t	0.038	0.229	0.973	0.111	0.585	0.804	0.090	0.118	0.989	0.113	-0.087	0.989

SODIUM PERBORATE 250-212μm

	t=250-210μm			t=210-180μm			t=180-150μm			t=150-125μm		
Component	1	2	3	1	2	3	1	2	3	1	2	3
% Variance	80	14	6	80	13	6	73	19	8	56	28	16
l	0.991	-0.131	0.034	0.991	-0.131	0.037	0.968	-0.249	-0.022	0.871	-0.489	0.053
w	0.127	0.811	-0.571	0.128	0.797	-0.590	0.220	0.890	-0.399	0.442	0.863	-0.120
t	0.044	0.570	0.820	0.048	0.589	0.806	0.119	0.381	0.917	0.013	0.131	0.991

The Electrical Sensing Zone Method (The Coulter Principle)

R. W. Lines

COULTER ELECTRONICS LTD., NORTHWELL DRIVE, LUTON, BEDS. LU3 3RH, UK

1 SUMMARY

The first types of particle and cell counters to reach the market employed optical sensing. This probably arose from the habits of viewing and counting such particles visually through a microscope. Indeed, one of the first two types was an automated microscope; the other was based on the use of a narrow beam of light focused through a narrow stream of particle-bearing fluid. The light beam in such a system is modulated by the passage of individual cells through the beam, and such effects are picked up by a photocell across the stream to produce electrical pulses for counting, and for size analysis if required.

These first optically based counters were quickly displaced for most applications by the simpler and more effective non-optical electrical sensing zone method which remains the dominant laboratory tool for such work.

Although the first optically-based counter designs were not commercially successful, one development of that time has been employed in virtually all optical counters since then. A fundamental requirement is the necessity of ensuring that the particle stream always intercepts the transverse light beam in precisely the same path for stable control of pulse amplitude and duration. This is accomplished by providing a small diameter sample-carrying stream into the centre of a larger sheath stream. This invention was due to Crosland-Taylor.[1] It is more commonly known today as "hydrodynamically focusing" the sample flow.

In those days, to obtain better particle sizing required a modified light path of small dimensions, an internal "aperture", but the laser did not then exist, and the requirements for an optical design exceeded the skill and patience available. Wallace H. Coulter had taken to exploring alternative ways of producing pulses, culminating in an electrical path of small dimensions.[2] This sensed particle volume, its sensitivity being dependent upon the ratio of the volume of the particle to the volume of the sensing zone. For example, 13 volts applied across a 100µm aperture of 13,000 ohms resistance generated a 1mA current and a particle volume ratio of 1:13,000 for a 5µm particle. This apparatus was the first to measure particle volume.

It has since been intensively developed for the particle counting, and the particle size distribution analysis, of almost all forms of finely divided particulate materials. The review which follows concentrates on this aspect of the principle as applied to non-biological particles.

However, the principle is that of the modulation of the impedance of an electrical current path of small dimensions, and this means not only by the application of a direct current, but also by an alternating field. Since around 1957 Coulter has known how to gather information from the interior of the particles electrically, however only recently has this begun to be exploited, for instance for biological cells.

Biological cells can be generally considered as little bags of electrolyte solution contained within very thin-walled membranes. At high applied electrical frequency, such as about 20-25 MHz, the membrane is mostly transparent, so if the applied aperture current is of the same frequency it is obvious that some of that current flows through the cells and not entirely around them. The resulting 'high frequency' pulse is slightly complex; it is a function both of particle volume and internal conductivity in relation to the conductivity of the surrounding fluid. In practice, a simultaneous direct current (d.c.) pulse produces a response primarily to particle volume. By dividing the high frequency (r.f.) signal by the d.c. signal, this quotient is quite a good measure of the cells' internal conduction, or resistivity. One interesting application is to separate white cells from whole blood, analyse with an aperture, and plot the pulses produced on a scatter diagram to give three distinct clusters representative of three distinct populations, i.e. inherently different salt concentrations. By challenging the cells with red cell lysing chemicals one can selectively alter the conductivity of at least one of the populations. This then presents four clusters, the fourth being from eosinophils. Further refinements give a fifth cluster, due to basophils.

That technology is the basis of one white cell blood counter. The use of d.c. and r.f. current gives measurements therefore of two parameters which did not exist before the aperture impedance method was invented.

The electrical sensing zone, or aperture impedance, method can be combined with other sensing means so that

additional parameters are available for more precisely
classifying the individual particles or cells. Optical
sensing can be combined in several fashions, such as light
absorption, light scatter, light polarisation and
fluorescence, which are all feasible and practical now
that reliable and more economical lasers are available.

2   INTRODUCTION

The principle of counting and sizing particles was
developed by Wallace H. Coulter in the late 1940s under
contract to the United States Navy as a method of counting
blood cells rapidly,[2,3] and was described in detail by
Mattern, Brackett and Olson in 1957.[4] Since that time,
COULTER® particle counters and other devices utilizing the
same principle have become established not only for counting
and sizing blood cells but also for measuring the particle
size distribution of any finely divided particulate matter
which can be suspended in an electrolyte solution. Over
97% of automated blood cell counters in use today employ
the aperture sensing system. Many hundreds, if not
thousands, of different industrial particulate materials
are characterised using the method. It is the only method
which measures a particle volumetrically, i.e. in three
dimensions. It has a lower limit of detection of below
0.3µm. The method is the subject of a British Standard,[5]
as well as many other national standards.

The principle, apparatus and operation for the
determination of particle size distribution has been
reviewed in many publications, for example Ullrich,[6]
Allen,[7-9] Rabinovitch,[10-11] and Allen and Marshall.[12]
Coulter Electronics Ltd. publishes two bibliography lists
(Industrial, including pharmaceutical; and Medical,
including haematological),[13,14] giving over 6,000
references to the uses of the various COULTER COUNTER®
models. Richardson-Jones[15] has provided a comprehensive
review of the application of the principle to blood cell
analysis.

3   THE COULTER PRINCIPLE

Figure 1 shows a schematic diagram of a simple form of
apparatus. Particles, suspended homogeneously at a low
concentration in electrolyte solution, are made to flow
through a small aperture (or orifice) in the wall of an
electrical insulator, which is commonly called the aperture
(or orifice) tube; the aperture creates the sensing zone.
In addition, a current path is established between two
immersed electrodes, across this aperture, setting a
certain base impedance to the electrical detection

---

*COULTER, COULTER COUNTER, and CHANNELYZER are registered
trademarks of Coulter Corporation.

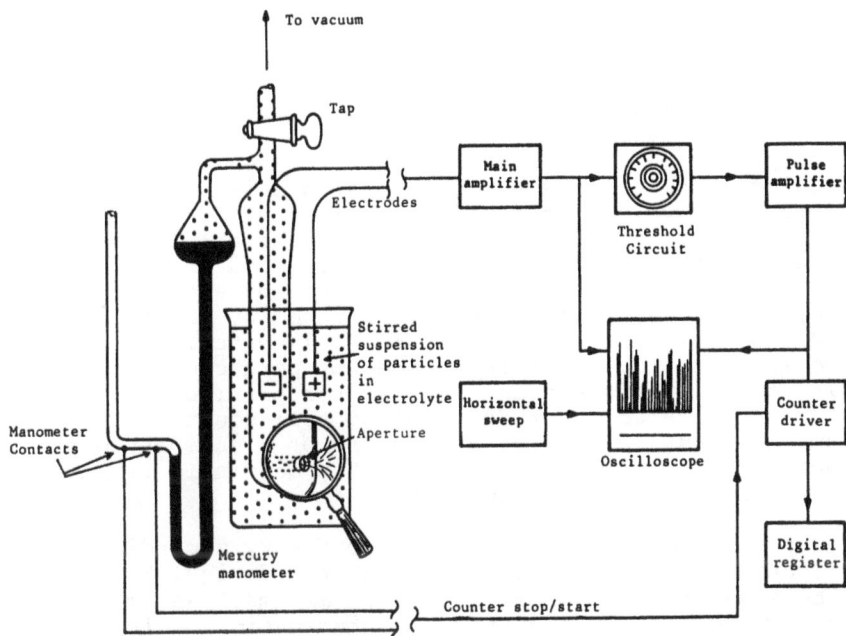

<u>Figure 1</u> Schematic drawing of a simple form of analyzer.

circuitry. A direct current is generally used. As each
particle enters the aperture, it has effectively displaced
a volume of electrolyte solution equal to its own immersed
volume, and the base impedance is therefore modulated by
an amount proportional to the displaced volume of the
particle. This results in an electrical pulse of short
duration being created by each particle; the height of the
pulse being essentially proportional to the volume of the
particle. The pulse may be measured for instance as the
change in resistance, current or voltage across the
electrodes.

The passage of a number of particles produces a train
of pulses which can be observed on an oscilloscope and
analyzed by counter and pulse height analyzer circuits to
produce a number against particle volume, or equivalent
spherical diameter, distribution. A volume or mass ("weight")
against size distribution can also be measured, calculated,
or computed; the "weight" percentage being possible if all
of the particles have uniform density or a known density
distribution across their size range.

Simple models have only one counter and size level
circuit, (and so are called single channel models); more
complex instruments can obtain number and/or mass (weight)
distributions automatically in up to 256 size channels

within a few seconds. Counting and sizing rates of up to
some 10,000 particles per second are possible, with each
pulse height being measured to within one or two percent.

Since most particulate materials are irregularly
shaped, the volumetric response is invaluable, as volume
is the only single measurement which can be made of an
irregular particle in order to characterize its size. In
biological applications the size response is usually left
calibrated in volume units (femtolitres, or 'cubic
microns'), but industrially it is conventional to report
the equivalent spherical diameter calculated from it.
This volumetric method makes no assumptions about particle
shape, and indeed is not greatly affected by particle
shape except in extreme cases like flaky materials, such
as some clays. To count the number of particles in a
known volume of suspension, such as for particulate
contamination studies or for a blood cell count, the sample
volume is accurately metered by means of a calibrated
"manometer". Figure 1 illustrates the original, simple,
mercury siphon and metering system.

The shape of the aperture tube can vary according to
application; for instance a very narrow design,[16] which
can be inserted into glass ampoules as small as 1ml
capacity, to allow the particle contamination of injectable
solutions to be measured.

Modern instrument designs have a range of extra
features, including embedded microprocessors, and various
data reduction handling and presentation methods, and the
manufacturers should be contacted for details. Their
volumetric sizing resolution, speed of data collection,
statistical accuracy of counting, freedom from any optical
response effects, and the reliability and simplicity of
calibration make these devices unsurpassed for providing
particle counting and size distribution analyses.

4 SIZING RESPONSE

The volumetric response is, in theory, absolute. For
instance, Kachel[17] stated that the electrical signal from
a Coulter-type aperture of length to diameter ratio in the
range of 1:1 to 2:1 depends on the particle and aperture
parameters:

$$U = V \rho_0 \, i \, f / \pi^2 \, R^4 \qquad \ldots\ldots\ldots\ldots\ldots\ldots\ldots (1)$$

Where U = amplitude of voltage pulse

   V = particle volume

   ρ_0 = electrolyte resistivity

   i = aperture current

f = particle "shape" factor

and   R = aperture radius

Particle resistivity has very little effect on voltage response, unless it is close to the resistivity of the fluid. If the particle resistivity changes from one millionth to one hundredth of that of the electrolyte solution there is less than 1% change in the response.[18] Colloidal charges are too small to have any effect.[18] The first such instruments from all sources were quite sensitive to electrolyte resistivity changes. This effect was eliminated from many later designs as a result of the principle and circuitry disclosed in 1966.[19]

Other simple, early, response theories (e.g. Coulter Electronics, Inc.,[18] Batch,[20] Allen,[7] and Grover et al.[21-23] predicted a dependence not only upon particle shape, but also that the response would increasingly deviate from linearity as the size of the particle approached the diameter of the aperture. It was thus established in the early literature that an upper limit of sizing accuracy lay at some 35-40% of the aperture diameter. This idea held for some years, even after Saranummi,[24] Kachel,[17] and Scarlett[25] independently developed more comprehensive theories which showed otherwise. Saranummi's approach predicted only a small dependency on the particle to aperture ratio, while Kachel's showed no such dependency, and Scarlett concluded that the response seemed to be due only to the volume of the particle, if second order effects were small. Those effects included the field potential distribution through and around the sensing zone, and are discussed later, (section 7).

The first experimental proof of virtual linearity of response, up to some 80% of the aperture diameter, came from Harfield and Knight[26] and Harfield, Wharton and Lines.[27] They used spherical particles of polymer latex and the COULTER COUNTER® model ZM. The experiment involved the use of different sizes of "mono-sized" latex particles measured by a range of different apertures, and therefore required no assumed or measured "real" sizes for the particles. The experiment has not been repeated for other particle shapes as no other series of suitable model particles appear to exist, so the linearity of response for other particle shapes has not yet been verified experimentally. It is reasonable to assume however, from all existing theory that no significant extra alinearity will exist for non-spherical shapes.

That particle shape has a basic effect upon instrument calibration was proposed in 1957[18] and has continued in later theories. It is however a constant effect and, as indicated by Harfield et al.,[26-28] is not dependent on the particle to aperture diameter ratio. That is, there

may be a different calibration factor for different shapes
of particles, but that one factor will hold over the full
measuring range of the aperture. The effect can be
studied, for instance, by calibrating the instrument with
a fraction of the material itself (see Section 5). Batch[20]
showed that the calibration factor varied little with
widely different shapes and resistivities. Eckhoff[29,30]
also used this "mass integration" calibration method to
determine the calibration factors for spherical particles
(glass beads) and angular particles (cement, and the raw
mix used in its preparation), and found no significant
variation between them.

Personal experience of rigid particles confirms that
moderately irregular particles like milled silica or
alumina have no measurable shape effect, and that only
grossly irregular shapes do, such as the flaky particles
of some clays and mica. These have been found to give
an apparent volume response double or even treble the
physical volume, i.e. the diameter calibration factor
could differ by up to 66% from the calibration factor
derived from spherical latex particles.

Rod-like particles appear to give no shape effect.
O'Connell and Martsch[31] microtomed wool fibres to a
constant length, and measured them by a COULTER COUNTER®
instrument, so that the "volume" response would reduce to
"cross-sectional area" response, from which fibre diameter
could be calculated and compared with microscopy. The
results were virtually identical.

Interestingly, just as fibre diameter can be measured
from the volume response of particles cut to a constant
length, fibre or particle length can be measured to some
degree of accuracy by observing for instance the length of
the particle-generated pulses under known conditions.[32,33]

The more voluminous reports which theorise a "shape"
factor have considered the response of red blood cells;
however, these cells are readily deformable, subject to
osmotic change, and can be conductive, so they are not as
easy to study as the rigid "industrial" particulate
materials. Gregg and Steidley,[34] Waterman et al.,[35]
Grover et al.,[21-23] Kachel,[36] and others have examined the
response of the electrical sensing zone method to blood
cells and have proposed a "form factor" of 1 for an elongated
object oriented along the axis of the aperture, 1.5 for a
sphere, and 2.8 for a disc perpendicular to the flow. The
analogy is that somehow the electrical path flows around
a streamlined envelope, or outline, of the particle,
thereby making the particle appear to be larger than it is.

Certainly, evidence presented from red blood cells
does support this view, but it is complicated by the above-
mentioned effects, as well as often to a shape change to

spherical or crenated morphometry when the cells are placed in an electrolyte diluent necessary for their analysis. All of this makes the expression "form factor" for red cells more meaningful than just "shape factor".

Marshall and Mehta,[37] Eckhoff,[38] Lloyd et al.,[39] and Lloyd[40] were forced to investigate the effect of shape in detail experimentally by using large scale mock-ups of the aperture, several centimetres across. If there is no error in such scaling up, their work indicates that shape does have a constant effect. Marshall and Mehta concluded that:

(1)   Pulse height was directly proportional to particle volume for spheres.

(2)   Pulse height was not directly proportional to particle volume for rod and disc shaped particles, but was modified by a shape factor, dependent on their orientation within the aperture.

(3)   Particle volume as seen by the sensing zone was greater than its physical volume for spheres, discs and short rods. [This observation supports the electrical flow path theory noted above].

Eckhoff's model showed that response is proportional to particle volume, including any re-entrant regions. For porous particles, response is proportional to displaced volume when the pores are parallel with the field.

With these effects in mind, it is recommended that for any particular, known, material the apparatus is self-calibrated when the instrument design allows.[5,18,41] This is a "primary" calibration method, as follows.

5 CALIBRATION

As already stated, the ideal method of calibration, when it can be performed, is that of mass integration (e.g.[5]), sometimes also called mass-balance. That is, the instrument is calibrated directly with the particles of material under test, using the standard gravimetric and volumetric methods of an accurate balance, pipettes and flasks. A size fraction of the particles under test, preferably not exceeding some 10:1 by particle diameter to be sure that they are all measured, is diluted to a known mass concentration in electrolyte solution. The total volume, in instrument units, of the particles measured in a known volume of suspension is related to the known mass concentration and the particles' immersed specific gravity (density), allowing the calibration factor to be calculated. In this way, the calibration procedure approaches an absolute reference method, as it eliminates any possible errors which may be caused by particle shape, porosity, conductance, or any other effect which may be present but cannot easily be isolated.

The diameter calibration factor (Kd) is calculated from

$$Kd = [6WV_m 10^{12}/\pi V_T \, \rho \, (\Delta n.V)]^{1/3} \quad \ldots\ldots\ldots\ldots(2)$$

where

W = the mass of sample in beaker (in g);

V_T = the volume of electrolyte solution in which W is diluted (in cm^3);

V_m = the manometer volume (in cm^3);

ρ = the immersed density of the powder (in g/cm^3);

Δn = the number of particles in a size interval;

and V = the arithmetic mean volume for that particular interval, in instrument units (e.g. product of threshold value, aperture current and attenuation).

It is of interest to note, then, that an electrical sensing zone instrument can be used to determine the immersed density of a powder, if the calibration factor is known. The approach has been used to obtain the wet densities of various clays.[42,43]

Of course not all particulate matter is homogeneous in terms of its density, for example river sediments, so the method is not universal. In these cases, and also for general convenience, it is usual instead to perform a "secondary" calibration method using a narrow size ranged polymer latex sample, whose mean, mode or some other "size" has been measured by another technique. The latex method is commonly used. Suitable particles are obtained from several sources including, in europe, the Community Bureau of Reference (B.C.R.),[44] and, in the U.S.A., the National Institute of Science and Technology (formerly the National Bureau of Standards).[45] Full traceability to those Reference Materials is recommended.

A common method of calibration is to choose latex particles having a mean diameter of between some 5 and 20% of the aperture diameter, and to determine the instrument settings (e.g. of current, attenuation and threshold) which halve the total particle count of the population of singlet particles. These settings are equated to the mean diameter of the latex particles. Doublets, caused by agglomeration and/or coincident particle passage through the sensing zone, are thereby prevented from causing any

error. This so-called 'half-count' method is simple and
very precise.

Models fitted with a multichannel pulse height analyzer
can be more quickly calibrated by relating the mode of the
generated size distribution to the assayed modal value of
that latex, as provided by the supplier.

Haematologically, for red blood cell sizing,
calibration is performed against a series of specimens of
normal fresh blood cells, taken into a known anticoagulant,
whose mean sizes have been determined by the reference
procedures of total red cell volume (centrifuged
haematocrit) and total red cell count (for example as
measured by an electrical sensing zone analyzer fitted with
an accurate manometer and used with accurate dilution
methods). The procedure may or may not be allowed to
account for any plasma still trapped within the packed bed
of red cells.

Whichever method of calibration has been chosen, it
should be specified when results are reported. The one
calibration will be valid for the whole range of size
measurement from any one aperture, or sensor. Note that,
with some more basic models, the size response is
dependent also upon the resistivity and temperature of the
electrolyte solution, in which case these two effects must
be eliminated by standardizing the electrolyte solution
composition and temperature, or by frequent re-calibration.

6   PRACTICAL CONSIDERATIONS

Particle Size Range

For routine practical purposes with most industrial
particulate materials, the overall size range commonly
considered for the electrical sensing zone method is about
0.5-200μm equivalent spherical diameter, although some
commercial instrument designs will provide results from
about 0.4-1200μm, and experimental designs have provided
results down to 0.1μm or below, and up to 1500μm or more.
More than one aperture (or sensor) size will be needed in
order to cover such a large size range.

The lower size limit of a sensor is set by the
electrical noise generated by the applied current upon the
electrically conducting fluid within the sensing zone
itself; as the aperture is, in effect, a somewhat noisy
resistor. This electrical noise level is often as low as
1.5% of the aperture diameter, e.g. 1.5μm with a 100μm
aperture. Coupled with the large sizes measurable with
each aperture (up to 80%), this means a dynamic measurement
range of some 50:1 by diameter from each sensor. Of course
not all instrument designs are such that this dynamic range
can be utilized, or are linear enough in their circuitry
to allow such a range to be measured with accuracy.

Commercially available apertures exist down to 15 or even 10μm in diameter, so that particle size distributions down to 0.3μm or below can be obtained.

An interesting experimental design using the Coulter principle was created by DeBlois and Bean.[46] They created a sensing zone by selecting an individual sub-micrometre pore etched in an irradiated polycarbonate plastic sheet; the material now known as Nuclepore® membrane. A typical aperture was some 0.5μm diameter and 3μm thick. Because it was very fragile, particles were made to pass through it more by means of their electrophoretic mobility rather than by a differential pressure. The aperture membrane only lasted for three weeks, but during that time they established that they could accumulate a size distribution of a total of 134 polystyrene latex particles of nominal diameter 0.091μm, obtaining a coefficient of variation from the mean diameter of 5.9%. The device was later used to count and size viruses and bacteriophage particles in the range of about 0.1-0.2μm diameter, and also to study simultaneously their electrophoretic and electro-osmotic velocities.[47,48]

The upper size limit of the method is usually set by the design's ability to suspend particles uniformly in the sample beaker. Various stirrer and container designs have been proposed and are used, and the electrolyte solution's density and/or viscosity may be increased in order to assist homogeneous suspension. McCave and Jarvis[49] found that, as the rate of sampling through a large sensing aperture (2000μm diameter) exceeded the rate of fall of sediment particles initially suspended in a large beaker, they could simply turn off the stirrer and obtain a particle size distribution directly when using a COULTER COUNTER® model T multichannel analyser.

Harfield et al.[50] approached the same problem by mounting the aperture wafer horizontally, and allowing (and encouraging) the particles to fall through the sensor. In this way they were able to measure 1.5mm steel balls. This device did not become commercially available.

Choice of Sensing Aperture

If the majority of the particles lie within a 50:1 diameter size range, the most suitable aperture will be chosen. For instance, a 30μm aperture will measure, on some instrument models, from about 0.5 to 20μm or more; a 140μm aperture from about 2 to 85μm or more.

If the particles cover a wider size range, two or more aperture sizes will be used, and the results overlapped to provide the full distribution analysis (e.g.[5]).

The smallest apertures may be more sensitive to external electrical noise, and are more prone to temporary blockage, such as from inadvertent contamination or agglomerated particles. Blockages are not normally common, but will occur from time to time depending on the particulate specimen and/or the care taken with the experimental conditions. Optics are provided to monitor the cleanliness of the aperture, as well as other indicators such as electronic blockage detectors, an oscilloscope display of pulses, and an audible counting rate. Blockages are usually easily cleared by applying momentary back-pressure to the aperture by means of a small piston, or by a rotation of the operating taps, and perhaps at the same time applying high current to the aperture to "boil" out the blockage. In case of difficulty, the aperture tube can be easily dismounted and cleared by the use of greater back-pressure, brushing, soaking in acid or by applying, carefully and at low power, ultrasonic energy for a fraction of a second.

Choice of Electrolyte Solution

The electrolyte solution should have a conductivity similar to 0.2-20% w/v sodium chloride in water, sufficient to give an aperture resistance between about 1 and 100 kilohms, and ideally around 5-40 kilohms. It must be compatible chemically with the sample material under test and allow proper dispersion (usually with the help of a surfactant and ultrasonic agitation) and suspension of the particles (with the help of a built-in stirrer, and thickening agents if necessary).

A great many materials are water-insoluble, or virtually so. In these cases, an electrolyte solution of about 10g/l sodium chloride in water (e.g. normal or isotonic saline) or 50g/l trisodium orthophosphate in water is commonly used. Sometimes chloride ions inhibit dispersion and the alkaline phosphate solution has advantages, for instance in dispersing silicas, clays and minerals. Many other solutions of inorganic salts in water have been used for particular purposes; for instance drug solubility studies can be carried out in dilute hydrochloric acid to simulate the contents of the stomach, or the common ion effect used to depress the solubility of some materials so that they can be more easily measured. Typical concentrations of particles for measurement are of the order of 10-100mg/100ml, so even low solubilities can be important, though it is the solubility rate which is the determining factor as to whether the analysis can be made at all. Multi-channel particle counters are invaluable for soluble materials as the total time of analysis is short (10-20 seconds). It has been found advantageous sometimes to remove the aperture illumination temporarily to prevent any small temperature rise in the electrolyte solution increasing the solubility of the particles.

Alternatively, organic electrolyte solutions may be used, such as 50g/l lithium chloride in methanol and 50g/l

ammonium thiocyanate in isopropanol, and many others are
available. The dielectric constant of the 'solvent' base
must be greater than about 10 to allow the dissolved salt
(or acid) to dissociate and give an acceptable aperture
resistance. Thus, liquids like petroleum spirit and
mineral oils cannot be used alone.

To measure very small particle sizes (e.g. 1µm and
below) with small apertures, it is often beneficial to
raise the conductivity of the electrolyte solution, where
this is not detrimental to the sample. An example is to
use 100g/l sodium chloride in water rather than 10g/l. As
well as improving the signal to noise ratio, this also
assists in preventing any 'electrical translucency' effects
due to the surface breakdown potential of conductive
materials such as metals (see following).

Large particles may not be adequately suspended, even
with the stirrer at the maximum rate which does not entrain
air bubbles, which would then be counted as particles. For
instance, 300µm particles of density 2.5g/ml may be
difficult to suspend. In these cases, it is customary to
add a thickening agent (e.g. up to 700ml/l glycerol or
1000g/l common sugar) to raise the viscosity and/or density
of the electrolyte solution, as well as to use a round-
bottomed sample beaker fitted with a vertical baffle and a
four bladed stirrer to maintain uniform suspension. Nienow
and Ang[51] and deNeve and Wuyts[52] have also published novel
approaches to measuring large or dense particles.

When selected, the electrolyte solution should be
filtered before use so that it is essentially free of
particles in the desired size range of measurement.
Commonly, 0.45µm or 0.22µm membrane filters are used for
that purpose.

Conducting Particles and Porous Particles

It might be expected that particles of conductive
materials could not be accurately measured by the
electrical sensing zone method. For this to be so, the
conductivities of the particle and of the electrolyte
solution would have to be very similar and there would
have to be no electrical boundary between them. It is a
fact that particles of highly conductive materials such as
copper (e.g. Ullrich[53]), silicon and carbon can be sized
correctly providing that the voltage applied across the
aperture does not allow the barrier potential at the
particles' surface to be exceeded. This is of the order
of some 5-20 volts across the aperture, depending on
material, according to Harfield.[54] The effects of the
Helmholtz double electrical layer and/or an oxide layer
or adsorbed gas film on the surface of the particles help
to render metal particles less conducting.

Particles which are porous and contain electrolyte

solution have been found to be sized as their solid volume, i.e. the current effectively enters the particles to give a sizing response not related to their 'envelope' volume. The effect of porosity, and the technique for filling porous particles so as to allow the measurement of 'envelope' volume, are detailed by Horák et al.[55] and Van der Plaats and Herps.[56]

In the case of cells enclosed by membranes, such as red blood cells, the linear response to cell volume changes at some 'breakdown' value of applied voltage (e.g. Groves[57]). Blood cell analysers are of course designed to operate below this breakdown potential.

Coincident Particle Passage

All types of sensing zone particle counters must be operated at concentrations low enough that essentially only one particle is present in the zone at any one time. Coincidence limits therefore vary according to the volume of the zone, (i.e. the size of the aperture), and according to the response of the electronic circuits. If two particles are in the zone at the same time, they may only be counted as one (Figure 2) and, in addition, if they are very close together they may be sized as one particle of volume equal to their sum (Figure 3), exactly as if they were physically joined. By keeping concentration levels relatively low, e.g. to 10^5 particles/ml for a 100µm electrical sensing zone aperture, this count "loss" is small and can be easily corrected mathematically. Under these conditions, any particle sizing error caused by coincidence is usually not detectable but, if necessary, the effects can be virtually eliminated by analyzing at a lower concentration for a longer time, or by using a smaller aperture.

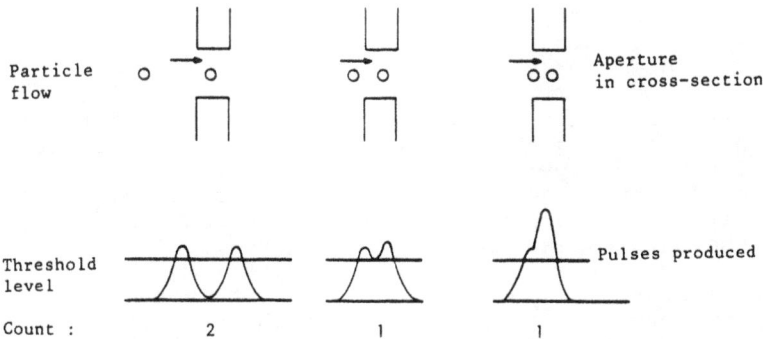

Figure 2   Effect of coincidence upon count recorded. (Two particles closely together in sensing zone can count only as one).

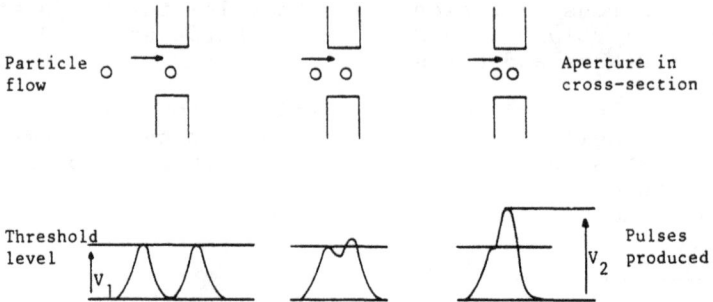

<u>Figure 3</u>   Effect of coincidence on size recorded. (Two
           particles of volume V_1 very closely together in the
           sensing zone can be sized only as one (V_2),
           which approaches the sum of their individual
           volumes).

Several coincidence theories exist, varying in
complexity, for instance [18,58-62] but, providing that the
concentration is kept fairly low, say to the so-called
"10% coincidence limit" (where one particle in 10 is not
being counted) they all give very similar corrections.   A
common expression[18] for the lost count due to coincidence
(n") is

$$n'' = p[n/1000]^2 \qquad\qquad \dots\dots\dots\dots\dots\dots(3)$$

where

$$p = 2.5\,[d/100]^3\,[500/V] \qquad \dots\dots\dots\dots\dots\dots(4)$$

p is the coincidence factor and n is the number of
particles counted in the metered volume V, in μl, using an
aperture of diameter d, in μm.

The factor of 2.5 was derived experimentally for an
aperture 75% as long as its diameter, a standard Coulter
configuration for apertures above 50μm.   It can be thought
of as the effective volume of the sensing zone being two
and a half times its physical volume; the zone extending
beyond the physical edges of the aperture itself.

Harfield, Lloyd and Cowan[63] cast doubt on the common
use of the Poisson distribution model for predicting
coincident occurrences and related the correction to the
particle transit times, sensing volume and circuit response.

In practice, coincidence factors are given by the
manufacturers for each aperture size and most modern

analyzers make the corrections automatically.

Precision of Measurements

In any type of particle counter, the number of particles in a metered volume of suspension will vary from the number in any other identically metered volume, as expected from statistical theory. A mean count of $\bar{n}$ particles per unit volume could be expected to vary according to the standard deviation of measurement, calculated as $(\bar{n})^{\frac{1}{2}}$. Thus a mean count of 100,000 particles can be obtained with a standard deviation of 316.2 particles, i.e. a precision (expressed as coefficient of variation, c.v.) of 0.316%.

Clearly, it is important that any particle counter/sizer should measure a large number of particles in order to achieve the greatest statistical confidence in the results. In electrical sensing zone models, the concentration should be sufficiently great, but not too great to exceed coincidence limitations. Larger measurement volumes may be preferred, at lower concentrations. Of course, in contamination studies where there are low numbers of particles per unit volume, ideal counting statistics may never be achieved.

The resulting precision of particle size distributions obtained from different total numbers counted has been evaluated by Lines[64] who showed that the average standard deviation of any point on a typical weight percentage size distribution was

2.12 wt% for a total distribution of 2,700 particles counted

0.74 wt% for a total distribution of 10,000 particles counted

and 0.33 wt% for a total distribution of 100,000 particles counted

Thus, a size distribution obtained from 100,000 particles gave a precision of significantly better than 1wt% over its whole range.

7   IMPROVEMENTS IN, AND DEVELOPMENTS OF, THE METHOD

Improving Sizing Accuracy

With standard apertures and sampling configuration, the electrical sensing zone method gives a much superior sizing accuracy compared with most other particle sizing methods. However, for very narrowly size ranged particles, such as polymer latexes, even greater sizing resolution can be achieved by making a few improvements.

    In the standard design configuration, e.g. Figure 1,
particles are drawn through the aperture from all
positions around it. Those particles passing centrally
into, through, and out of the sensing zone generate pulses
which are essentially proportional to their displaced
volume (Figure 4, A). However, those particles which do
not pass centrally through the electrical zone produce a
slightly larger 'artefact' pulse than expected for their
displaced volume (e.g. Thom et al.,[65] Thom,[66] Grover et al.,
[21] Kachel,[36] Atkinson and Wilson,[67] and Wilson[68]) the error
being most for those particles passing closest to the edge
of the aperture, (Figure 4,B,C). This causes the size
distribution to be skewed towards to the coarser sizes, the
effect being most significant and hence most noticeable on
very narrow size distribution materials (e.g. [67,28]). In
extremely narrow size distributions, the error may even
show as a second 'particle' population (Figure 5, broken
line plot). On wider size range samples, such as most
powders, suspensions and emulsions (wider than some 3:1 by
diameter) the effect is not significant or observable.

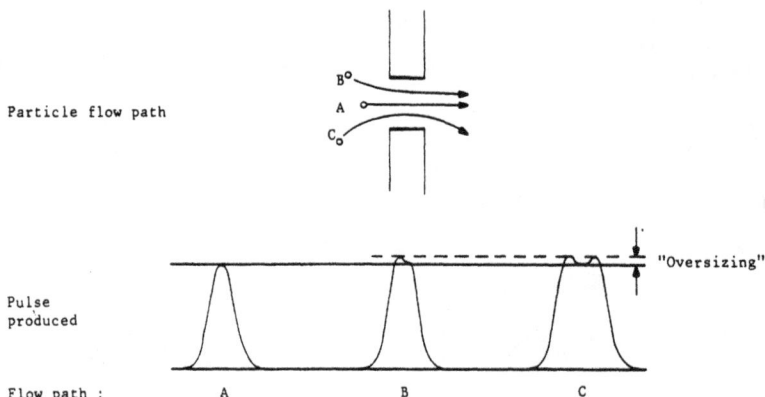

Particle flow path

Pulse
produced

Flow path :   A   B   C

"Oversizing"

Figure 4   Pulse shape (and height) depends on particle
           flow path through the aperture.

    For increased sizing accuracy, when needed, many of
these 'artefact' pulses can be screened out electronically
(by examining each pulse and then "editing" the unwanted
ones out of the analysis), or the particles can be
physically prevented from flowing non-axially by
"hydrodynamically focusing" them into the sensing zone,
(Figure 6). In both of these ways, extreme size
distribution accuracy can be obtained if it is required.
Figure 5, solid line plot, shows the same latex sample,
but analyzed from "edited" pulses. The "editing" approach
(e.g. Hogg[69]) is also useful in removing some of the
electronic noise, allowing small particles to be resolved
more clearly.

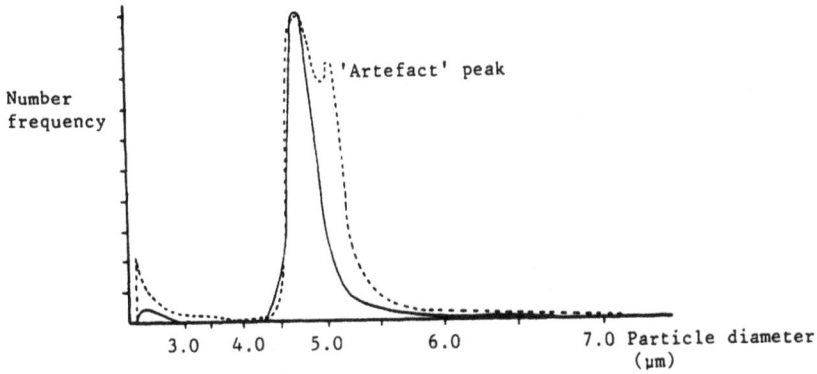

Figure 5 Very narrow range latex size distribution. [Broken line : standard aperture and sample flow, without editing. Solid line : with editing applied).

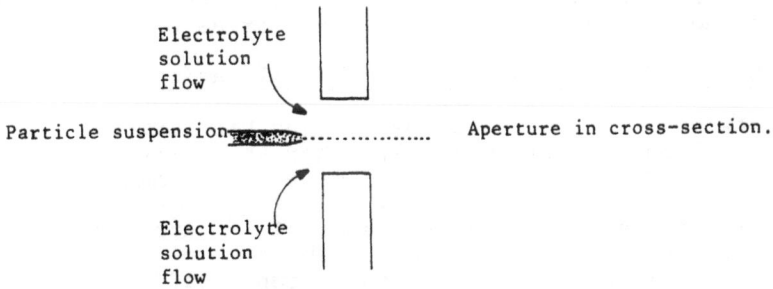

Figure 6 Hydrodynamic focusing of particle flow.

Thom [65,66] introduced hydrodynamic focusing as a means of improving the sizing response of the electrical sensing zone method. Harfield and Cowan,[70,71] Göransson,[72] and Merkus et al.[73] have reported results from an improved experimental hydrodynamic focusing device. Göransson[72] also compared his results from narrow size-ranged materials with the "editing" method. The use of electronic "editing" instead of the excellent method of Thom is to avoid complicating the sample flow path with its susceptibility to collecting debris.

Measuring Contaminants in Molten Metals

The electrical sensing zone method requires a conducting liquid within the sensor. Simensen[74] and

Guthrie and Doutre[75] have dipped an aperture tube made of
boron nitride into molten metals, such as aluminium, zinc
and lead, applied a high current across the resulting
sensing zone by way of iron electrodes, and measured the
insoluble particles, or inclusions, within the molten
metal. Currents of 20-60A were used, along with apertures
of up to 2mm diameter.

Continuous Monitoring of Cell Cultures

It has been shown[76] that cells can be efficiently
grown, in continuous culture in a bio-fermenter, on the
surface of polymer beads. Only the electrical sensing
zone method allows the growth to be monitored to the
required degree of sensitivity, as the 'bead' size
increases due to the attached, growing, cells. At the end
of the process, the cells must be removed from the beads'
surface and harvested. Coulter Electronics has cooperated
in producing a unique autoclavable, self contained,
sampling, diluting, and counting probe to be fitted
permanently to a bio-fermenter.[77] The signals are
processed by standard COULTER COUNTER® and CHANNELYZER®
instrument circuitry. It is possible that this new
technology may allow patients' own skin cells to be rapidly
cultured, to be re-applied to large areas of burns.

On-Line, In Situ, and On-Stream Applications

As well as the molten metals and bio-fermenter
applications described above, other specially designed
devices have been described to make direct measurements.

Barnett et al.[78] and Lines et al.[79] introduced an on-
line monitor system which comprised a sampling device, a
control unit and a printer, to be connected to a
conventional COULTER COUNTER® instrument. It was used for
continually monitoring the particulate contamination levels
within an electrolyte solution, such as normal saline. The
sampling device included three apertures, any one of which
was used at any one time while the others were held in
reserve in case of blockage. At pre-selected time
intervals, samples of solution were passed through the
active sensor and particle counts determined at various
pre-set size levels.

Submersible and self-contained sensor devices to
determine the numbers and sizes of individual particles
directly in sea-water at various depths have been reported.[80-82] These towed devices contained the necessary pumps
and sensing apertures, and passed their signals to on-
board COULTER COUNTER® instruments for real-time
processing.

Single Particle Measurement

A novel approach has recently been used at the University of Oslo by Berge et al.[83-87] They constructed an electrical sensing zone device with a pressure reversal mechanism, which enabled one particle to be measured by repeated transit through the aperture. In this way they have studied single particle dissolution and particle flow dynamics,[83,87] transit time and electrokinetic effects,[86] the dissolution of air bubbles,[85] and the pulse height response to particles transitting close to the walls of the aperture.[84] In the latter work, they found a maximum error in pulse height with respect to particle volume of 10%. This is a 3.3% diameter oversizing for those particles passing closest to the aperture walls; c.f. Figure 4.

Internal Measurements of Particles

So far, the Coulter principle has been illustrated when using an applied direct current (d.c.) across the aperture. This enables particle volume to be measured with very high precision and accuracy. Recently, Coulter Electronics has begun exploiting the use of alternating current (or radio frequency, r.f.). This has the ability to generate a response dependent upon the composition of particles, and so can help to identify them (e.g. Coulter and Hogg[88]). One such application is to use r.f. to obtain information within blood cells and, in combination with a laser for further identification purposes, a unique triple transducer has been developed to allow the routine quantification of different types of white blood cells. This is now commercially available,[89,90] and will be further developed. It is also expected that the combined use of r.f. and d.c. aperture technology will have future benefits elsewhere in particle size analysis applications.

REFERENCES

1. P.J. Crosland-Taylor, Nature, 1953, 171, 37.
2. W.H. Coulter, U.S. Patent 2,656,508, 1953.
3. W.H. Coulter, Proc. Natl. Electr. Conf., 1956, 12, 1034.
4. C.F.T. Mattern, F.S. Brackett and B.J. Olson, J. Appl. Physiol., 1957, 10, 56.
5. British Standard 3406: Part 5:1983; British Standards Institution, London, 1983.
6. A.O. Ullrich, Instr. Soc. America Conf., New York City, 1960.

7. T. Allen, in 'Proc. Conf. Particle Size Analysis, Loughborough', Society for Analytical Chemistry, London, 1966, p. 110.
8. T. Allen, 'Particle Size Measurement', Chapman and Hall, London, 1st ed., 1968.
9. T. Allen, 'Particle Size Measurement', Chapman and Hall, London, 4th ed., 1990, p. 445.
10. F.M. Rabinovitch, 'Conductimetric Method of Dispersion Analysis', Chemical Publishing House 'Khimiya', Leningrad, 1970.
11. F.M. Rabinovitch, 'The Application of Conductimetric Particle Counters to Medicine', Meditzina, Moscow, 1972.
12. T. Allen and K. Marshall, 'The Electrical Sensing Zone Method of Particle Size Measurement', University of Bradford, 1972.
13. Industrial Bibliography, Coulter Electronics Limited, 1989.
14. Medical and Biological Bibliography, Coulter Electronics Limited, 1990.
15. A. Richardson-Jones, 'Advances in Hematological Methods : The Blood Count', O.W. van Assendelft and J.M. England, eds., CRC Press, Inc., Boca Raton, Fla., Chapter 5, 1982, p. 49.
16. R.W. Lines, J. Pharm. Pharmac., 1967, 19, 701.
17. V. Kachel, 'Flow Cytometry and Sorting', M.R. Melamed et al., eds., Wiley, New York, 1979, p. 64.
18. 'Theory of the Coulter Counter-Bulletin T1', Coulter Electronics, Inc., 1957.
19. W.H. Coulter, W.R. Hogg, J.P. Moran and W. Claps, U.S. Patent 3,259,842, 1966.
20. B.A. Batch, J. Inst. Fuel, 1964, 37, 455.
21. N.B. Grover, J. Naaman, S. Ben-Sasson, and F. Doljanski, Biophys. J., 1969, 9, 1398.
22. N.B. Grover, J. Naaman, S. Ben-Sasson and F. Doljanski, Biophys. J., 1969, 9, 1415.
23. N.B. Grover, J. Naaman, S. Ben-Sasson and F. Doljanski, Biophys. J., 1969, 12, 1099.
24. N. Saranummi, Techn. Res. Centre of Finland Report No. 10, 1975.
25. B. Scarlett, 'Proc. Conf. Partec (Nuremberg)', 1979, p. 682.
26. J.G. Harfield and P. Knight, 'Proc. Conf. Particle Size Analysis 1981, Loughborough', Wiley-Heyden, Chichester, 1982, p. 151.
27. J.G. Harfield, R.A. Wharton and R.W. Lines, Part. Charact., 1984, 1, 32.
28. J.G. Harfield and R.A. Wharton, 'Proc. Conf. Partec (Nuremberg)', 1984, p. 661.
29. R.K. Eckhoff, J. Sci. Instr., 1967, 44, 648.
30. R.K. Eckhoff and P. Soelberg, Betongtekniske Publikasjoner, 1967, 7, 1.
31. R.A. O'Connell and R.J. Martsch, Textile Res. J., 1962, 32, 581.
32. R.B. Valley and T.H. Morse, U.S. Patent 3,441,848, 1969.

33. R.H. Berg and R.F. Karuhn, U.S. Patent 4,290,011, 1981.
34. E.C. Gregg and K.D. Steidley, Biophys. J., 1965, 5, 393.
35. C.S. Waterman, E.A. Atkinson, B. Wilkins Jnr., C.L. Fisher and S.L. Kimzey, Clin. Chem., 1975, 21, 1201.
36. V. Kachel, J. Histochem. Cytochem., 1976, 24, 211.
37. K. Marshall and R.G. Mehta, M. Sc. Thesis, University of Bradford, 1969.
38. R.K. Eckhoff, J. Sci. Instr., (J. Phys. E.). 1969, 2, 973.
39. P.J. Lloyd, B. Scarlett and I. Sinclair, 'Proc. Conf. Particle Size Analysis 1970, Bradford', Society for Analytical Chemistry, London, 1970, p. 276.
40. P.J. Lloyd, 'Proc. Conf. Partec (Nuremberg)', 1979, p. 694.
41. C.M.L. Atkinson and R. Wilson, 'Proc. Conf. Particle Size Analysis 1981, Loughborough', Wiley-Heyden, Chichester, 1982, p. 185.
42. J.W. Gartner, Ph.D Thesis, University of S. Florida, 1978.
43. J.W. Gartner and K.L. Carder, J. Sed. Petrol., 1979, 49, 631.
44. Community Bureau of Reference-BCR, Rue de la Roi 200, B-1049 Brussels, Belgium; suppliers of particulate reference materials, quartz powders and latex suspensions.
45. National Institute of Standards and Technology, Standard Reference Materials Program, Room 204, Building 202, Gaithersburg MD 20899, U.S.A.; suppliers of particulate reference materials, glass beads and latex suspensions. (Formerly, National Bureau of Standards - NBS).
46. R.W. DeBlois and C.P. Bean, Rev. Sci. Instr., 1970, 41, 909.
47. R.W. DeBlois and R.K.A. Wesley, J. Virology, 1977, 23, 227.
48. R.W. DeBlois, C.P. Bean and R.K. Wesley, J. Coll. Interf. Sci., 1977, 61, 323.
49. I.N. McCave and J. Jarvis, Sedimentology, 1973, 20, 305.
50. J.G. Harfield, B. Miller, R.W. Lines and T. Godin, 'Proc. Conf. Particle Size Analysis 1977, Salford', Heyden, London, 1978, p. 378.
51. A.W. Nienow and H.M. Ang, Lab. Practice, 1972, 21, 495.
52. R. DeNeve and L. Wuyts, Pharm. Weekblad, 1976, 111, 220.
53. W.J. Ullrich, 'Modern Developments in Powder Metallurgy, Vol. 1, Fundamentals and Methods', Plenum Press, 1966, p. 125.
54. J.G. Harfield, 'Proc. Conf. Particle Size Analysis 1981, Loughborough', Wiley-Heyden, Chichester, 1982, p. 165.

55. D. Horák, L. Peška, F. Švec and J. Štamberg, Powder Technol., 1982, 31, 263.
56. G. van der Plaats and H. Herps, Powder Technol., 1984, 38, 73.
57. M.R. Groves, I.E.E.E. Trans. Biomed. Engrg., 1980, BME-27, 364.
58. M. Wales and J.N. Wilson, Rev. Sci. Instr., 1961, 32, 1132.
59. L.H. Princen and W.F. Kwolek, Rev. Sci. Instr., 1965, 36, 646.
60. L.H. Princen, Rev. Sci. Instr., 1966, 37, 1416.
61. H. Bader, H.R. Gordon and O.B. Brown, Rev. Sci. Instr., 1972, 43, 1407.
62. P.J. Lloyd, 'Proc. First European Symp. Particle Size Measurement, Nuremberg', 1975, p. 325.
63. J.G. Harfield, P.J. Lloyd and M.P. Cowan, 'Proc. Conf. Particle Size Analysis 1988, Guildford', J. Wiley and Sons, Chichester, 1988, p. 121.
64. R.W. Lines, Powder Technol., 1973, 7, 129.
65. R. Thom. A. Hampe and G. Sauerbrey, Z. ges. exp. Med., 1969, 151, 331.
66. R. Thom, Vergleichende Untersuchungen zur Elektronischen Zellvolumen - analyse, A.E.G. Telefunken, Ulm, W. Germany, 1972.
67. C.M.L. Atkinson and R. Wilson, Powder Technol., 1983, 34, 275.
68. R. Wilson, Part. Charact., 1984, 1, 37.
69. W.R. Hogg, U.S. Patent 3,668,531, 1972.
70. J.G. Harfield and M.P. Cowan, 'Proc. Conf. Particle Size Analysis 1988, Guildford', J. Wiley and Sons, Chichester, 1988, p. 151.
71. M.P. Cowan and J.G. Harfield, Part. Part. Syst. Charact., 1990, 7, 1.
72. B. Göransson, Part. Part. Syst. Charact., 1990, 7, 6.
73. H.G. Merkus, H. Liu and B. Scarlett, Part. Part. Syst. Charact., 1990, 7, 11.
74. Chr. Simensen, Personal communication, (1985).
75. R.I.L. Guthrie and D.A. Doutre, 'Proc. Intnl. Seminar on Refining and Alloying of Liquid Aluminium and Ferro-Alloys', Trondheim, Norway, August 26-28, 1985, p. 147.
76. O. Hanotte, C. Dejaiffe, D. Dubois, P. Vanderpoorten, F. Menozzi and A.O.A. Miller, Biology of the Cell, 1987, 59, 175.
77. D. Thebline, J. Harfield, O. Hanotte, D. Dubois and A.O.A. Miller, Conf. 8th ESACT-32nd OHOLO, Tiberias, Israel, April 6-10th, 1987.
78. M.I. Barnett, E. Sims and R.W. Lines, Powder Technol., 1976, 14, 125.
79. R.W. Lines, J.G. Harfield, W.M. Wood and B.V. Miller, Intnl. Symp. "In Stream Measurements of Particulate Solid Properties", Bergen, August 22-23, 1978.
80. W.S. Maddux and J.W. Kanwisher, Limnol. Oceanogr., 1965, 10, R162.
81. R.W. Lines, Conf. Oceanology International '69, Brighton, 1969.

82. C.M. Boyd, Neth. J. Sea Res., 1973, 7, 103.
83. L.I. Berge, J. Feder and T. Jøssang, Rev. Sci. Instr., 1989, 60, 2756.
84. L.I. Berge, T. Jøssang and J. Feder, Meas. Sci. Technol., 1990, 1, 471.
85. L.I. Berge, J. Coll. Interf. Sci., 1990, 134, 548.
86. L.I. Berge, J. Coll. Interf. Sci., 1990, 135, 283.
87. L.I. Berge, J. Feder and T. Jøssang, J. Coll. Interf. Sci., 1990, 138, 480.
88. W.H. Coulter and W.R. Hogg, U.S. Patent 3,502,973, 1970.
89. A. Richardson-Jones, 'Evaluation of the COULTER® VCS Automated Differential Counter', Coulter Electronics, Inc., 1988.
90. D.F. Barnard, S.A. Barnard, A.B. Carter, K.G. Patterson, A. Yardumian and S.J. Machin, Clin. lab. Haemat., 1989, 11, 255.

A New Perspective on Particle Sizing by the Coulter Principle: Single Particle Dynamics

L. I. Berge, J. Feder and T. Jøssang

DEPARTMENT OF PHYSICS, UNIVERSITY OF OSLO, P.O. BOX 1048 BLINDERN, 0316 OSLO 3, NORWAY

1 INTRODUCTION

Since the invention of a means to count and size particles suspended in a conducting solution was patented by Coulter in 1953,[1] the Coulter principle has become widely used for counting and sizing particles. The Coulter Counter was first used to count and size red blood cells.[2] Kubitschek[3] modified a Coulter Counter for studying bacteria. Later, a large variety of systems have been studied by numerous workers.[4,5] DeBlois and Bean[6] extended the Coulter principle by introducing a new kind of pore made by the etched particle track process.[7] Using submicron pores, they reported the routine detection of 0.09 μm diameter particles, which represented a qualitative advance in sensitivity.

We present a novel extension of the resistive pulse technique (Coulter principle) for studying single particle dynamics. A unique pressure reversal technique has been developed, where the pressure gradient may be reversed shortly after a particle exits the pore (aperture) such that the same particle re-enters the pore at regular intervals. This way, single particle dynamics may be studied over long periods of time. By using long pores as apertures, end effects are minimized, the pulses have well defined shapes[8] and the pulse height, form and width are related to the particle volume, shape and transit time. In long cylindrical pores, a pressure drop sets up a Poiseuille flow, where the effect of radial migration (particle motion transverse to streamlines) has been utilized to study off–axis particle response. The pressure reversal technique can also be used to study single particle dissolution and growth processes. This is a new application of the Coulter principle and measurements of dissolution and growth of gas bubbles and crystal dissolution are used to illustrate the potential of this new technique.

2 EXPERIMENTAL

The experimental arrangement has previously been described[9] in detail. The central part is a Plexiglas cell with two chambers (~ 1 cm^3) connected by a pore. The particles to be analysed are suspended in the electrolytic solution which fills the cell. Particles flow through the pore by pressure drive. Each chamber contains a silver, silver–chloride electrode connected to a constant voltage source. When a particle enters the pore, the resistance between the electrodes increases and consequently the current decreases. The current is amplified using a current amplifier with a voltage output. The output signal is connected to a trigger circuit for pressure reversal and to a waveform analyser, which is interfaced to a personal computer. In pressure reversal mode, the pressure is reversed shortly after a particle exits the pore and the same particle re–enters the pore at regular intervals. This way, single particle dynamics may be studied over long periods of time. Particles have been reversed for many hours.

The frequency of pressure reversal has mostly been 2 Hz. It is varied using a delay circuit, by adjusting the time a particle spends outside the pore between reversals. The pressure reversal technique should in principle be usable with any pore length, but a reasonable pressure drop across the pore is required. In order for the pressure reversal technique to function satisfactorily, the signal to noise ratio must be sufficient to eliminate malfunction. Also, the particle concentration must be kept low. At large concentrations, a reversed particle is easily lost due to interactions with other particles.

3   RESPONSE OF THE PARTICLE ANALYSER

The basis for the resistive pulse technique is the resistance pulses which are detected every time a particle passes through the pore. The principal theoretical problem is to determine the increase in resistance of a conducting circular cylinder when an insulating particle is inserted far from the ends. Maxwell[10] obtained an expression for the effective resistivity of a dilute suspension of small insulating spheres in a solution of known resistivity. Smythe[11] analysed the problem of flow around a spheroid in a circular tube considering also large particle sizes. In the limit of small spheres, Smythe's numerical results confirm the derivation by Maxwell. Golibersuch[8] introduced a shape factor which is a function of particle shape and orientation.

On this basis, the response of our particle analyser may be expressed as[9]

$$\Delta E = \frac{EG}{R_e} \left(\frac{V}{f_e v S(d/D)} + 1 \right)^{-1}. \tag{1}$$

Here ΔE is the measured voltage change at the output of the current amplifier due to the presence of a sphere of volume v inside a pore of volume V. Further, E is the constant voltage across the pore, G is the current amplifier gain (volts/ampere), R_e is the pore resistance in the absence of particles and $S(d/D)$ is a size correction factor based on the calculations by Smythe, where D and d are pore and sphere diameter, respectively. The size correction factor has been given a simple empirical form,[12] $S(d/D) = (1 - 0.8(d/D)^3)^{-1}$. The shape factor is $f_e = f_b + (f_a - f_b) \cos^2 \alpha$ for a spheroid, where $f_a = 1/(1 - D_a)$ and $f_b = 1/(1 - D_b)$. D_a and D_b are the demagnetization factors[13,14] for a

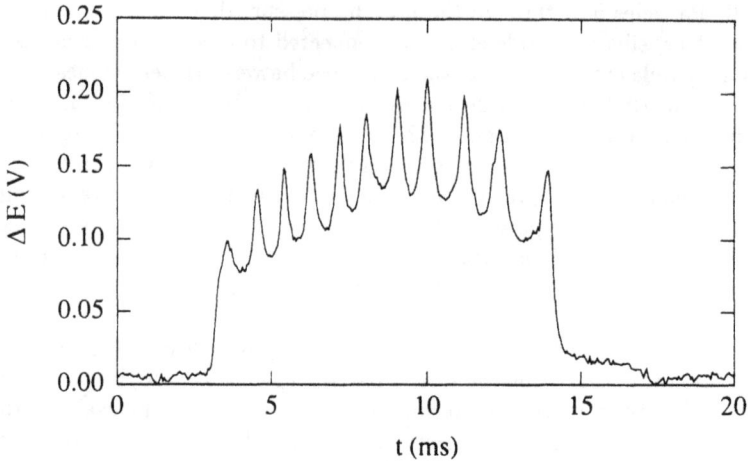

Figure 1: The measured pulse from a rotating crystal particle (Rochelle salt) which is transported in Poiseuille flow through a 9 μm diameter and 540 μm long glass capillary. A measured pulse height of 0.15 V corresponds to a particle diameter of about 3 μm. The particle makes 5 complete revolutions during the transit through the pore.

field applied parallell and perpendicular, respectively, to the axis of revolution. The demagnetization factors are given in terms of the ratio of polar diameter to equatorial diameter,[13] and α is the angle between the field and the axis of revolution. For a sphere, $f_e = 1.5$.

4   RESULTS AND DISCUSSION

Pulse Amplitude

As noted before, the pulse amplitude is not uniquely determined by the particle volume. Shape and orientation of aspherical particles also contibute to the measured amplitude.[8] This is easily demonstrated in long pores and Figure 1 shows the measured pulse from an approximately 3 μm crystal particle (Rochelle salt) which is transported by pressure drive through a 9 μm diameter and 540 μm long pore. The pulse amplitude varies significantly as the particle rotates in the shear field. The undulations are superimposed on a rounded pulse which depicts the structure of the pore, the local pore diameter decreases inwards in the pore.

In long cylindrical pores, a pressure drop sets up a Poiseuille flow (parabolic velocity profile) and particles suspended in the solution are transported through the pore with transit times depending on the radial positions at which the particles enter the pore. In Poiseuille flow, solid spheres migrate to an off–axis

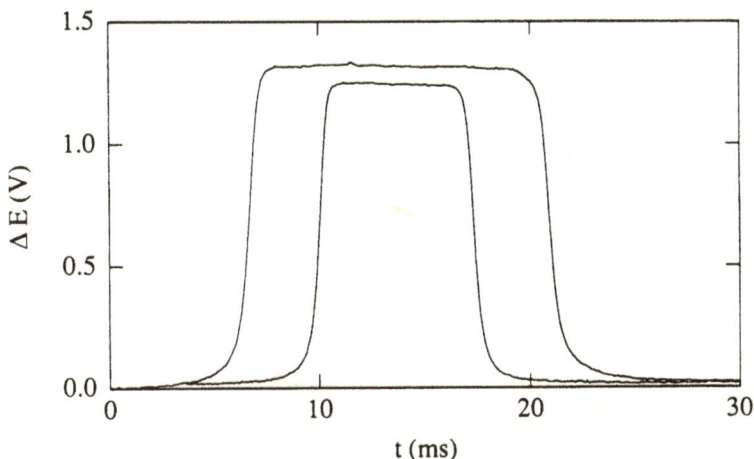

Figure 2: Two measured pulses from a 25 μm diameter sphere which is reversed back and forth through a 70 μm diameter and 570 μm long glass capillary. The particle migrates radially in Poiseuille flow from the axis (short transit time) towards the wall (long transit time) and the pulse height increases when the particle moves closer to the wall. Approximately 20 reversals separate the two realizations.

equilibrium position[15,16] and this non–linear hydrodynamic effect has been utilized to study how the measured pulse amplitude from a single particle flowing back and forth through a pore increases when the particle migrates closer to the pore wall.[17] Figure 2 shows two recorded pulses from a reversed 25 μm diameter polystyrene sphere flowing through a 70 μm diameter and 570 μm long glass capillary. The pore is very uniform and the pulses rise rapidly to a constant amplitude value. The pulse width is well defined and it can be related to the flow characteristics of the system.[16,18]

The pulse width and height increase as the particle migrates towards the wall and the two pulses correspond to particle positions close to the pore axis and wall. There are approximately 20 reversals between the two realizations shown in the figure. For all particle and pore sizes studied, the increase in pulse height is found to be less than 10%.[17] This is considerably less than predicted by the off–axis upper limit theory by Smythe.[19] The variations in recorded pulse heights for particles of the same size entering the pore at different radial positions is related to artifact peaks in particle size distributions measured with Coulter Counters, caused by particles passing through the aperture very close to the peripheral edge.[20]

Dissolution and Growth of Gas Bubbles

The pressure reversal technique extends the concept of particle sizing to also include single particle dynamics. As a supplement to studying the time evolution

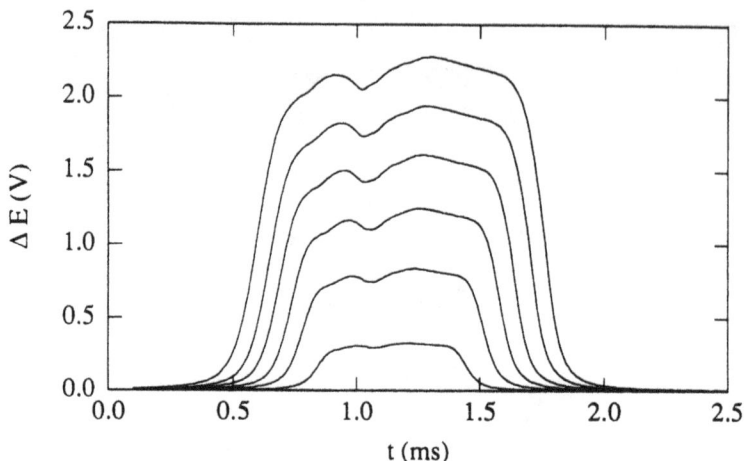

Figure 3: Six measured pulses from the same air bubble dissolving in a 46 wt% sucrose/0.1 M NaCl solution using a 22 μm diameter and 110 μm long mica pore. The largest pulse height corresponds to a bubble diameter of about 20 μm.

of particle size distributions[21,22] one can follow the dissolution or growth process of single particles. We have studied air bubbles[23] with diameters in the range from 22 to 3 μm, a size range which is not easily accessible by other methods. Bubbles have been produced by shaking. Typical transit times through the pore have been a few milliseconds.

When a bubble is reversed back and forth through the pore at low frequencies ($\sim$ 2 Hz), the bubble spends most of its time outside the pore practically at rest. The gas molecules in the bubble enter the liquid phase and diffuse away and the governing equation for the dissolution process is the diffusion equation.[24] The diffusion coefficient for air in the solution is inversely proportional to the viscosity of the solution. Sucrose added to the electrolyte increases the viscosity and as a consequence the diffusion coefficient decreases. Thus, the bubble lifetime may be sufficiently increased to make single small bubbles accessible to measurements. In pure water, bubbles in this size range would dissolve in less than 10 s.

Several recorded pulses from the same air bubble dissolving in an undersaturated 46 wt% sucrose/0.1 M NaCl solution at 26°C are shown in Figure 3. The bubble was reversed twice every second. All pulses are for the same flow direction and the pulse form reveals that the 22 μm equivalent diameter and 110 μm long mica pore is not completely smooth. The pulse width decreases with bubble size as expected.

Theory predicts that the surface area of the bubble should decrease linearly with time until surface tension becomes important, enhancing the dissolution

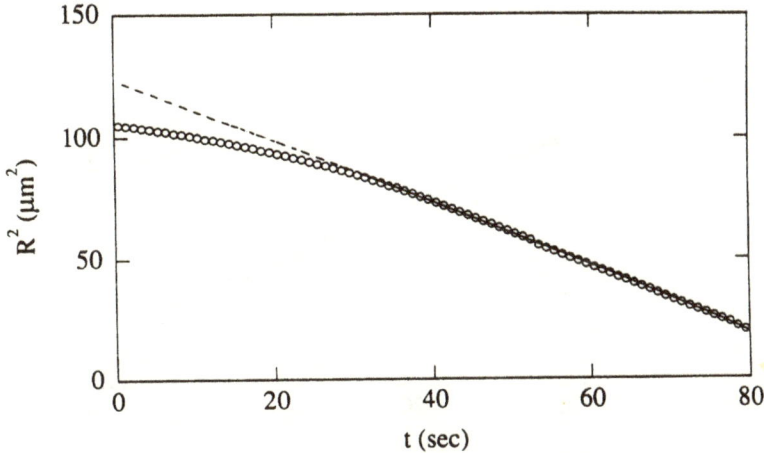

Figure 4: The measured radius squared (open circles) as a function of time for the dissolving air bubble referred to in Figure 3. The fitted curve (Eq. (2), including surface tension) to the last part of the measurement is also shown. Note the deviation from the expected behaviour for large bubble sizes due to bubble deformation.

process and causing the linear relationship to become more and more curved. The time it takes for a bubble of radius R_0 to shrink to a radius R (neglecting corrections in the initial phase before the gas concentration reaches a steady state distribution) is given by[23,24]

$$t = \frac{R_0^2}{2\alpha(1-f)\mathcal{D}} \left(1 - \left(\frac{R}{R_0}\right)^2\right)(1 - \mathcal{F}(\gamma)) , \qquad (2)$$

where α is the ratio of the dissolved gas concentration for a saturated solution and the gas density inside the bubble, f is the solution saturation ratio, $\mathcal{D}$ is the diffusion coefficient of the gas in the liquid, γ is the surface tension, and $\mathcal{F}(\gamma)$ is a function of surface tension.[24]

The measured dissolution curve for the case in Figure 3 is shown in Figure 4. When the bubble first enters the pore, its diameter is close to the pore diameter. Every second data point (for clarity) has been plotted as an open circle and a fitted curve to the last part of the measured curve is also shown. The fitted parameter was $f = 0.6$, while $\gamma = 75$ dyn/cm, $\mathcal{D} = 200$ $\mu m^2/s$, and $\alpha = 0.0061$ were held fixed at their expectation values.[23] The bubble diameters have been calculated from the measured pulse heights assuming a spherical bubble. This assumption is clearly not fulfilled for the largest bubble sizes and the observed deviation between measurement and theory can be explained in terms of bubble deformation. A prolat spheroid with a ratio $m = 4/3$ of the axis of revolution to the equatorial axis explains the deviation at time $t = 0$, since a prolat ellipsoid gives a smaller increase in resistance compared to a sphere of equal volume.

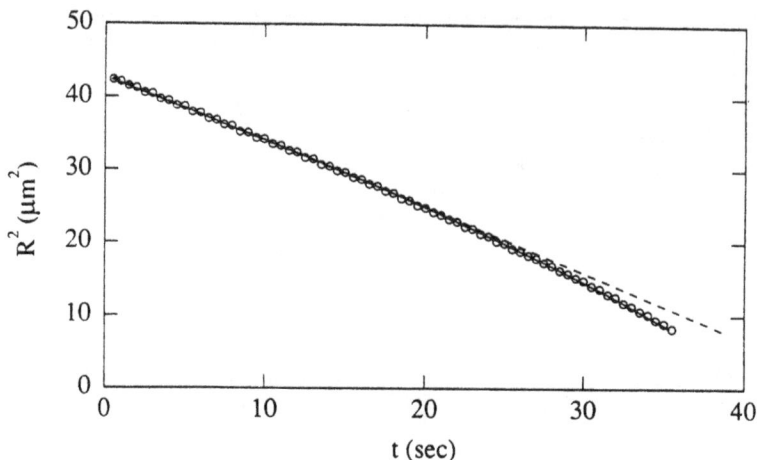

Figure 5: The measured radius squared (open circles) as a function of time for an air bubble dissolving in a 46 wt% sucrose/0.33 M NaCl solution using a 15.9 μm diameter and 104 μm long mica pore. The solid curve is the fit to Eq. (2) with $\gamma = 75$ dyn/cm and the broken line is a fit to the first part of the data neglecting surface tension. Note the deviation from the linear behaviour for small bubble sizes due to surface tension.

Figure 5 shows measured and fitted curves for the radius squared R^2 as a function of time t for an air bubble dissolving in an undersaturated 46 wt% sucrose/0.33 M NaCl solution at 26°C. The bubble was reversed back and forth through a 15.9 μm equivalent diameter and 104 μm long mica pore two times each second. The measured data points in Figure 5 fit the above equation very well (solid curve) with $f = 0.83$ and otherwise the same parameters as before. Also, a linear fit to the first part of the curve is shown (broken line). We clearly see that the first linear part crosses over to a curved region when the bubble has become so small that the increased pressure inside the bubble due to surface tension is significant. Proteins (BSA) added to the solution have a large stabilizing effect and the rate of dissolution is drastically reduced when the bubbles become small. We hope to go further down in size to see to what extent microbubbles are stabilized by organic material always present in water.

A different system is shown in Figure 6 which illustrates the growth of a protein (BSA) coated chlorine bubble produced electrolytically in a 0.15 M NaCl solution close to the entrance of the pore. The bubble was reversed at a frequency of about 80 Hz using a 27 μm diameter and 540 μm long glass capillary. The pulse height was measured 3 times each second. Instead of the expected linear relationship between surface area and time, we observe a much more rapid growth, $R^2 \sim t^3$. This is not well understood, but it may be an enhancement which is due to the high frequency of pressure reversal.

4. 'Coulter Counter Medical and Biological Bibliography,' Coulter Electronics Limited, Northwell Drive, Luton, Bedfordshire, LU3 3RH, England, 1990.
5. 'Coulter Counter Industrial Bibliography,' Coulter Electronics Limited, Northwell Drive, Luton, Bedfordshire, LU3 3RH, England, 1989.
6. R.W. DeBlois and C.P. Bean, Rev.Sci.Instrum., 1970, 41, 909.
7. P.B. Price and R.M. Walker, J.Appl.Phys., 1962, 33, 3407.
8. D.C. Golibersuch, J.Appl.Phys., 1973, 44, 2580.
9. L.I. Berge, J. Feder and T. Jøssang, Rev.Sci.Instrum., 1989, 60, 2756.
10. J.C. Maxwell, 'A Treatise on Electricity and Magnetism', Clarendon, Oxford, 1904, 3rd ed., Vol. I.
11. W.R. Smythe, Phys.Fluids, 1964, 7, 633.
12. R.W. DeBlois, C.P. Bean and R.K.A. Wesley, J.Colloid Interface Sci., 1977, 61, 323.
13. E.C. Stoner, Philos.Mag., 1945, 36, 803.
14. J.A. Osborn, Phys.Rev., 1945, 67, 351.
15. G. Segré and A. Silberberg, J.Fluid Mech., 1962, 14, 136.
16. L.I. Berge, J.Fluid Mech., 1990, 217, 349.
17. L.I. Berge, T. Jøssang and J. Feder, Meas.Sci.Technol., 1990, 1, 471.
18. L.I. Berge, J.Colloid Interface Sci., 1990, 135, 283.
19. W.R. Smythe, Rev.Sci.Instrum., 1972, 43, 817.
20. D.A. Elkington and R. Wilson, in 'Particle Size Analysis 1985', ed. P.J. Lloyd, Wiley, Chichester, 1987, p. 509.
21. R. Smither, J.Appl.Bact., 1975, 39, 157.
22. M. Marković and L. Komunjer, J.Cryst.Growth, 1979, 46, 701.
23. L. I. Berge, J.Colloid Interface Sci., 1990, 134, 548.
24. P.S. Epstein and M.S. Plesset, J.Chem.Phys., 1950, 18, 1505.

Integrated Granulometric Techniques for Particle Swelling Characterization

C. Caramella, F. Ferrari, M. C. Bonferoni and M. Bertoni

DEPARTMENT OF PHARMACEUTICAL CHEMISTRY, UNIVERSITY OF PAVIA,
VIA TARAMELLI 12, 27100 PAVIA, ITALY

1 INTRODUCTION

Water uptake capacity (swelling in bulk) and particle swelling (intrinsic swelling) may be considered the physical properties of tablet disintegrants that are more commmonly related to disintegrant efficiency [1-3].

For the evaluation of swelling in bulk both static methods, like hydration capacity[4] and sedimentation volume [5] and dynamic methods, like water uptake of powder bed [6,7], swelling of pure disintegrant tablets[8] and dilatometry[9] have been proposed.

The methods for the evaluation of intrinsic swelling, i.e. individual particle volume increase, are mostly based on the microscopic observation of particles. This presents the obvious advantage of allowing the direct measurement of the increase in disintegrant particle dimensions due to water absorption [10-11].

Recently an improvement of the microscopic method for evaluating the swelling characteristics of tablet disintegrants has been proposed[12]. The authors also considered the changes in particle shape that may take place upon swelling. All the microscopic methods, however, present some limitations in their applicability to routine swelling measurements, especially when irregularly shaped materials or materials characterized by small variations of the particle linear dimensions are considered.

Given the difficulties of evaluating particle volume increase from microscopic data, alternative instrumental methods, in particular methods based on the Coulter Counter principle, have been proposed in the literature[13-15].

The Coulter Counter method provides facilities for both counting and sizing particles and its counting principle has a volume basis, which should allow the exact calculation of the particulate volume. In particular the method we proposed[15] was essentially based on the comparison between the total particulate volume of a disintegrant as measured in a swelling and in an inert medium.

This method provides a rapid and reproducible means for measuring particulate volumes of many kinds of materials in various media and was especially advised in the case of limited swelling materials[15,16].

The accuracy of the total particulate volume measurements, which is likely to depend on the volume computation method, was further investigated and assessed on dry polystyrene latices[17].

On the other hand, considering the operating principle of the Coulter Counter, which is based on electrical conductivity differences between the suspending medium and the suspended particles, it could be expected that, if the particle conductivity changes on swelling owing to electrolyte absorption, the response of the Coulter Counter could be impaired.

This assumption was verified in a previous work[15] where the results obtained with the Coulter Counter were validated with optical microscopy. The validation procedure showed that the method was not applicable to strongly swelling materials containing highly hydrophilic ionizable moieties, due to the changes in particle conductivity that may occur owing to electrolyte absorption.

Therefore, although the Coulter Counter method allows an accurate evaluation of particle volume diameter in inert media as checked by mass balance evaluation, in some cases it is not suitable to measure particle swelling. In these cases, the Coulter Counter method needs to be integrated with other granulometric techniques.

Given these premises, in the present work some tablet disintegrants were examined using both the Coulter Counter and other instrumental techniques, in particular optical instrumental methods capable of measuring swollen particle dimensions independently of their changes in conductivity.

Two different types of optical instruments were employed: a light blockage apparatus and a laser diffraction apparatus.

Although optical methods permit us to overcome the problem of changes in conductivity, they provide for particle sizing on the basis of equivalent diameters different from the volume diameter and, unlike the Coulter Counter, do not allow the exact calculation of particulate volume from particle size distributions.

Therefore, additional reference methods were employed, which could provide a direct evaluation of the particle swelling, even in those experimental conditions that could bias the instrumental responses.

In particular, optical microscopy analysis, which allows the evaluation of particle shape factors and the assessment of their possible change in differing media, was performed both in inert and in swelling media.

Furthermore, a modified centrifugation method was devised, which allowed the correct evaluation of the amount of water actually retained by the swollen particles.

The feasibility of the integrated approach for particle swelling measurements has been assessed on sieved fractions of the following disintegrants: Sodium starch glycolate, characterized by spheroidal shape, β Cyclodextrin polymer, characterized by sub-angular shape, and Croscarmellose sodium, a fiber-like material.It could be argued that the swelling behaviour of sieved fractions may differ from that of the unfractionated material. While this aspect deserves further investigation, in the present paper the employment of sieved (middle to narrow size) fractions instead of the commercial grades was justified by the fact that sieved fractions exhibit a rather well defined log-normal particle size distribution that may be easily characterized by means of the relevant parameters (median size and standard deviation)[15].

2 EXPERIMENTAL

<u>Materials</u>

The disintegrants examined were:
- Sodium starch glycolate (Explotab® ; E.MENDELL Co., USA - New York): granulometric fractions 44-53 μm and 53-63 μm;
- β Cyclodextrin polymer (CHINOIN Pharm.& Chem.Works, HUNGARY - Budapest): granulometric fractions 44-53 μm and 53-63 μm.
- Croscarmellose sodium (Ac-di-Sol®; FMC Co.,USA-Philadelphia, Pa.): granulometric fractions 44-53 μm and 53-63 μm.

The granulometric fractions were obtained by wet sieving with ethyl ether, having previously checked that the treatment did not modify particle size and shape.

The true density and water contents were determined as previously described[15].The true density was 1.57 g/cm^3 for Explotab®, 1.39 g/cm^3 for β Cyclodextrin polymer and 1.59 g/cm^3 for Ac-di-Sol®.

The suspending media employed were: isotonic saline containing 0.05% w/v of formalin; distilled water; isopropyl alcohol; 5% w/v NH_4SCN in isopropyl alcohol. All suspending media were filtered using 0.8 μm and subsequently 0.45 μm MF filters (MILLIPORE CORP., USA-Bedford,MASS.) for aqueous media and 0.8-0.4 μm polycarbonate filters (Nucleopore, USA-Pleasanton, CA.) for organic media.

<u>Methods</u>

<u>Microscopic analysis.</u>Optical microscopy measurements were performed in inert medium (isopropyl alcohol) and swelling medium (isotonic saline) by means of a transmitted light microscope (Microstar 120, Reichert-Young Amer. OPT., USA-Buffalo, N.Y.), fitted with a Polaroid camera (3.5 x 4.5) and with standard ocular micrometer.

The swelling index was calculated as previously described[15].

Furthermore, the elongation ratios of the examined materials both in inert and swelling media were calculated, according to Prasad[18].

<u>Coulter Counter</u>.A COULTER Counter Mod. TA II, 16 channels (COULTER ELECTR. LTD., GB-Luton), equipped with PCA was employed.

Calibration was effected on 140 and 280 μm orifice tubes using standard latex suspensions (COULTER ELECTR. LTD., GB-Luton) of nominal sizes 18.3 and 46.5 μm, respectively, according to two different procedures recommended in the instruction manual[19] : the volume calibration which is performed without the PCA equipment by using an exactly weighed quantity of standard material (Method A); the number calibration which is performed with the PCA equipment (Method B).

Particle size distribution of the examined materials was determined both on a volume (Method A) and on a number/volume basis (Method B) both in inert (isopropyl alcohol) and in swelling media (isotonic saline). From the differential number distribution supplied by the instrument, differential and cumulative volume distributions were calculated by means of a suitable program run on an IBM AT personal computer.

Mass balance evaluation was effected on exactly weighed samples by direct volume measurements (Method A) and by calculating the total particulate volume according to Nystrom[20] (Method B).

HIAC device. An HIAC/ROYCO model size 3000, 6 channels (Pacific Scientific Co., USA - Silver Spring, MD.) fitted with an HR60H sensor, with standard size range of 1-120 µm, flow rate 10 ml min^{-1} and supplied by the manifacturer already calibrated with standard spherical materials was employed.

Particle size distributions of the examined disintegrants were determined both in inert (isopropyl alcohol) and in swelling media (isotonic saline).

The size range considered was between 10 and 120 µm: the number particle size distribution was obtained using a six equally-spaced (10 µm) channel resolution.

The instrument supplies the differential number distribution, from which differential and cumulative volume distributions were obtained by means of the same program already mentioned for the COULTER Counter.

Laser particle sizer. Laser diffraction measurements were performed with a Laser Particle Sizer ANALYSETTE 22, 31 channels (FRITSCH GMBH, FRG-Idar-Oberstein) equipped with a liquid dispersion unit.

Particle size distributions of all the examined disintegrants were determined both in inert medium (isopropyl alcohol) and in isotonic saline. For Explotab® 44-53 µm measurements were also carried out in distilled water. The measuring range was between 5 and 722.8 µm subdivided into 31 logarithmically-spaced channels.

Particle size distribution is directly provided on a volume basis. The instrument supplies both the differential and the cumulative volume distributions.

For all the instrumental methods employed, geometric mean diameter (dg') and geometric standard deviation (σg) were obtained on a volume basis[21] from log probability plots of cumulative volume distributions.

Modified centrifugation method. Hydration capacity measurements were effected in swelling media as described by Kornblum[4]. One gram of material was thoroughly dispersed in 20 mL of suspending medium and then centrifuged using a refrigerating super speed centrifuge (Sorvall RC 5B, Ing. Terzano & C., I-Milan). The speed of the centrifuge was progressively increased from 270 to 23500 g until the weight of the sediment did not change any more. Each sample was maintained for 15 min at the following speed conditions: 270, 1475, 5900, 13300, 17300, 23500 g. The amount of water eventually retained by the material was then calculated by weighing the wet sediment and subtracting the weight of dry material. It represents the volume increase ratio, as obtained by centrifugation, of the material itself.

3 RESULTS

Microscopic analysis. The elongation ratios, both in inert and in swelling media, together with the swelling indexes are given in Table 1. Although the differences in elongation ratios between inert and swelling medium were significant in most cases, the changes in particle shape upon swelling were not dramatic either for Explotab® or for β Cyclodextrin polymer fractions, thus indicating a rather isometric swelling. Concerning Ac-di-Sol®, which consists of fiber-like particles that mainly swell in radial direction[15], the different behaviour of the two fractions as far as the elongation ratios are concerned is in line with their different propensity to uncoil upon swelling. In the case of Ac-di-Sol® the fiber-like shape of the particles, coupled with their anisometric swelling, is likely to bias any instrumental measurements. A rather good agreement was found for Explotab® and Ac-di-Sol® fractions between the swelling index values here found and those already reported in the previous work[15].

Table. 1 Elongation ratios (mean ± S.D. n=100) and swelling indexes of the materials examined.

Disintegrant	Isopropyl alcohol	Isotonic saline	t test	Swelling index
Explotab®				
44-53 μm	1.31 (±0.21)	1.60 (±0.44)	S	2.21
53-63 μm	1.33 (±0.20)	1.51 (±0.27)	S	2.09
β Cyclodextrin polymer				
44-53 μm	1.37 (±0.30)	1.49 (±0.40)	S	1.98
53-63 μm	1.39 (±0.25)	1.52 (±0.47)	S	2.03
Ac-di-Sol®				
44-53 μm	3.43 (±1.34)	2.43 (±0.81)	S	1.41
53-63 μm	4.09 (±1.78)	4.16 (±1.85)	NS	1.38

S = significant difference (P=0.05)
NS = not significant difference (P=0.05)

Coulter Counter and HIAC. In Table 2 the particle size distribution parameters (geometric mean diameter and geometric standard deviation) of the examined materials, as obtained with the Coulter Counter either in inert or in swelling medium, are given. Swelling indexes that were calculated according to[15] are also given in the table. Concerning measurements in inert medium, the mass balance evaluation showed that in all cases the total particulate volume measured with Method A was not significantly different from the true sample volume as calculated from the true density. This suggests a complete mass recovery with the Coulter Counter and confirms the suitability of the method for measurements in inert medium.

Using Method B, the total particulate volume measured was slightly overestimated (about 10%) as previously assessed[17]. However, this overestimation was not likely to impair the size distribution, thus advising the use of Method B for routine analysis.

Concerning measurements in swelling medium, the inadequacy of Coulter Counter method for strongly swelling materials, already pointed out[15] and ascribed to electrolyte uptake, is confirmed by the present results.

In fact, as can be appreciated from the data reported in Table 2, the increase in mean diameters occurring in isotonic saline (as expressed by swelling index) is definitely underestimated with respect to microscopic observation. This is especially evident in the case of Ac-di-Sol® fractions.

Table 2 Geometric mean diameters (dg') and geometric standard deviations (σg) of the particle size distribution obtained with the Coulter Counter in inert and swelling media and relevant swelling indexes.

	Isopropyl alcohol		Isotonic saline		Swelling index
	dg'	σg	dg'	σg	
Explotab®					
44-53 μm	46.4	1.26	58.1	1.29	1.25
53-63 μm	53.7	1.50	71.8	1.46	1.34
β Cyclodextrin polymer					
44-53 μm	47.3	1.27	65.0	1.18	1.37
53-63 μm	63.3	1.22	79.6	1.18	1.26
Ac-di-Sol®					
44-53 μm	53.3	1.66	34.9	1.52	0.65
53-63 μm	49.6	1.47	39.7	1.55	0.80

Similar results were obtained with the light blockage method (HIAC), whose inadequacy for swelling measurements can be explained by too small a difference in refractive index between aqueous medium and swollen particles [22].

Laser particle sizer. In Figure 1 (a and b) a typical example of particle size distribution as obtained with the laser diffraction apparatus in inert and in swelling medium is given. The shift of the mode of the distribution in swelling medium is evident.

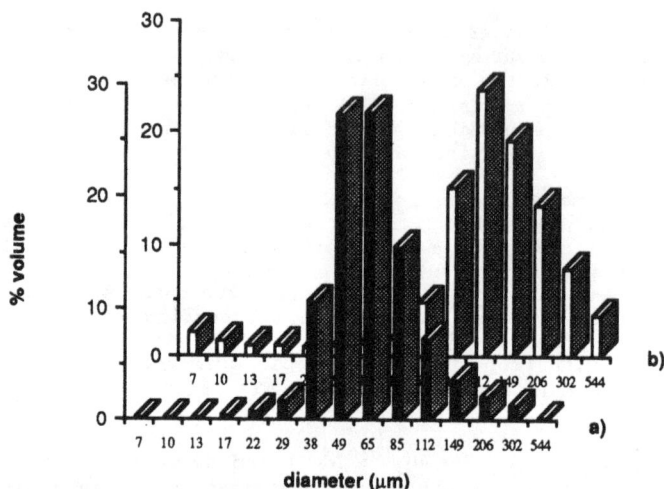

Figure 1 Particle size distributions of β Cyclodextrin polymer (granulometric fraction 44-53 μm) in isopropyl alcohol a) and isotonic saline b) as obtained with laser diffraction apparatus.

The geometric mean diameters and geometric standard deviations of the particle size distributions of Explotab® and β Cyclodextrin polymer fractions are given in Table 3. The results relative to Ac-di-Sol® fractions are not reported since their particle size distribution in inert medium was rather broad, possibly owing to the critical fiber-like shape, and did not fit any distribution model.

Table 3 Geometric mean diameters (dg') and geometric standard deviations (σg) of the particle size distribution obtained with the laser diffraction apparatus in inert and swelling media and relevant swelling indexes.

Disintegrant	Isopropyl alcohol		Isotonic saline		Swelling index
	dg'	σg	dg'	σg	
Explotab®					
44-53 μm	57.9	1.60	130.7	1.74	2.26*
53-63 μm	65.0	1.62	138.5	1.74	2.13
β Cyclodextrin polymer					
44-53 μm	61.1	1.51	121.5	1.76	1.99
53-63 μm	80.5	1.51	158.1	1.62	1.96

* The swelling index obtained in distilled water was 2,75.

The swelling indexes, calculated as the ratio of the mean diameters in swelling and in inert medium are also given. Although obtained on the basis of the equivalent diameter provided by the laser diffraction method and not on the basis of a volume diameter, swelling indexes are in good agreement with those obtained by microscopic analysis.

By raising to the third power the swelling index (which represents a linear increase ratio) the volume increase ratio of the four fractions was calculated on the assumption (supported by microscopic analysis) of an isometric swelling.

<u>Modified centrifugation method</u>. The results obtained for Explotab® and β Cyclodextrin polymer are illustrated in Figure 2. For both samples the test was performed in isotonic saline; for Explotab®, besides isotonic saline, distilled water was also employed. Maize starch was also tested as reference limited swelling material, given its well known swelling properties.

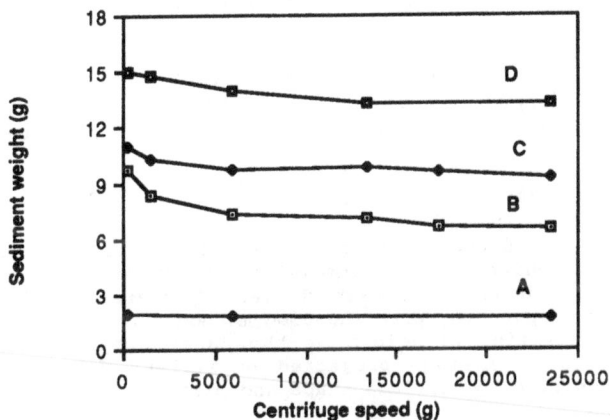

<u>Figure 2</u> Decrease of the sediment weight of the samples as a function of increase in centrifugation speed.
A) Maize Starch; B) β Cyclodextrin polymer; C) Explotab® in isotonic saline; D) Explotab® in distilled water.

It is assumed that when a plateau is attained, the interparticulate medium has been completely squeezed out and only the swelling medium inside particles is present.

As anticipated in the experimental part, from the limit constant value of the sediment weight the volume increase ratio of the materials is calculated. The relationship between the volume increase ratio obtained by centrifugation and that obtained by laser diffraction method is given in Figure 3.

The data relative to maize starch, whose volume ratio had been previously determined[17], were also considered in the regression. A significant correlation is found between the two ratios.

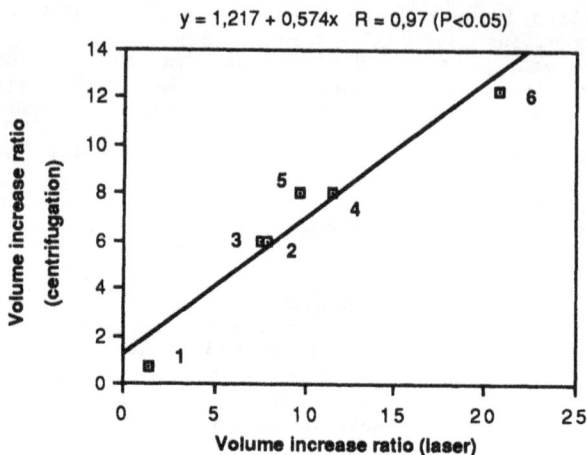

Figure 3 Relationship between volume increase ratio (by centrifugation) and volume increase ratio (by laser diffraction).
1) Maize Starch; 2) β Cyclodextrin polymer 44-53 μm; 3) β Cyclodextrin polymer 53-63 μm; 4) Explotab® 44-53 μm; 5) Explotab® 53-63 μm; 6) Explotab® 44-53 μm in distilled water.

4   CONCLUSIONS

For Explotab® and β Cyclodextrin polymer the laser diffraction method provides quite acceptable measurements of particle volume increase, even though a certain degree of overestimation is observed with respect to centrifugation method. This is possibly linked to the sizing principle of the laser apparatus that, being based on an average projected equivalent diameter, is certainly sensitive to particle shape and particle shape changes upon swelling.

Besides that, the laser apparatus provides a useful tool for the routine determination of particle size distribution of such materials.

Concerning fiber-like materials, like Ac-di-Sol®, optical microscopy seems to be the only applicable method for the evaluation of particle swelling, which was to be expected on the basis of its particle shape and swelling characteristics.

Acknowledgements

The authors wish to thank COULTER Scientific S.p.A.(Milano, Italy) for partially supporting the paper, MICROCHEM (Fiorenzuola D'Arda, Italy) for kind cooperation in laser diffraction measurements and Prof. V. Prodi (LAVORO e AMBIENTE, Bologna, Italy) for helpful suggestions.
Work partially supported by a grant of C.N.R.

REFERENCES

1. P. Couvreur, J. Gillard, H.G. Van der Schrieck, M. Roland, J.Pharm.Belg. 1974,29,399.

2. J.Ringard, A.M. Guyot-Hermann, J.Pharm.Belg. 1978,33,99.

3. P.H. List, U.A. Muazzam, Pharm.Ind. 1979,41,1075.

4. A.S. Kornblum, S.B. Stoopak, J.Pharm.Sci.,1973,62,43.

5. R. Shangraw, A.Mitrevej , M. Shah, Pharm.Technol.,1980,4,49

6. A.M. Guyot-Hermann, J.Ringard, Drug Dev. Ind. Pharm., 1980, 6, 511.

7. H.V. Van Kamp, G.K. Bolhuis, A.H. De Boer, C.F. Lerk, L.Lie-A-Huen, Pharm. Acta Helv.,1986,61,22.

8. D.Gissinger, A.Stamm, Drug Dev. Ind. Pharm.,1980,6,511.

9. S.Erdos,A.Bezegh, Pharm.Ind.,1977,39,1130.

10. E.Rudnic, C.T.Rhodes,S.Welch,P.Bernardo, Drug Dev.Ind.Pharm., 1982,9,87.

11. A.Mitrevej, R.G.Holembeck, Proc.Pharmaceutical Technology Conference, New York, 1986, p.211

12. L.S.C. Wan, K.P.P. Prasad, Acta Pharm.Technol.,1990,36,20.

13. P.Paronen,M.Juslin,K.Kasnanen,Drug Dev.Ind.Pharm.1985,11,405.

14. K.J.Steffens,P.H.List,V.Muazzam, Acta Pharm.Technol., 1980,26, 254.

15. C.Caramella,P.Colombo,G.Bettinetti,F.Giordano,U.Conte, A.La Manna, Acta Pharm. Technol.1984, 30, 132.

16. C.Caramella, F.Ferrari, U.Conte, 'Particle Size Analysis 1988' Wiley & Sons, New York, 1988, p.167.

17. C.Caramella,F.Ferrari,M.C.Bonferoni,Part.Part.Syst.Charact., 1990, 7, 131.

18. K.P.P. Prasad, L.S.C. Wan, Pharm. Res., 1987, 4, 504.

19. Instruction Manual for Coulter Counter Model TA II. Technical literature Coulter Electronics Ltd, 1976, p. 45 and p.75

20. C. Nystrom, J. Mazur, M. I. Barnett, M. Glazer, J. Pharm. Pharmacol.,1984, 37, 217.

21. A. Martin,J.Swarbrick,A. Cammarata, 'Physical Pharmacy', Lea & Febiger, Philadelphia, 1983, p. 500-501.

22. L.S. Golden, 'Particle Size Analysis 1988', Wiley & Sons, New York, 1988, p. 335.

Testing Mass Calibration with the Coulter Counter

M. M. Figueiredo, M. G. Rasteiro and M. F. Ferreira

DEPARTAMENTO DE ENGENHARIA QUÍMICA, UNIVERSIDADE DE COIMBRA,
LARGO MARQUÊS DE POMBAL, 3000 COIMBRA, PORTUGAL

ABSTRACT

This paper concerns the calibration methods of the Coulter
Counter, namely the mass integration method. These methods
have been approached from a theoretical and experimental
point of view.

Tests were performed using an accurate counter
(Coulter Counter ZM) and also a multichannel analyser
(Channelyzer 256).

1 INTRODUCTION

The Coulter Counter, through its various models, has
become an established and well evaluated method of
counting and sizing particles. The Coulter principle, or
better, the electrical sensing zone method, has been the
subject of many publications[1], including textbooks[2] and
the British Standard 3406[3]. The main advantage of this
method is that it yields directly a number distribution of
the equivalent volume diameter.

The most common method of calibrating these
instruments is by using spherical particles (normally
latex spheres) of uniform size. However a more fundamental
approach consists of using, for calibration, the particles
of the material under test. This is known as the mass
integration method of calibration, or simply, mass
calibration. This method is recommended by the British
Standards and is considered superior to the latex
calibration. Being, however, rather time consuming it is
not used for routine purposes.

The present paper concerns the calibration methods of
the Coulter Counter. Emphasis is put on the mass
integration method, namely on the equations normally used
to calculate the mass calibration constant. Experimental
results comparing both methods of calibration are also
presented.

2   CALIBRATION METHODS

As the response of the Coulter is essentially proportional to particle volume, the threshold scale provides a relative scale of individual particle volume. To calibrate the instrument means to convert the arbitrary threshold scale into equivalent spherical diameter, that is, to determine the calibration constant according to

$$K_d = \frac{d}{\sqrt[3]{T}} \qquad (1)$$

where

K_d = calibration constant based on the diameter
d = equivalent particle diameter
T = threshold value.

As mentioned above, this can be done in two different ways:

a) using polymer latex particles - <u>latex calibration</u>;
b) using the material under investigation - <u>mass (or self) calibration</u>.

For the mass integration method it is necessary to prepare a suspension of accurately known concentration of a narrow size range fraction of the material under test[3]. The mass calibration constant can, then, be evaluated from the ratio of the real volume of particles, calculated from the particle concentration for a metered volume of suspension, to the measured threshold values, being given by:

$$K_d = \left[\frac{6}{\pi} \frac{V_m}{V_s} \frac{W}{\rho_s} \frac{1}{\Sigma \Delta n \overline{T}} \right]^{1/3} \qquad (2)$$

where

V_m = volume of metered suspension (manometer volume)
W = total weight of particles
V_s = total volume of suspension
ρ_s = density of particles
Δn = number of particles in a size interval
$\overline{T}$ = average threshold corresponding to Δn.

Particular attention should be given to this equation which often appears in the literature in the form

$$K_d = \left[\frac{6}{\pi} \frac{V_m}{V_s} \frac{W}{\rho_s} \frac{1}{\Sigma \Delta n \overline{V}} \right]^{1/3} \qquad (3)$$

where $\overline{V}$ is, sometimes, not well defined[3,4] and can be incorrectly identified to a "real" volume value. This could be rather confusing, especially for the Coulter ZM users, since this model computes automatically volume values and not threshold values. These have to be hand calculated for each combination of the instrument settings.

A more suitable form of Eq.(2), in terms of volume values, has been proposed[5]. Obviously, these volumes have to be related to a prior calibration constant, which could well be the latex calibration constant. K_d can, then, be calculated as

$$ K_d = K_{do} \left[\frac{V_m}{V_s} \; \frac{W}{\rho_s} \; \frac{1}{\Sigma \Delta n \bar{V}_o} \right]^{1/3} \qquad (4) $$

where

K_{do} = initial estimate of the calibration constant
$\Sigma \Delta n \bar{V}_o$ = total volume of particles computed by the instrument when using K_{do}

The differences between Eq.(3) and (4) are obvious, being therefore apparent the errors that can be committed if volume values are used in the former.

3 EXPERIMENTAL

Experimental tests were carried out with irregular glass particles in order to compare both calibration techniques. The orifice tube selected was a 140 μm tube which was calibrated with a 20 μm nominal size latex (Coulter Electronics Ltd.) using the so called "half count method"[4]. For the mass calibration technique, a sieve fraction of the material under test, in the range of 37-44 μm, was used and a suspension of known concentration ($4-8 \times 10^{-5}$ g/ /cm^3) was carefully prepared. The density of the glass particles (Schott Duran 50) was measured to be 2.23 g/cm^3.

The analysis was performed, according to the British Standard recommendations, in the single threshold mode, with coincidence correction for, at least, 15 size classes.

This method of calibration is, however, rather tedious, especially if a considerable number of size classes is to be used. It was then decided to link the Coulter ZM to a Coulter Channelyzer 256 in order to speed up the analysis. The channelyzer was connected to a PC compatible which provided the data in a suitable form for direct use in the above equations. The channelyzer was calibrated using the same latex as for the ZM, following the instruction manual calibration procedure. The suspensions for the mass calibration technique were prepared as previously described. The channelyzer was operated in manometer mode with 256 channels for higher resolution.

4   RESULTS AND DISCUSSION

The first set of experiments was performed with the Coulter ZM. The mass calibration constant was calculated from Eq.(4) using the latex calibration constant as K_{do}. Although this is not strictly necessary it is more logical since it gives a direct indication of the deviation between both constants.

As mentioned, it was later decided to use the Channelyzer 256. Preliminary tests were carried out in order to compare the results obtained with the two instruments in terms of size analysis. The agreement found for the size distribution curves was perfect. The following step was to test the counting efficiency of the channelyzer. For that, The Coulter ZM was operated in a dual threshold mode with size ranges corresponding to the positions of the channelyzer cursors. The ZM was used as reference as it is considered an accurate counter. The counting differences were found to be less than 2%, with lower channelyzed countings. This result encouraged the final step: to test the accuracy of the channelyzer for mass calibration. The equation used to calculate the mass calibration constant was, as before, Eq.(4).

The results for both the ZM and the Channelyzer are presented in Table 1. The values for K_d(mass) are mean values of at least three separate experiments. The standard deviations are given in brackets.

The values of Table 1 were used to determine the particle size distribution of the 37-44 μm sieve fraction, being the results summarized in Table 2, in terms of d_{10}, d_{50} and d_{90}.

Refering to the results obtained with the Coulter ZM, a good agreement was found between the two calibration techniques. Also the size analysis fall well within the sieve cut. Similar results have been reported for BCR reference materials[6]. The differences between the two calibration constants, although minor, could be explained by the existence of a small fraction of fine particles which is not detected by the system.

Table 1   Comparison of the calibration constants using the latex and the mass calibration methods

Instrument	K_d(latex)	K_d(mass)	Dev.(%)
ZM	23.24	23.78(0.04)	2.3
Channelyzer 256	16.34	17.17(0.06)	5.1

Table 2   Results of the particle size analysis using the
            latex and the mass calibration constants

instrument	Latex calib.			Mass calib.		
	d_{10}	d_{50}	d_{90}	d_{10}	d_{50}	d_{90}
	(μm)	(μm)	(μm)	(μm)	(μm)	(μm)
ZM	36.72	42.24	48.10	37.61	43.20	49.20
Channelyzer 256	36.70	42.15	47.50	38.50	44.25	49.37

As for the channelyzer the deviations are larger, as
expected. The reason for that is probably related to the
accumulation of two effects: the loss of counts in the
channelyzing process, experimentally verified, and the
presence of small particles outside the tube measuring
range, as mentioned above.

5 CONCLUSIONS

The calibration methods of the Coulter Counter have been
discussed in some detail in the present work. A more
suitable equation has been used for the calculation of the
mass calibration constant.

For the material under test, the differences
experimentally found for the latex and the mass
calibration constants, were not appreciable. However the
mass integration method is potentially more accurate, as
it takes into account the properties of the material under
analysis, being recommended as a primary calibration
procedure. This method requires great care in the
preparation of the suspensions and also the guarantee that
all particles are counted. Its main drawback is the
considerable time needed for the calibration. Hence, some
tests were also performed using a channelyzer which,
despite of being a high resolution and user friendly
instrument, was found to be not accurate enough for mass
calibration purposes.

REFERENCES

1.   Coulter Electronics Ltd., 'Coulter Counter -
      - Industrial Bibliography', 1987
2.   T. Allen, 'Particle Size Measurement', Chapman &
      Hall, London 1990
3.   British Standards Institution, 'British Standard
      Methods for Determination of Particle Size of Powders',
      BS3406, Part 5, 1983
4.   Coulter Electronics Ltd., 'Instruction Manual for the
      Coulter Counter Model ZM', Issue H, 1985
5.   M.M. Figueiredo, M.G. Rasteiro, C. Santos and
      C. Monteiro, submitted for publication, Part.
      Charact., 1991

6.   C.M.L. Atkinson   and   R. Wilson,   Proc. Particle Size
      Analysis, 1981.   Ed. N.G. Stanley-Wood and T. Allen.
      Wiley Heyden Ltd., 1982.

The Use of a Mass Balance in the Coulter Counter Technique

Henk G. Merkus[1], Ed H. L. Jansma[1], Brian Scarlett[1]
and Marguerida Figueiredo[2]

[1] DEPARTMENT OF CHEMICAL ENGINEERING, DELFT UNIVERSITY OF TECHNOLOGY,
P.O. BOX 5045, 2600 GA DELFT, THE NETHERLANDS
[2] DEPARTAMENTO ENGENHARIA QUÍMICA, UNIVERSIDADE DE COIMBRA, LARGO
MARQUÊS DE POMBAL, 3000 COIMBRA, PORTUGAL

Abstract

A distinct advantage of the electrical sensing zone technique is that it
allows setting up a mass balance for the measurement, since a volume related
signal is measured with high resolution. The mass balance can be used for
two purposes. The first and older one uses the mass balance to calculate
appropriate calibration constants for those materials where latex
calibration leads to systematic errors, such as porous and conducting
particulates, the so-called mass calibration procedure.
The second application of the mass balance is to calculate the amount of
sub-sized material which can not be measured in the orifice used.
Prerequisite for this application is, of course, that the counting
efficiency over the whole measuring range is 100%. It is particularly useful
for characterisation of e.g. pigments.

1. Introduction

The electrical sensing zone method, also known as Coulter Counter
method, has since its development in 1953 [1] been subject to numerous
studies and applications. As a result the technique is well described, as
well in textbooks [2, 3] as in manuals accompanying the instruments [4], and
accepted as a Standard Method [5, 6, 7]. A schematic drawing of the instrument is given in figure 1.

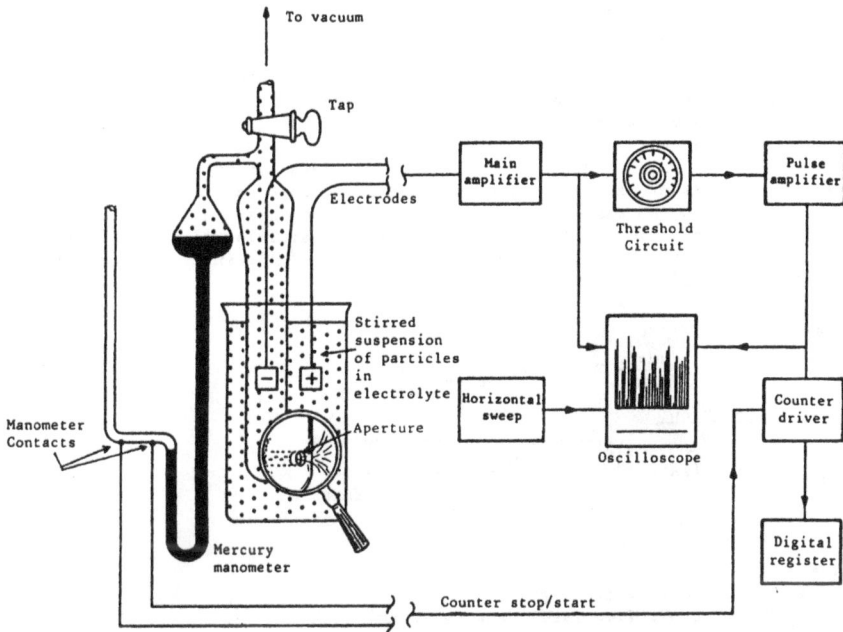

Figure 1. Schematic drawing of an electrical sensing zone instrument.

It is based on the principle that particles, suspended at low concentration in an electrolyte solution, are detected upon their passage through a small orifice in an insulating wall by the modulation of an electrical field existing within the orifice. This electrical modulation is sensed as a voltage pulse for each particle, the height of which is proportional to the volume of the particle. This is very convenient since a measurement yields directly a number distribution of equivalent volume diameters. This distribution is, in turn, easily converted to a volume-based distribution of equivalent volume diameters. Since the volume of a particle is directly related to its mass through the density, this distribution is identical to a mass distribution and, thus, most relevant to characterize a lot of particulate material. Therefore, in our opinion standardisation and calibration of particle size measuring techniques and characterisation of reference materials should be based on the equivalent volume diameter of particles. For these types of applications the utmost accuracy is required, both in the size measurement and in the number count. As the electrical sensing zone technique has the potential of high accuracies, it appears to be a logical candidate to be used as a basis for the above calibration work. A drawback of the technique, however, is that it offers no direct indication of the amount of sub-size material. This is especially important in the size analysis of powders such as pigments containing an appreciable amount of submicron particles and/or having very wide size distributions. In the latter case sometimes a solution can be found in classification of the material, followed by analysis of the fractions in different orifices, the so-called multitube analysis. Of course, the classification step is an extra - often time consuming - step and should fulfil the requirement that it does not selectively "absorb" material of any size. For the former case, however, where an appreciable amount of submicron particles is present, the setting up of a mass-balance over the whole measurement seems the only way to give an indication for the amount of material below the lower detection limit of the technique. Here too, high accuracy of the analytical result for the size analysis is required. This paper will discuss the error sources related to this mass balance and procedures to minimize them.

2. The Mass Balance and its Error Sources

The procedures for setting up a mass balance and for mass calibration of the Coulter Counter technique are identical. A weighed amount of sample of known density is suspended in a known volume of electrolyte, a known portion of which is analysed. From these data the total volume of particles in the analysed volume of electrolyte can be calculated and compared with the measured volume of particles. For a mass (or volume) balance this measured volume of particles is calculated by means of a prior calibration, generally with an appropriate latex standard. For mass calibration, on the other-hand the ratio of measured and calculated particle volume is used to correct the prior calibration constant. The following equations can be derived:

mass balance:

$$W = 10^{-12} \cdot \rho_s \cdot \frac{Vsusp}{Vman} \cdot \Sigma \, (n_i \cdot \hat{V}_i)_{latex} + \Delta \qquad (1)$$

mass calibration:

$$K_d(mass) = K_d(latex) \sqrt[3]{\frac{W \cdot Vman}{\rho_s \cdot Vsusp} \cdot \frac{10^{12}}{\Sigma(n_i \cdot \hat{V}_i)_{latex}}} \qquad (2)$$

where:

W — sample weight, g

ρ_s — sample (immersed) density, g/cm³

V_{susp} — total suspension volume, cm³

V_{man} — analysed volume of suspension, cm³

N_i — number of particles in size class i

$\hat{V}_i$ — mean particle volume of size class i, 10^{-12} cm³ (fl) (obtained with prior latex calibration)

Δ — error in mass balance, or sub-size fraction weight, g

K_d — (diameter) calibration constant for instrument, as measured either from prior latex calibration or by mass calibration.

The following error sources can be designated in the procedure for mass balancing:

(a)   non-representative sampling of the particulate material. Generally only small samples are weighed - some tens to hundreds of mg - in order to keep the concentration in the suspension low. Especially in the case of broad size distributions and free flowing powders there are chances for segregation in the dry state, and thus for sampling errors. In that case rotating sample splitters are required. Moreover, it is advisable to use somewhat larger sample weights (hundreds of mg) and to apply an extra dilution step of the liquid suspension while stirring adequately. For fine powders, which are non-free-flowing, and narrow size distributions sampling errors due to segregation can generally be disregarded.

(b)   analysis of a non-representative fraction of the total suspension. One reason for this problem may be the instability of the suspension. Generally addition of a specified dispersant at a given concentration and application of limited times of ultrasonication favour deagglomeration and stabilise the suspension formed. A second reason may be a faulty design or positioning of the suspension beaker and the wrong positioning of the stirrer in the beaker. Generally these types of errors will be excluded by an experienced operator. An easy indication for the occurrence of these problems is given by bad reproducibility of subsequent measurements from the same beaker with suspension.

(c)   inaccuracy of the size axis of the instrument.
It is assumed in the total procedure that the conversion of pulse height to particle size is straightforward, or in other words, independent of the exact location of particle passage through the orifice. This is not true, however. e.g. in [8] we demonstrated that the coefficient of variation of the size distribution of an ultra-narrow latex standard decreases from about 3% to about 1% when applying hydrodynamic focussing of the sample stream to the centre of the orifice instead of the normal sampling stand. This broadening effect in standard orifices may lead to an apparent change of the particle size if the calibration is not performed properly [9]. But even with proper calibration it leads to some 20% overestimation of the particles volume. Another error source is related to the properties of the particles; viz. for conducting or porous particles much smaller pulse heights are obtained than corresponding to their envelop size [10-12]. In these special cases special care has to be given to the calibration.

(d)   inaccurate counting of the instrument.
There are two reasons possible for too high count rates. Firstly, there may be excessive electronic noise resulting in countable pulses. And secondly, recirculation of particles through the electrical field in front of or especially behind the orifice will also result in countable pulses. Built-up of high particle concentrations in the measuring probe at the back side of the orifice is especially conceivable in case of fairly large particles of high density. Both these reasons generally result in extra pulses in the small size region. A blank experiment with clean electrolyte after the measurement should reveal these extra counts and offer compensation possibilities. An example is given in figure 2, in which a comparison is also given with the hydrodynamic focussing sampling stand, in which extra electrolyte is used to sweep the particles away from the back side of the orifice (cf.ref.8).

On the other hand, the counting efficiency may be lower than 100% through losses in the digitisation or channelyzing stage. These type of difficulties are much more difficult to reveal, since they are not always apparent in the size distribution.
This counting efficiency has been the main subject of the studies described here.

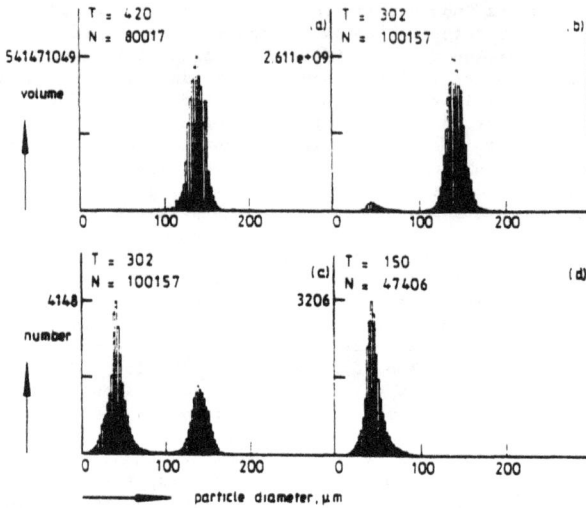

Figure 2. Coulter Counter analysis of 139 μm glass beads,
          (a) sampling stand with hydrodynamic focussing
          (b) normal sampling stand; volume based distribution
          (c) normal sampling stand; number based distribution
          (d) blank analysis after b/c; number based distribution, showing
              particle built-up behind orifice.

3. Counting efficiency

The counting efficiency in the channelyzing stage was tested by comparing raw counts and channelyzed counts from a COULTER Multisizer during the analysis of BCR-70. An example is shown in figure 3 for a 32 channel operation of the Multisizers I and II.

Data clearly indicate that a loss of counts occurs during channelyzing here, especially in the channels for smaller sizes. By consequence, the number distribution shows too low figures for the smaller sizes. The consequences for the volume/mass distribution of these type of powders with a broad size distribution, however, are negligible. Only if a significant part of the mass of the particulate material is counted in these channels, significant losses will occur. Furthermore it should be remarked that the introduction of a new electronic board in the Multisizer II resulted in a significant improvement.

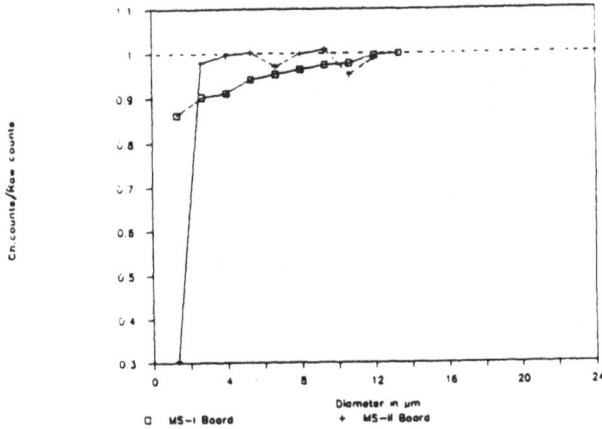

Figure 3. Comparison of raw to channelyzed counts in Multisizer I and II.

The overall relative counting efficiency of the COULTER COUNTER was tested with trimodal latex samples, applying different orifices. An example is shown in figure 4, where a mixture of nominally 8,3; 18,6 and 45,2 μm latexes were used in 100 and 200 μm orifices respectively.

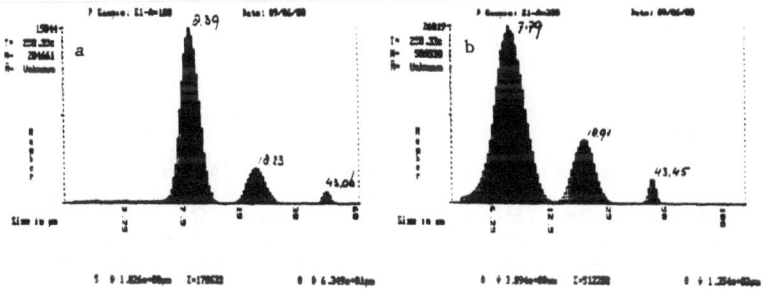

Figure 4. COULTER COUNTER analysis of a 8,3/18,6/45,2 μm latex mixture using a 100 (a) and 200 μm (b) orifice.

The results are summarized in Table 1 as given below:

mode value standard, μm	counted size range, μm	mean ratio of counts, 200/100 μm orifice
8,3	5,4 - 12,7	0,74
18,6	12,7 - 32,6	0,95
45,2	32,6 - 63,5	1,00 (normalised)

Differences in the measuring volumes of both orifices are corrected for through normalisation of the largest size peaks. Of course, this normalisation includes the assumption that count losses for that peak in the 200 μm orifice (counted size range = 16,3 - 31,7% of orifice diameter) and in the 100 μm orifice (counted size range = 32,6 - 63,5 % of orifice diameter) are either negligible or identical. Since the middle sized peak, however, indicates that only 5% (extra) losses occur in the size range of 6,4 - 16,3% of the orifice diameter, this assumption is probably correct. On the other hand the ratio obtained for the smallest sized peak indicates that some 30% extra count losses occur for the size range of 2,7 - 6,4% of the orifice diameter. Apparently recognition of particle pulses in that lower size region relative to the orifice size is difficult and the lower limit of detection, generally given as some 2% of the orifice size, is not a hard limit, above which all particles are counted.

Figure 4 also indicates the modal sizes found for the three peaks with the two orifices. It is clear that also some size shifts occur, which amount to several channel widths. This will also highly influence the mass balances, since the volumes of the particles are related to the third power of their diameter.

Final tests of the mass balance were performed with BCR-70 standard using a 70 μm orifice. Bad results were obtained so far: the mass calculated from the analyses generally amounted some 80% of the input material weight. Moreover, the reproducibility of this value was rather low: its standard deviation was about 5%. These results are inacceptable if one wants to use the analyses for certification of reference materials or to calculate the sub-size fraction from such a balance. On the other hand, the cumulative size distributions agree reasonably well with the calibration data of BCR-70. Thus, probably significant count losses occur over the total measuring range to about the same degree. At this time we are still searching for the major error source(s) in these mass balances and investigating whether application of hydrodynamic focussing of the sample to the orifice will improve accuracy and precision.

4. Sub-size amount.

The second application of the mass balance is to calculate the amount of sub-size material which escapes measurement as their pulses are small and drowned in the electronic noise for a given orifice. The calculation proceeds according to equation (1) given in chapter 2. Prerequisite for such a calculation is, of course, that the counting efficiency over the whole measuring range is 100%.

This application is very fruitful for e.g. pigments. These materials generally have their particle size in the range between 0,2 - 30 μm where an optimum scattering behaviour results for visible light (wavelength 0,4 - 0,8 μm). An example for three yellow pigments is presented in figure 5.

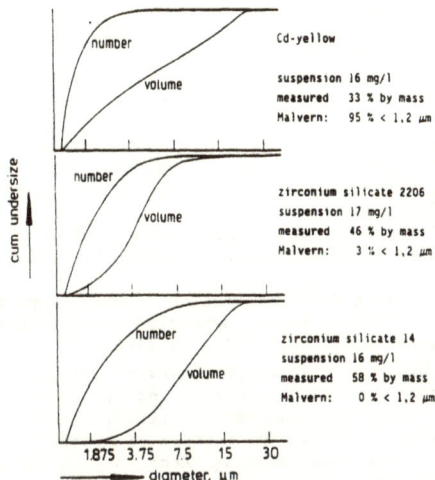

Figure 5. Particle size analysis for three yellow pigments by COULTER Multisizer and Malvern 2600

For comparison also the amount of material smaller than 1,2 μm is given, resulting from a forward light-scattering measurement with a Malvern 2600. This latter technique clearly indicates large differences between the three pigments. Especially for the CdS pigment the particle size of the majority of the material is below 1 μm. This is much less clear if one looks at the volume distribution of the COULTER COUNTER analysis. On the other hand, the number distribution of this sample is very steep at small sizes, indicating a very high number of counts in the low channels. And although these measurements were made at a time where the mass balance for BCR-70 material was still too low (some 60%), it is clear that there is a distinct deficit on the mass balance for this pigment. In contrast, the yellow zirconium silicate sample 14 does not show any submicron material in the Malvern forward light scattering analysis. Also the course of the cumulative volume distribution found in the COULTER Multisizer and its mass balance (as compared with the BCR-70) indicate the same conclusion, although the number of counts in the lower channels is still appreciable. These results clearly show the need of a mass balance with high precision and accuracy, when analyzing these types of materials.

5. Conclusions

The work presented clearly indicates that the electrical sensing zone technique offers a good potential both for certification of reference materials and for analysis of materials such as pigments having a significant amount of sub-sized material, provided that a mass balance can be offered with high precision and accuracy. The target should be no systematic deviations in the mass balance with a precision in mass of 3%.

6. References

(1)   W.H. Coulter: Means for Counting Particles Suspended in a Fluid. US Patent 2.656.508 (1953)
(2)   T. Allen: Particle Size Measurement. Chapman and Hall, London, New York, 4th edition, (1990) pp. 455-482
(3)   P.W. Helleman: The Coulter Electronic Particle Counter (1971)
(4)   Coulter Electronics Ltd: Reference Manual for the Coulter Multisizer (1986)
(5)   British Standards Institution, British Standard Methods for Determination of Particle Size of Powders, BS 3406: Part 5, Recommendations for Electrical Sensing Zone Method (the Coulter Principle) (1983)
(6)   L'Association Française de Normalisation (AFNOR): Particle Size Analysis in an Electrolytic Suspension using the Resistance Variation Counter. X 11-670 (1979)
(7)   L'Association Française de Normalisation (AFNOR): Apparatus for Grain Size Analysis of Particles in Suspension in an Electrolyte; Method based on Resistance Variations. X 11-671 (1981)
(8)   H.G. Merkus, H. Liu and B. Scarlett: Improved Resolution and Accuracy in Electrical Sensing Zone Particle Counters through Hydrodynamic Focussing. Part. Part. Syst. Charact. $\underline{7}$ (1990) 11-15.
(9)   D.A. Elkington and R. Wilson: The Effects of Artefact Peaks on the Calibration of Electrical Sensing Zone Instruments with Reference Latices. Particle Size Analysis 1985, P.J. Lloyd ed., Wiley 1987, pp. 509-528.
(10)   G. van der Plaats and H. Herps: A Study on the Sizing Process of an Instrument Based on the Electrical Sensing Zone Principle, Part 1; The Influence of Particle Material. Powder Technol. $\underline{36}$ (1983) 131-136.
(11)   G. van der Plaats and H. Herps: A Study on the Sizing Process of an Instrument based on the Electrical Sensing Zone Principle, Part 2; The Influence of Particle Porosity. Powder Technol. $\underline{38}$ (1984) 73-76.
(12)   B. Göransson and E. Dagerus: Image Analysis as a Tool for Calibration of Electrical Sensing Zone Instruments in the Size Measurement of Porous Spherical Particles. Particle Size Analysis 1988; ed. P.J. Lloyd; 1988, Wiley & Sons, pp. 159-166.

Two Phase Flow Techniques as Applied in Characterization

Kurt Leschonski

INSTIT. FÜR MECH. VERFAHRENSTECHNIK, TECHNISCHEN UNIVERSITÄT CLAUSTHAL,
LEIBRUZSTRASSE 15, 3392 CLAUSTHAL-ZELLERFELD, GERMANY

1 Introduction

The modern trend of Particle Size Analysis seems to be a one way road to optical methods of all kinds. The range of possibilities is enormously wide, ranging from the direct optical measurement of single particles from the light scattered to the measurement of a mixture of all particles from the sum of their diffracted light. The latter technique, the so-called diffraction pattern analysis, having at present a privileged position amongst industrially applied methods. The list of optical methods is long. I shall not try to summarize them again. There are plenty of examples at this Conference.

Optical methods yield in principle results of high reproducibility if the withdrawal of a representative sample has been performed with care. The results obtained may however differ between instruments, due to systematical differences in the translation of the physical principle to the instrument by its manufacturer. The rapid determination of particle size distributions over at least two decades of particle size being the biggest advantage of optical methods. With diffraction pattern analysis, nowadays the time for the determination of a size distribution dropped to less than a minute. The results are plotted with respect to the diameter of a sphere. This diameter represents an equivalent diameter which is either calculated from the basic physical dependency or has been obtained from calibration. With optical methods one obtains the diameter of a sphere of equivalent light scattered, diffracted or extinguished.

Due to the fact that theory presupposes spherical particles, the particles, however, are irregular in shape, differences must be observed between a geometrical particle dimension and the equivalent diameter measured. It depends on the purpose of the measurement whether the information obtained is acceptable as it is, or the result achieved with the optical method has to be matched to the physical property one is interested in.

In many cases of application it is not sufficient to measure a size distribution with respect to an arbitrary physical particle property, but to the one which is as closely as possible related to the desired product property. In most cases it would anyhow be better, to measure the product property directly and not the particle size distribution of the product. Only if the dependency between product property and particle size distribution is known, can conclusions be drawn between both. Sugar in chocolate feels gritty between one's teeth if its coarsest particles are bigger than 15 μm. The strength of concrete depends on the size distribution of its binder, that is cement, and so on.

Plenty of examples may be given. It is, however, not always the strength of concrete, the covering power of a paint, the permeability of a packing, etc. which is controlled by product particle size, other physical properties may also be of importance, for example those, which can be used for the design of machines where particles travel in a flow. In the design of a classifier, a cyclone separator, a centrifuge and other two phase flow systems not the size but the stationary settling rate of the particles in the fluid controls their movement. Particle trajectories are based on settling rate and not on size alone. The design related to practice of two-phase flow systems should therefore rather be based on the distribution of stationary settling rates than on optical particle properties.

Methods which measure the product size distribution from its distribution of stationary settling rates should therefore not to be taken as obsolete in particle size analysis. Critics might object at this point, that size distributions are in general only needed approximately and differences in equivalent diameters are not that important in actual practice, in spite of the fact that even industrial users nowadays ask for the highest possible reproducibility. It might also be said that life is too short to spend too much of it in the measurement of settling rate distributions of small particles.

It is the aim of this paper to summarize some of the existing possibilities and to show that one of the biggest disadvantages of settling rate measurements, the long duration of an analysis, can be overcome in principle. It is possible to outwit physics and to measure settling rate distributions of fine particles in a very short time. It is even possible and highly desirable to combine settling rate measurements with optical methods. If only differences between products have to be measured, it is indeed irrelevant which method is used, the highest possible reproducibility is then more important than anything else.

2 Settling Rate

The stationary settling rate of a single, small, spherical particle in an infinite, stationary fluid under the influence of gravity can be calculated from Eqns. 1 and 2. It has been assumed in these equations that the flow around the particle is laminar and *Stokes* law of resistance may be applied.

$$w_g = \frac{(\rho_p - \rho_{fl})gx_w^2}{18\eta} \qquad\qquad \text{Re} < 0.25 \qquad\qquad (1)$$

$$w_g = \frac{\rho_p g x_w^2}{18\eta} \qquad\qquad \text{Re} < 0.25 \qquad\qquad (2)$$

One realizes from these equations that the stationary settling rate includes a number of important properties which control the particle movement. It is directly proportional to the density difference between particle and fluid, the gravity constant and the square of the equivalent settling rate diameter. It is inversely proportional to the viscosity of the fluid. The stationary settling rate therefore not only depends on the size and density of the particle but also on the material data of the fluid. Eqn. 2 represents the movement of a particle in a gas. The settling rate is, however, also dependent on particle shape. If one introduces a shape factor, which describes the deviation of the irregularly shaped particle from a sphere, for example, with the so-called sphericity, ψ, /1/, Eqn 1 is replaced by Eqn. 3

with particle size being described by its volume diameter, x_V:

$$w_g = \frac{(\rho_p - \rho_n)gx_V^2}{18\eta}\sqrt{\psi} \qquad\qquad Re < 0.25 \qquad\qquad\qquad (3)$$

From these· equations one may calculate the time necessary for a particle of a given size to travel a certain distance in a fluid. Spherical particles of 1 μm diameter require, for example, 3.09 hours in water, or 2.08 minutes in air to travel a distance of one centimeter. A reduction of particle size to 0.1 μm decreases the settling rate by a factor of 100. Even if one could neglect other factors, such as *Brownian* motion, convection currents and the like, it would therefore not be advisable to use gravity sedimentation for the determination of small particles. No wonder that gravity sedimentation methods have almost disappeared from the market. Even if the height of fall is drastically reduced, the time for one analysis is too long to be accepted nowadays in the routine analysis of particle size distributions.

A considerable reduction of the time necessary for one analysis is obtained, however, if the analysis is performed in a centrifugal field. Due to the fact that the settling rate in a centrifugal field is increased over the stationary gravity settling rate by the ratio of centrifugal acceleration, a, to the gravity acceleration, g:

$$w_a = w_g \frac{a}{g} \qquad\qquad\qquad (4)$$

the time for the analysis is reduced. Since, however, at the same time the desire to measure even smaller particles also increases, even centrifugal methods cannot be used for the almost instantaneous measurement of small settling rates. On the other hand, high speed centrifuges have been and still are used to measure the molecular weight of particles. But these measurements never were and never will be performed under the above time restrictions.

One might conclude, therefore, that the time for sedimentation methods is over. This, however, is only true at first sight. There are physical principles available which permit the measurement of settling rate distributions without having the drawbacks of sedimentation techniques.

3 The Measurement of Settling Rate Distributions in Cross Flow Systems

The principle of a cross flow system is shown in Fig. 1, which may either be used for the measurement of settling rate distributions, the technical classification in fluids, or the separation of particles from fluids. It consists, for example, of a rectangular channel which is flown through with a fluid from left to right and an average fluid velocity. The feed particles enter the flow with a certain ini-

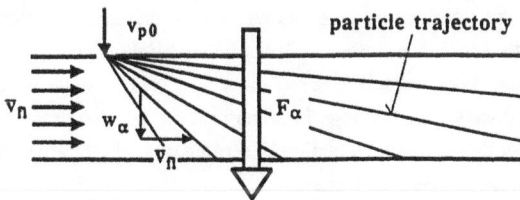

Fig. 1: Principle of a cross flow system

tial velocity, v_{p0}, at a point near the beginning of the channel through a slit, small in the direction of the flow. The particles move across the flow under the influence of a force perpendicular to the flow lines. These forces may either be a gravity, centrifugal or electrical force, or diffusion. The velocity of the cross flow movement can be calculated from the force balance between the drag force and the forces mentioned. In the case of gravity and centrifugal forces the particles settle in the flow as they did in a stationary fluid. Therefore Eqns. 1 to 4 describe the particle velocity perpendicular to the flow. As long as the particles are small, they also travel from left to right with the velocity of the fluid.

If one introduces the particles into the flow with a negligible initial velocity, v_{p0}, the particle trajectories are straight lines if the flow around the particles is laminar. One therefore obtains a fan of settling rate dependent particle trajectories. Cross flow systems, contrary to the classical sedimentation methods,therefore already have the substantial advantage of permitting the measurement of individual settling rate classes. Since these settling rate classes are represented by their trajectories, the amounts present in each class can be determined independently, for example, from the particles deposited and collected at the wall opposite the particle entrance.

Typical examples for instruments of this type are the so-called aerosol centrifuges, as developed and described by *K.F. Sawyer* and *W.H. Walton* /2/, *C.H. Keith* and *J.C. Derrick* /3/, *A. Goetz* and *J.R. Stevenson* /4/, *W. Stöber* and *U. Zessack* /5/, *W. Kast* /6/, *W. Stöber and H. Flachsbart* /7/ and others. One distinguishes between spiral channel centrifuges and the so-called conifuge and its derivatives. In a spiral channel centrifuge, for example, the rectangular channel forms a spiral in a disk. A comparatively long channel can be accommodated in a disk of small diameter. Due to the spiral the centrifugal force increases with channel length, which helps to separate even small particles from the flow. The particles are collected on one side of the channel at the end of the analysis and the amount of particles present in each settling rate class is determined independently, for example, microscopically.

I shall not deal any further with aerosol centrifuges in this paper, even though these instruments have been developed to extremely high standards and have a high potential for the measurement of small cut sizes. In the end uncontrollable secondary flow problems, the problem of not being able to measure the amount of particles deposited more economically, the high purchase price and other simpler two phase flow techniques, such as impactors, have diminished their chances of survival. Only a spiral centrifuge is to my knowledge still available commercially.

Gravity and centrifugal spectrometers, as described in Fig. 1, still suffer from the fact that the settling rates are the same as in conventional sedimentation techniques. Therefore the height of the channel must be the smaller, the smaller the particles to be measured are and the shorter the channel is which has been used for the separation of the particles from the flow. In principle, however, spectrometer-like systems have the highest potential for the rapid determination of settling rate distributions, but they should be operated in a different way, as will be shown later.

The so-called electro precipitators, with electrical forces acting perpendicular to the flow, have been developed by *K.T. Whitby* and *W.E. Clark* /8,9/. A temperature gradient across the flow is applied in thermo precipitators, with diffusion being the driving force across the flow.

All these systems have been developed for the determination of atmospheric aerosol size distributions. They are therefore limited to extremely small solids concentrations. This simplifies on the one hand their theoretical precalculation, because the particles will not interact with the flow, they are, however, difficult to apply in the routine analysis of aerosols produced from powders.

4 The Measurement of Distributions from the Accelerated Particle Movement

4.1 The Accelerated Particle Movement in a Flow of Constant Velocity

If one introduces a spherical particle with negligible initial velocity in a gas or liquid flow of constant and higher velocity, the particle is accelerated. The particle approximately assumes the velocity of the flow after a certain acceleration distance. This distance can be calculated for small *Reynolds*-numbers that is within the validity range of *Stokes* law of resistance. With a system, as shown in Fig. 2, the acceleration distance, z, has been calculated with respect to the relative approach of the particle velocity to the gas velocity, Δc. With:

$$\Delta c = \frac{v_p - v_{p0}}{v_a - v_{p0}} \tag{5}$$

one obtains:

$$z = \frac{w_g v_a}{g}\left[\ln(1-\Delta c)^{-1} - \frac{v_a - v_{p0}}{v_a}\Delta c\right] \tag{6}$$

or:

$$z = s_0\left[\ln(1-\Delta c)^{-1} - \frac{v_a - v_{p0}}{v_a}\Delta c\right] \tag{7}$$

s_0 represents the so-called stop distance, which, for laminar particle movement, can be calculated from:

$$s_0 = \frac{w_g v_a}{g} \tag{8}$$

One realizes: under laminar flow conditions, the acceleration distance, z, is directly proportional to the stationary settling rate, w_g, or the stop distance, s_0. All other variables are constant for a given layout. It has

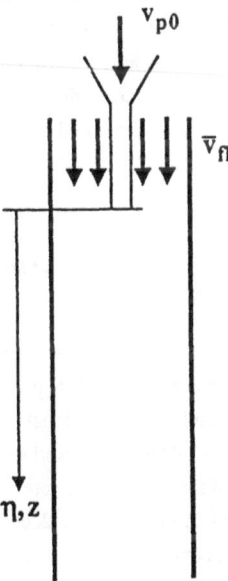

Fig. 2: Particle acceleration in a flow of constant velocity

been assumed in this system that the particles enter the flow of velocity, v_a, with the constant initial velocity, v_{p0}. Fig.3 shows the relative velocity increase, Δc, with respect to the acceleration di-

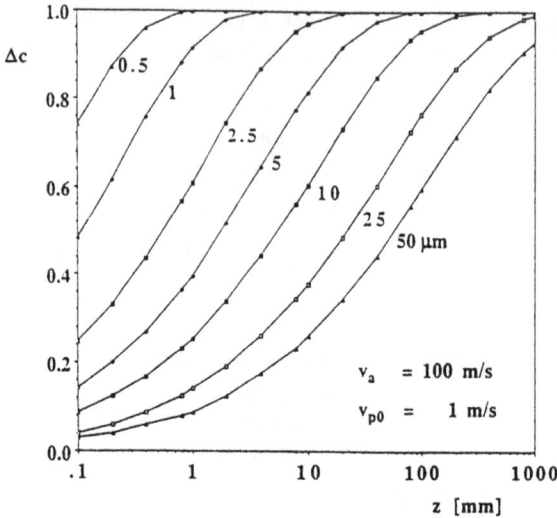

Fig. 3: Relative velocity increase, Δc, with respect to acceleration distance, z

stance, z. These curves have been calculated with a law of resistance which is not limited to small *Reynolds*-numbers /10/. Eqn. 9 has been used, which is valid for *Reynolds*-numbers: Re $< 2 \cdot 10^5$:

$$c_d = \frac{24}{Re} + \frac{4}{\sqrt{Re}} + 0.4 \qquad (9)$$

One realizes from Fig. 3 that small particles attain, for example, 99% of the gas velocity after a very short distance. A 0.5 μm particle, for example, needs approximately 0.8 mm, a 1 μm particle 4 mm to attain this relative velocity increase. The situation improves with coarser particles. The particle sizes, given in Fig. 3, are equivalent settling rate diameters. They have been calculated from the stationary settling rate, which controls the particle acceleration, for the sake of better demonstration. This system can be used for the measurement of settling rate distributions, if at a certain distance the increase of the particle velocity is measured for each individual particle. This has been used in an accelerated gas flow, as shown in 4.3.

The system as explained may, however also, be used at higher solids concentrations.

4.2 The Measurement of Accelerated Particle Movement in a Gas Flow of Constant Velocity at High Solids Concentrations

As *P. Bernutat* /11/ showed in 1968, the different acceleration behavior of small particles in a gas flow of constant velocity can be used for the measurement of settling rate distributions. The particle acceleration caused by the drag of the gas flow consumes energy which results in an additional pressure drop on top of the pressure drop caused by the transport of the flow. *P. Bernutat* showed that it is possible to calculate the settling rate distribution of the particles from the course of the pressure drop with tube length. The system which resembles, in principle, the layout, shown in Fig. 2, corresponds to a carefully designed and operated acceleration zone of a pneumatic duct. In order to obtain measurable pressure drops, this system needs a certain, non negligible mass throughput. It can therefore only be used for technical aerosols, for example, produced from a powder.

The basic equation may be derived as follows: If one accelerates a certain mass flow rate, m_s, of particles of a certain settling rate, $w_{g,i}$, in a tube of constant cross section, A, a momentum balance yields the additional pressure drop, $\Delta p_{j,i}$, at the distance, η_j, from the entry point of the particles:

$$\Delta p(\eta_j, w_{g,i}) = \Delta p_{j,i} = \frac{\dot{m}_s}{A} \Delta c_{j,i} \qquad (10)$$

$\Delta c_{j,i}$ represents the velocity increase of the particle of settling rate, $w_{g,i}$, over the entry velocity, v_{p0}, at the tube length η_j.

$$\Delta c_{j,i} = v_{p_{j,i}} - v_{p0} \qquad (11)$$

With the solids loading, μ:

$$\mu = \frac{\dot{m}_s}{\rho_a v_a A} \qquad (12)$$

one obtains:

$$\Delta p_{j,i} = \mu \, \rho_a \, v_a \, \Delta c_{j,i} \qquad (13)$$

Each individual settling rate class adds to the overall pressure drop with a fraction of $\Delta p_{j,i}$ which is represented by the relative amount of particles, $dQ_{3,i}$, present in this settling rate class:

$$dQ_{3,i} = q_{3,i} dw_{g,i} \qquad (14)$$

One therefore obtains:

$$d\left(\Delta p_{j,i}\right) = \mu \, \rho_a \, v_a \, \Delta c_{j,i} \, dQ_{3,i} \qquad (15)$$

Integrating this equation over all particles sizes yields the pressure drop, Δp_j, at a given tube length, η_j:

$$\boxed{\int_0^{\Delta p_j} d\left(\Delta p_{j,i}\right) = \mu \, \rho_a \, v_a \int_0^1 \Delta c_{j,i} \, dQ_{3,i}} \qquad (16)$$

or, written as a sum:

$$\boxed{\Delta p_j = \mu \, \rho_a \, v_a \sum_{i=1}^{n} \Delta c_{j,i} \, \Delta Q_{3,i}} \qquad (17)$$

Eqn. 17 represents a linear set of equations, with: $1 < j,i < n$.

If one measures Δp_j at different tube lengths, η_j, the relative amounts of particles, $\Delta Q_{3,i}$, of the corresponding settling rate classes, $w_{g,i}$, may in principle be calculated from the linear set of equations, described by Eqn. 17. The different velocity increases, $\Delta c_{j,i}$, for particles of a certain settling rate at position, η_j, can be calculated from a force balance. Eqn. 17 represents a so-called first order *Fredholm* integral equation, well known from several optical methods, for example, diffraction pattern analysis. The problems encountered in the solution of this equation /12/ occur here as well. We suffered from this at the beginning of 1970, when we tried to set up an instrument according to *P. Bernutat*'s idea. At that time we did not succeed for a number of reasons. To obtain a mathematical solution was possible in principle, because some of the algorithms presently used were already available, however, the efficiency of the computer which we used parallel to the analytical system was by no means comparable to modern personal computers. Furthermore problems arose in the dispersion and the controlled feeding of fine particles, some of which we have solved in the meantime.

I should like to point out, however, that the system has successfully been applied in 1948 by *W. Barth* /13/ in the measurement of the mass flow rate or the solids concentration of coal particles when feeding them to the burners of a power plant. He used a tube which was long enough to accelerate all particles to the velocity of the gas. The *Fredholm* equation, which may also be written in a dimensionless form:

$$\boxed{\frac{\Delta p_j}{\rho_a v_a^2} = Eu_j = 2\mu \sum_{i=1}^{n} \frac{\Delta c_{j,i}}{v_a} \Delta Q_{3,i}} \qquad (18)$$

then turns into:

$$Eu = 2\mu \qquad (19)$$

4.3 The Measurement of Settling Rate Distributions in an Accelerated Flow

Another possibility which can be used for the measurement of settling rate distributions originates from the particle behavior in an accelerated gas flow. The gas and the particles are, in this case, accelerated in a converging nozzle. The acceleration of the particles is settling rate dependent. Fine particles follow the flow almost immediately, while, over a given distance, coarse particles obtain only part of the gas velocity. The particle velocities obtained at the orifice of the nozzle therefore differ, similar to the situation described in Fig.3. Since an unambiguous dependency exists between settling rate, acceleration distance and particle velocity, the measurement of the particle velocity at a given distance close to the orifice of the nozzle yields information on the settling rates or the equivalent settling rate diameters of the particles. In existing systems the particle velocity is therefore measured at a fixed distance from the nozzle exit, for example, by the time of flight between two narrow laser beams. Experimental and theoretical evidence of this technique has been given by a number of authors, some of whom are listed in Refs. 14 to 21. The dependency between particle velocity measured and size or settling rate of the particle can either be determined theoretically or by calibration. The settling rate distribution obtained is a number distribution.

There are to my knowledge two instruments on the market which use this principle, firstly the APS 33 of TSI, St. Paul, Min. and secondly, to my knowledge since 1989, the API Aerosizer of Malvern Instr. Ltd, Worcestershire, U.K. Fig. 4 shows the main part of the principle set up of the APS 33, as given in the TSI advertising leaflet.

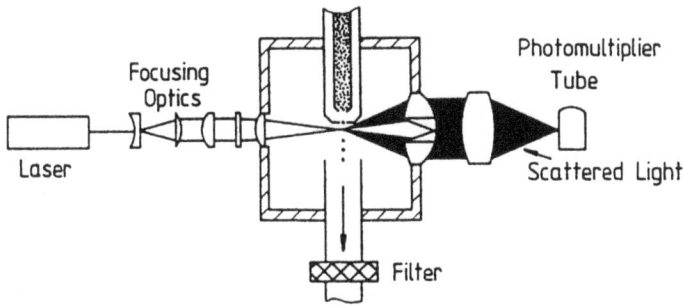

Fig. 4: Principle set up of the APS 33 of TSI

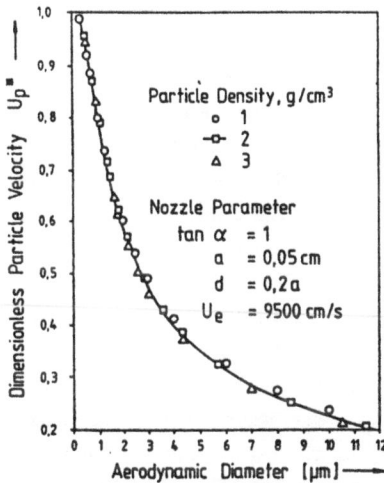

Fig. 5: Calculated particle velocity with respect to aerodynamic diameter, x_{ae}

Fig. 5 shows the calculated particle velocity as a function of the so-called aerodynamic diameter, as published by *J.C. Wilson* /18/ in 1978 and *J.C. Wilson* and *B.Y.H. Liu* /19/ in 1980.

The aerodynamic diameter, x_{ae}, represents the equivalent settling rate diameter of particles with the density of water that is 1 g/cm^3. It is a very popular equivalent diameter in aerosol physics. Due to the fact that the true particle density is not taken into account, or better replaced by the density of water, the aerodynamic diameter is directly proportional to the square root of the stationary settling rate. The abscissa of Fig. 5 could therefore be replaced by a square root settling rate scale.

Fig. 4 shows that the aerosol enters the flow in front of the jet through a separate tube. The aerosol flow is surrounded by a clean air stream. During the acceleration process, the aerosol stream is reduced in cross sectional area and it is also focussed in the center of the jet. This system, also used in other counting techniques, is called: hydrodynamic focussing. The particles to be measured therefore leave the jet in a thin stream and in the center of the jet. This simplifies the detection of the particles in the time of flight measurement. The method can only be applied at low solids concentrations. The lowest detectable particle size seems to be of the order of 0.5 μm. The size range to be covered and the resolution depend on the geometrical and the flow boundary conditions chosen.

A dimensionless representation of the calibration curve has been given by *B.T. Chen*, *Y.S. Cheng* and *H.C. Yeh* /21/ in 1985. They plotted the Stokes-number, St, as a function of the ratio of particle velocity and gas velocity at the exit of the nozzle.

4.4   The Measurement of Settling Rate Distributions in a Curved Flow

4.4.1   The Measurement in a Stagnation Point Flow

Another two phase flow system, which can be used for the comminution /22/ and the measurement /23-27/ of particles in a gas stream originates from the movement of particles in a stagnation point flow. If, as shown in Fig. 6, a high speed gas stream, formed in a jet of diameter, d, hits a plate of circular cross section, distance, s, apart from the orifice of the jet, a stagnation point flow is formed. The flow lines are differently curved, with the curvature being the more pronounced, the smaller the distance, s, is. Most of the particles, which have been accelerated to the exit velocity of the gas flow, will not be able to follow the flow lines, due to their inertia. The coarser ones will therefore collide with the surface of the

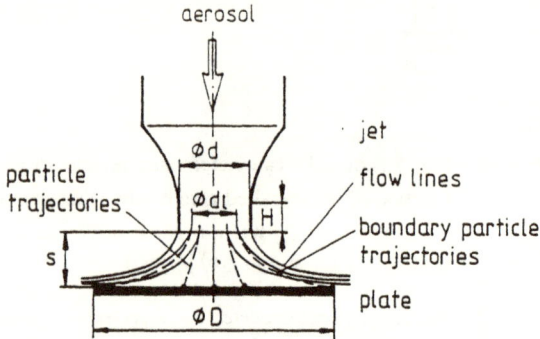

Fig. 6: Set up of an impactor

plate, that is the target. Particles of the same settling rate class will be evenly distributed in the cross section of the orifice of the jet. Depending on their radial position only part of these particles will reach the surface of the plate. A cross section exists with a certain limiting diameter, d_l, from which particles will reach the plate of diameter, D. Those particles in the outer ring between d_l and d will by-pass the plate and remain in the flow. Therefore a so-called grade efficiency can be calculated for each settling rate class, which represents the amount of particles collected on the plate, to those initially present in the cross section of the orifice. The grade efficiency can be calculated from Eqn. 20:

$$T = \left(\frac{d_l}{d}\right)^2 \tag{20}$$

Therefore the deposition of the particles with respect to their settling rate can be described by their grade-efficiency curve, $T(w_g)$. The precalculation of the limiting diameter, d_l, is based on trajectory calculations. These particle trajectories are calculated from a force balance at the particle, the assumed stagnation point flow, a drag law and the relevant geometrical and flow boundary conditions. *C.N. Davies* and *M. Aylward* /23/, *W. E. Ranz* and *J.B. Wong* /24/, *T.T. Mercer* and *H.Y. Chow* /25/ and others have calculated the particle movement in a frictionless flow. The most comprehensive calculations have been performed, however, by *R. Roeber* /26/ in 1957 and *V. A. Marple* /27/ in 1970. *V. A. Marple* showed, that it is advisable to plot the grade efficiency, T, against the *Stokes*-number, St. The *Stokes*-number is directly proportional to the stationary settling rate of the particles:

$$St = \frac{v_a w_g}{dg} = \frac{v_a \rho_p x_w^2}{18 \eta d} Cu \tag{21}$$

The *Cunningham* correction /28/, or slip correction, must be taken into account, as soon as the particles become small in comparison to the mean free path of the gas molecules. This correction has always to be taken into account if small particles, for example, with diameters smaller than 1 µm, travel in a gas.

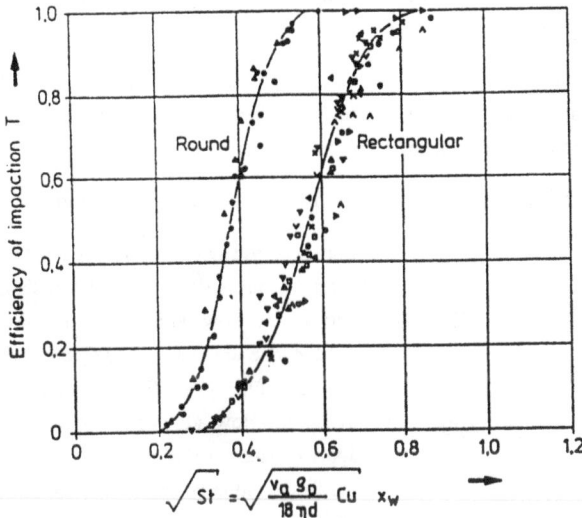

Fig. 7: Impactor grade efficiency curves /24/

V.A. Marple /27/ showed that the grade efficiency, T, depends on a number of variables. In dimensionless terms T is a function of the *Stokes*-number, St, the *Reynolds*-number, Re, and several geometrical dimensions. Fig. 7 shows theoretical and experimental results as obtained by *W.E. Ranz* and *J.B. Wong* /24/ in 1952. This figure shows the grade efficiency curves for a jet of circular and rectangular cross section. The square root of the *Stokes*-number is plotted on the abscissa, which is therefore directly proportional to the equivalent settling rate diameter.

Due to the fact that the grade efficiency curves are not ideal that is do not form a step function, which jumps from zero to one at a certain *Stokes*-number, the corresponding point of the settling rate distribution curve can only be determined correctly if the amounts of misplaced material in the fine and the coarse fraction are the same. Fig. 8 shows the situation. The amount collected on the plate is represented by the bell shaped curve on the right for the coarse fraction. It represents the residue. The vertical line must be shifted to a point, x_a, until the areas I and II, are equal. The residue measured on the plate can then be plotted with respect to x_a, the analytical cut size /29/. The me-

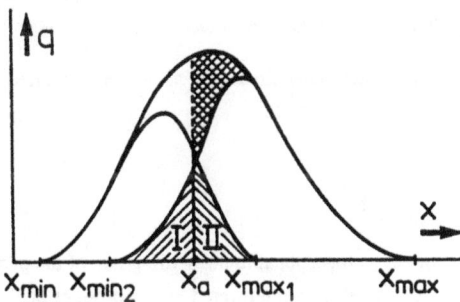

Fig.8: Analytical cut size /29/

dian of the grade efficiency curve should only be used in the description of a technical classification process.

The classification performed at a single plate subdivides a given feed material into a fine and a coarse fraction. The coarse fraction is collected on the plate and therefore removed from the flow, the fine fraction is still entrained in it. If more points of a distribution curve have to be determined, several impactor stages have to be arranged in series, with the cut sizes decreasing in the direction of the flow. One obtains a so-called cascade impactor.

Its overall behavior is similar to a sieve, where successive size classes are collected between sieves. In both cases the amounts of the particles collected have to be determined by weighing, for example. Single stage impactors as well as cascade impactors exist and are built in a number of ways.

4.4.2 The Measurement of Settling Rate Distributions in a Flow Round a Bend

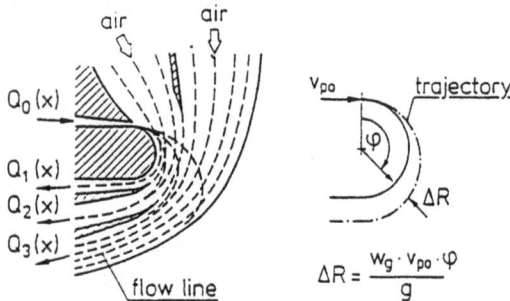

Fig.9: Coanda cross flow system /30,31/

$$\Delta R = \frac{w_g \cdot v_{po} \cdot \varphi}{g}$$

If an aerosol stream, formed in a jet, adheres to one side of a wall which is curved, the flow will stick to the wall surface, due to the so-called *Coanda*-effect (see Fig. 9). Without additional means the flow adheres to the wall for approximately 135 degrees. Only then the flow gradually starts to break away from it and to form vortices. It is possible to keep it there, if approximately at this point the boundary layer is withdrawn through the wall. *H. Rumpf, K. Leschonski* and *K. Maly* /30, 31/ have investigated this system in 1974 and later in a technical classifier for the production of fine powders. A 10 cm wide classifier of this type had a cut point of approximately 2 μm and throughputs up to 1500 kg/h.

In this system, one again obtains curved flow lines and settling rate dependent centrifugal forces are exerted on the particles entrained in the flow. The particles therefore form a settling rate dependent fan of trajectories, with the smallest particles travelling closest to the curved wall.

An estimate of the radial distance travelled by a particle at a certain angle, φ, can be calculated from Eqn. 22, assuming potential flow:

$$\Delta R = \frac{w_g v_{p0}}{g} \varphi = \frac{\rho_p x_w^2 v_{p0}}{18\eta} \varphi = s_0 \varphi \tag{22}$$

The radial distance travelled by the particle is directly proportional to the settling rate of the particle, w_g, its initial velocity, v_{p0}, and the angle, φ. ΔR can also be interpreted as the product of the so-called stop distance, s_0, and the angle, φ:

$$s_0 = \frac{w_g v_{p0}}{g} \qquad\qquad Re < 1 \qquad\qquad (23)$$

The stop distance has been calculated from the equation of motion, using:

$$c_d = \frac{21}{Re} + \frac{6}{\sqrt{Re}} + 0.28 \qquad\qquad 0.1 \le Re \le 4 \cdot 10^3 \qquad\qquad (24)$$

for the drag coefficient. The results are shown in Fig. 10 with respect to the equivalent settling rate diameter, x_w, and the initial particle velocity, v_{p0}. With an initial velocity of 100 m/s one obtains a stop distance of $s_0 = 621$ μm if the particle size is 1 μm. The stop distance drops to 8 μm if the particle size is further reduced by a factor of ten.

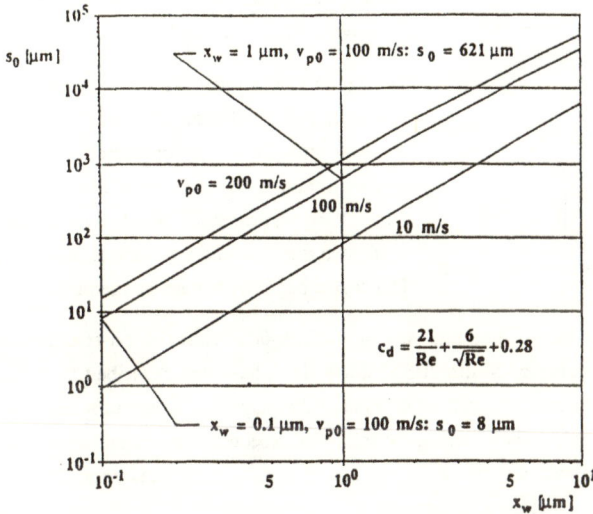

Fig.10: Stop distance, s_0, with respect to equivalent settling rate diameter, x_w, and initial particle velocity, v_{p0}

In general air classifiers are operated in the laminar flow range. The flow velocities in *Coanda* type systems must, however, be much higher, in order to obtain an appreciable radial particle displacement. It can be seen from Fig. 10 that the air and particle velocities should be of the order of 100 m/s and above. The particle movement therefore takes place in turbulent flow. In order to reduce the influence of turbulent mixing on the particle trajectories, the particle movement should take place in an extremely small classification zone.

This system no longer suffers from the drawbacks of the sedimentation techniques discussed at the beginning of this paper. The settling rate controls the trajectory that is the radial distance, ΔR, travelled by the particle. The residence time is controlled by the velocity of the curved flow lines, it is of the order of milliseconds and in most cases even smaller. With a constant aerosol feeding system one therefore obtains a stationary set of particle trajectories. Under these circumstances a size distribution may easily be obtained if one measures the amount, or better the concentration, of particles in each individual settling rate class. The first possibility is represented in Fig 9. The fan of particles is subdivided into settling rate classes by several blades and the particles present in each settling rate class are separated from the gas streams and collected, for example, in cyclones or filters. The amount of particles present in each settling rate class can then be obtained by weighing or other suitable methods. The *Coanda* system then represents a multi blade classifier, which can also be used for the production of narow settling rate classes.

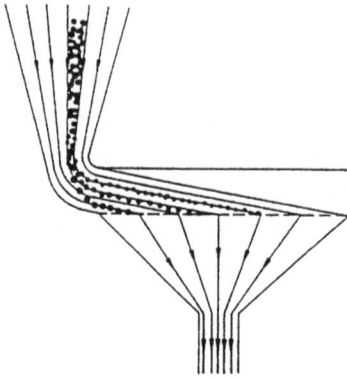

Fig.11: The principle of the Inspec/Inpac (Palas GmbH)

An instrument which uses the flow round a 90 degree bend has been described by *V. Prodi, F. Belosi* and *A. Mularoni* /32/. The instrument called Inspec or Inpac is manufactured by Palas GmbH, Karlsruhe, Germany. The basic principle is shown in shown in Fig 11.

The second possibility uses the system as described in Fig. 9. The fan of particles is analysed with respect to the solids concentrations of the individual settling rate classes from the extinction of a travelling photometer. *M. Heuer* /33/ has investigated the flow and the particle behavior in a 180 degree bend, with water as medium.

Another example, with water as carrier medium, was the *A. B. Holland-Batt* /34/ on-stream particle size analyser which under the name of Mintek/RSM slurry sizer, has been described by *B.F. Osborne* /35/. This instrument consisted of a curved channel of the shape of a single turn helix. A size dependent separation of the particles when leaving the helix was expected, the concentration distribution of which could be analysed with a ß-gauge. Due to non ideal flow conditions within the helical channel the instrument could not fulfil its expectations.

5 The Decelerated Particle Movement perpendicular to a Flow of Constant Velocity

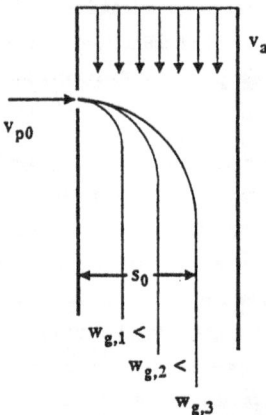

Fig.12: The principle of cross flow classifier

If one introduces particles into a gas or a liquid flow of, for example, rectangular cross section, the particles are fanned out in the flow. With the system as schematically shown in Fig. 12., the particles enter the flow with a high initial velocity, v_{p0}, and travel perpendicular to the flow. They are continuously decelerated within the flow and finally come to a halt. The distance travelled is the stop distance, s_0, already explained and shown in Fig. 10. One again obtains a fan of settling rate dependent particle trajectories. Due to the fact, that the fluid and the initial particle velocities are comparatively high, in gas, for example, both, v_a and v_{p0}, should be higher than 20 m/s, the residence times of the particles within the classification zone are small. The fan of particles is again stationary and can be used in the technical classification of powders but also for the measurement of settling rate distributions.

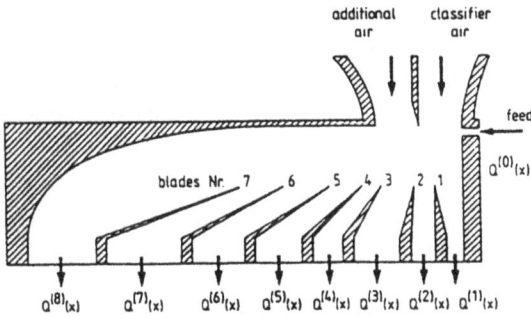

Fig.13: The multi-blade cross flow classifier /36,37/

K.-L. Metzger and *K. Leschonski* /36,37/ have developed a so-called multi-blade classifier, the principle of which is shown in Fig. 13. The feed material is subdivided in this multi-blade cross flow classifier into 8 to 11 settling rate classes, if 7 or 10 blades are used. The relative amount of particles collected in each settling rate class is determined by weighing.

This system has been developed by *K. Leschonski, K.-L. Metzger* and *U. Schindler* /38/ into an on-line analyser capable of handling throughputs of dry powders of several kg/h. The relative solids concentration has been determined with a travelling photometer. The extinction of a white light beam has been used to measure solids concentration. Due to the fact that the settling rate depending fan of particles may either be determined theoretically or by experiment, *Lambert-Beer*'s law of extinction can be used for the measurement of the volume related solids concentration. The analysis of the settling rate distribution of the feed material was performed in less than a minute.

This example shows, that it is possible in principle, to use the settling rate behavior of small particles in a flow as a measure of particle size and to use optical techniques for the determination of concentration that is the relative amount of individual settling rate classes.

6 Conclusions

This paper has mainly been written aiming at a rehabiltation of settling rate as a valuable and most useful physical property in the characterisation of particles. In parts I have done this before /39,40/. Its drawbacks in conventional sedimentation methods, mainly their extremely long sedimentaion times for one analysis, can in principle be overcome in two phase flow systems. The particle position in a flow can be used to characterise its settling rate and with continuous feeding of a stream of particles a stationary fan of the distribution of settling rates is set up in the measuring zone. From this fan of particles the settling rate dependent relative amounts of the density distribution can be determined. The measurement can be performed almost instantaneously and continuously depending on the method used for the measurement of the relative quantities present in each settling rate class. The most useful systems are those were the particles are fanned out in space forming a spectrometer type arrangement. If time is most important, optical methods, such as extinction measurements, should be used for the determination of the quantities. Cross flow principles have succesfully and commercially been applied, apart from technical classification systems, only in the analysis of atmospheric aerosols. The analysis of aerodispersions of higher solids concentrations originating from powders has been neglected, at least commercially.

I hope this paper, even though dealing with techniques and systems well known to most of the experts, will not be regarded as a second brew of already well known techniques. My intention was to remind some of you of the potential of two phase flow systems in characterisation. I did not intend to suggest to you to forget about true optical systems. Fortunately a fairly successful diffraction pattern analyser has been developed at my Department, which should me render unsuspicious /12/.

7 References

/1/	H. Wadell	J. Geology, 40(1932)443-451.
/2/	K.F. Sawyer, W.H. Walton	J. Sci. Instr. 27(1950)272-276.
/3/	C.H. Keith, J.C. Derrick	J. Coll. Sci. 15(1969)340-356.
/4/	A. Goetz, J.R. Stevenson	APCA Proc.,Technical Conf. San Francisco (1957)228-267.
/5/	W. Stöber, U. Zessack	Zentralblatt f. Biol. Aerosolforschung, 13(1969)263-281.
/6/	W. Kast	Staub-Reinhaltung Luft, 21(1961)215-223.
/7/	W. Stöber, H. Flachsbart,	Environ. Sci. Technol. 3(1969)1280-1296.
/8/	K.T. Whitby , W.E. Clark	Tellus 18(1966)573.
/9/	K.T. Whitby	"Fine Particles, Aerosol Generation, Measurement, Sampling Analysis" B.Y. H. Liu, Academic Press (1976)581-624.
/10/	J. Mühle	Dissertation TU Berlin (1969).
/11/	P.Bernutat	Deutsche Offenlegungsschrift 1810711, 1968.
/12/	M. Heuer, K. Leschonski	Part. Charact. 2(1985)7-13.
/13/	W. Barth	Ingenieur-Archiv 16(1948)3/4, 147-152.
/14/	B. Dahneke	Nature 54(1973)244.
/15/	W. J. Yanta	NBS, Special Publ. No. 412 (1974).
/16/	M. H. Schwartz, R.P. Andres	J. Aerosol Sci. 7(1976)281.
/17/	F.N. Weber	Bull. Am. Phys. Soc. 22(1977), No. 10, Talk FD-15.
/18/	J.C. Wilson	Ph. D. Thesis, Univ. Min. (1978).
/19/	J. Wilson, B.Y.H. Liu	J. Aerosol Sci. 11(1980)139-150.

/20/ J.K. Agarwal, R.J. Renwarz,   J. Aerosol Sci. 13(1982)222-223.
      G.J. Sem

/21/ B.T. Chen, Y.S. Cheng,   Aerosol, Sci. and Tech. 4(1985)89-97.
      H.C. Yeh

/22/ K. Leschonski, U. Menzel   1. World Congress Particle Technology, Nürnberg
                                          (1986), Part II, 297-323.

/23/ C.N. Davies, M. Aylward   Proc. Phys. Soc. 64(1951)889-911.

/24/ W.E. Ranz, J.B. Wong   Ind. and Engng. Chem. 44(1952)1371-1381.

/25/ T.T. Mercer, H.Y. Chow   J. Coll. and Interface Sci. 27(1968)75.

/26/ R. Roeber   Staub-Reinhaltg Luft, (1957)41-100, 273-296, 418-449.

/27/ V.A. Marple   Ph.-D. Thesis, Univ. Min. (1970).

/28/ M. Knudsen, S. Weber   Ann. Physik 36 (1911)981-994.

/29/ K. Leschonski   Ullmann´s Encyklopädie, Vol. II, (1972)35-42. Verlag
                                          Chemie, Weinheim.

/30/ H. Rumpf, K. Leschonski   Deutsches Bundespatent.
      K. Maly

/31/ K. Maly   Dissertation Univ. Karlsruhe (1976).

/32/ V. Prodi, F. Belosi, A. Mularoni   Proc. of the 4th European Symp. Particle Characterization,
                                          Nürnberg 1986.

/33/ M. Heuer   Diss. Technische Univ. Clausthal, (1987)

/34/ A. B. Holland-Batt, M.G. Fleming   8th Int. Min. Proc. Congress, Leningrad (1968).

/35/ B.F. Osborne   ISA Trans 10(1971)379-385.

/36/ K.-L. Metzger, K. Leschonski   1. Europ. Symp. Particle Charac., Dechema Monogr. 79,
                                          Part B (1976)77-94., Verlag Chemie, Weinheim.

/37/ K.-L. Metzger   Dissertation TU Clausthal (1977).

/38/ K.Leschonski, K.-L. Metzger,   Proc. Salford Conf. Particel Size Analysis, Heyden, Lon-
      U. Schindler   don, (1978)227-237.

/39/ K. Leschonski   Part. Charact. 1(1984)7-13.

/40/ K. Leschonski   Part. Charact. 3(1986)99-103.

Statistical Analysis of Cyclone Grade Efficiency Measurements

J. G. Bernard[1], J. Andries[1] and B. Scarlett[2]

[1] DELFT UNIVERSITY OF TECHNOLOGY, LABORATORY FOR THERMAL ENGINEERING
MEKELWEG 2, 2628 CD DELFT, THE NETHERLANDS
[2] DELFT UNIVERSITY OF TECHNOLOGY, PARTICLE TECHNOLOGY GROUP, JULIANALAAN 136,
2628 BL DELFT, THE NETHERLANDS

ABSTRACT

A set of 18 grade efficiency measurements were made for a cyclone. Three different inlet velocities and three dust concentrations were chosen. Each condition was tested twice. By statistical tests the reliability of the experiments was investigated and a procedure to analyse grade efficiency data is proposed. The results show that 12 out 18 experiments could be used when considering only the cyclone mass efficiency. In total 8 resulted in useful grade efficiency curves.
The size analysis data appeared to be the most critical part of the whole procedure.

1. INTRODUCTION

Particle size measurements must always be related to their purpose, particularly the accuracy necessary in different regions of the particle size range. The determination of the grade efficiency of separation equipment is particulary difficult because it represents, essentially, the difference in two particle size distributions. At least one of the distributions must be accurately known at the ends of the distribution. In this paper we report on the determination of the grade efficiency of a cyclone.

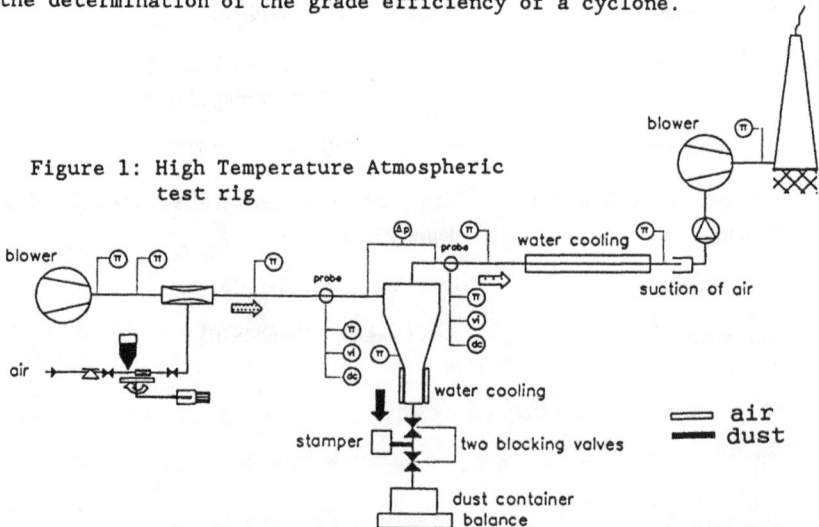

Figure 1: High Temperature Atmospheric test rig

The High Temperature Atmospheric (HTA) test rig (Fig. 1) is fitted to control most of the experimental parameters in an accurate way. The cyclone which was tested, is a commonly used industrial cyclone, shown in Fig. 2, which causes considerable experimental complications.

2. EXPERIMENTAL PERFORMANCE

A controlled amount of dust is fed to the system at a fixed rate. After the experiment the dust which has been collected, is weighed and this determines the mass efficiency of the cyclone The system also contains two probes (Figure 1). One to measure the dust concentration at the inlet of the cyclone (probe 1) and one at the outlet of the cyclone (probe 2). The samples which were obtained are denoted feed, probe 1 (cylone precollector and filter), probe 2 (only a filter) and catch. They were analyzed by a Coulter Multisizer in 128 channels using a logarithmic scale.

3. ERRORS IN GRADE EFFICIENCY CALCULATION

One appropiate way to express the efficiency of a cyclone is in terms of a grade efficiency curve. In this case the efficiency is calculated for every particle size class and the resulting curve is an S-type curve with efficiency values between 0 and 1. In the test rig under investigation we can apply three models of grade efficiency calculation methods:

Model 1: based on the probe 1 and probe 2 samples:

$$\eta_i = 1 - (1-\eta_m) . f2(i)/f1(i) \tag{1}$$

Model 2: based on the catch and probe 1 samples:

$$\eta_i = \eta_m . fc(i)/f1(i) \tag{2}$$

Model 3: based on the probe 2 and the catch samples:

$$\eta_i = 1 - (1-\eta_m) . f2(i)/((1-\eta_m)f2(i) + \eta_m . fc(i)) \tag{3}$$

η_i is the grade efficiency for size i, $fc(i)$, $f1(i)$ and $f2(i)$ are the relative mass fractions for that size of catch, probe 1 and probe 2. η_m is the mass efficiency. The error in the grade efficiency curves ($mc(i)$, $m1(i)$ $m2(i)$ are the absolute mass fractions!) can be calculated according to:

1) $\delta\eta_i = -\dfrac{\delta(m2(i)m1(i) - \delta(m1(i)m2(i))}{m1(i)^2}$ \hfill (4)

2) $\delta\eta_i = \dfrac{\delta(mc(i)m1(i) - \delta(m1(i)mc(i))}{m1(i)^2}$ \hfill (5)

3) $\delta\eta_i = -\delta(m2(i)/(m2(i)+mc(i))$

$\quad +\dfrac{(\delta m2(i))+\delta mc(i))m2(i)}{(m2(i) + mc(i))^2}$ \hfill (6)

4. SAMPLING [1]

Determination of the cyclone grade efficiency needs reliable measurement of the size distribution at both the inlet and outlet of the cyclone. The inlet distribution is determined by probe 1. Probe 1 samples a rather broad size distribution. In principle one should aim for isokinetic conditions. Another complication is that the probe occupies a significant area in the duct which has a rather small diameter due to which a velocity profile is present. The flowlines are supposed to deviate away from the probe opening due to which the velocity at the probe opening might be considerably lower than it should be. This means that in the case that in fact an optimal sample velocity has been chosen smaller particles have already been moved to a position where it is too far out of range to be captured by the probe again. This is mainly caused by the tendency of the small particles to follow the flow. Due to that an undersampling of fine particles is probable by probe 1.

Probe 2, at the outlet of the cyclone, causes another problem as it is positioned in a slightly swirling flow, due to the cyclone outlet. We carefully investigated the assumption that at probe 2 coarser particles are segregated by taking samples at several positions in the tube cross section. The sampled concentration was found to be quite similar at all positions, average 0.1009 g/m^3 and $\sigma=0.0025$ g/m^3.

The catch can also contain an error due to breakage and attrition, processes which are very difficult to trace.

If the possible errors in the particle size distributions are used to assess the possible errors in the grade efficiency curves we obtain the results of Table 1.

Type 1: probe 1 and probe 2

size class	probe 1	probe 2	grade eff
fine	<	·	<
middle	>	·	>
coarse	>	·	>

Type 2: probe 1 and catch

size class	probe 1	catch	grade eff
fine	<	>	>
middle	>	<	<
coarse	>	<	<

* Figure 2: cyclone dimensions

Type 3: Combination of probe 2 and catch as inlet in combination with catch or probe 2.

size class	pr2+catch	catch	probe 2	grade eff
fine	>	>	·	>
middle	<	<	·	<
coarse	<	<	·	<

TABLE 1: Effects of errors in particle size distributions on the grade efficiency. The signs > and < indicate the direction of the error. The grade efficiency curve should be shifted in the opposite direction.

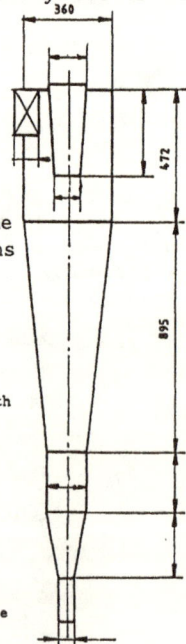

Assuming a relative error in the fines region of +5% for the catch and -5% for the inlet (size distribution) the error in the grade efficiency for the three grade efficiency models can be estimated by using equations (4)-(6). The total efficiency is 90%, and the calculated grade efficiency for the particular particle size is 80 %, the mass fraction feed is 0.01, the catch 0.008 and probe 2 0.002. This means that the absolute error in the inlet is -0.0005 and in the catch +0.0004 and at probe 2 an error of 0.0001. The sign of the errors is valid in the small size ranges. The same calculation is made for the size with 20 % efficiency; probe 2 = 0.008, error 0.0001 and catch = 0.002, error 0.0004. The absolute errors in the size distributions remain constant. The errors in the grade efficiency are listed in Table 2:

TABLE 2: Influence of errors in size distributions on grade efficiency curve.

	grade eff=80%		grade eff=20%	
gr.eff model:	error pr2=0	error pr2=5%	error pr2=0	error pr2=1%
model 1	-1.0.%	-2.0%	-4.0%	-5.0%
model 2	+8.0 %	+8.0%	+5.0%	+5.0%
model 3	+0.8 %	-	+3.2%	+3.0%

The errors for models 1 and 2 are clearly the highest. Model 3 has a smaller error. Model 3 offers the best potential for the calculation of the grade efficiency curve. It is less sensitive to systematic errors since they act in the same direction.
A way to express the quality of the experiments is by the ratio:

$$((1-\eta_m)f2(i) + \eta_m fc(i))/f1(i) \tag{7}$$

In the case that there are no disturbances due to breakage or sampling this ratio should be one. It appears that for particles smaller than about 2 μm the ratio is larger than one. This normally results in a negative efficiency for the fine particles. There are two ways in which this can happen: oversampling of fines by probe 2 or undersampling at probe 1. Probe 2 contains some of the products due to breakage which are not found at probe 1. The effects of the swirl are not likely as concluded earlier. On the other hand there can be a systematic undersampling as mentioned earlier. Model 3 combines two size distributions with relatively more fines than probe 1. The numerator is the same as in the models 1 and 3, which means that the grade efficiency for the fine particles should always be higher than with the model 3. From simple comparision of the grade efficiency data for both models we can conclude that in the normal case the error in grade efficiency is only present in the size range below about 2 μm. Above this size the models 1 and 3 fit very well together.

5. REPRODUCIBILITY

The reproducibility of experiments can be influenced by the following factors:
1) non ideal sampling with the probes
2) size analysis errors

A procedure to determine the statistical reliability of a set of experiments is as follows.

1) the mass efficiency is not comparable to experiments which were made under the same conditions: F-test
2) the size distribution of probe 1 or probe 2+catch differs too much from the average: Student t-test
3) the size distribution of the catch or probe 2 deviates from comparable measurements: F-test

6. F-TEST OF VARIANCE [2]

If a set of experiments in accordance to a certain pattern has been made it is possible to use these experiments to calculate the standard deviation for all the experiments. The experiments which were made are listed in Table 3.

TABLE 3: Matrix of the experiments: mass efficiencies of the experiments.

	flow [%]			
conc. [%]	50 %	75 %	100 %	code
33%	0.914 0.926	0.935 0.933	0.948 0.949	1
50%	0.936 0.932	0.942 0.941	0.958 0.954	2
100%	0.942 0.935	0.954 0.950	0.962 0.950	3
code	A	B	C	

Pairs of experiments are coded A1.1, A1.2, A2.1 etc..

In the case that at one condition more than one experiment is made it is possible to compare the variance of the multiple experiments at one condition with the variance for the whole set of experiments by using the F-test:

$$F(f_1, f_2) = \sigma_1^2 / \sigma_2^2 \qquad (8)$$

Here f_1, f_2 are the degrees of freedom for the denominator and numerator. Using an F-test we stated the Nul-hypothesis that a group of experiments must be discarded if the variance within the group is significantly larger than the variance calculated for the whole set of experiments.
For this reason all experimental values are normalized to one distribution by dividing the mass efficiencies by the average for a group. The variance in a group was determined as a fraction, so that the variances can be compared. It is now decided on the basis of the significance levels listed in Table 4 whether the Nul-hypothesis is valid or not. If the Nul-hypothesis is not significant the pair of experiments is not a good duplication within the limits determined by the variance for the complete set of experiments.
First the standard deviation for the 18 experiments was calculated, and on the basis of the F-tests results. The pair A1 was discarded. After a second F-test the pair C3 was also

TABLE 4: Results of the F-test.

exp	eff	average	std dev	norm std	norm eff	F18 *)	F16	F14
A1	0.914	0.920	0.0084	0.00915	0.9935	6.18	-	-
	0.926				1.0064	c)		
A2	0.936	0.935	0.0028	0.003025	1.0021	0.68	1.28	1.98
	0.933				0.9978	a)	a)	a)
A3	0.942	0.939	0.0052	0.005575	1.0039	2.30	4.33	6.70
	0.935				0.9961	a)	b)	c)
B1	0.935	0.934	0.00014	0.001514	1.0011	0.17	0.32	0.49
	0.933				0.9990	a)	a)	a)
B2	0.941	0.942	0.00071	0.000751	0.9995	0.04	0.08	0.12
	0.942				1.0005	a)	a)	a)
B3	0.950	0.952	0.0028	0.002971	0.9979	0.65	1.24	0.90
	0.954				1.0021	a)	a)	a)
C1	0.948	0.949	0.00049	0.000521	0.9996	0.02	0.04	0.08
	0.949				1.0004	a)	a)	a)
C2	0.958	0.956	0.0028	0.002958	1.0021	0.64	1.22	1.88
	0.954				0.9979	a)	a)	a)
C3	0.962	0.957	0.0088	0.009246	1.0065	6.31	11.9	-
	0.950				0.9934	c)	d)	

*) Nul-hypothesis: a) not sigificant b) probably significant
c) not significant d) highly significant

discarded. The standard deviation then decreases from 0.003679 to
0.002677 and 0.002154 respectively. This means that the criterion
for the reliability of duplication is sharpened by calculating the
standard deviation from fewer experiments, since the F-values only
slightly increase.
Of all the pairs of experiments only A2, B1, B2, B3, C1 and C2 are
suitable for further comparision.
The second step is the comparision of the inlet size distributions.
There are two inlet size distributions which might be useful: the
probe 1 size distribution and the combined (probe 2+catch) size
distribution. The probe 2+catch distribution was used.

7. STUDENT t-TEST [2]

The Coulter Multisizer used 128 size channels. Each channel
represents a certain size in which a number of particles are
counted from which a total volume for the contents of the channel
can be calculated. The volumes are normalised, so each channel
represents a certain mass fraction. In the case of undisturbed
separation behaviour in the cyclone, low concentration and exactly
the same conditions, the grade efficiency for a certain particle
should be the same. In that case the contribution of that particle
size fraction to the total mass efficiency should be exactly the
same.
Errors in sampling or the use of different batches of material may,
however, cause the inlet size distributions to deviate from the
mean. The Nul-hypothesis which we will test is that a size
distribution can be rejected by comparision with the average.
The probability that a value of the mass fraction deviates from an
average distribution is determined by a certain t-value, depending
on the number of degres of freedom.
If the t-value exceeds the 10% value there is no reason to reject
the size distribution data in comparision with the average
distribution. Between 5 and 10 % there is some doubt about the
significance of the Nul-hypothesis. Usually if the t-value
exceeds a value of 2 the size distribution is not comparable and
the results cannot easily be used to compare with other
experiments.

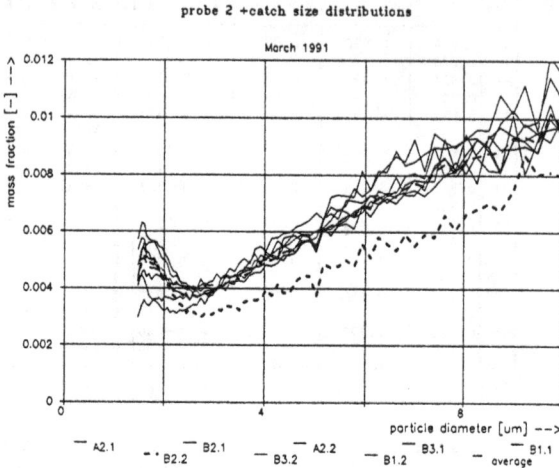

probe 2 +catch size distributions

March 1991

mass fraction [-] --->

particle diameter [um] -->

— A2.1   — B2.1   — A2.2   — B3.1   — B1.1
  - · B2.2   — B3.2   — B1.2   — average

*Figure 3: Inlet size distributions for 8 selected experiments.
           There is a clear difference for experiment B2.2.

        In total 8 size distribution curves (Figure 3) were useful for
further comparision. First of all the average mass fraction and the
standard deviation was calculated for each size range. The variance
was used in an F-test and compared with the variance for a group.
The Nul-hypothesis was used again to decide whether to discard a
group on the basis of the fact that its variance is significantly
larger than the variance for the whole set.
The F-test and t-test results are shown in Figures 4 and 5.
If this procedure is followed the F-test shows that A2, B1 and B3
are acceptable sets of experiments. B2 has too large an F-value and
cannot be used directly. Using the t-test A2 and B3 are very
acceptable, B3 is only acceptable up to 4 μm. Refering to the
grade efficiency curves A2, B1 and B3 show very reproducible
results. The fact that the B1 inlet size distribution is slightly
different above 4 μm does not matter very much since the grade
efficiency has already reached a value over 95 % for that size.
It appears that for experiment B2.2 both the flow and the
concentration were slightly lower than for B2.1. The
reproducibility for the B2 set appears to be worse than for the
other three.

8. GRADE EFFICIENCY CURVES

        The results of the grade efficiency calculations using
equation (3), model 3, are shown in Figure 6.
The criterion can be used less rigorously.For instance if we use
the F-value for 5 % as a critical value, which is 5.59, it is not
necessary to reject any of the experiments A2, B1, B2 and B3.
Considering the scale of the installation this can be justified.

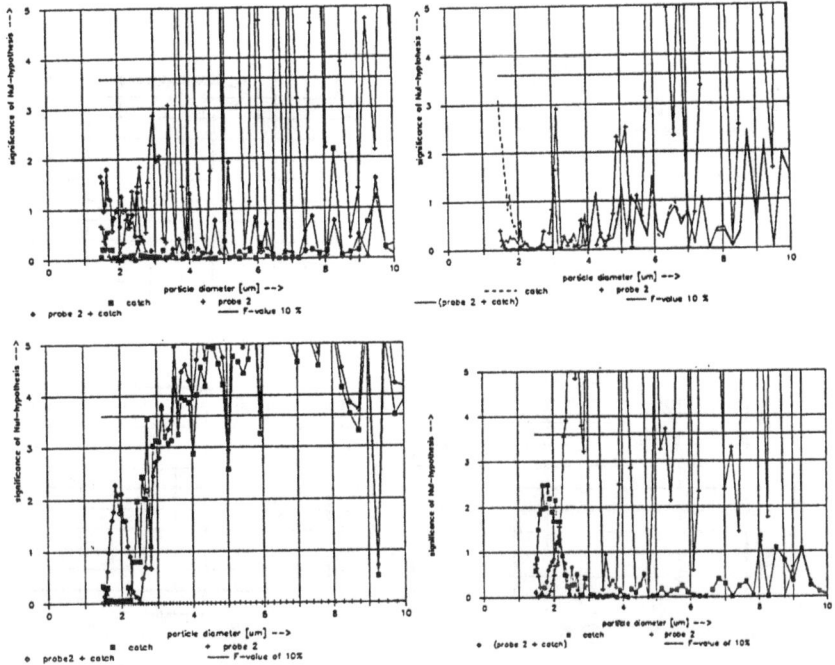

*Figure 4: results of the F-test on the probe 2+catch inlet size distributions for the experimental pairs: A2, B1-3.

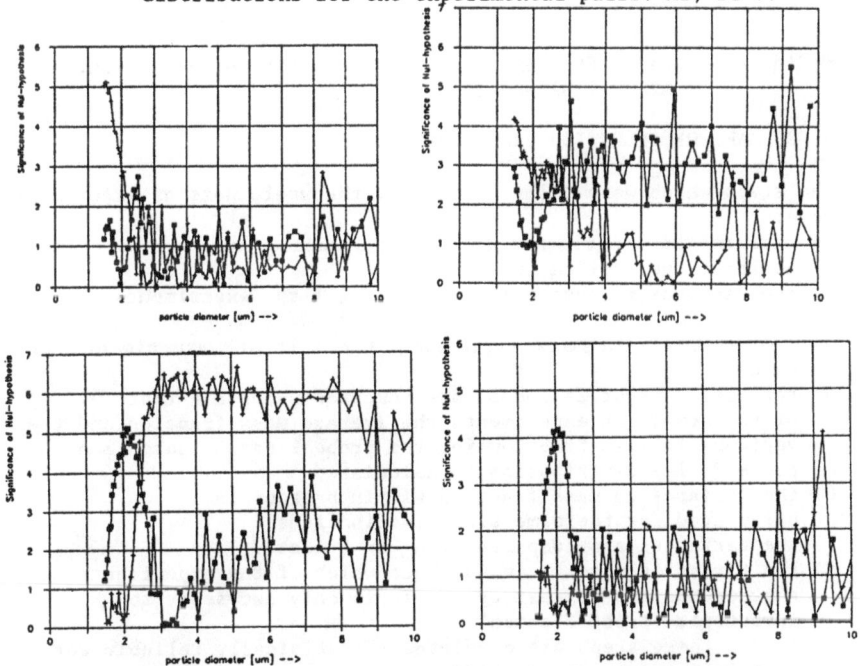

*Figure 5 results of the t-test on the probe 2+catch inlet size distributions for the experimental pairs: A2, B1-3.

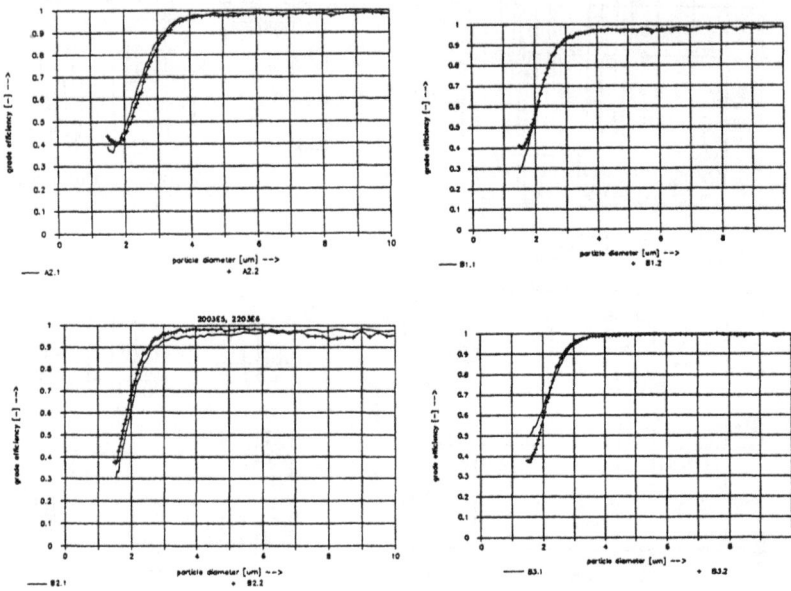

* Figure 6: Grade efficiency data for the pairs of experiments
           A2, B1, B2 and B3.

9. SUMMARY OF SELECTION PROCEDURE

The steps which were followed to test the usefulness of experiments
were:
1) perform the experiments in random sequence; of each experiment
   at least one duplication is made
2) the variance in mass efficiency for all the experiments is
   calculated
3) the variance in mass efficiency for multiple experiments is
calculated
4) the variances of 2) and 3) are compared using a F-test
5) of the remaining experiments the average mass fraction and the
   variance in mass fraction for the probe 2+catch, catch and
   probe 2 size distribution is calculated
6) the variance in mass fraction within a group is
   calculated for the same size distributions
7) the variances are compared using an F-test
8) the t-test is used to investigate which of the experiments of
   a group has the largest error. It is only necessary to
   investigate probe 2+catch
9) if all these steps are complete, a statisically reliable set of
   experiments can be selected.

10. DISCUSSION AND CONCLUSIONS

The interpretation of (cyclone) efficiency data is a complicated problem in which statistical analysis of the results together with an appropriate statistical procedure are essential.
The number of possible error sources, due to the complex way the experiments are carried out, is very large. As a consequence not all the results can be used. In this paper we suggest one method of selecting experiments from a set to minimise the error. As a consequence it might be possible to discriminate between results within a set in which the variance between the experiments is clearly higher. For instance for A1, the grade efficiency curves are very well reproducible, except above 7 μm. Only the mass efficiencies are significantly different. After the application of some cross multiplication we must consider the result of A1.2 as the best.
As the best model for grade efficiency calculations we applied model 3 in which the catch and probe 2 sample should be known. This model resembles the procedure proposed by Hermann and Leschonski [3], as a correction of the mass balance is applied in this calculation model as well. It appears from our experience that the size analysis of the probe 2 sample must be carried out very accurately. Enough sample should be collected on the filter in order to perform reliable Coulter Counter analysis.
The grade efficieny results appear to be reproducible above 2 μm. Firstly this is caused by the sensitivity of the grade efficiency towards very small fluctuations in the size distribution curves. Secondly the Coulter Multisizer is too sensitive for noise in the range from 1.4-2.0 μm, when using a 70 μm tube, which affects the size distribution results. This problem can be avoided by applying exactly the same concentration in the Coulter analysis.

11. REFERENCES

[1] Bernard, J.G., Andries, J., Scarlett, B., Pitchumani B., Sampling of Fine Particles from Turbulent Gas Streams, Proc. of 2nd World Congress Particle Technology, Kyoto, Jimbo, G.(ed), (Sept 1990)
[2] Davies, O.L., Statistical Methods in Research and Production, 2nd Ed., Oliver and Boyd, London (1954), 51-70.
[3] Herrmann, H.,Leschonski, K., Einfluss der Beruecksichtigung von Fehlern der Partikelgrossenverteilung bei der Ermittlung von Trennkurven, (Influence of errors in particle size distributions on the averaging of grade efficiency curves), 2nd European Symposium on Particle Characterisation, Nurnberg, (Sept 1979).

12. ACKNOWLEDGEMENTS

We acknowledge the assistance of Messrs. K.J.F. Bloos and D. Triezenberg who carried out most of the experiments reported in this paper.
This project is financially supported by European Community (contract number EN3F-0028-NL-CDF)), the Dutch Ministery of Economic Affairs (NOVEM contract 20.36.-0110.10), TNO-KRI, Van Tongeren and Stork Boilers.

A New Method and Apparatus for Monitoring of Particle Size Distributions in Industrial Processes

L. Svarovsky[1] and J. Svarovsky[2]

[1] DEPARTMENT OF CHEMICAL ENGINEERING, UNIVERSITY OF BRADFORD, BRADFORD BD7 1DP, UK
[2] BRADFORD GRAMMAR SCHOOL, BRADFORD BD9 4JP, UK

1 INTRODUCTION

This paper describes a new method for monitoring of mean particle size and equivalent spread of the size distribution in process streams. Measurement of mean particle size is important in process control, quality control or in setting up of processing plants but the present method can also monitor the width of the distribution, which is often also of interest.

The concept of equivalence is well known and accepted in particle size measurement, and the paper applies this concept to the measure of the spread of the distribution. It characterises the actual distribution of particle size in the slurry by an equivalent, log-normal distribution described by a simple formula with two numerical parameters, the geometric mean (as a measure of the mean size) and the geometric standard deviation (as a measure of the distribution spread). The equivalence is by separation efficiency in a dynamic separator such as a hydrocyclone or a sedimenting centrifuge.

The underlying theory is fully developed in the paper, together with its application to a specific case of monitoring of very fine particle size using a small diameter hydrocyclone. Experimental results obtained with a specially-developed and fully instrumented test rig are given and the usefulness of the method is clearly demonstrated.

Although the test rig developed to demonstrate the validity and use of the new method can be used as an "off-the-shelf" instrument, it would require a continuous sample taken from the stream whilst making sure that the sample is representative of the particle size distribution of the solids suspended in the flow. The main strength of the new method, however, is in its ability to make use of an existing separator in the process so that the whole of the process stream can pass through the separator and thus avoiding any sampling errors.

2 GRADE EFFICIENCY OF A DYNAMIC SEPARATOR

For a separator with a size-dependent performance the grade efficiency varies with particle size, and a graphical representation of this is called the grade efficiency curve.

The effect of flow splitting (or "dead flux") in applications with appreciable and dilute underflow, as for example with hydrocyclones, is to modify the shape of the grade efficiency curve making it appear that the performance of the separator is better than would be expected from its ability to separate particles of different properties. An example is given in Figure 1 where a typical grade efficiency curve of a hydrocyclone is plotted. The curve does not start from the origin (as it should for inertial separation) but has an intercept, the value of which is usually equal to the underflow-to-throughput ratio R_f. This is because the very fine particles simply follow the flow and are split between the underflow and the overflow in the same ratio as the fluid. The R_f ratio is defined as the fraction of the volumetric feed rate which turns up in the underflow, i.e. the underflow rate divided by the feed rate.

In order to remove the effect of flow splitting from the efficiency definition so that it describes only the true "centrifugal efficiency", the grade efficiency is "reduced" as in the following equation:

$$G'(x) = \frac{G(x) - R_f}{1 - R_f} \qquad (1)$$

This forces the curve to pass through the origin as indicated by the second curve, $G'(x)$, in Fig.1. The reduced grade efficiency curve can, for some separators, be approximated by an analytical expression such as the one used in this method - see eqn.2 given on the following page.

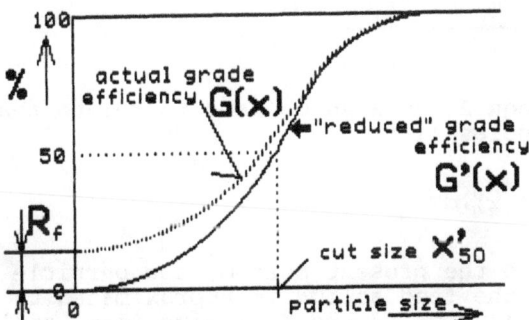

Figure 1 Grade efficiency curves for a hydrocyclone

The size corresponding to 50% on the reduced grade efficiency curve G'(x) is referred to as the "reduced cut size" - see Fig.1. Most mathematical descriptions of the performance of hydrocyclones or sedimenting centrifuges are in terms of the reduced cut size.

3 THE THEORY

The reduced grade efficiency curves of some dynamic separators, including hydrocyclones and sedimenting centrifuges, can be fitted by a cumulative log-normal function in the following form:

$$G'(x) = 0.5 + 0.5 \ erf \left[\frac{\ln x - \ln x'_{50}}{\sqrt{2} \ \ln \sigma_s} \right] \tag{2}$$

where the erf function is defined as:

$$erf(z) = \frac{2}{\sqrt{\pi}} \int_0^z e^{-t^2} \ dt \tag{3}$$

and erf(z) can be evaluated using tables, series or analytical approximations. Note that x'_{50} and σ_s can be determined from a plot of the grade efficiency curve in a log-probability graph paper.

Once the grade efficiency curve is known for a given set of operating conditions, the total efficiency E_T (recovery of solids into the underflow) expected with a particular of a cumulative particle size distribution $F(x)$ feed can be predicted using the following relationship :

$$E_T = \int_0^1 G(x).dF \tag{4}$$

The effect of dead flux on total efficiency is eliminated in the same way as for grade efficiency (eqn.1), i.e.

$$E'_T = \frac{E_T - R_f}{1 - R_f} \tag{5}$$

Note that eqn.2 can also be used to relate the reduced efficiencies, i.e.

$$E'_T = \int_0^1 G'(x) \ dF \tag{6}$$

According to the present method, the particle size distribution of the feed solids is approximated by the log-normal distribution in the following form, analogous to eqn.2 (as cumulative fraction undersize):

$$F(x) = 0.5 + 0.5 \text{ erf} \left[\frac{\ln x - \ln x_g}{\sqrt{2} \ \ln \sigma_g} \right] \qquad (7)$$

The total reduced efficiency E'_T can be predicted from $G'(x)$ and $F(x)$ by integration (eqn.6) and this, using eqns 2 and 7 leads to the following formula[1]:

$$E'_T = 0.5 + 0.5 \text{ erf} \left[\frac{\ln x_g - \ln x'_{50}}{\sqrt{2} \ \sqrt{(\ln^2 \sigma_g + \ln^2 \sigma_s)}} \right] \qquad (8)$$

The total reduced efficiency E'_T can also be evaluated directly from the feed solids concentration c and the solids concentration in the overflow c_o using another formula[1]:

$$E'_T = 1 - \frac{c_o}{c} \qquad (9)$$

It is the combination of equations 8 and 9, by eliminating E'_T, which forms the foundation of the present method:

$$1 - \frac{c_o}{c} = 0.5 + 0.5 \text{ erf} \left[\frac{\ln x_g - \ln x'_{50}}{\sqrt{2} \ \sqrt{(\ln^2 \sigma_g + \ln^2 \sigma_s)}} \right] \qquad (10)$$

If the response of the separator to the operating conditions, in terms of the cut size x'_{50} and the standard geometric deviation of the reduced grade efficiency σ_s, is known (from tests and/or theory) and the two concentrations c and c_o are monitored, the two parameters chracterising the particle size distribution in the feed, x_g and σ_g are the only unknowns in the above equation. Two sets measurements (e.g. at two different flowrates) are, therefore, required to allow the calculation of the two unknowns.

4 PRACTICAL USE OF THE METHOD

The sequence of the measurement, preferably controlled by a computer, is to take the readings of c and c_o at one pressure drop (result set 1), then switch to another pressure drop and repeat the measurement (set 2). Fig.2 shows a schematic diagram of the measurement positions, with all of the readings being capable to be taken and logged by a computer. The concentration readings are from two separate and suitably calibrated densitometers such as gamma gauges, vibrating U-tube density meters or ultrasonic devices. It is also possible to use just one densitometer and re-route periodically and alternately the feed and overflow streams through it. This is best done by solenoid or motorised valves, controlled by the computer (see an example in Fig.3 described in section 5).

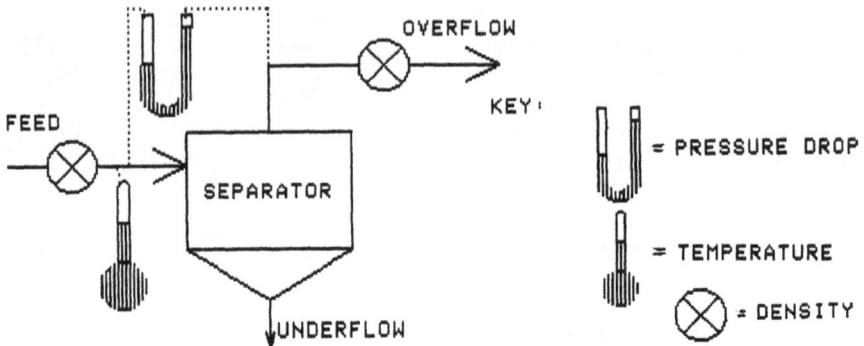

Figure 2 A schematic diagram of the measurement positions

The evaluation of results is done with a computer which uses a mathematical model of the separator function and the following simple algorithm for the evaluation of the two parameters x_g and σ_g:

1. Assume σ_g and evaluate x_{g1} from eqn.10 and result set 1
2. change Δp and calculate x_{g2},
3. compare x_{g1} and x_{g2}, and if they are not the same, keep changing σ_g and re-calculating the values of x_g until they become the same, i.e.

$$x_{g1} = x_{g2} = x_g$$

4. print or display the final values of x_g and σ_g, use these as control signals, plot the whole distribution (eqn.7) or calculate and print specific points on it (such as % less than 5 micron etc.).

Given the simplicity of the above calculations and the speed of personal computers, the calculations are done virtually instantaneously. The only delay in the measurement is through the necessity of switching to another pressure drop and the need to establish steady-state operation after the change. This is quite short (a few seconds) in hydrocyclones but in centrifuges, it may take up to a minute or more.

The equipment, therefore, continuously cycles through the measurement sequence, constantly updating the values of x_g and σ_g and, if required, averaging them over a pre-determined period of time.

There are two ways in which the present method can be used in practice. In one alternative, a sampling stream is taken from the process stream and passed through a specially built and instrumented test rig such as the one shown schematically in Fig.3 and further described in section 5. The required step change in pressure drop (or flowrate, if a centrifuge is used) can be easily introduced by using a solenoid or motorised valve, or by changing the speed of the supply pump, with the change initiated by the controlling computer.

The other, second alternative avoids the sampling problem altogether. It makes use of an existing separator in the process itself so that the whole of the process stream passes through the separator and participates in the measurement. In some instances, it is not necessary to introduce any artificial changes in pressure drop (or flowrate) but use the normal variations in the operating conditions that often occur in industrial practice.

In either of the alternatives, no dilution is necessary and, furthermore, the separator itself can be tested using the same instrumentation as installed for the monitoring of particle size. This in effect relies on the same basic equation (eqn.10) but this time, x_g and σ_g are known for the test suspension fed to the separator and x'_{50} (the cut size) and σ_a (the standard deviation of the grade efficiency curve) are determined from the tests.

5 THE EXPERIMENTAL APPARATUS

In the experimental rig shown in Fig.3, the change from one operating pressure to another is achieved by switching on or off the solenoid valve (sv) in the by-pass line 1. This is, however, only possible when the sample stream to be monitored contains only very fine particles which do not separate significantly in the T-junction 2 which splits the pumped flow into the hydrocyclone feed line 3 and the by-pass line 1. In the case of coarser slurries, this arrangement would not be acceptable and the change in the operating pressure drop might better be achieved by changing the speed of the supply pump.

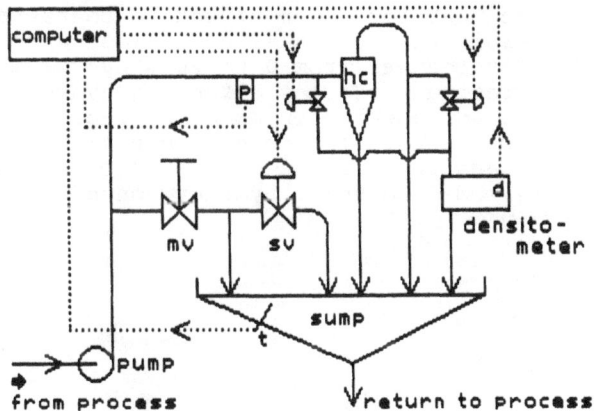

Figure 3 A schematic diagram of the experimental setup
d = density measurement (vibrating U-tube)
p = pressure drop measurement (piezo-electric)
t = temperature measurement (thermocouple)
hc= hydrocyclone
mv= manual valve
sv= solenoid valve (controlled by computer)

It may be noted from Fig.3 that only one densitometer is used and the two streams to be measured, the sample stream 4 from the feed stream 3 and the sample stream 5 from the overflow 6, are switched through it alternately, with a delay in between to allow each sample stream to reach the sensor head of the instrument. Note also that both of the streams taken through the densitometer have to be well de-aerated and this is achieved by venting the lines and routing them in such a way that de-aeration takes place by gravity.

6 RESULTS OF TESTS

Fig. 4 shows the particle size distribution, by mass, of the solids used in the tests (chalk), as obtained with the Ladal Pipette Centrifuge and the Andreasen Pipette Method. As can be seen from the log-probability plot, the distribution is very nearly log-normal and thus suitable for testing the performance of the hydrocyclone. Further-more, the medium size of the chalk (3.9 microns) is close to the range of cut sizes expected from the hydrocyclone (2 to 4 microns) which is also a desirable feature for effective separator testing.

An IBM-compatible microcomputer was used to control the rig and log the data, using a specially-developed program which included the necessary algorithms mentioned previously. Two versions of the program were developed, one for testing the separator with a known solid and another for using the separator for on-line monitoring of the median size and the geometric standard deviation.

The most important effect to be tested in hydro-cyclones is that of the feed solids concentration. The range covered was from 0 to 20 % by volume. The rest of the operating variables affecting hydrocyclone performance are grouped together in dimensionless groups. A model based on the use of such groups has been published before[2] and the present method is based on an adaptation of this model to small hydrocyclones.

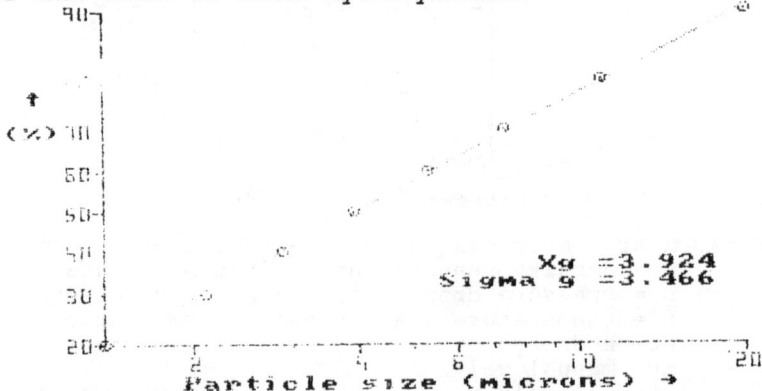

$X_g = 3.924$
$Sigma\ g = 3.466$

Figure 4 Mass distribution of the test powder (chalk)

The main equation relating the relevant dimensionless groups is as follows:

$$Stk'_{50} \sqrt{Eu} = k_1 \exp(k_2 c) + k_3 \qquad (11)$$

Euler number Eu is a pressure loss factor based on the static pressure drop across the cyclone:

$$Eu = \Delta p / (\rho v^2 / 2) \qquad (12)$$

Stk'_{50} is the Stokes number (for cut size x'_{50}) defined as:

$$Stk'_{50} = x'_{50}{}^2 (\rho_s - \rho) v / (18 \mu D) \qquad (13)$$

where ρ_s and ρ are densities of the solids and of the liquid respectively, μ is liquid viscosity and D is cyclone diameter, x'_{50} is the reduced cut size (see the Nomenclature for the definitions).

Both of the above equations use the superficial velocity in the cyclone body as the characteristic velocity, i.e.

$$v = 4Q / (\pi D^2) \qquad (14)$$

The left-hand term in eqn.11 is a dimensionless group that has its theoretical basis in the turbulent two-phase flow theory for hydrocyclones[2] whilst the right-hand side is a semi-empirical expression for the effect of feed concentration, which has its roots in the theory of hindered settling[3].

The test rig described in section 5 was run with the chalk slurry at several solids concentrations ranging up to 20 % by volume, with duplicated measurements at each concentration. Fig.5 shows the results and the best fit curve obtained by the minimum sum of squares method. This provided the constants for equation 11, with the following values (if c is in %): $k_1 = 0.083419$, $k_2 = 0.22359$ and $k_3 = 1.1335$. Eqn.11 can then be used as a model for the hydrocyclone performance in particle size measurement of unknown slurries, using the same rig as used in testing the hydrocyclone.

Fig.6 gives an example of such measurement on a wheat starch slurry, with the full line representing the "equivalent", mono-modal distribution substituted for the actual, bi-modal distribution as measured by Coulter Counter and shown as points in Fig.6 (as % by volume).

7 CONCLUSIONS

The advantages of the method and apparatus of the present method can be summarised as follows:

1. no dilution of the feed stream is necessary,
2. a personal computer can be used to automatically control the measurement, to log and evaluate the data,
3. existing separators used in the process can be employed and only relatively simple instrumentation is needed,
4. sampling problems are reduced or completely eliminated,
5. the same equipment is used for the calibration as for the actual on-line monitoring,
6. testing of separators using the method is greatly simplified and sped up, compared to conventional methods, and can itself be used in research and development.

REFERENCES

1. L. Svarovsky (ed), 'Solid-liquid Separation', 3rd edition, Butterworths, London, 1990, Chapter 2, p.61 & p.67
2. L. Svarovsky, 'Hydrocyclones', Holt Rinehart and Winston, London, 1984, Chapters 4 and 9, p.55 & 119
3. L. Davies, D. Dollimore and G.B. McBridge, Powder Technology, 1977, 16, 45

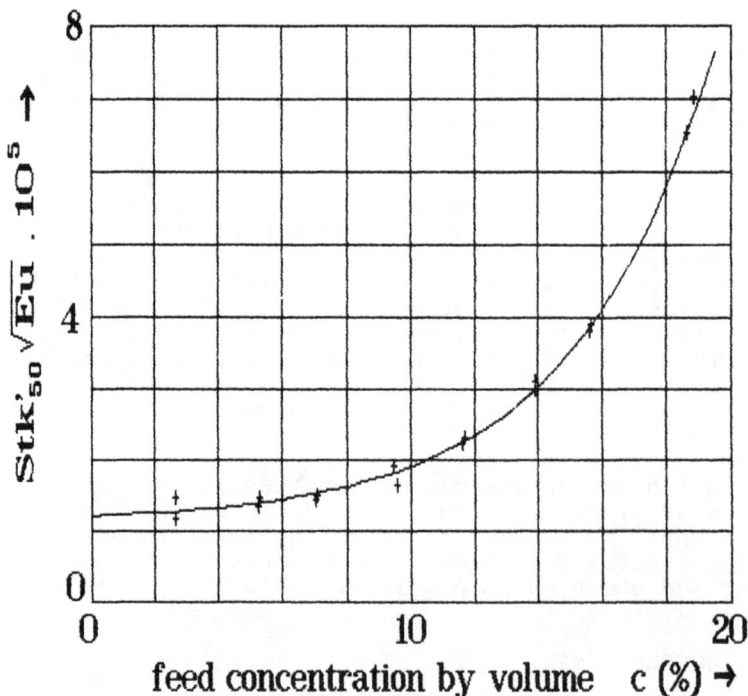

Figure 5 Test data obtained with a 10 mm hydrocyclone

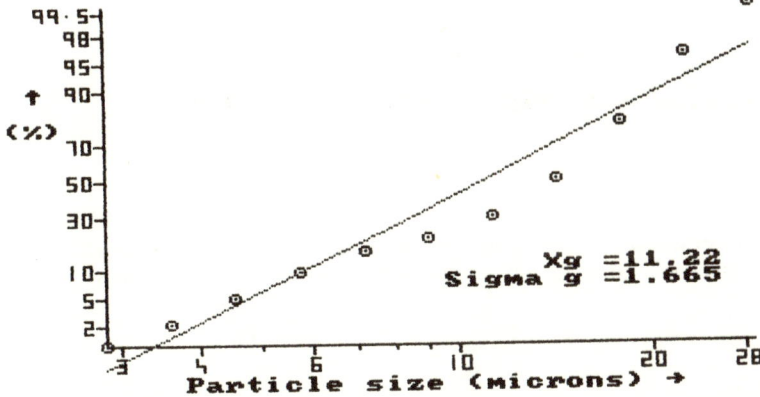

Figure 6 Example of measurement on a wheat starch slurry

NOMENCLATURE

c	is mass concentration of solids in the feed
c_o	is mass concentration of solids in the overflow
D	is the internal diameter of a hydrocyclone
E_T	is the total coarse efficiency (or recovery)
E_T'	is the reduced total efficiency defined in eqn.5
Eu	is Euler number defined in eqn.12
F(x)	is the cumulative percentage undersize in the feed
G(x)	is the actual grade efficiency function (curve)
G'(x)	is the reduced grade efficiency function, eqn.1
R_f	is the underflow-to-throughput ratio (by volume),
Q	is the volumetric flowrate of the feed
Stk'_{50}	is Stokes number defined in eqn.13
v	is characteristic velocity defined in eqn.14
x	is particle size as a variable
x_g	is the mass median of the feed solids [$F(x_g) = 0.50$]
x_{50}'	is a the reduced cut size [$G'(x_{50}') = 0.50$]
Δp	is static pressure drop across the hydrocyclone
μ	is liquid viscosity
ρ	is liquid density
ρ_s	is solids density
σ_g	is the geometric standard deviation of F(x)
σ_s	is the geometric standard deviation of G'(x)

Evaluation of Particle Size Analysis Data Regarding Reproducibility and Accuracy

E. Heidenreich and M. Stintz

TECHNISCHE UNIVERSITÄT DRESDEN, INSTITUT FÜR MECHANISCHE, VERFAHRENSTECHNIK UND SYSTEMVERFAHRENSTECHNIK, MOMMSENSTR. 13, O-8027 DRESDEN, GERMANY

1 INTRODUCTION

There are three major reasons for the growing importance of information about reproducibility and accuracy in particle size analysis.

Automation in Chemical Engineering needs on-line-information about the state of disperse systems which should be able to distinguish between random and systematic deviations from a certain norm.

It is advisable to investigate the behaviour of a disperse system under different conditions (fluid or dispersing treatment), or to measure different charac-teristics of the same particles. The combination of different particle size analysis data also requires individual information about their reliability.

Some on-line-measurement devices give relative infor-mation only about changes of particle size distributions or of their integral parameters. This fact is caused by the dependence of the calibration function on other chemical or physical characteristics of the disperse system.

It is intended to give a basis for discussion in the field of computer-aided interpretation of PSA results. The operator should be encouraged to choose such a way of data presentation that physical interpretation of deviations is possible.

2 STATISTICAL EQUATIONS

Most authors prefer the standard deviation of the cumulative distribution function for the description of the reproducibility of measurements[1,2].

After n repetitive measurements of the cumulative distribution function Q of a certain particle size x the standard deviation $s_Q(x)$ from its mean $\bar{Q}(x)$ can be calculated to

$$s_Q(x) = \sqrt{\frac{1}{n-1} \cdot \sum_{i=1}^{n} (Q_i(x) - \bar{Q}(x))^2}$$

(1).

The same method can be used to describe the reproducibility $s_q(x)$ of the density distribution function $q(x)$ with

$$s_q(x) = \sqrt{\frac{1}{n-1} \cdot \sum_{i=1}^{n} (q_i(x) - \bar{q}(x))^2}$$

(2).

Because of high values differ with high absolute deviations the standard deviation s_q should be related to the mean density $\bar{q}(x)$, represented by the variation coefficient

$$v_q(x) = \frac{s_q(x)}{\bar{q}(x)}$$

(3).

Different quantity of the distribution, number or volume, and different presentation, density or cumulative, give different values of the reproducibilty of the same measuring results.

3   EXPERIMENTS

On the example of the statistical errors of dry sample splitting of Aluminium oxide alternative methods of evaluation should be tested.
About 200 g were split   by a rotary riffler into 10 samples. After that every sample was suspended in water and after ultrasonic treatment measured by three very different methods:

-single particle light extinction measurement of about
 400.000 particles or 45 mg solid in 100 g filtered
 water,

-laser diffraction measurement of about 500 mg solid in
 a closed loop with an ultrasonic bath,

-wet sieving of about 5 g solid on single test sieves
 with the help of an ultrasonic agitator below 100 μm.

As a unit of the single particle measurement software on-line calculation of all interesting statistic parameters was integrated.
By this way the operator is able to improve sample preparation and sample feeding referring to its effect on the result during the analysis.

Figures 1-4 show direct hardcopys of the PC-screen.

For comparison between the analysis methods the results
of laser diffraction and wet sieving were imported in a
commercial program for matrix calculation and presentation in figures 5-8.

4   RESULTS

To get an impression of the particle size distribution
and its deviations the ten number density distributions
and cumulative distributions of the split samples,
measured by optical single particle extinction counting,
are plotted in figure 1.

Because most processes in Chemical Engineering can be
better characterized by mass or volume distributions,
figure 2 presents the converted data of the same method.
This kind of quantity shows a small amount of coarse
material greater than 200 μm in every sample.

Figures 3 and 4 show the mean values of the density
function $\bar{q}(x)$ by number and by volume. Additionally the
according standard deviations s_q and variation
coefficients v_q are plotted.
After magnification by the factor 100 it is obvious
that the standard deviation firstly depends on the
amount of particles in a certain particle size interval.
On the other hand the particle size interval with
relative errors lower than 15 % can be found
-between 6 and 25 μm regarding the number density
 distribution and
-between 30 and 90 μm regarding the volume density
 distribution.

For convenient comparison of the different measuring
methods upper figures on the following pages contain the
statistical evaluation of the volume density distribution function of the split samples.

Figure 5 shows the equivalent evaluation of the laser
diffraction measurements. The higher amount of material
under investigation improves the reproducibility.
For example the particle size range with relative errors
below 15% is expanded from 30...90 μm to 25...100 μm.
On the other hand the density function is smoother and
broader, the relative minimum around 160 μm is erased by
the diffraction pattern calculations and the broader
particle size intervals.

Figure 6 can be used to calculate the pure splitting
error for different particle sizes. It shows the
statistical parameters of ten repetitive measurements of
only one sample in a closed loop.
With exception of the 70 μm class splitting was the
dominant source of statistical errors. Its influence can

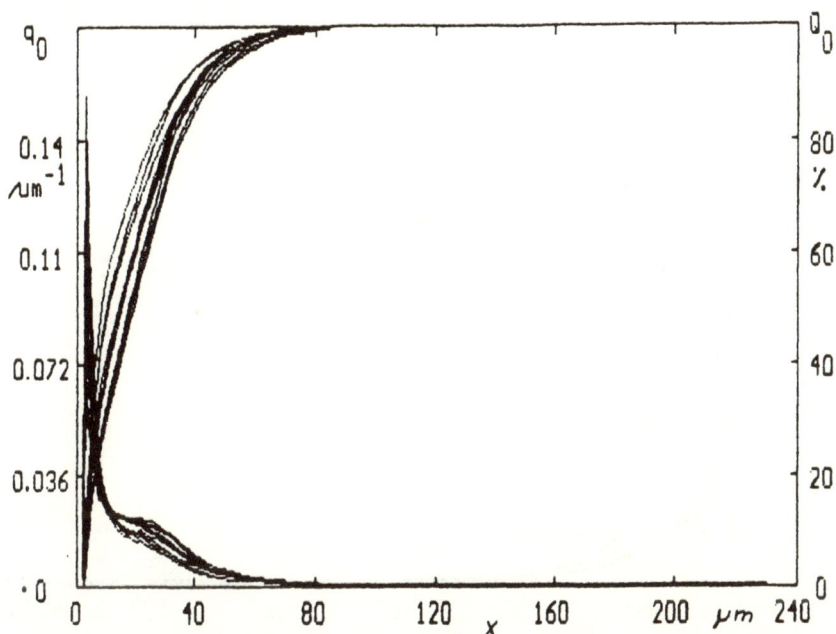

Figure 1   Number density and cumulative distribution functions of 10 samples of Aluminium oxide split by a rotaty riffler, analysed by optical single particle extinction method

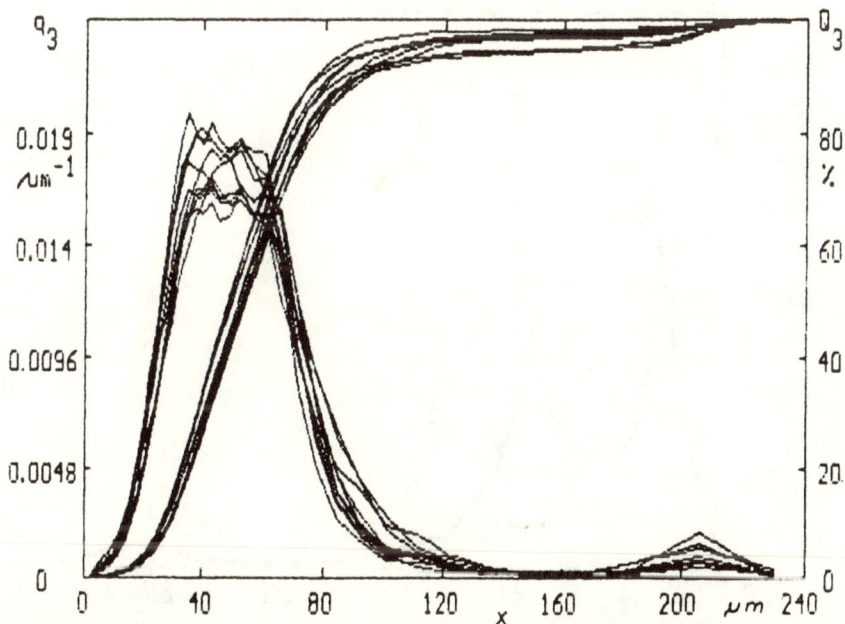

Figure 2   Volume density and cumulative distribution functions of 10 split samples

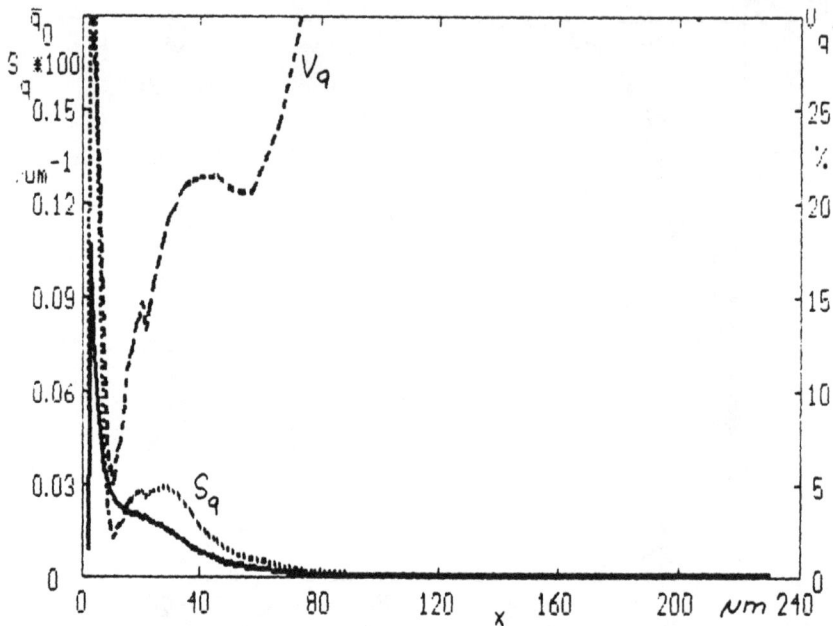

Figure 3   Mean values of the number density distribution function, standard deviation s_q and variation coefficient v_q according to figure 1

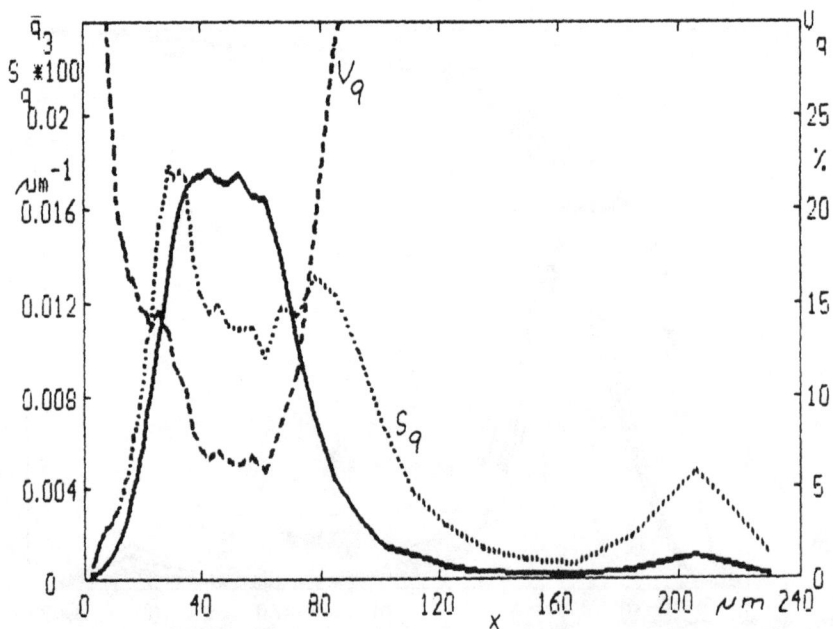

Figure 4   Mean values and statistical data of volume distribution functions according to figure 2

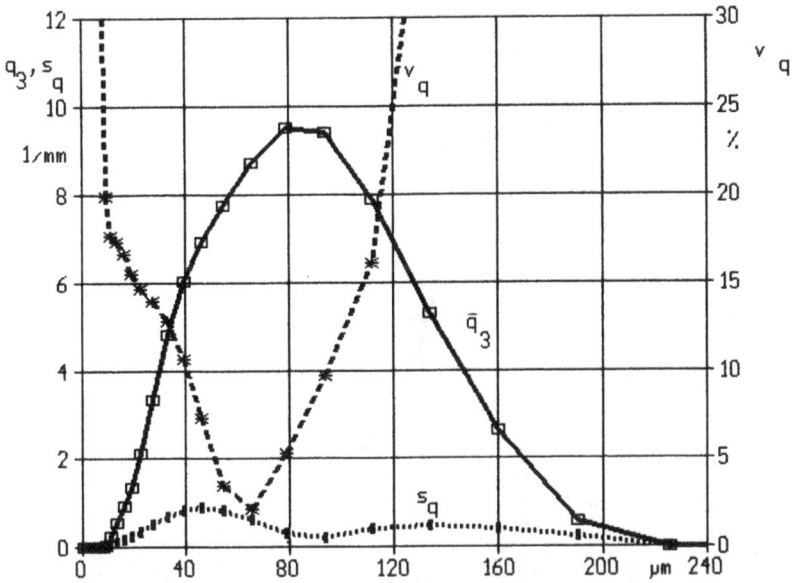

Figure 5 Laser diffraction measurements of the same 10 split samples, mean volume density distribution function q(x), standard deviation s_q and variation coefficient v_q

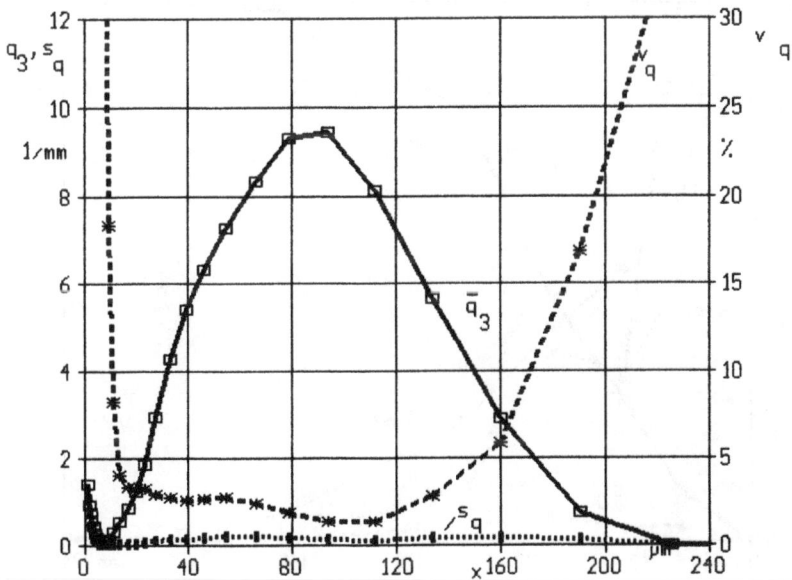

Figure 6 Mean values and statistical data of 10 repetitive measurements of only one sample by laser diffraction measurement

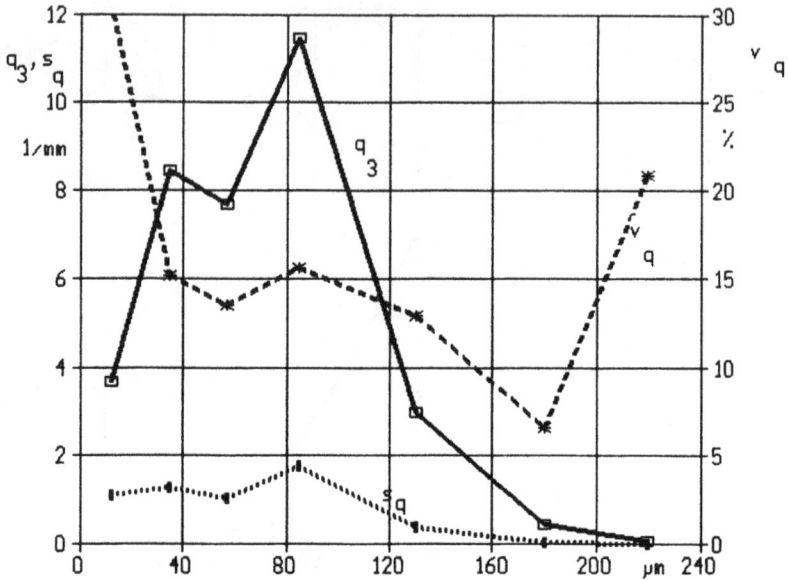

Figure 7 Mean volume density distribution function q(x)
of 10 split samples, measured by wet
sieving, standard deviation s_q and variation
coefficient v_q of the density function

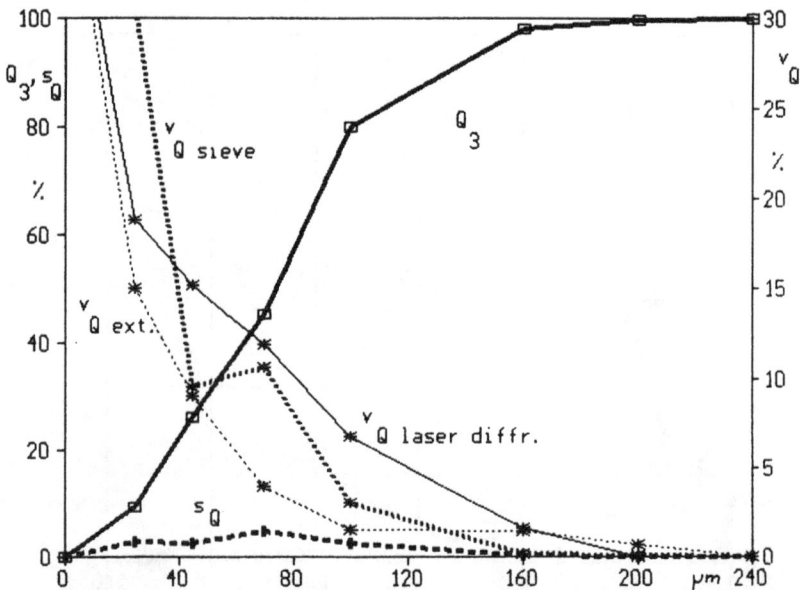

Figure 8 Mean cumulative distribution function Q(x) and
its statistical data of wet sieving according to
figure 7, added by variation coefficients of
laser diffraction and single particle extinction

be quantified by the difference between the variation coefficient for a certain particle size in figure 5 and the pure measuring variation coefficient in figure 6.

The volume density distribution of the ten split samples measured by wet sieving is evaluated in figure 7. Relative errors below 15% can be achieved in the range between 35 and 200 μm.
The high amount of analysed material of about 1...4 g delivers reliable results in coarse particle size ranges with a small relative amount of particles.

Figure 8 answers the question for the suitable kind of presentation of particle size analysis data for statistically evaluation.
The relative errors or variational coefficients of cumulative distribution functions increase from zero at the maximum particle size to high values of more than a hundred percent at the minimum size class.
These relations are inversed for the cumulative oversize distribution function (1-Q).

The standard deviation both of the cumulative distribution and of the density distribution decreases to values of zero for minimum and maximum particle size classes without respect to any reproducibility.

5 CONCLUSIONS

The calculation and presentation of variation coefficients of density distribution functions can be used as a powerfull tool to

-evaluate the quality of laboratory work,
-improve the methods of sampling, splitting, dispersing and feeding of samples,
-select the suitable measuring method to characterize a certain particle size range with a certain reliability,
-quantify the limits for interpretation of particle size analysis data with respect to process modelling or investigating property functions.

REFERENCES

1. K. Leschonski and K. Legenhausen, 'Comparison of Errors of Different Instruments for Particle Size Analysis and their Influence on the Characterization of Classification Process', 1. World Congress Particle Technology, Nürnberg, 1986, Preprints, Part I, 235-254.
2. T. Allen and R. Davies, 'Evaluation of Instruments of Particle Size Analysis', 4. European Symposium Particle Characterization, Nürnberg, 1989, Preprints, Part I, 17-46.

A Review of Sedimentation Methods of Particle Size Analysis

T. Allen

PARTICLE SCIENCE AND TECHNOLOGY CENTER (PARSAT), E. I. DU PONT DE NEMOURS AND
COMPANY, WILMINGTON, DE 19714–6090, USA

1 INTRODUCTION

The settling behaviour of particles in fluids is of wide
industrial relevance. In an ideal situation "tailored"
powders would be produced, eg powders having the right
surface characteristics for their designated end use,
powders which did not attrit during handling and powders
having a desired size distribution. In the real world
powders frequently have to be modified, often by
classifying them in order to obtain a desired size
distribution. The quality of a classifier is governed
by its ability to accurately separate desired from
undesired material and this is given quantitatively by
its grade efficiency curve.

Measured size distributions depend not only on the
physical dimensions of the particles but also on the
method of size analysis used. Size distributions by the
Coulter Principle will only agree with sedimentation
data if the particles are spherical. Indeed the
difference is a measure of particle shape. Since
classifiers separate particles on the basis of their
Stokes sizes a sedimentation method of size analysis
should be used to determine their grade efficiency.
Sedimentation analyses are also applicable to many other
industrial situations.

It has also been found that many end-use
properties, such as pigment gloss and hiding power, can
also be predicted from size distributions by
sedimentation techniques. A combination of
gravitational and centrifugal sedimentation covers a
wide size range and a wide range of measurement
techniques are possible making the technique very
popular.

Sedimentation techniques are based on the settling
behaviour of a single sphere, under gravity, in a fluid
of infinite extent. Many experiments have been carried

out to determine the relationship between settling velocity and particle size and a unique relationship has been found between drag factor and Reynolds Number. This relationship reduces to a simple equation, the Stokes equation, which applies at low Reynolds Numbers, relating settling velocity and particle size. Thus at low Reynolds Numbers the settling velocity defines an equivalent Stokes diameter which, for a homogeneous spherical particle, is its physical diameter.

Table 1 Principles of Sedimentation Techniques

Suspension Type	Measurement Principle	Force Field
Homogeneous	Incremental	Gravitational
Line Start	Cumulative	Centrifugal

Homogeneous, incremental gravitational sedimentation	Homogeneous, cumulative, gravitational sedimentation
Andreasen pipet	Oden Balance
Leschonski pipet	Svedberg and Rinde
Fixed depth pipet	automatic recording
Side-arm pipet	sedimentation beam
Wagner photosedimentometer	balance
EEL photosedimentometer	Cahn Balance
Bound Brook photosedimentometer	Gallenkamp balance
Seishin Photomicrosizer	Mettler H20E balance
Ladal Wide Angle Scanning Photosedimentometer	Sartorious Recording Sedibel balance
Paar Lumosed	Palik torsion balance
ICI X-Ray sedimentometer	Kiffer continuous weighing chain link balance
Ladal X-Ray sedimentometer	Rabatin and Gale spring balance
Micromeretics Sedigraphs 5000 & 5100	Shimadzu balance
Quantachrome Microscan X-Ray sedimentometer	ICI sedimentation column
Hydrometers	BCURA sedimentation column
Divers	Fisher Dotts apparatus
Suito specific gravity balance	Decanting
	β-Back-scattering

456

Particle Size Analysis

Line-Start, incremental, gravitational sedimentation MSA Analyzer	Line-Start, cumulative, gravitational sedimentation Werner and Travis method Granumeter Micromerograph MSA Analyzer

Homogeneous, incremental centrifugal sedimentation
 	
Simcar centrifuge
Ladal Pipet centrifuge
Ladal X-Ray centrifuge
 	
Du Pont/Brookhaven
    Scanning X-Ray
    Centrifuge
Kaye Disc Photocentrifuge
Coulter Photofuge
Technord photocentrifuge
Horiba cuvet
    photocentrifuges
    (CAPA 300, 500, 700)
Seishin cuvet
    photocentrifuge
Shimadzu cuvet
    photocentrifuge
	Homogeneous, cumulative, centrifugal sedimentation
	
	Alpine centrifuge
	Hosokawa Mikropul
	    Sedimentputer

	Line-Start, incremental, centrifugal sedimentation
	
	Joyce-Loebl disc
	    photocentrifuge
	Brookhaven disc
	    photocentrifuge

	Line-Start, cumulative, centrifugal sedimentation MSA Analyzer

Sedimentation techniques can be classified according to the principles outlined in Table 1. Some of these techniques will be discussed here; a fuller discussion can be found in my book[1].

In the homogeneous, incremental, gravitational technique the solids concentration (or suspension density) is monitored at a known depth below the surface for an initially homogeneous suspension settling under gravity. The concentration will remain constant until the largest particle present in the suspension has fallen from the surface to the measurement zone (Figure 1). At the measurement zone the system will be in a state of dynamic equilibrium since, as particles leave the zone, similar particles will enter it from above to replace them. The concentration will then fall since there will be no particles larger than Stokes diameter present above the measurement zone. Thus the concentration will be of particles smaller than the Stokes diameter and a plot of concentration against Stokes diameter is, in essence, the size distribution.

In the homogeneous, cumulative, gravitational technique the rate at which solids settle out of suspension is determined for an initially homogeneous suspension settling under gravity (Figure 2). This technique is typified by the sedimentation balance in which the balance pan can be in the suspension (Figure 3) or in a clear liquid (Figure 4). With the former set-up correction has to be made for the particles which do not fall on the pan; errors are also introduced since the particle free zone below the pan leads to convection currents. The latter technique also suffers from problems due to the motion of the pan as particles settle on it.

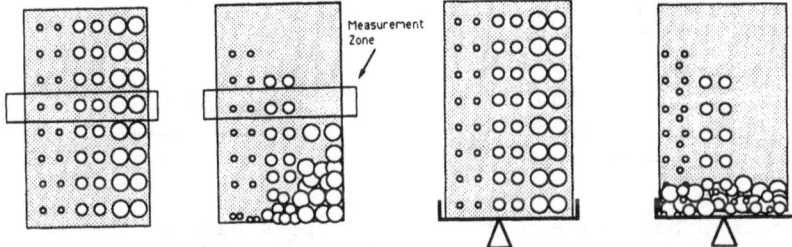

Figure 1 Homogeneous, incremental, gravitational sedimentation

Figure 2 Homogeneous, cumulative, gravitational sedimentation

With the incremental, gravitational, line start technique (Figure 5) the suspension is floated on top of a container of clear liquid and, provided the particles fall independently, the largest particles present in the suspension will reach the measurement zone first and the measured concentration will be of the size band in the measurement zone. This technique can also be used in the cumulative mode (Figure 6)

In homogeneous, incremental, centrifugal techniques matters are more complex since the particles move in radial paths hence the measured concentration is less than the original concentration (Figure 7). This problem does not occur with centrifugal line-start method at a fixed measurement radius.

In the paper presented here I intend to describe some of the methods for sedimentation particle size analysis in current use. Although I do not intend to cover operating procedures it would be remiss of me not to mention the two factors which, more than anything else, lead to incorrect analyses. The first is incorrect sampling; analyses are carried out on from a tenth of a gram up to a few grams and these samples must be representative of the bulk for the analyses to be meaningful. The second is dispersion: It has been rightly said that the most important factor in obtaining accurate sedimentation data is dispersion – the second most important factor is dispersion and the third is also dispersion!

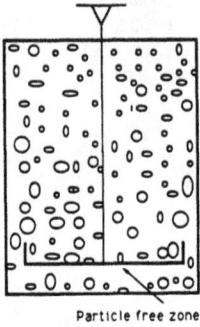

Figure 3 Balance Pan in Suspension

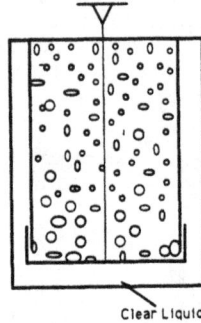

Figure 4 Balance Pan in Clear Liquid

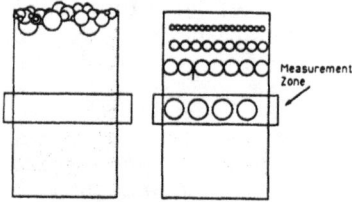

Figure 5 Line Start, Incremental Gravitational Sedimentation

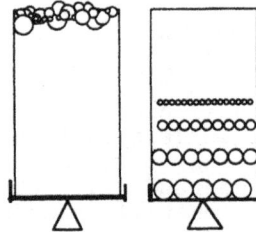

Figure 6 Line Start, Cumulative Gravitational Sedimentation

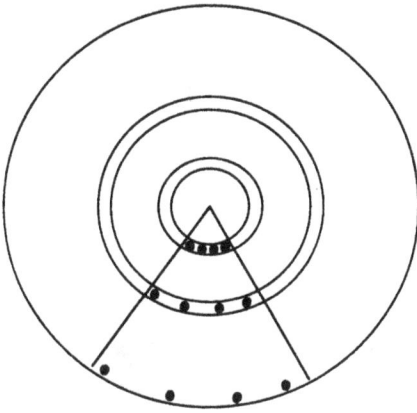

Figure 7   Homogeneous,
Incremental, Centrifugal
Sedimentation.
(The Radial Dilution
Effect)

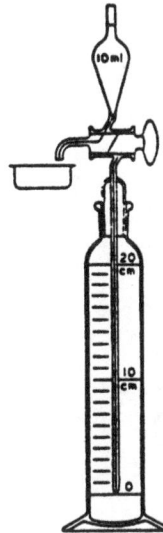

Figure 8   The Andreasen
Pipet

2   HOMOGENEOUS INCREMENTAL GRAVITATIONAL SEDIMENTATION

The Pipet Method

In the pipet method (Figure 8), concentration changes occurring within a settling suspension are followed by drawing off definite volumes, at pre-determined times and known depths, by means of a pipet. The method was first described in 1922 by Robinson[2] who used a normal laboratory pipet.   Various modifications were later suggested which complicated the operating procedure or the apparatus[3].   Andreasen was the first to leave the pipet in the sedimentation vessel for the duration of the analysis.   The apparatus described by Andreasen and Lundberg[4] is the one in general use today.

Although, theoretically, errors can be reduced by the use of more complicated construction and operation it is highly debatable as to whether this is worthwhile for routine analyses since conventional apparatus is reproducible to ± 2% if operated with care[5].

This technique is a standard procedure since both the Stokes diameter and the mass undersize are determined from first principles. The method is versatile, since it can handle any powder which can be dispersed in a liquid, and the apparatus is inexpensive. The analysis is however time consuming and operator intensive.

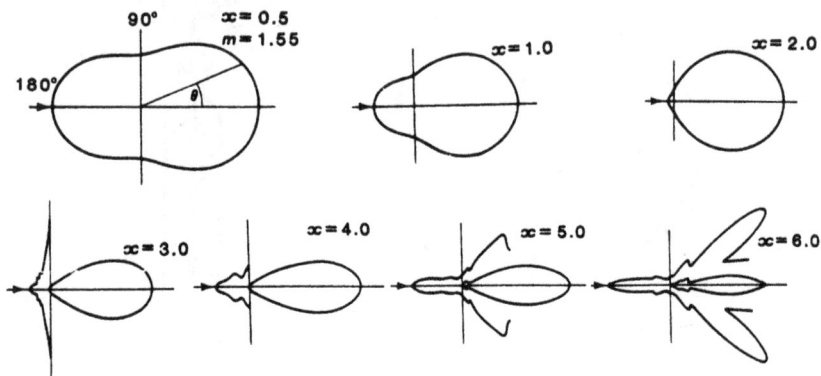

Figure 9   Polar light scattering diagrams[6]. The outer curve magnifies the inner by a factor of 10 in order to show fine detail. [$x = \dfrac{\pi D}{\lambda}$] where D = particle diameter and λ the wavelength of light.

The Photosedimentation Technique

The photosedimentometer combines gravitational settling with photo-electric measurement. The principle of the technique is that a narrow horizontal beam of parallel light is projected through the suspension at a known depth on to a photocell. Assuming an initially homogeneous suspension, the attenuation at any time will be related to the undersize concentration.

Superficially, the attenuation is related to the random projected areas of the particles. The relationship is more complex than this however due to the breakdown in the laws of geometric optics. For small particles an amount of light flux, equal in magnitude to that incident upon the particle, is bent away from the forward direction (Figure 9). As the particle size decreases the diffracted light is contained in an increasing solid angle so, no matter how small the light detector, with large particles most of the diffracted light is accepted; this amount reduces until the particle size is about the same as the wavelength of light when the diffracted light is mainly back-scattered. For partially transparent particles some of the incident light is absorbed and some refracted to cause interference in the transmitted beam.

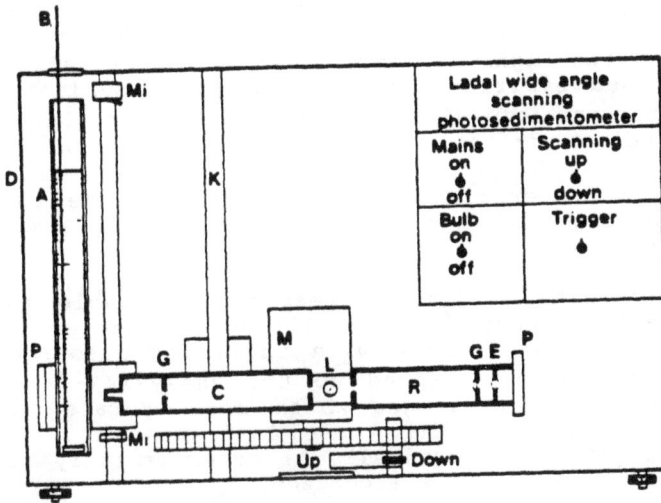

Figure 10 The Ladal Wide Angle Scanning Photosedimentometer
[A, sedimentation tank; B, stirrer; C, collimator; D, light proof box; E, variable aperture; G, lenses; K, drive screw; L, light source; M, motor; Mi, microswitches; P, photocells].

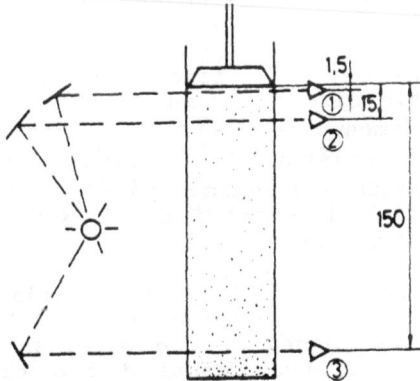

Figure 11 The Paar Lumosed [depth of beams are marked in mm]

It cannot be assumed that each particle obstructs the light with its geometric cross-sectional area since complex diffraction, scattering, interference and absorption effects have to be considered. These effects are compensated for by inclusion of an extinction coefficient (K) in the equation making the apparent area K times the geometric area.

Early experimenters[7,8] were either unaware of or neglected this correction. Some research workers used monochromatic light and determined K theoretically[6,9] others used empirical calibration by comparison with some other particle sizing technique. Rose and Lloyd[10] attempted to define a Universal Calibration Curve; Allen[11,12] (Figure 10) designed a Wide Angle Scanning Photosedimentometer (WASP) which accepted the forward scattered light so that K was constant down to a size of around 3 μm.

Commercial equipment presently available ranges from the PAAR Lumosed (Figure 11) which operates in the gravitational size range with three light sources at different depths to speed up the analysis[13] and a range of photocentrifuges which also operate in the centrifugal mode. With these instruments a K factor obtained either theoretically or experimentally can be inserted in the software algorithm.

X-Ray Sedimentation

A natural extension to the use of white light is to use X-rays in which case the X-ray attenuation is directly proportional to the atomic mass, ie the mass undersize for a homogeneous suspension, of material in the beam.

Brown and Skrebowski[14] first suggested the use of X-rays for particle size analysis and this resulted in the ICI X-ray Sedimentometer[15,16]. Other instruments were developed by Kalshoven[17] (1966) and Oliver et al[18]. In 1970 Allen and Svarovsky[19-21] developed an instrument in which the traditional X-ray tube was replaced by an isotope source.

Several commercial instruments utilizing these principles were developed. The Micromeretic's Sedigraph 5000 (Figure 12) was based on the paper by Kalshoven. The Quantachrome Microscan reduced the time for an analysis by a factor of about two and the Sedigraph 5100 was designed as a faster version of the 5000. Allen and Svarovsky's design was incorporated in the LADAL X-ray Sedimentometer (Figure 13).

Hydrometers and Divers

The changes in density of a sedimenting suspension may be followed with a hydrometer (Figure 14), a method still used in the ceramic industry. The method is open to several objections not least being the high concentration required in order to obtain accurate readings.

Resolution is defined as:

$$\frac{\Delta D}{D} = \frac{\Delta h}{2h}$$

where Δh is the depth of the measurement zone with a mid-point at a depth h below the surface and ΔD is the difference in size between a particle leaving the measurement zone and a particle entering it from above; D is the Stokes diameter for a particle at depth h.

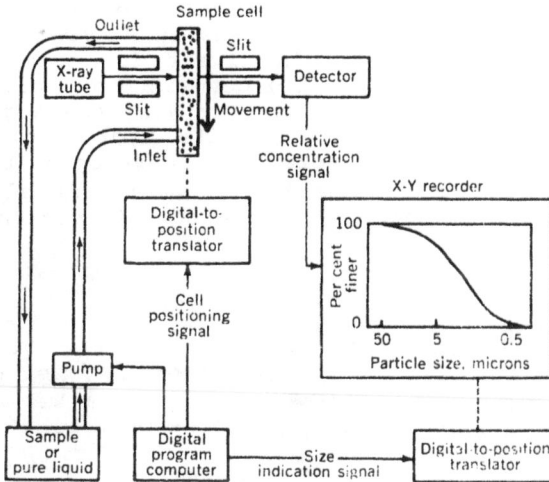

Figure 12   The Micromeretics Sedigraph

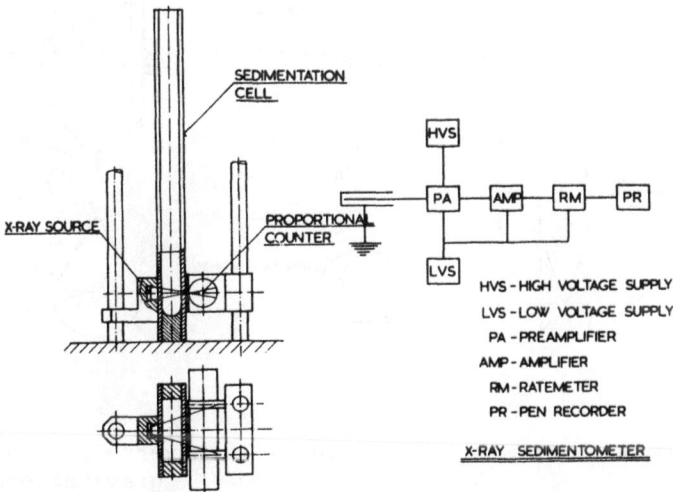

Figure 13   The Ladal X-Ray gravitational sedimentometer

The resolution is particularly poor for the hydrometer method of size analysis where Δh is of the same magnitude as h. Despite these objections the instrument is useful for control purposes with wide size range continuous distributions.

Divers (Figure 15), overcome many of the objections associated with the hydrometer technique. These miniature hydrometers were developed by Berg[22] for use with both gravitational and centrifugal sedimentation but have never been widely used. Basically, divers are small objects of known density which are immersed in the suspension so that they find their density level. Berg's divers for example, were hollow glass containers which contained mercury to give the desired density. The density was then adjusted to the desired value by etching with hydrofluoric acid. Various modified divers were later developed, the final ones, by Kaye and James[23], being metal coated polythene spheres which were located with search coils.

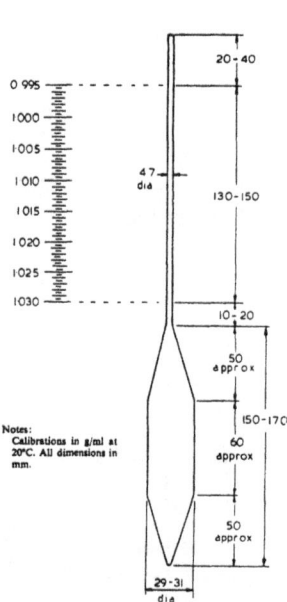

Figure 14 Hydrometer

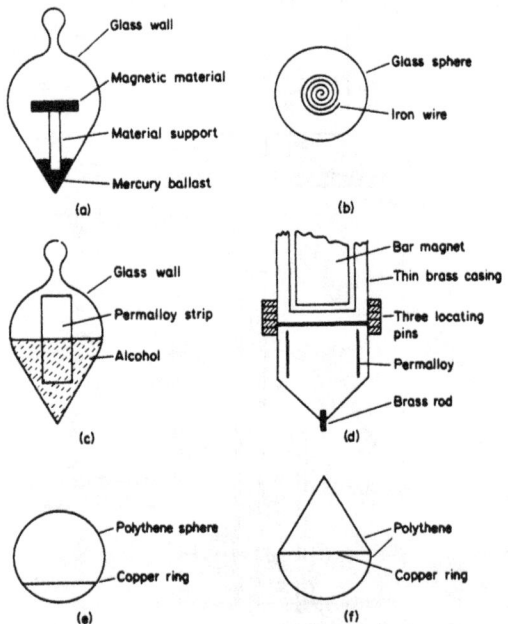

Figure 15 Divers [(a) and (b) Berg[22] gravitational and centrifugal (c) and (d) Jarret and Heywood[7] (e) and (f) Kaye and James[23]] (see Reference 1)

3 HOMOGENEOUS CUMULATIVE GRAVITATIONAL SEDIMENTATION

Introduction

The principle of this method is the determination of the rate at which particles settle out of a homogeneous suspension. This may be done by extracting the sediment and weighing it; allowing the sediment to fall on to a balance pan or determining the weight of powder still in suspension by using a manometer or pressure transducer.

One problem associated with this technique is that the sediment consists both of oversize (greater than Stokes diameter) and undersize particles so that the sedimentation curve of amount settled (*P*) against time (*t*) has to be differentiated to yield the weight (*W*) larger than Stokes diameter:

$$W = P - t\frac{dP}{dt}$$

Several balance systems, based on this equation, have been described.

Balances

In the Gallenkamp balance[24,25] the pan is placed below a sedimentation chamber with an open bottom and the whole assembly is placed in a second chamber filled with sedimentation liquid so that all the powder falls on to the pan. The weight settled is determined from the deflection of a torsion wire and either the run continues until all the powder has settled out of suspension or a second experiment is carried out to determine the supernatant fraction. Problems also arise during the charging operation with leakage into the clear water reservoir and particle adhesion to the premixing tube.

In the Sartorious balance[26-28] the pan is suspended in the suspending liquid and a correction has to be applied for the particles which fall between the rim of the pan and the sedimentation vessel. In this instrument, when 2 mg of sediment has deposited, electronic circuitry activates a step by step motor which twists a torsion wire to bring the beam back to its original position. A pen records each step on a chart. The manufacturers suggest that about 8% of the powder does not settle on the pan. Leschonski[29] and Leschonski and Alex[30] report losses of between 10% and 35% depending on the fineness of the powder and the difference is attributes to the pumping action of the pan as it rebalances. Leschonski modified the instrument by placing the pan at the bottom of a

sedimenting column surrounded by a second column of
clear liquid so that all the powder settled on to the
pan. This eliminated powder losses and resulted in more
accurate analyses[31]

The manufacturers of the Cahn micro-balance make
available   an   accessory   to   convert   it   into   a
sedimentation   balance[30,32].   The balance pan is
immediately below the sedimentation cylinder in order to
eliminate convection currents. Shimadzu also make a
beam balance[33] which operates using a simple
compensating system prone to considerable error.

Sedimentation Columns

Sedimentation columns (ICI, BCURA) have also been
described in which the sediment is extracted, dried and
weighed.   A   full   description   of   these   and   other
sedimentation columns may be found in[1].

4 LINE-START INCREMENTAL GRAVITATIONAL
   SEDIMENTATION

Photosedimentation

The Horiba cuvet photo(centri)fuge has been
operated in this mode but the method is subject to the
objection that the cuvet lies vertically under gravity
at the start of the analysis so that the layer is not
sitting on top of the fill liquid.

5 LINE-START CUMULATIVE GRAVITATIONAL
   SEDIMENTATION

Introduction

If the powder is initially concentrated in a thin
layer floating on the top of a suspending fluid, the
size distribution may be determined by plotting the
fractional weight settled against the free falling
diameter.

Methods

Marshall[34] was the first to use this principle.
Eadie and Payne[35] developed the Micromerograph, the
only method in which the suspending fluid is air.
Brezina[36,37] developed a similar water based system
the Granumeter which operated in the sieve size range,
and was intended as a replacement for sieve analyses.

The Werner and Travis methods[38,39] also operate
on the layer principle but their methods have found

little favor due to the basic instability of the system; a dense liquid on top of a less dense liquid being responsible for a phenomenon known as "streaming" in which the suspension settles *en masse* in the form of pockets of particles which fall rapidly through the clear liquid leaving a tail of particles behind.

Whitby[40] eliminated this fault by using a clear liquid with a density greater than that of the suspension. He also extended the size range covered by using centrifugal settling for the finer fraction. The apparatus enjoyed wide commercial success as the (Mines Safety Appliances) MSA Particle Size Analyzer although it is less widely used today[41]. The MSA analyzer can be operated in the gravitational mode although it is more usually used in the centrifugal mode.

The line-start technique has also been used to fractionate UO_3 particles by measuring the radioactivity at the bottom of a tube, the settled powder being washed out at regular intervals without disturbing the sediment[57].

6 HOMOGENEOUS, INCREMENTAL CENTRIFUGAL SEDIMENTATION

Introduction

Gravitational sedimentation techniques have limited worth for particles smaller than about a micron due to the long settling times required. In addition most sedimentation devices suffer from the effect of convection, diffusion and Brownian motion. These difficulties may be reduced by speeding up the settling process by centrifuging the suspension.

One of the complications that arises is that particle velocity is not only dependent on particle size as in gravity sedimentation but also depends upon the radial position of the particle. The radial velocity may be written as:

$$v = \{\frac{ln\{\frac{r}{S}\}}{t}\}$$

where t = time
          r = radial position of the particle
          S = radial position of the surface of the
                suspension.

In long arm centrifuges $(r-S)$ is made much smaller than r or S so that the velocity may be assumed constant.

Correction is also necessary for radial dilution effects. Particles, of narrow size range centered on x_r, originating from radius r and occupying an initial volume of $h_r \Delta_r$ will occupy a volume of $h_R \Delta_R$ at the measurement radius R. The relationship between the concentration at the starting radius (the initial concentration in the suspension) and the concentration at the measurement zone is given by:

$$\frac{F(x_r)}{C(x_r)} = \frac{R\Delta R}{r\Delta r} \text{ which can be shown to equate to } (\frac{R^2}{r^2})$$

and it can be shown that:

$$\int_0^{x_{max}} F(x)dx = \int_{r=R}^{r=S} \int_0^{x_{max}} (\frac{R}{r})^2 Q(x)dx$$

This is solvable for fixed R by an equation developed by Kamack[42,43], for a variable surface radius (S) by a modified equation developed by Allen and Svarovsky[44] and for variable R by an equation developed by Allen[45].

The Simcar Pipet Disc Centrifuge

The Simcar centrifuge was developed by Slater and Cohen[46] to conform with Kamack's theory and was, in essence, a centrifugal version of the gravitational pipet. One problem with this instrument was the amount of suspension required (about two and a half liters of liquid containing about 5g of powder) so that the liquid level did not alter appreciably during an analysis. The amount removed at each withdrawal was variable and about 40ml, hence the error due to assuming a constant liquid level (S) increased as more samples were withdrawn. Since no

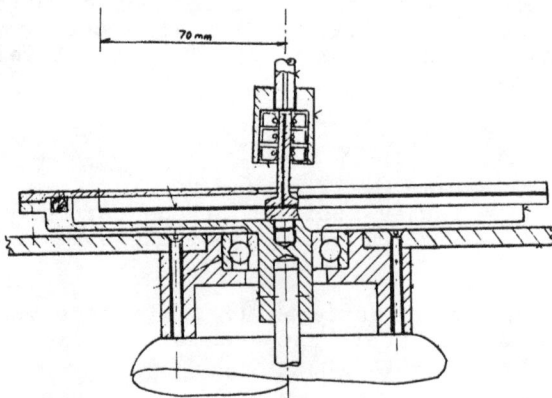

Figure 16   Line diagram of the Ladal Pipet Centrifuge

correction was available for the fall in surface with each extraction it was inadvisable to withdraw more than 4 samples in each analysis. The analysis was carried out in duplicate to give 8 points on the distribution curve, a single evaluation from about 5μm to 0.2μm taking up to a full day.

Concentrations were determined by drying and weighing. Using the modified equation to take into account the changing liquid level would allow more data points to be taken in a single run.

The Ladal Pipet Disc Centrifuge

This pipet (Figure 16) was designed by Allen and Svarovsky[44] to operate with a reduced volume of suspension (150 ml) and a modified Kamack equation to reduce the measurement time down to about an hour. At 500 rpm the size range for quartz is approximately 8 μm to 0.8 μm and these sizes are halved if the speed is doubled.

The Ladal X-Ray Disc Centrifuge

This is an extension of the Allen and Svarovsky[48] X-ray gravitational sedimentometer (Figure 17). The X-rays are generated by an isotope source and, after passing through the suspension, are detected by a scintillation counter. The signal is then processed to generate the size distribution.

The attenuation is proportional to the mass concentration at the measurement radius which has to be converted to the size distribution using the Kamack equation. A size range of about 8:1 is covered in about an hour.

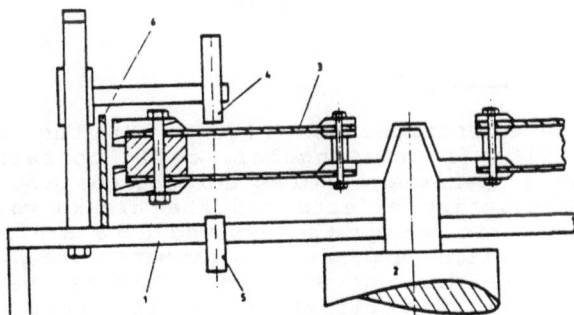

Figure 17   The Ladal X-Ray Centrifugal Sedimentometer [(1) Base plate; (2) Electric Motor; (3) Disc centrifuge bowl; (4) Isotope source; (5) Detector; (6) Safety ring.

The Du Pont Scanning X-ray Disc Sedimentometer

This instrument was designed by the writer[45,49] to fill a need for fast, reproducible sedimentation analyses in the sub-µm size range. The heart of the instrument is a hollow, X-ray transparent, disc which, under normal operating conditions, contains 25ml of of suspension at a concentration of around 0.2% by volume.

The speed is selectable in the range 750 to 6000 rpm. The default condition is for the source and detector to remain stationary for 1 minute at a radial position of 48.00 mm and then to scan towards the surface. Total run time is normally 8 minutes. A commercial version is available from Brookhaven as the BI-XDC (Figure 18).

The instrument can operate in the gravitational or centrifugal mode and the analyses can be blended to cover a total size range of 0.05 µm to over 100 µm. A size range of 15:1 is covered in a standard 8 minute analysis.

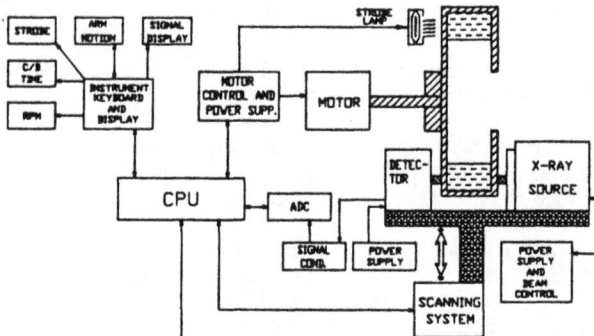

Figure 18   The Du Pont/Brookhaven X-Ray Scanning
                           Centrifuge

Cuvet Photocentrifuges

In these instruments (Figure 19) the disc is replaced with a rectangular cell containing a homogeneous suspension. Unless corrections are applied for radial dilution effects and the breakdown in the laws of geometric optics the derived data are suitable only for comparison purposes. Instruments are available from Horiba, Seishin and Shimadzu. They can be run in the gravitational, centrifugal or gradient mode. In the gradient mode the centrifuge accelerates over the analysis time to reduce the measurement time. The simpler instruments operate at constant speed and an analysis can take 45 minutes which can be reduced to a few minutes in the more sophisticated versions.

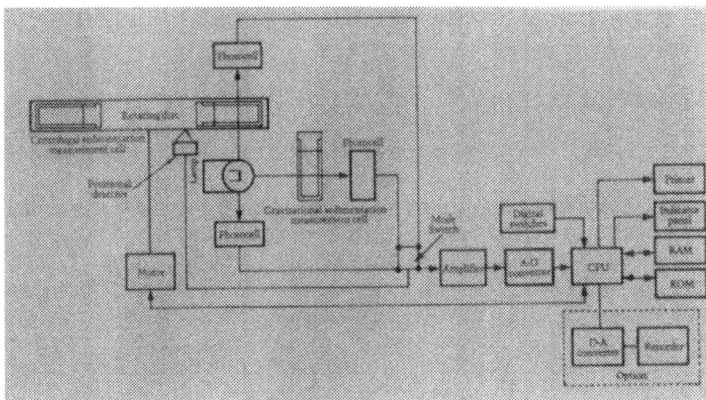

Figure 19   Block diagram of the Horiba cuvet
photocentrifuge

7   HOMOGENEOUS, CUMULATIVE CENTRIFUGAL
SEDIMENTATION

Methods

The Alpine sedimentation centrifuge is a long-arm
centrifuge of diameter 400 mm with a 50 mm high
measuring cell.   The rate at which sediment settles out
is determined by measuring pressure changes at the
bottom of the cell using a diaphragm arrangement.

The Hosokawa Mikropul Sedimentputer[50,51] has a
sealed suspension in a cell which is then rotated.   as
the particles settle out the center of gravity changes,
which creates an in balance which causes the cell to
vibrate (Figure 20).   By detecting the amplitude and
angular velocity of the vibration the size distribution
is obtained.

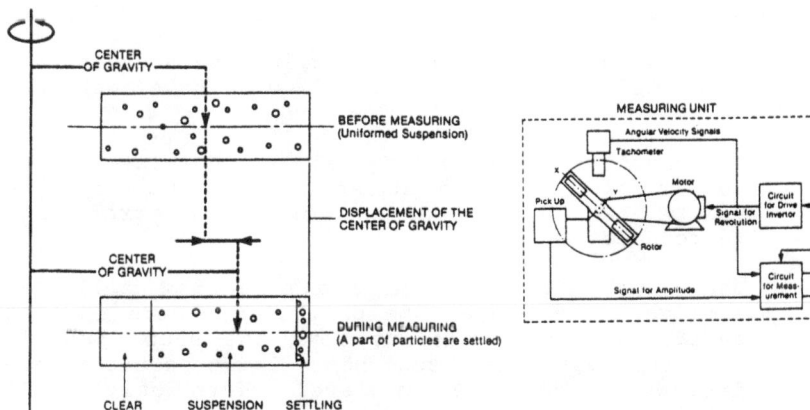

Figure 20   The MikroPul Sedimenputer {20.1 The settling
of particles causes the center of gravity to shift 20.2
Schematic of the system}

8 LINE-START, INCREMENTAL CENTRIFUGAL SEDIMENTATION

Disc Photocentrifuges.

The first disc photocentrifuge was developed by Kaye[52]. In this instrument concentration changes within a suspension are followed using a white light beam. The instrument is usually used in the line start mode and the data evaluated as a weight distribution using theory developed by Treasure[53]. The instrument can also be used with a homogeneous suspension. The technique is available in the Joyce Loebl and the Brookhaven Disc Photocentrifugal Photosedimentometer (Figure 21).

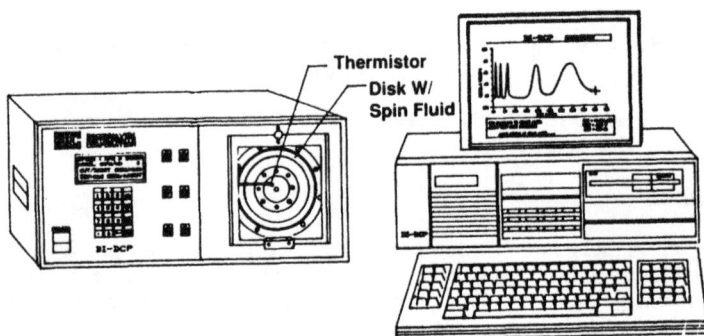

Figure 21   The Brookhaven disc photocentrifuge

It must be stressed that the raw curves are not size distributions and calibration is required to convert to absolute values[54].

9 LINE-START, CUMULATIVE CENTRIFUGAL SEDIMENTATION

MSA Analyzer

The MSA analyzer described earlier operates in this mode. Objections that can be levelled at this technique are:

- The amount of settled material is determined by the height of the sediment and, since the settled volume is not independent of size, errors are introduced.

- The lower part of the sedimentation cell has sloping walls hence some particles adhere to this section and others slide down the sloping walls into the measurement zone hence large particles are frequently found at a level where only small particles should be present.

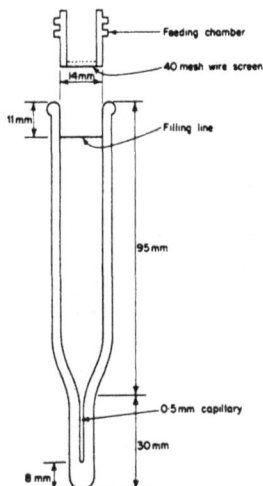

Figure 22 MSA special centrifuge tube

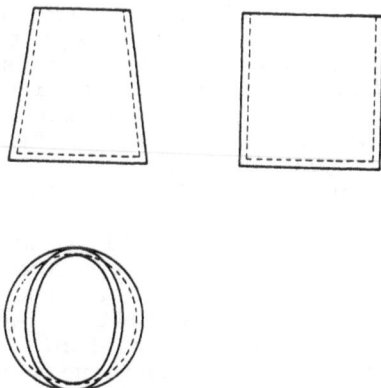

Figure 23 Berg's conoidal centrifuge tubes

One would expect that the large particles would carry some small particles with them so that the sediment would not be stratified into size bands but would contain small particles mixed in with the large ones.

Zwicker[55] found the method highly unsatisfactory and recommended that it should no longer be used.

The problems associated with the use of cylindrical tubes ie particles striking the walls of the tube and sticking or agglomerating and reaching the bottom more rapidly than freely sedimenting particles has been recognized for many years. Berg's solution[22] was to construct sector-or conoidal shaped tubes (Figure 23).

10   CONCLUSIONS

During the past 30 years I have specialized in the field
of powder characterization and in this talk I have
highlighted contributions, made by my colleagues and I,
in sedimentation techniques. As an academic I favored
the gravity pipet since the capital costs were low and
labor was cheap. The Ladal pipet centrifuge grew out
of a need for a Simcar type instrument which needed less
than 2 gram of powder. The Ladal Photosedimentometer
originated in the work done for a Masters Degree under
the supervision of Brian Kaye. The Ladal X-Ray systems
were natural extensions of these apparatuses.   In my
present position in industry, speed and running costs
are more important than capital costs, and the Scanning
X-Ray Centrifuge was developed with these needs in mind.
The relatively high resolution and improved accuracy
were an additional bonus.

The pressures in industry are to introduce on-line
size analysis systems, preferably methods that can
operate without pre-dilution since this may modify the
suspension.   The drive is always to finer and finer
powders hence sub-μm, on-line size analysis of
concentrated suspensions would be ideal. Failing this,
rapid off-line analyses of concentrated suspensions
would be acceptable. Accuracy is not always important;
detecting changes which may affect powder handling or
final product may be all that is necessary.

Despite this, there will always be a demand for
laboratory instruments that are more than mere
difference meters and this is the niche into which
sedimentation techniques fall. The large number of
techniques highlighted in Table 1 indicate past
importance and this will continue with the introduction
of faster, more accurate methods.

REFERENCES

1.   T. Allen, Particle Size Measurement, 1990, Chapman
      & Hall,   Fourth Ed.
2.   G. W. Robinson, J. Agr. Sci., 1922, 12 (3), 306-321.
3.   A.H.M. Andreasen, Kolloid Beith., 1928, 27, 405
4.   A.H.M. Andreasen, and J.J.V. Lundberg, Ber. Dt.
      Keram.   Ges., 1930, 11(5) 312-323
5.   T. Allen, Powder Technology, 1969, 2(3) 132-140
6.   V. Vouk, Ph D Thesis, 1948, London University
7.   B.A. Jarrett, and H.Heywood, Br. J. Appl.
      Phys., Suppl. No   3, 1954
8.   V.T. Morgan, Symp. Powder Metallurgy., 1954, Iron
      and Steel Institute, preprint, Group 1, 38-43
9.   P. C. Lewis, and G.F. Lothian, Br. J. Appl. Phys.,
      1954,   Suppl. No 3 , S571

10.   H.E. Rose, and H.B. Lloyd, <u>J.Soc. Chem. Ind</u>. 1946,
      <u>65</u>, 52
11.   T. Allen, Powder Technology, 1969, 2(3) 141-153
12.   T. Allen, Proc. <u>Particle Size Analysis Conference</u>,
      1970,   Bradford, Pub Soc. Anal. Chem. 1971
13.   G. Staudinger, M. Hangl, and P. Peschtl,<u>Proc.
      Partec</u>, 1986,   Nurenberg
14.   J.F. Brown, and J.N. Skrebowski, <u>Br. J. Appl.Phys</u>.
      1954,   Suppl. No 3 S27.
15.   S. G. Conlin, et al <u>J. Sc. Instrum</u>.,1967, <u>44</u>, 606-
      6
16.   G.Nonhebel, ed <u>Gas Purification Processes</u>, 1964,
      Newnes.
17.   J. Kalshoven, <u>Proc. Conf. Proc. Particle Size
      Analysis</u>,   1967, Soc. Anal. Chem., London
18.   J.P. Oliver, G.K.Hicken, and C. Orr, <u>US Patent,</u>
      1969,   3,449 567
19.   T. Allen, (1970) <u>Br. Pat</u>. 1764/70 3
20.   T. Allen, and L. Svarovsky, <u>J. Phys. E</u>, 1970, <u>3</u>,
      458- 460
21.   T. Allen, and L. Svarovsky,Proc. <u>Particle Size
      Analysis Conference</u>, 1970, Bradford, Pub Soc.
      Anal. Chem. (1971)
22.   S. Berg, <u>Ingen Vidensk</u>., 1940, Skr. B., no
      2.Phys.,Suppl.   No 3, S27
23.   B.H. Kaye and G.W. James, <u>Br. J. Appl. Phys</u>., 1962,
      <u>13</u>,   415
24.   W. Bostock, <u>J. Scient. Instrum.</u>, 1952, <u>29</u>, 209
25.   Cohen, L. <u>Instrum. Pract</u>., 1959 <u>13</u> 1036
26.   D, Bachman and H. Gerstenberg, <u>Chem. Ing. Tech</u>.,
      1957,   <u>8</u>.589
27.   D. Bachnan <u>Dechema Monograph</u>, 1959, <u>31</u>, 23-51
28.   H. Gerstenberg, <u>Dechema Monograph</u>, 1959, <u>31</u>, 52-60
29.   K. Leschonski, <u>Staub</u>, 1962, <u>22</u>, 475-486
30.   K. Leschonski, and W. Alex, <u>Proc. Int. Symp.
      Particle Size Analysis</u>, Bradford,1972, Soc. Anal.
      Chem.
31.   S.T. Pretorius, and W.G.B. Mandersloot, <u>Powder
      Technol</u>,   1967, <u>1</u>, 23-27
32.   B.H. Kaye, and R. Davies, <u>Proc. Conf. Particle Size
      Analysis</u>, ed. Groves & Wyatt-Sargent, Soc.
      Anal. Chem.,1972   207-222,
33.   E. Suito and M. Arakawa, <u>Bull. J. Chem. Res</u>.,1950,
      Kyota   University, <u>23</u>, 7
34.   C.E. Marshall, <u>Proc. R. Soc</u>., 1930 <u>A126</u>, 427
35.   F.A. Eadie, and R.E., Payne, <u>Iron Age</u>,1954 <u>174</u> 99
36.   J. Brezina, J. Sediment, Petrol.,1969, 1627-31
37.   J. Brezina, <u>Proc. Conf. Particle Size Analysis</u>
      1970, Publ.   Soc. Anal. Chem
38.   D. Werner,<u>Trans. Farad. Soc</u>., 1925 <u>21</u>, 381
39.   P.M. Travis, <u>ASTM Bull</u>.,1940,   29, 102
40.   K. T. Whitby, <u>Heat Pip. Air Condit</u>.,1955 Jan. Part
      1, 231;   June Part 2, 139

41.   K. T. Whitby, A.B.Algren and J.C.Annis, 1958, ASTM
      Spec. No 234, 117
42.   Kamack, H.J.,Anal. Chem., 1951, 23, 6, 844-850
43.   Kamack, H.J., Br.J. Appl. Phys.,1972, 5, 1962-1968
44.   T. Allen, and L. Svarovsky, Dechema Monogram, 1976
            Nurenberg, 1975, Nos 1589-1615, 215-221
45.   T. Allen, Proc. Int. Symp. Particle Size Analysis,
      1991   to be publ.Anal. Div. Royal Soc. Chem.
46.   C. Slater, and L. Cohen, J. Scient. Instrum.,1962,
      39,   614
48.   T. Allen, and L. Svarovsky, Proc. Soc. Anal Chem.,
      1972,
      9, 2, 38-40
49.   T,. Allen and L. Svarovsky, J. Powder
      Technology,1974,
      10, 1/2, 23-28
50.   N. Kaya, T. Yokoyama, M. Arakawa, and N. Kona No 4
            1986 Japan
51.   N. Kaya, T. Yokoyama, M. Arakawa, and N. Yazawa,
      Proc.2   Partec,1986, Nurenberg, April
52.   B.H. Kaye, British Patent 895 222, 1962
53.   C.R.G. Treasure, Tech. Paper No 50, 1964, Whiting
      and Industrial Powders Research Council
54.   T. Allen, Powder Technology.,1988 50, 3, 193-200
      Loughborough, Anal. Div. Royal Soc. Chem.
55.   J.D. Zwicker, Powder Technol.,1972 6 133-138
56.   P. Imris and H. Landsperky, Silikaty, 1956, 9, 4,
      327

Factors Causing Apparent Size Distributions in Sedimentation Analysis

C. Bernhardt

RESEARCH INSTITUTE OF MINERAL PROCESSING, FREIBERG, SAXONY, GERMANY

1 INTRODUCTION

Numerous disturbances of the sedimentation motion of particles are frequently encountered in incremental sedimentation analysis. Some of these disturbances lead to the
fact that the true distribution is not measured, but only
an apparent one. This finds expression in the fact that
a particle system presumed to be monodisperse is measured
as an apparently polydisperse system. Factors contributing
to this effect are:
1. Factors caused by the particles
- Brownian motion of the particles
- deviation from the spherical shape
- heterogeneity of the solid
2. Factors caused by the device:
- horizontal and vertical limitations of the vessel
- extension of the measuring gap
We are now interested in the question, what apparent distributions are caused by these effects and finally also
measured, if the starting distribution is polydisperse already from the beginning. This problem can be solved by
the strict application of the "concept of the apparent
distribution", allowing to pass always from the monodisperse case on the polydisperse one.
What is the principle of this concept? For this purpose
we regard Fig. 1.
The upper part of the figure shows the principle of the
incremental sedimentation analysis: The disperse solid
shall be homogenously distributed in a sedimentation vessel at the time t = 0. A sensor is situated at the depth
h_M, which measures the solid concentration there as a function of time. In the ideal case, the highest particle layer
of a monodisperse system is just reaching the measuring
plan at the time $t = t_M$. This leads to the concentration
jump at this point, shown below. Consequently, the resulting cumulative distribution, represents monodispersity.
In reality, however, after the time $t = t_M$ has passed,
there is a more or less diffuse border of particles near
the plane or it looks like it. In the two cases, the
measured course of concentration is as shown in the figure.

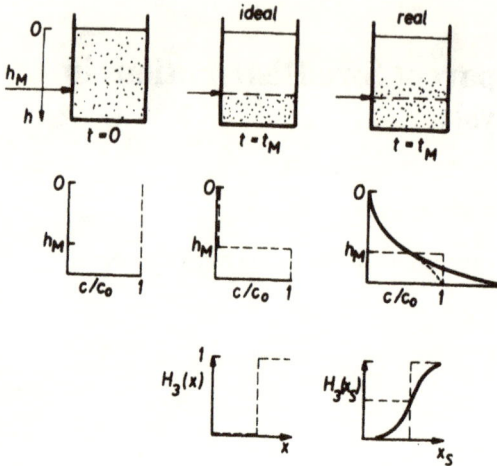

Fig. 1:
Ideal and real
sedimentation

The cumulative distribution calculated from it shows an
apparent polydisperse distribution, which does not exist
at all in reality.

2   DIFFUSION BY BROWNIAN MOTION

As known fluid molecules perform a thermal motion, whose
kinetic energy is transfered to the suspended particles.
This leads to non-directional motion of the particles themselves. For the sedimentation analysis only the influence
of the diffusion mechanism in the direction of sedimentation is relevant, since in a multiparticle system with ho-
mogenous concentration distribution over the cross section
of the vessel merely the concentration changes along the
sedimentation height have an effect on the measuring result. Mathematically the problem is treated by the solution
of the Fick's differential equation

$$\frac{\partial c}{\partial t} = D \frac{\partial^2 c}{\partial h^2} - U \frac{\partial c}{\partial h} \tag{1}$$

where c = particle concentration; D = coefficient of
       Diffusion; U = sedimentation velocity; h = sedi-
       mentation height†

The solution of the Fick's equation for a sedimentation
vessel with a suspension level and with a bottom has been
found by Meason et al. /1/ and later by Fürth /2/ nearly
70 years ago. Their equations are complicated and very
difficult to treat, and therefore frequently you find
in the literature /3, 4, 5/ the more simple solution for
an infinite long vessel.
Fig.2 illustrates the occurring differences by an example.

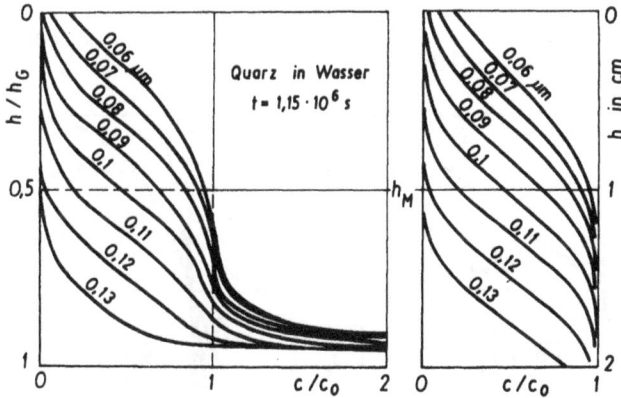

Fig. 2
Concentration distribution in the sedimentation vessel; at the left: exact values, at the right: approximation

There the relation is shown between the relative solid concentration and the sedimentation path for quartz particles in water. The measuring plane is in the middle of the vessel. After the sedimentation time for a particle of 0.1 µm size has passed, concentration distributions are found, which only near the measuring plane and also near the suspension level are described with sufficient accuracy by the approximation equation. As shown in /6/, the latter, however, is only applicable, if the particles are greater than 0.1 µm. If only the mechanisms in the middle of a sufficiently big sedimentation vessel are considered, then the   theory gives the following expression for the measured apparent distributions of monodisperse particles:

$$H_3(x_s) = \frac{1}{2} - \phi_0(z) \; ; \quad z = \frac{x^2 - x_s^2}{x} \cdot \left[x \, h_M \, \frac{\pi}{12} \, \frac{\Delta \rho \, g}{kT} \right]^{1/2} \quad (2)$$

with ϕ_0 :Gaussian error function integral; x:particle size; x_s :apparent particle size; h_m :level of the measuring plane; $\Delta\rho$:density difference; g: acceleration; k:Boltzmann constant; T: temperature

The greater the argument z is, the smaller are the effects of the diffusion phenomena on the measuring result. This shall be illustrated by two examples (see Fig. 3).

The left graph shows the enormous importance of the sedimentation path: measurements at low h_M - despite of lower measuring time - lead to considerably broader apparent distributions than at higher h_M. This is a factor restricting

the use of modern devices, by which narrow distributions
are to be measured. The right graph reflects the effect
of the particle size on the measuring result.

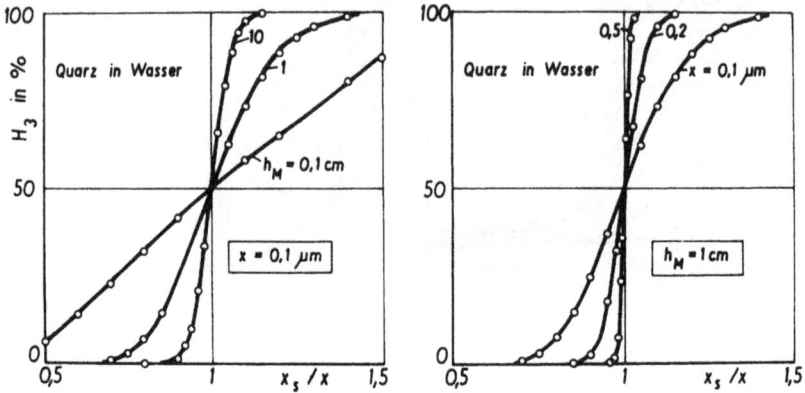

Fig. 3: Apparent distributions of monodisperse quartz
        particles in water

For polydisperse solids with the true distribution density $h_3(x)$ the measured (apparent) density is

$$H_3(x_s) = \int_0^{x_{max}} \left[\tfrac{1}{2} - \phi_o(z) \right] h_3(x)\, dx \qquad (3)$$

In order to estimate the errors existing between the true
and the measured distribution, it has been repeatedly proposed to use one of the known distribution functions for
the distribution density and to integrate equ. (3) with it
/3, 4, 5/. Several of the results announced in /4/ and /5/
are presented in Fig. 4

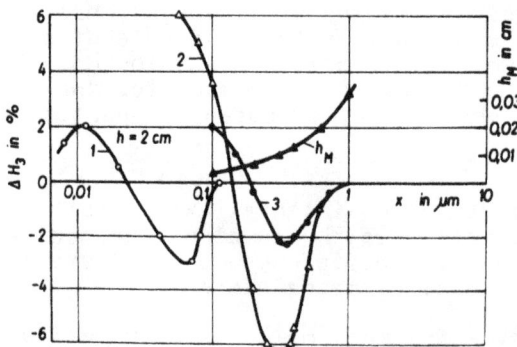

Fig. 4:
Measuring errors
caused by diffusion;
curve 1 according to
/4/, curve 2 and 3
according to /5/ with
simultaneous decrease
of the height h_M (curve
h_M)

For the true distribution $H_3(x)$ logarithmic normal distributions were used and the differences $\Delta H_3 = H_3(x_s) - H_3(x)$

were calculated. It becomes evident that the measuring errors on the whole remain small. Obviously, the reason for this is that there is a certain symmetry of the concentration distribution of the individual particle sizes; and that, therefore, by the integration effects on the measuring result can only be remarked in so far as this symmetry is incomplete.

From this view, measurements in the gravitational field up to 0.1 µm and in the centrifugal field up to 0.01 µm are the limits, up to which measurements by incremental sedimentation analysis can be performed with an admissible error.

3 DEVIATION FROM THE SPHERICAL SHAPE

The application of the sphere equivalence principle in sedimentation is always failing, when the products of mechanical processes leading to changed particle shapes have to be characterized by sedimentation analysis. Then it becomes necessary to include the particle shape in the consideration.

Particles of a uniform shape deviating from the sphere, settle at different velocity depending on their angle position. In the left part of Fig. 5 such a particle system is schematically represented in its initial state (t=0). After the time $t = t_1$, the situation is similar to the right part of the sketch: the particles settling in longitudinal direction passed a longer way than the particles of $\alpha < 90°$.

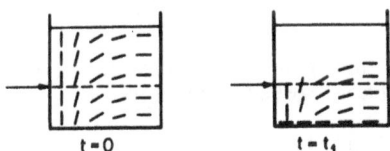

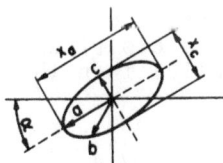

$$p = \frac{c}{a} = \frac{x_c}{x_a}$$

Fig. 5:
Sedimentation of non-spherical particles

In this way, in the measuring plane indicated by an arrow, a concentration change with time takes place, which is reflected in the measuring result as an apparent distribution. In order to be able to describe this apparent distribution, a geometric body must be found, by which the occurring particle shapes can sufficiently accurately be approximated and the equations of motion of which are known. Such a body is the ellipsoid with the semi-axes a, b, c and in particular the special case of the spheroid (a=b >c for plates and b=c< a for rods), whose equations of motion have been known for 80 years /7/. Using this theory, one obtains for the apparent distribution of monodisperse particles

$$H_3 \ (x_s/x_{s,50}) = \frac{1}{\pi} \ arc \ cos \ Q_{P/S}; \ mit \ Q_{P/S} = \frac{2 \ \xi_s^2 - q_c - q_a}{q_c - q_a} (4)$$

$$\xi_s^2 = \frac{x_s^2}{x_a \cdot x_c} \quad for \ plates; \qquad \xi_s^2 = \frac{x_s^2}{x_c^2} \quad for \ rods$$

The quantities q_a and q_c are complicated expressions,which only still depend on the axial ratio p = c/a /7,8/.
Some examples of this apparent distribution relating to its median are represented in Fig. 6

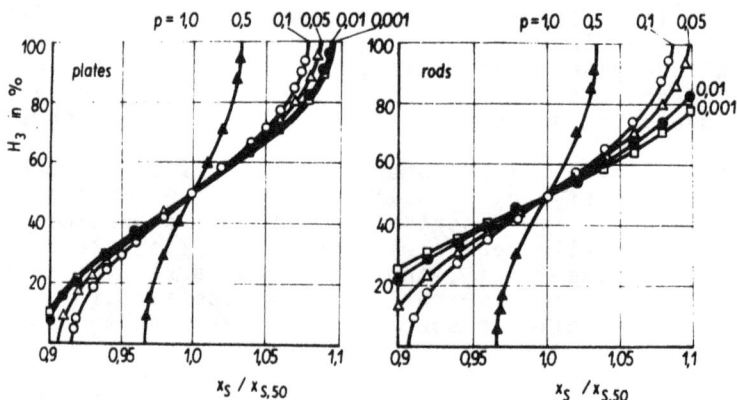

Fig. 6: Apparent size distributions of monodisperse
        spheroids with constant axial ratio p

The apparent distribution of plates are more narrow than those of rods. Moreover, very thin plates (p<0.01) lead to practically constant distributions.
Now let us assume that polydispersity extends to the particle size x_a: particle size x_c be distributed in such a way that always p = const. is valid; then the apparent distribution is:

$$H_3(x_s x_a) = \frac{1}{\pi} \int_{x_{a,min}}^{x_a} h_3(x_a) \ arc \ cos \ Q_{P/S} \ dx_a \qquad (5)$$

In Fig. 7 apparent distributions of several p = c/a = x_c/x_a are represented, where equ. (5) is numerically integrated using GGS-laws for the x_a-distribution.
The upper part shows the apparent distributions for the case, in which at constant distribution of x_a the plates become more and more thinner (delamination).
In the lower part the differences are represented between the apparent distribution and that of the volume-equivalent sphere diameter. It becomes evident that in general the distribution of x_v does not lead to a sufficiently accurate estimation of the distribution of x_s.

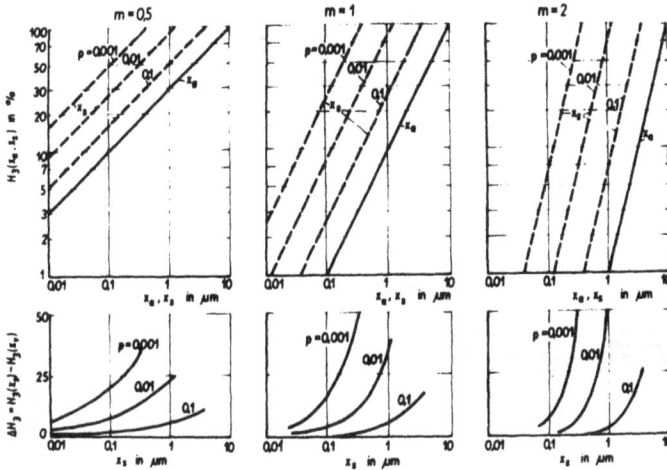

Fig. 7.: Apparent distributions of polydisperse plate-shaped particles

The effect of solids heterogeneities on the measuring result shall be mentioned here only for the sake of completeness. The theory and experimental results are described in /9/. The relationships are very complex, as the sedimentation analysis of solids mixtures, consisting of several components, is influenced by the following heterogeneities: differences in density, in particle size distribution and differences in the characteristics having an effect on the concentration measurement (attenuation of light or X-ray radiation). A more detailed description would be beyond the scope of this paper.

4 HORIZONTAL AND VERTICAL LIMITATIONS OF THE VESSEL

It has been known for a long time that walls and other vessel limitations have a slowing-down effect on the sedimentation velocity of individual particles as a function of the distance (comprehensive description see /10/). Monodisperse particles, moving between two parallel walls, lead to an apparent distribution of

$$H_3(x_s) = 1 - \left[1 - \frac{9}{8L} \cdot \frac{x^3}{x^2 - x_s^2} \right]^{1/2} \tag{6}$$

with L = distance of walls.
Fig. 8 shows such distributions for a wall distance of 0.5 cm, as it can be met indeed in modern sedimentation analysis instruments.
In the case of polydisperse material, the integration of equ. (6) and the application of a GGS distribution leads to differences with respect to the true distribution, as they are shown in Fig. 9.
In the calculated example, the differences ΔH_3 reach a value of about more than 2 % only with narrow particle size distributions ($m \gtrless 2$) and then only within the coarse particle size range. For a sedimentation vessel closed at

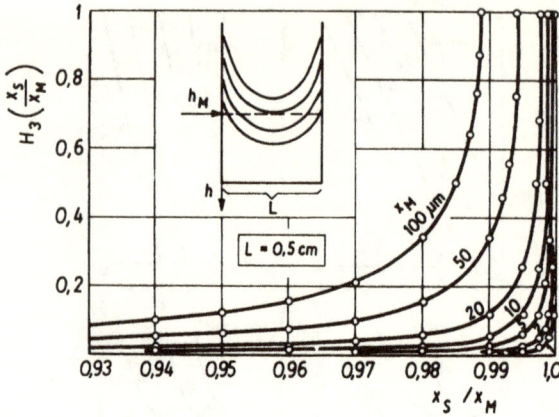

Fig. 8:
Apparent distributions of monodisperse material with wall effect

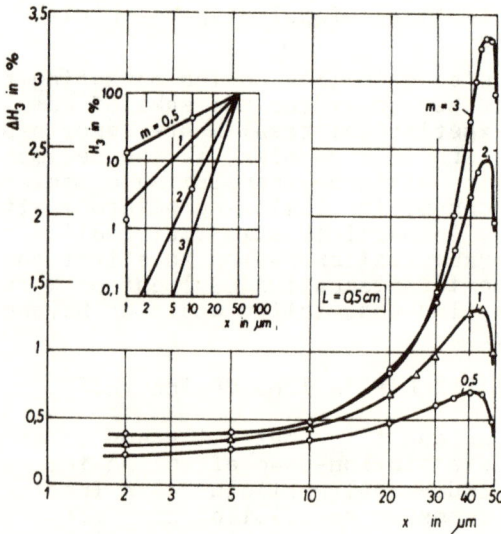

Fig. 9:
Errors due to the wall effect at polydispersity

the top, the apparent distribution of a monodisperse material is approximately given by /6, 10/:

$$H_3(x_s) = \frac{0.25 \; x/h_M}{1 + 0.25 \; x/h_M - x_s^2/x^2} \qquad (7)$$

As also with modern apparatuses no values over $x/h_M = 0.01$ are reached, this influence can be neglected on the rule.

5 EXTENSION OF THE MEASURING GAP

The theory of the incremental sedimentation analysis is
based on the principle that the solids concentration is
measured as a function of time in a measuring plane. In
practice, however, there is always a measuring volume,
since both the entrance slit and the exit slit of the
measuring radiation have a finite extension. An apparent
distribution is obtained by the fact that the concentration
limit needs a certain time to pass from the upper margin
to the lower margin of the gap. The apparent distribution
of a monodisperse solid is then /6/:

$$H_3(x_s) = \frac{1}{2} - \frac{h}{\Delta h} \left(\frac{x^2}{x_s^2} - 1 \right) \qquad (8)$$

with Δh = height of the measuring gap,
     h = sedimentation depth

Some examples are shown in Fig. 10.

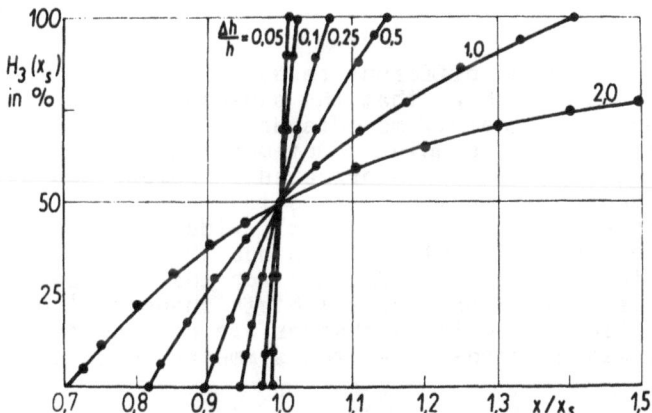

Fig. 10:
Apparent dis-
tributions of
monodisperse
material for
different
sizes of the
measuring gap

The influence is only important with very large size of
the measuring gap. In commercial devices with very small
measuring cells, the ratio of $\Delta h/h$ is usually not greater
than about 0.2. Also in this case the apparent distributions are still very narrow. The error between the true
and the apparent distribution with polydisperse material
can again be calculated by integration of equ. (8). In
Fig. 11 the error $f(x_s) = [H_3(x) - H_3(x_s)]/H_3(x_s)$ is given
for different GGS distributions (parameter $\bar{m}$).
It can be seen that even for the extreme values of $\Delta h/h=1$
the deviation of the measured (apparent) distribution from
the actual one is in general negligible. This is an important conclusion with respect to the design of measuring devices.

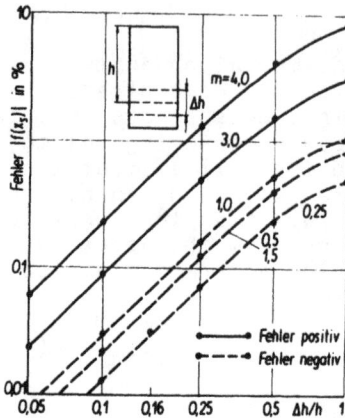

Fig. 11:
Influence of the measuring
gap dimension on the measured
result

6   CONCLUSION

The investigation of the different reasons leading to apparent distributions has shown that the sedimentation analysis of monodisperse or nearly monodisperse solids involves
some sources of errors excluding sometimes its application.
At least its results must be interpreted in comparison with
other methods under the mentioned aspects.
With polydisperse solids of sufficiently wide distributions,
the effects are less apparent. This is due to the fact that
in most cases the factors of influence for the respective
particle size cause more or less symmetric changes of particle concentrations, and these changes are only so far reflected in the measured result as the symmetry is incomplete.

REFERENCES

/1/ Meason, M.; Weaver, W.: The settling of small partic-
    les in a fluid. Physical Rev. Sev. 2, 23 (1924), p.
    412 - 426
/2/ Fürth, R.: Über ein Problem der Diffusion im Schwere-
    feld. Zeitschr. f. Physik 40 (1926), p. 351 - 363,
    45 (1927), p. 83 - 85
/3/ Berg, S.: Determination of particle size distribution
    by examining gravitational and centrifugal sedimenta-
    tion to the pipett method and with divers. Symp. PSA,
    June 1958, Boston, ASTM STP 234 (1959), p. 143 - 171
/4/ Chung, H. S.; Hogg, R.: The effect of Brownian motion
    on particle size analysis by sedimentation. Powder
    Techn. 41 (1985) 3, p. 211 - 216
/5/ Allen, T.: Sedimentation techniques of particle size
    measurement. Conf. PSA Sept. 1985, Bradford, Proceed.
    p. 24 - 45
/6/ Bernhardt, C.: Granulometrie - Klassier- und Sedimen-
    tationsmethoden. Leipzig 1990

/7/ Gans, R.: Wie fallen Stäbe und Scheiben in einer
     reibenden Flüssigkeit? Sitzungsber. d. math.-phys.
     Kl., Königl. Bayer. Akad. d. Wiss. Bd. 41 (1911),
     S. 191 - 203
/8/ Bernhardt, C.: Sedimentation of nonspherical par-
     ticles. 4. Europ. Symp. Partikelmeßtechnik, Nürn-
     berg 1989, Proceed. p. 87 - 111
/9/ Bernhardt, C.: Particle size analysis of hetero-
     geneous powders by sedimentation. Part. Characteri-
     zation 1 (1984) 3, S. 121 - 126

/10/ Happel, J.; Brenner, H.: Low Reynolds number hydro-
     dynamics. Leyden 1973

Ultrasonic Spectrometry: On-line Particle Size Analysis at Extremely High Particle Concentrations

U. Riebel

INSTITUT FÜR MECHANISCHE VERFAHRENSTECHNIK, UND MECHANIK, POSTFACH 6980, UNIVERSITÄT KARLSRUHE (TH), D-7500, GERMANY

1   INTRODUCTION

Most of the known methods of on-line particle size measurement can be applied at low particle concentrations only. At higher concentrations, serious systematic errors will occur due to specific particle interaction mechanisms. So e.g. with counting methods such as the Coulter-Counter or optical particle counters, the concentration limit is set by the condition that the statistical probability of particle coincidences in the measuring volume must be kept very low [1]. With laser diffraction methods, the onset of multiple scattering leads to an over-estimation of the fines content [2]. Roughly speaking, the application of the conventional particle sizing techniques is mostly restricted to particle concentrations below 1 % by volume, while a majority of possible applications in on-line process control is characterized by much higher particle concentrations.

Ultrasonic spectrometry is a new method of particle size analysis [3,4] based on ultrasonic extinction measurements at multiple frequencies. Due to some inherent features of the measuring principle, which will be discussed in the following, measurements by ultrasonic spectrometry can be extended to particle concentrations as high as 20 % by volume and more.

2   FUNDAMENTALS OF ULTRASONIC SPECTROMETRY

With low particle concentrations, the extinction of ultrasonics in a suspension of particles conforms to Lambert-Beer's law:

$$E_f = - \ln\left(\frac{I}{I_o}\right)_f = \Delta l \cdot C_{Pf} \int_{x_{min}}^{x_{max}} K <f, x> \cdot q_2 <x> dx \qquad (1)$$

with E   extinction, $- \ln (I / I_0)$

I_0, I ultrasonic intensities received in the absence / presence of particles

f   ultrasonic frequency

Δl length of ultrasonic path through the suspension

C_{Pf} particle concentration expressed in particle projected area / suspension volume

K   extinction efficiency

$q_2 <x>$ density distribution of particle projected area

x   particle diameter

The dependence of the ultrasonic extinction efficiency of the particles, K_{ext}, on particle size x and ultrasonic frequency f can be expressed as a function of the non-dimensional particle size parameter $\sigma = \pi x / \lambda$, where λ is the ultrasonic wavelength, see **Fig. 1**.

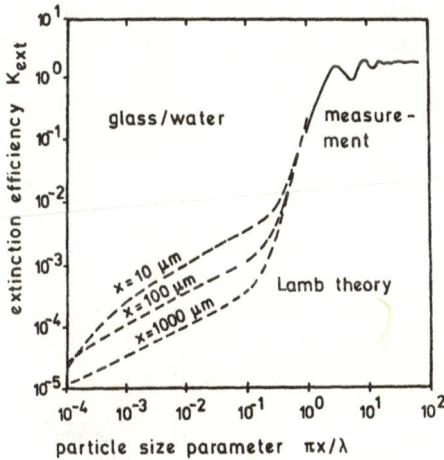

Fig. 1: Ultrasonic extinction efficiency of glass particles as a function of the particle size parameter $\pi x / \lambda$.

The determination of particle concentration and particle size distribution in a suspension proceeds from a set of ultrasonic extinction measurements at various frequencies f_i, i = 1...M . Based on a discretized version of Eq. (1),

$$E_i = \Delta l \, C_{Pf} \sum_{j=1}^{N} K_{ij} \, q_{2j} \, \Delta x_j \qquad (2)$$

a system of linear equations is established and solved numerically to obtain the particle concentrations $C_{Pf} \cdot q_{2j} \Delta x_j$ for N discrete particle size intervals Δx_j.

As a whole, the procedure of measurement and mathematical evaluation is closely analogous to laser diffraction measurements in the Fraunhofer domain [5], but the problems arising from the numerical instability of the linear equation system are even more severe. Thus the Phillips-Twomey-Algorithm (PTA), successfully applied in laser diffraction spectrometry, will perform poorly in the presence of systematic errors, as they may arise from unknown particle shape or particle material. A non-linear iterative procedure however, the relaxation method, has proven to yield excellent results even under difficult conditions.

3   EXPERIMENTAL SET-UP

The experimental set-up used for ultrasonic spectrometry is sketched in **Fig. 2**. The suspension sample to be investigated circulates in a closed loop with a volume of 1.2 litres. It is passed through the ultrasonic measuring cuvette with ultrasonic emitter and receiver and through an optical cuvette for reference measurements of optical extinction. The actual configuration of the system allows measurements within the frequency range from 0.5 to 90 MHz; the extinction may be as high as 30 to 60 dB, depending on the frequency.

For particle size analyses, ultrasonic extinction is measured for a number of different frequencies consecutively. The whole procedure of measurement, including frequency selection, extinction measurement and mathematical evaluation is automated. One measurement of particle size distribution and particle concentration takes about 2 to 5 minutes, depending on the resolution and the reproducibility required. Some cumulative area density distributions measured by ultrasonic spectrometry, compared to the results obtained from a microscopic evaluation, are shown in **Fig. 3**. Broad and narrow distributions are clearly distinguished, owing to the high resolution and reliability of the method.

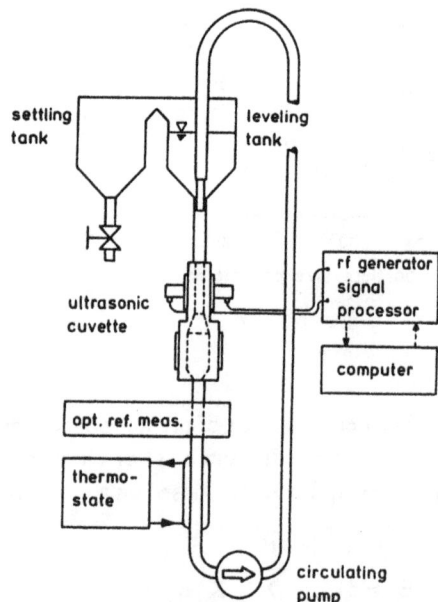

Fig. 2: Experimental set-up.

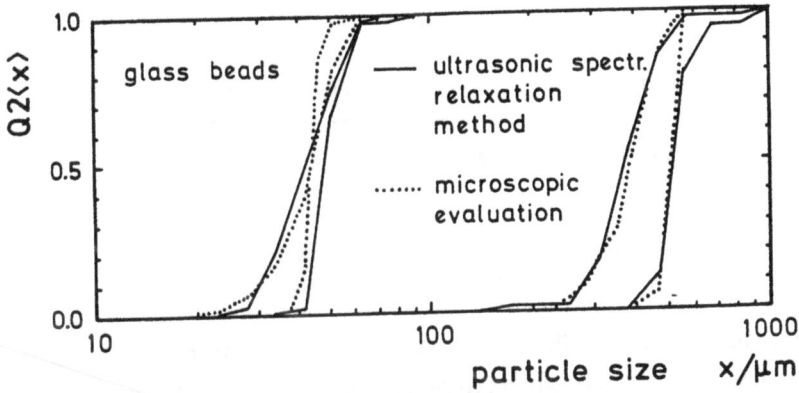

Fig. 3: Some particle size distributions measured by ultrasonic spectrometry, compared to results of microscopic evaluation.

4   EXTINCTION IN HIGHLY CONCENTRATED SUSPENSIONS

Lambert-Beer's law is fully valid in the limit of low extinction and low particle concentration only. With increasing particle concentration, particles will interact by various mechanisms leading to deviations from Lambert-Beer's law. Here, mainly two out of the large number of particle-particle-interactions are of interest and will be discussed in more detail: multiple scattering and steric interactions.

Multiple Scattering
The occurrence of multiple scattering phenomena with extinction measurements at high concentrations is related to the fact that the extinction efficiency of particles is generally higher than the absorbtion efficiency [6], $K_{ext} \geq K_{abs}$. In case that a beam of coherent or collimated radiation of intensity $I_{c,o}$ is incident upon a disperse system, the coherent intensity I_c will decay proportionally to K_{ext}, while the total intensity I_t will decay only proportionally to K_{abs}:

$$ -\ln\left(\frac{I_c}{I_{c,o}}\right) \sim K_{ext} \cdot C_{Pf} \cdot \Delta l \tag{3} $$

$$ -\ln\left(\frac{I_t}{I_{c,o}}\right) \sim K_{abs} \cdot C_{Pf} \cdot \Delta l \tag{4} $$

Since I_c decays faster than I_t, a background of incoherent / uncollimated radiation will build up, with an intensity I_i corresponding to the difference of I_c and I_t (see **Fig. 4a**):

$$I_i = I_t - I_i \tag{5}$$

Moreover, as a rule, the sensitivity of the receiver will be inferior for the incoherent part of radiation in comparison to the coherent part, which can be expressed by a relative sensitivity β_i ($\beta \leq 1$):

$$\beta_i = \frac{\text{sensitivity for incoh. / uncoll. part of radiation}}{\text{sensitivity for coh. / coll. part of radiation}} \tag{6}$$

Even though I_c and I_t will each go down exponentially with increasing particle concentration or Δl, the intensity received, I_r, will hence show a non-exponential behaviour (see **Fig. 4b**):

$$I_r \sim I_c + \beta_i I_i = e^{-K_{ext} C_{Pf} \Delta l} + \beta_i \left(e^{-K_{abs} C_{Pf} \Delta l} - e^{-K_{ext} C_{Pf} \Delta l} \right) \tag{7}$$

Eq. (7) can give a fairly good description of experimental results on extinction with multiple scattering, but is of little use in practical applications, since K_{abs} remains, in general, unknown.

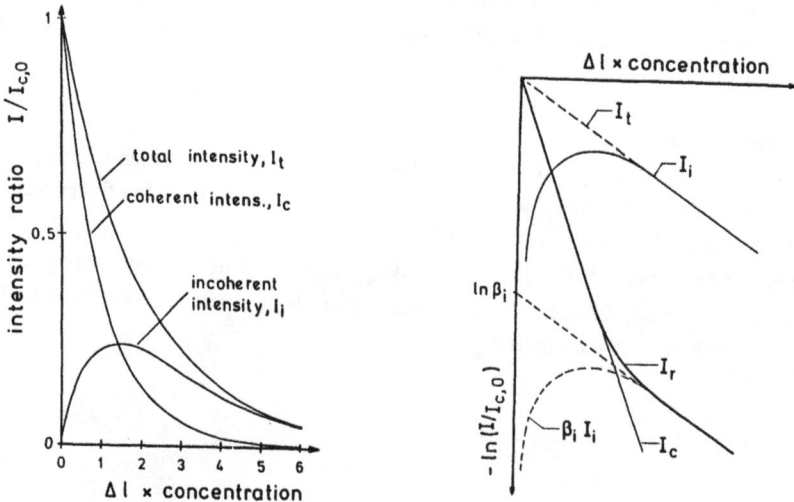

Fig. 4: Multiple scattering effects on received intensity, I_r.

Hence for practical concentration measurements, β_i must be reduced to a minimum in order to maximize the validity range of Lambert-Beer's law. With optical or x-ray extinction measurements, this is achieved by reducing the receiver's aperture angle [7], but it appears to be difficult to reduce β_i below about $\beta_i = 10^{-2}....10^{-3}$. With piezoelectric ultrasonic receivers however, β_i is extremely low, since the electric output from the receiver is proportional to the integral of all mechanical stresses over the sensor volume. Hence, upon incidence on a plane piezoelectric receiver, only a plane wave can produce a significant signal, whereas the stresses produced by scattered or multiply scattered waves will cancel out statistically, see **Fig. 5**.

As a consequence, β_i is extremely low for sufficiently large ultrasonic receivers (in the order of $\beta_i = 10^{-2}...10^{-4}$), and ultrasonic extinction measurements can be extended to extremely high particle concentrations without disturbances from multiple scattering.

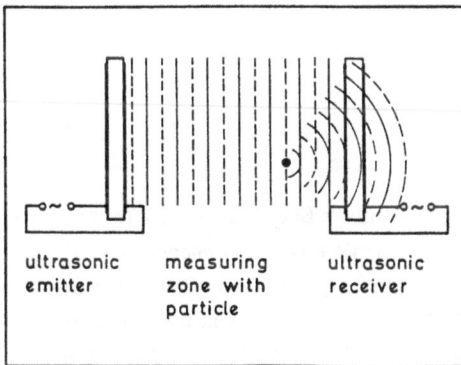

Fig. 5: Detection of plane and scattered waves by an ultrasonic receiver.

Steric Interactions

Steric interactions arise from the particle volume, which induces certain suspension structures at high particle concentrations. This is most evident for the case of monodisperse spherical particles, which will arrange into a lattice-like structure in the limit of high particle concentrations.

The influence of steric interactions on the extinction of radiations is illustrated by **Fig. 6** for a 2-dimensional model suspension. Whereas non-interacting particles (as they are presumed in the derivation of Lambert-Beer's law) will interpenetrate each other with a probability increasing with concentration, sterically interacting particles will arrange and hence expose more projected area to the radiation than non-interacting particles would do. As a consequence, steric interactions will lead to an increase of extinction as compared with Lambert-Beer's law.

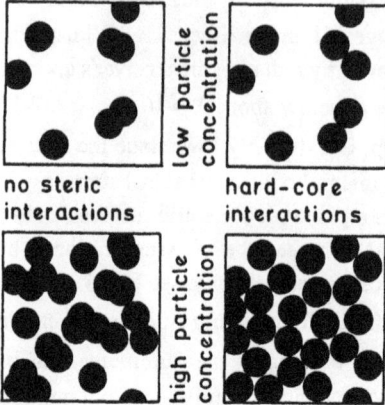

Fig. 6: Steric interactions in a 2-dimensional model suspension.

Theoretical and experimental investigations [3,8] on extinction with steric inter-actions have been limited, so far, to suspensions of monodisperse spherical particles. The experimental findings for ultrasonic extinction can be described by

$$E = E_{L-B} \cdot F < c_v >$$

$$= E_{L-B} \cdot \frac{\ln (1 - A c_v)}{- A c_v} \qquad (8)$$

where E_{L-B} is the extinction according to Lambert-Beer's law (Eq. 1 for monodis-perse particles), $F < C_v >$ is a function of the particle volume concentration C_v, and A is an adaptation parameter depending on some still unknown structural properties of the suspension (see **Fig. 7**, $A = 1.8...2.9$). Eq. (8) has been proven to be valid for concentrations up to 20 to 30 % by volume. In the limit of low concentration, $C_v \to 0$, $F< C_v > = 1$ is found, that is, Lambert-Beer's equation is found to be a special case of the extinction equation (Eq. 8).

Fig. 7: $F< C_v >$ for various suspensions.

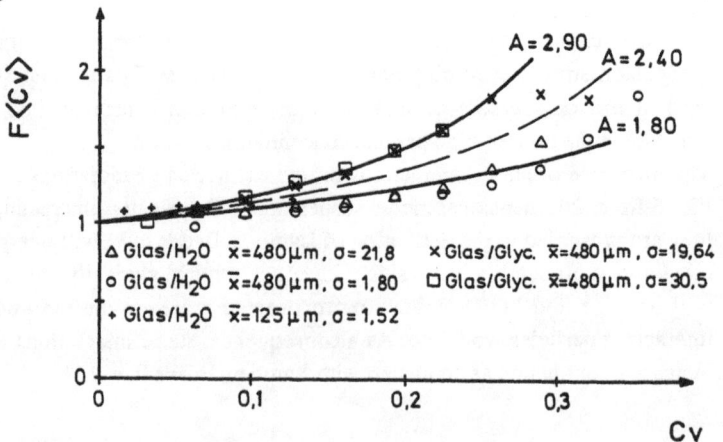

Interference of Multiple Scattering and Steric Interactions

Whereas in optical extinction measurements multiple scattering is so dominant that it masks steric interaction effects totally, both multiple scattering and steric interaction effects can be observed with ultrasonic extinction measurements in highly concentrated suspensions. Due to an extremely low β_i, ultrasonic measurements will in general show the effect of steric interactions first, and only later on, with a further increase of concentration, multiple scattering will be observed additionally. A typical result is shown in **Fig. 8**, whereby the contributions of coherent (I_c) and incoherent intensity (I_i) to received intensity (I_r) have been determined separately. In the range of medium particle volume concentrations C_v, I_r is lower than predicted from Lambert-Beer's law, which can be attributed to a dominance of steric interactions, whereas with a further increase of volume concentration, multiply scattered intensity begins to dominate and I_r becomes higher than predicted by Lambert-Beer's law.

Fig. 8: Extinction in concentrated suspensions of glass beads ($x = 480\ \mu m$) in water, measured at a frequency of 25 MHz.

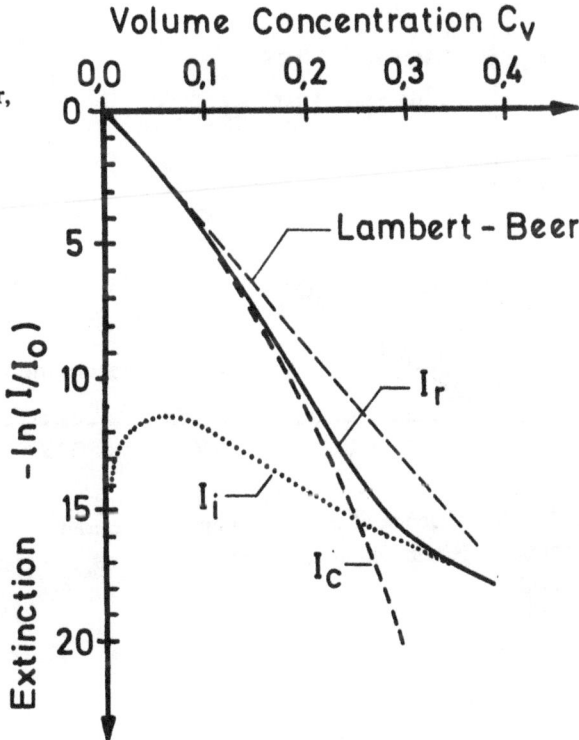

5   PARTICLE SIZE MEASUREMENTS AT HIGH CONCENTRATION

In default of a more thorough knowledge of steric interactions, including especially steric interactions between particles of different size, the evaluation of particle size measurements by ultrasonic spectrometry is based, so far, on the linear equation system (Eq. 2). Thus, the shortcomings of Lambert-Beer's law in the range of high particle concentrations are reflected in the possibility of systematic measurement errors. As **Fig. 9** shows for a series of measurements with increasing particle volume concentration, the particle concentration C_{Pf}, as determined by ultrasonic spectrometry, tends to be higher than the theoretical value, $C_{Pf} = 1,5 \, / \, x \, C_v$, especially when the Phillips-Twomey-Algorithm is used for evaluation of the results. The particle size distribution measured remains almost unaffected by concentration effects (**Fig. 10**), as long as the particle size distribution is not too broad. With broader distributions, there is a slight tendency to over-estimate the fines content.

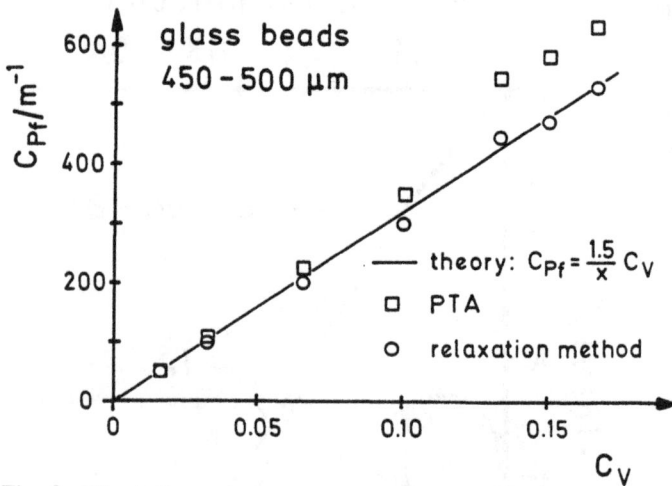

Fig. 9: The influence of particle volume concentration Cv on the particle concentration as determined by ultrasonic spectrometry, whereby Lambert-Beer's law was used.

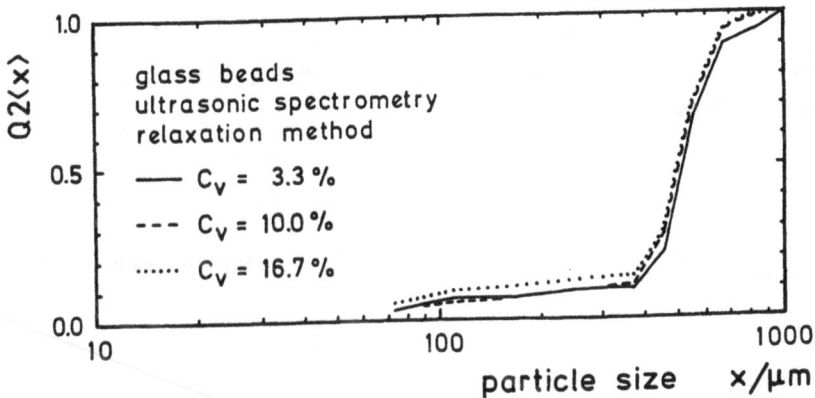

Fig. 10: The influence of particle volume concentration on particle size distributions determined by ultrasonic spectrometry.

6 CONCLUSION

Thanks to a low sensitivity to disturbances from multiple scattering, particle size measurements by ultrasonic spectrometry can be extended to extremely high particle concentrations. An evaluation of the measurements based on Lambert-Beer's law can provide results of sufficient quality for particle concentrations up to about 20 % by volume. An extension of ultrasonic measurements to even higher concentrations will require a more thorough understanding of steric interactions, including steric interactions between particles of different size.

7 REFERENCES

1. J. Raasch, H. Umhauer: Der Koinzidenzfehler bei der Streulicht-Partikelgrößen-Zählanalyse. Fortschr. -Berichte der VDI-Zeitschriften Reihe 3, Nr. 95, VDI-Verlag, Düsseldorf 1984.
2. A. Bürkholz, R. Polke, Part. Charact. 1 (1984), 153-160.
3. U. Riebel: Die Grundlagen der Partikelgrößenanalyse mittels Ultraschall-Spektrometrie. Thesis, Karlsruhe 1988.
4. U. Riebel, F. Löffler, Part. Part. Syst. Charact. 6 (1989), 135-143.
5. M. Heuer, K. Leschonski, Part. Charact. 2 (1985), 7-13.
6. A. Ishimaru: Wave Propagation and Scattering in Random Media, Vol. I Academic Press, San Diego, 1978.
7. K. Leschonski, T. Boeck, Part. Charact. 2 (1985), 81-90.
8. U. Riebel: An Estimate on some Statistical Properties of Extinction Signals in Dilute and Concentrated Suspensions of Monosized Spherical Particles. Part. Part. Syst. Charact. 8 (1991), in press.

The Du Pont/Brookhaven Scanning X-Ray Disc Centrifuge (BI-XDC)

T. Allen

PARTICLE SCIENCE AND TECHNOLOGY CENTER (PARSAT), E. I. DU PONT DE NEMOURS AND COMPANY, WILMINGTON, DE 19714–6090, USA

1   ABSTRACT

This instrument was designed to fill an industrial need for fast, reproducible, accurate analyses of inorganic powders in the sub-micron size range.

The heart of the instrument is a hollow, x-ray transparent, disc which, under normal operating conditions, contains 20 ml of suspension at a volume concentration of around 0.2%.

It is necessary to key in the operating variables such as powder density and temperature (for a water based suspension); for other liquids their density and viscosity are required. The instrument will then display the operating range covered by the pre-selected centrifuge speed.

Maximum x-ray transmission is determined daily by the injection of clear suspending liquid into the spinning disc. The suspension is injected into a stationary disc to determine maximum opacity; the disc is then spun at any pre-selected speed between 750 and 6000 rpm and the run started. The nominal run condition is for the source and detector to remain stationary for 1 minute at a radial position of 48mm, reproducible to ± 4μm, and then to scan towards the surface at a constant radial velocity. Total run time is 8 minutes but this, again, is selectable.

The BI-XDC can operate in the gravitational or centrifugal mode and the analyses can be merged to cover a total size range of 100 to 0.01μm.

2   INTRODUCTION

The BI-XDC consists of a disc with associated electronics. The disc is mounted vertically onto the shaft of an electric motor with a digitally variable speed between 750 and 6000 rpm.

For both gravitational or centrifugal sedimentation the instrument can be run in a stationary mode or held for a preset time at a fixed radius followed by scanning.

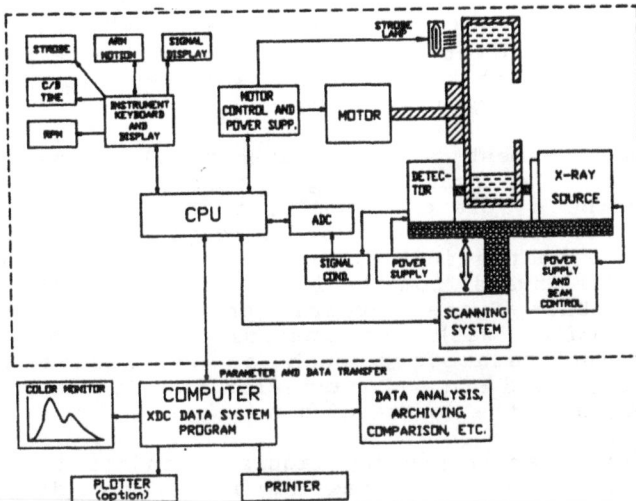

Diagram 1 Line Diagram of the Du Pont/Brookhaven BI-XDC

Nominal disc dimensions and running conditions are given below:

Inner radius of disc cavity= 50.8mm
Initial measurement radius = 48.8mm
Disc thickness = 10mm
Beam dimensions are: length 5.0mm, width 0.254mm

For a suspension volume of 20ml the surface radius will be 44.1mm hence, for an analysis time of 8 minutes, the measurement will be carried out at a fixed radius for 1 minute and the beam will then scan to reach 1mm below the surface at a scan rate of 4.7mm in 7 minutes. Under these default conditions, assuming valid data 15 seconds after injection of the suspension, a size range of 15:1 will be covered; increasing the run time by a factor of 2 increases the measured range by a factor of $\sqrt{2}$. In any analysis the scan rate depends therefore on the volume of suspension used and the pre-selected analysis time. In the gravitational mode 12 to 25 ml of suspension is required whilst in the centrifugal mode between 15 and 25ml may be used.

3   MEASUREMENT PROCEDURE

It is recommended that the instrument be switched on 20 minutes prior to use in order to obtain stable operating conditions. Prior to analysis the run conditions are set by keying into the computer program the operating variables such as powder and liquid densities and liquid density and viscosity if the suspending liquid is not water. For water the viscosity is determined from the operating temperature determined with a thermister situated close to the opening of the centrifuge bowl.

The measurement arm is automatically moved into position. The location program uses an optical switch to define a reference position for the measurement arm. The arm is first moved through a pre-determined distance to the initial measurement radius which is 2mm + 4μm inward from the inner edge of the disc. The BI-XDC Data System program on the host computer communicates with the instrument sending it the operating parameters such as centrifuge speed, scanning speed etc and retrieving from it the raw data collected in the course of a measurement.

5 ml of clear suspending liquid is injected into the disc via a syringe for maximum intensity reading: The program will, upon request, record this reading which normally remains stable throughout an operating session. Subsequent runs can access this reading, obviating the need to perform this step for every sample in an operating session.

The most important step in any sedimentation particle size analysis is the preparation of the suspension which must be carried out in such a way that a well dispersed suspension results. 5ml of this suspension is injected into the stationary disc and a reading taken of the maximum attenuation. The program will, on request, record this reading so that the 100% and 0% attenuation levels are now recorded.

The disc is now spun at a pre-selected speed, the rest of the suspension is added and the run commences. At completion of the analysis the arm containing the x-ray source and detector retracts into the body of the equipment so that the disc can be emptied.

For inorganic powders having a high atomic mass a volume concentration of 0.2% gives adequate attenuation but this needs to be increased with powders having a lower atomic mass and a volume concentration as high as 1% is necessary with quartz.

The above procedure ensures that, even if there is coarse or fine powder outside the measurement range,

the amount of this unmeasured fraction is known.   A
second analysis under different operating conditions
can then be carried out to analyze the unmeasured
fraction.   If necessary this can be carried out in the
gravitational mode and the two analyses merged in the
instrument.

Data are presented in both tabular and differential and
cumulative graphical mode and up to 5 measurements can
be compared in a single presentation.

The instrument measures mass undersize concentration
and this can be converted to a number distribution.
The errors involved in this conversion are considerable
hence the precision and reproducibility of this
distribution is much lower than for the original data.

4   THEORY

For a homogeneous suspension the attenuation of the x-
ray beam is proportional to the mass concentration of
the suspension at the measurement radius. The size of
the largest particle present in the suspension can be
calculated using Stokes equation and the mass
concentration undersize can be determined using the
Kamack equation[1-3].

The diameter (D_m) of the largest particle in the x-ray
beam is given by Stokes equation:

$$D_m^2 = \frac{18\eta \ln(\frac{r_i}{S})}{(\rho_s - \rho_f)\omega^2 t_i}$$

where: h   =   viscosity of the liquid medium
           r_i =   radial position of source and
                         detector
           S   =   radial position of suspension
                         surface
           t_i =   measurement time
           ω =   radial velocity of the centrifuge
                         disc
           ρ_s =   density of powder
           ρ_f =   density of liquid

Q_i, the measured suspension concentration at radius r_i
and time t_i, is determined from the following equation:

$$I_t = I_0 \exp(-BQ_i)$$

where I_t is the measured intensity of the emergent
               beam with suspension in the disc

I_0 is the measured intensity of the emergent beam
         with clear liquid in the disc.

B   is a constant

D   (The x-ray density) is defined as $D = \log_{10}$

$\dfrac{I_0}{I_t}$

The mass fraction smaller than D_m is given by:

$$F(D_m) = \int_0^i \left(\frac{r_i}{S}\right)^2 dQ$$

Kamack solved the above equation for a fixed measurement
radius (r) and this solution was later extended[4] to a
variable surface radius (S_i).

The solution for the above equation, for a variable
measurement radius r_i, may be expressed by the following
iterative equation:

$$F_i = \frac{1}{2}(y_i - y_{i-1,i})Q_i + \sum_{j=1}^{i-1} \left[\frac{y_i - y_{i-1,i}}{y_{j+1,i} - y_{j,i}} - \frac{y_i - y_{i-1,i}}{y_{j,i} - y_{j-1,i}}\right]F_j$$

Where:

y_i $= \left(\frac{r_i}{S_i}\right)^2$

$y_{j,i}$ $= y_i^{(\frac{D_i}{D_1})^2}$

r_i = the radial distance from the center of
            rotation to the measurement beam

$y_{i,i}$ $= y_i$

$y_{0,i}$ $= 1$

D_i is the diameter of the largest particle
            in the measurement beam at radius r_i and
            time t_i

n   is the number of data points selected i is
            an integer from 1 to n

F_i is the mass percentage of the powder
            smaller than D_i

5   EXPERIMENTAL RESULTS

As an extension of work carried out earlier[5,6]
experimental data with BCR 66 were quantified by the
introduction of two definitions:

Accuracy (A) is a measure of how closely the measured data reflects the standard BCR silica data:

$$A = 100\frac{(x-x_S)}{x_S}$$

where x and x_S are the measured and standard sizes at the percentiles.

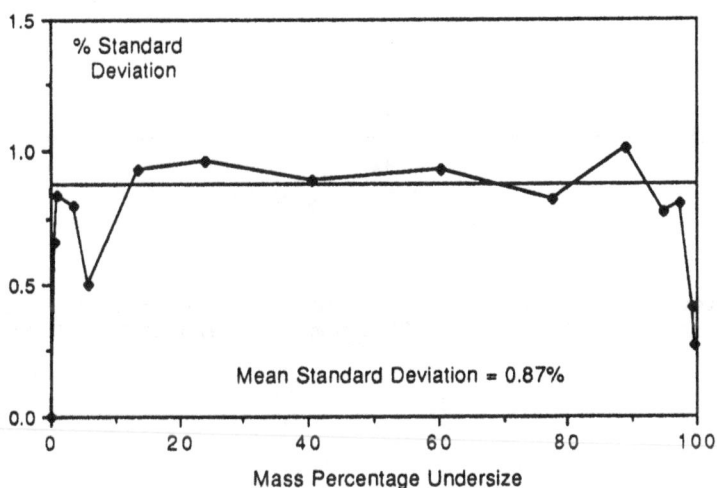

Figure 1   Reproducibility of the Du Pont/Brookhaven X-Ray Scanning Disc Centrifuge (BI-XDC) with BCR 66

Table 1   Repeat Analyses of BCR 66 using the Du Pont/Brookhaven X-Ray Scanning Disc Centrifuge

Size	Percentage by Mass Undersize (P)						Mean	S.D.
(μm)	Run 1	Run 2	Run 3	Run 4	Run 5	Run 6	($\overline{P}$)	(s)
0.211							0.0	0.00
0.262	0.1	1.2	0.0	0.4	1.5	0.0	0.5	0.66
0.326	0.1	1.8	0.3	1.1	1.8	0.0	0.9	0.83
0.404	2.9	3.3	3.2	3.7	4.9	2.7	3.5	0.79
0.502	5.8	5.5	5.4	6.4	6.5	5.4	5.8	0.50
0.624	12.2	13.9	13.2	13.5	15.0	13.2	13.5	0.93
0.775	22.7	24.2	23.3	25.2	25.0	24.1	24.1	0.96
0.962	39.0	40.8	41.0	40.7	41.4	41.4	40.7	0.89
1.195	59.0	60.2	61.3	59.5	61.1	60.9	60.3	0.93
1.484	76.2	78.7	77.8	77.8	77.5	77.9	77.7	0.82
1.843	87.7	90.7	89.0	89.2	88.3	89.2	89.0	1.01
2.288	94.1	96.3	94.7	95.1	94.4	94.9	94.9	0.77
2.842	96.7	98.8	96.9	97.3	96.7	97.5	97.3	0.80
3.529	99.0	99.9	98.8	99.4	98.9	99.4	99.2	0.41
3.772	99.7	100	99.4	100	99.5	100	99.8	0.27
Mean Standard Deviation							$\sum\frac{\overline{S}\Delta P}{100}$	0.87

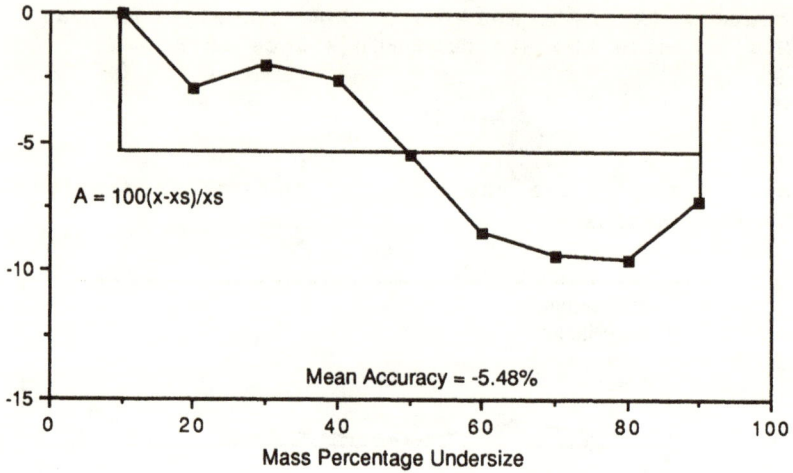

Figure 2   Deviation of Brookhaven/Du Pont X-Ray Scanning
Disc Centrifuge (BI-XDC) from Standard BCR 66 Data

Table 2   Deviation of Data using the Brookhaven/Du Pont
X-Ray Scanning Disc Centrifuge from Standard BCR 66 Data

Percentage by Mass Undersize	Standard Size (x_S)	X-Ray Centrifuge Size (x)	Accuracy $A= \dfrac{(x-x_S)}{x_S}$
10	0.57	0.57	0.00
20	0.74	0.72	-2.91
30	0.86	0.84	-1.96
40	0.98	0.96	-2.53
50	1.13	1.07	-5.44
60	1.30	1.19	-8.45
70	1.47	1.33	-9.38
80	1.70	1.54	-9.57
90	2.04	1.89	-7.26
Mean Deviation from Standard			-5.48

Reproducibility is defined as the average standard deviation for 6 measurements on different representative samples from the same parent suspension:

$$\overline{s} = \sum_{0}^{100} \frac{s\,\Delta P}{100}$$

$100\overline{s}$ being the area under the error curve.

Reproducibility data for BCR 66 are presented in Figure 1 & Table 1, and give a mean value of 0.87% which compares well with the Sedigraph 5100 value of 0.85%. Total run time was 10 minutes as opposed to 1 hour for the Sedigraph analyses (For complete data contact [6])

Accuracy data for BCR 66 are presented in Figure 2 & Table 2; deviation from the standard data is zero at the 0.57μm level and rises to about -10% at the 2μm level to give an average value of -5.48%. The comparable Sedigraph values are between 3.61% and 7.11% depending on the scanning speed. However the data lie within the error limits for the standard powder.

Results are compared with the Sedigraph since the BI-XDC was designed to extend the size range of the X-Ray absorption technique and reduce analytical time. The cumulative and relative size distributions for BCR 66 are presented in Figures 3 & 4 with a comparison with Sedigraph data.

A test was carried out to determine the effect of changing the operating variables on accuracy using BCR 66. The 6 standard runs were carried out at a disc speed of 1000 RPM and a run time of 8 minutes. Two further runs at 900 and 1100 RPM gave similar data. In order to test the validity of the scanning theory 2 runs were carried out without scanning and these gave similar data. The data are presented in Figure 5

The blending capability of the instrument was tested by carrying out a gravitational analysis of BCR 70 down to 2μm and determining the sub-2 μm distribution by centrifugal analysis. The data are presented in Figures 6 & 7 and the deviation from standard data is shown in Figure 8 and Table 3 with a mean value of 6.28%.

The reproducibility was determined for a ceramic powder to give a mean standard deviation of 0.79%, Figure 9 and Table 5. The run conditions are set down in Table 4. The analysis was also repeated using double the scan time to give a total run time of 11 minutes to determine whether this affected the results. The cumulative and relative data are presented in Table 6 and Figures 10 & 11.

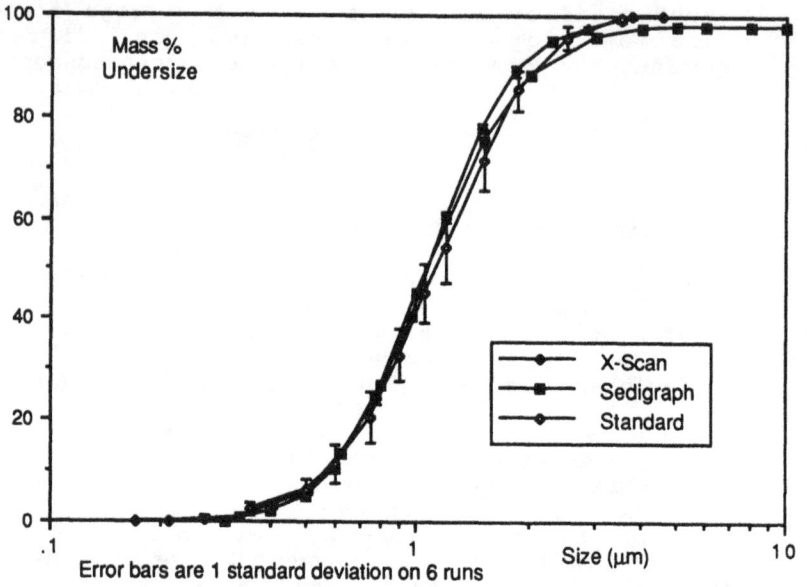

Figure 3 Comparison Between the BI-XDC, Sedigraph 5100
and BCR 66 Standard Data:Cumulative Undersize Mass
Distribution

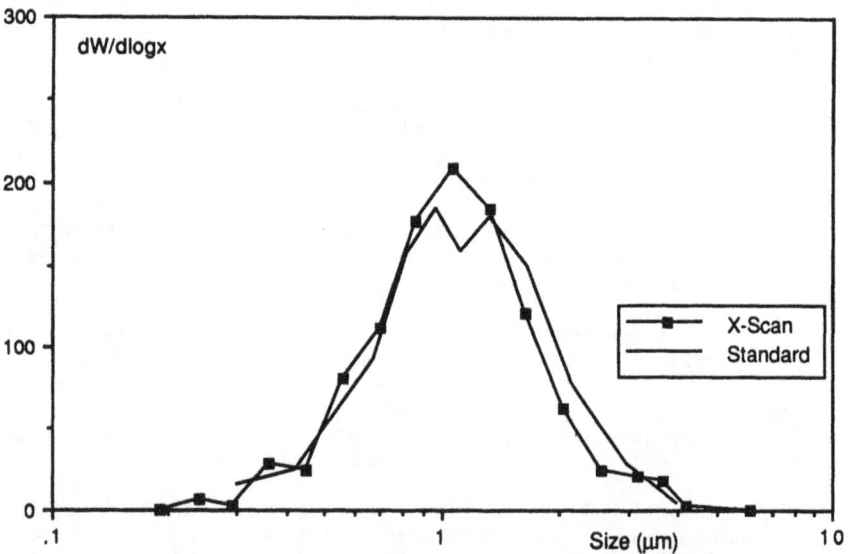

Figure 4   Comparison Between the BI-XDC and BCR 66
Standard Data Relative Mass   Distribution

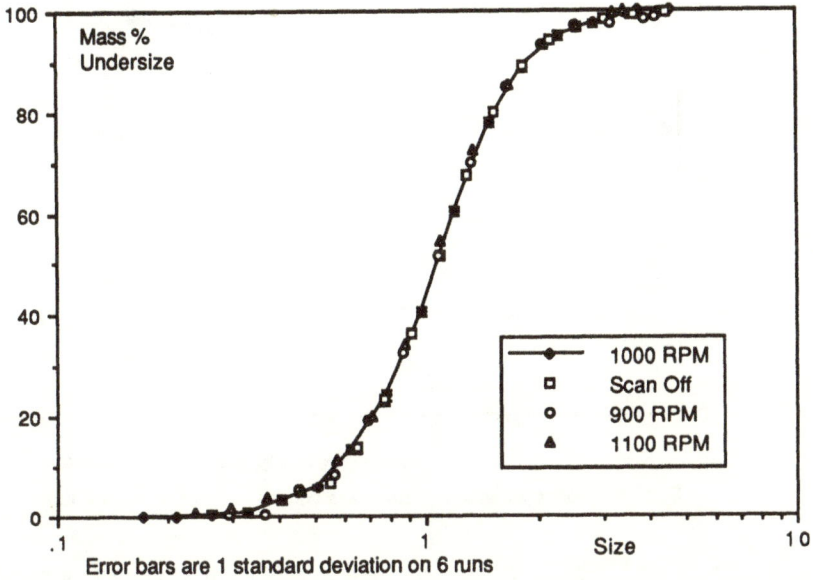

Figure 5   Effect of Varying the Operating Conditions
on the Generated Size Distributions of BCR 66 with the
BI- XDC

Table 4 Run Conditions for Analysis of a Ceramic
Powder

Run
Parameters   _Baseline_
                                                          Parameters

Disc Speed:	1502.04 RPM	High Concentration	0.514 V
Particle Density:	4.25 g/cm^3	Low Concentration	1.363 V
Spin Fluid	Water		
Spin Fluid Volume	20.0 ml		
Spin Fluid Density	0.9972 g/cm^3		
Stop Time*	6 minutes		
Temperature	24.0 ^{o}C		
Spin fluid Viscosity	0.9097 cp		

* 1 minute stationary followed by a 5 minute scan.

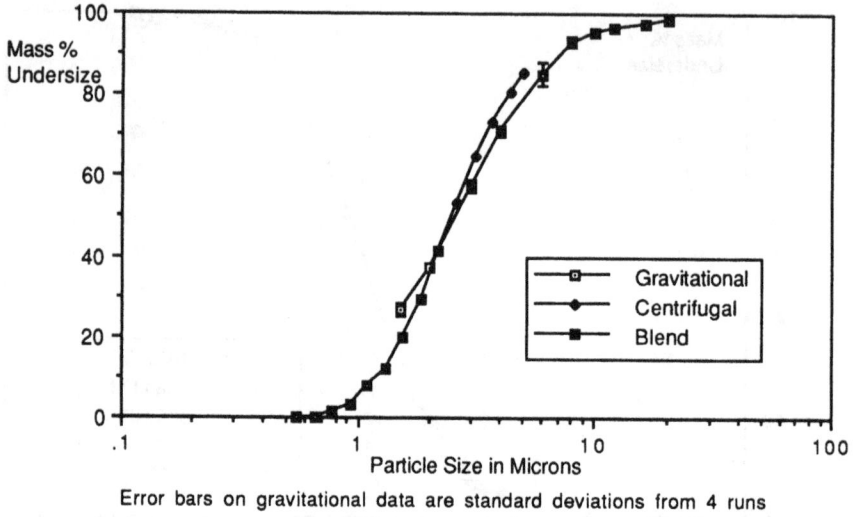

Error bars on gravitational data are standard deviations from 4 runs

Figure 6   Blending Gravitational and Centrifugal BCR
70 Data with the BI-XDC

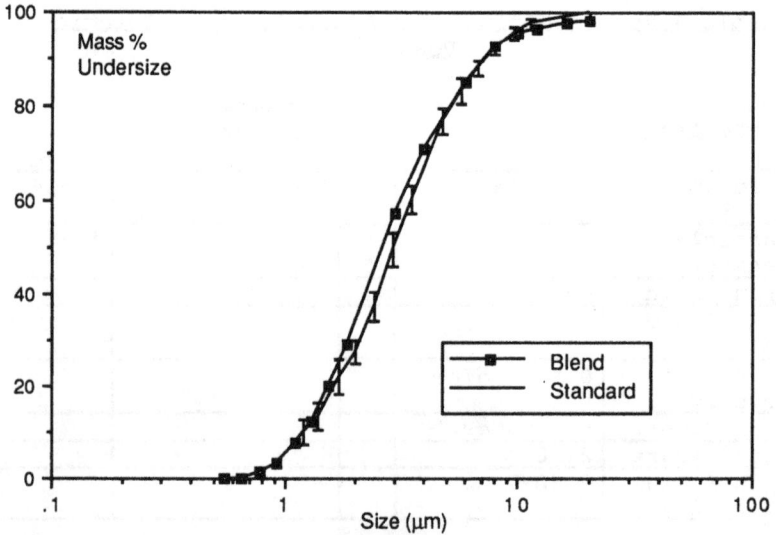

Figure 7   Comparison between the BI-XDC and BCR 70
Standard Data

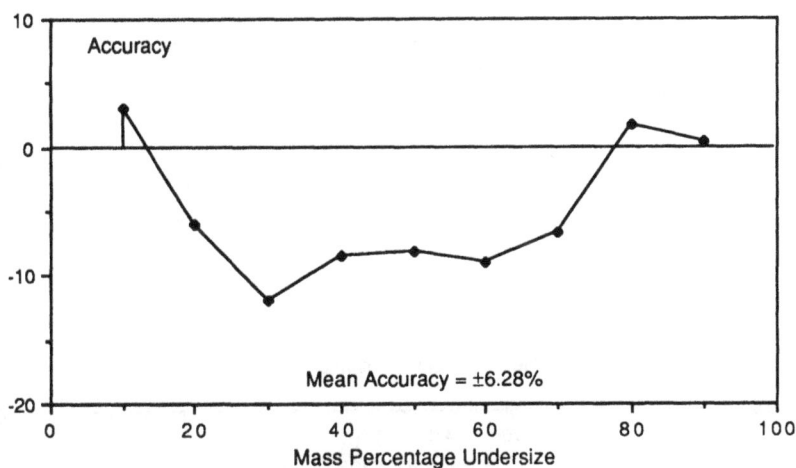

Figure 8   Deviation of Brookhaven/Du Pont BI-XDC from
Standard BCR 70 Data

Table 3   Deviation of Data, using the Brookhaven/Du
Pont BI-XDC from Standard BCR 70 Data

Percentage by Mass Undersize	Standard Size (x_S)	X-Ray Centrifuge Size (x)	Accuracy $A = \dfrac{(x - x_S)}{x_S}$
10	1.20	1.237	3.08
20	1.65	1.551	−6.0
30	2.10	1.848	−12.0
40	2.50	2.29	−8.40
50	2.93	2.69	−8.19
60	3.49	3.18	−8.88
70	4.20	3.92	−6.67
80	5.20	5.29	1.73
90	7.25	7.28	0.41
Mean Deviation from Standard			±6.28

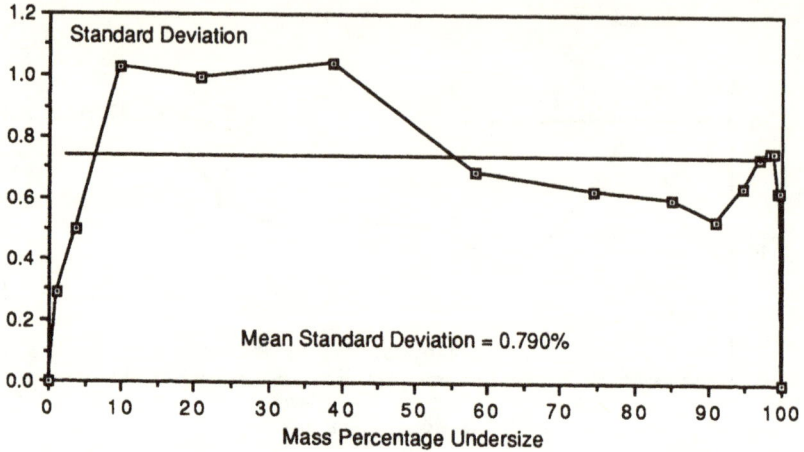

Figure 9   Reproducibility of Brookhaven/Du Pont BI-XDC
with a Ceramic Powder

Table 5   Repeat Analyses of a Ceramic Powder using the
Du Pont/Brookhaven BI-XDC

Size	Percentage by Mass Undersize (P)						Mean	S.D.
(μm)	Run 1	Run 2	Run 3	Run 4	Run 5	Run 6	($\bar{P}$)	(s)
0.133							0	0
0.158	1.4	1	1.2	1.1	1.7	0.9	1.2	0.29
0.188	3.6	3.5	4.0	4.2	4.2	2.9	3.7	0.50
0.224	9.2	9.0	11.0	10.2	10.1	8.1	9.6	1.03
0.266	20.3	20.3	21.8	21.3	21.4	19.1	20.7	0.99
0.316	37.8	38.5	40.2	39.2	39.0	37.3	38.7	1.04
0.376	57.3	58.4	59.3	58.5	58.3	57.7	58.3	0.69
0.447	73.9	74.8	75.0	73.7	73.6	74.8	74.3	0.63
0.532	85.0	85.3	85.4	84.6	83.8	85.2	84.9	0.60
0.633	91.3	91.4	91.6	90.6	90.2	91.1	91.0	0.53
0.752	94.3	95.5	94.9	94.8	93.6	94.8	94.7	0.64
0.895	96.1	97.5	97.5	97.5	95.9	96.9	96.9	0.74
1.064	96.9	99.0	98.7	98.2	97.8	98.6	98.2	0.76
1.265	98.0	100	99.1	98.8	98.3	99.6	99.0	0.76
1.505	98.9	100	99.8	99.6	98.5	100	99.5	0.63
1.572	99.1	100	99.8	100	98.5	100	99.6	0.63
							100	0
Mean Standard Deviation							$\frac{\Sigma \bar{S} \Delta P}{100}$	0.79

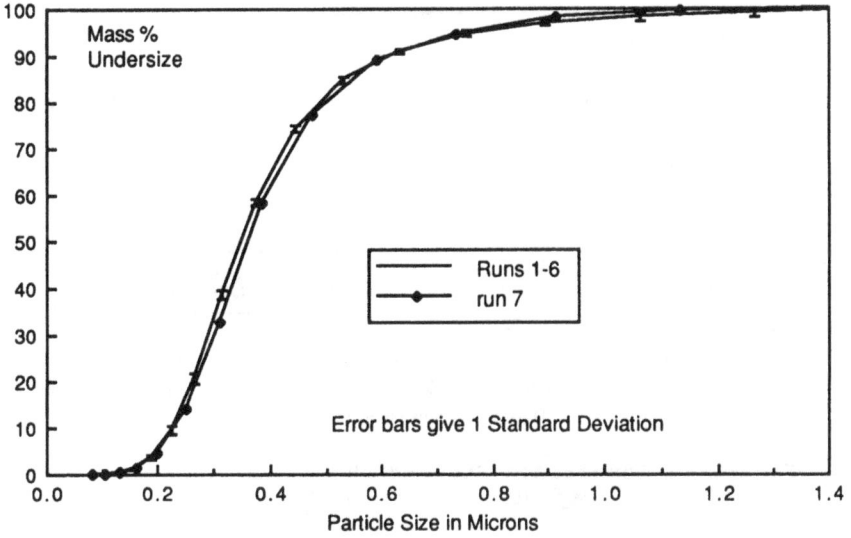

Figure 10   Cumulative Percentage Undersize by Mass of
a Ceramic Powder by BI-XDC

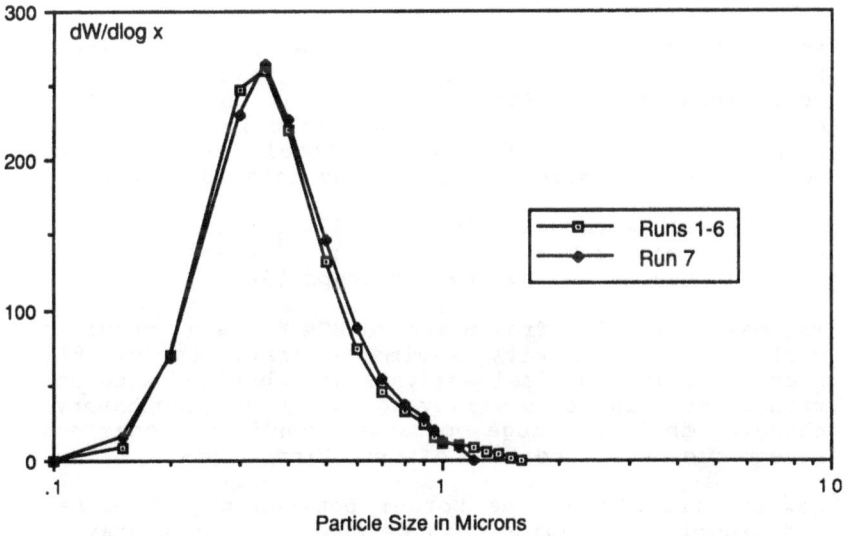

Figure 11   Relative Percentage by Mass of a Ceramic
Powder by BI-XDC

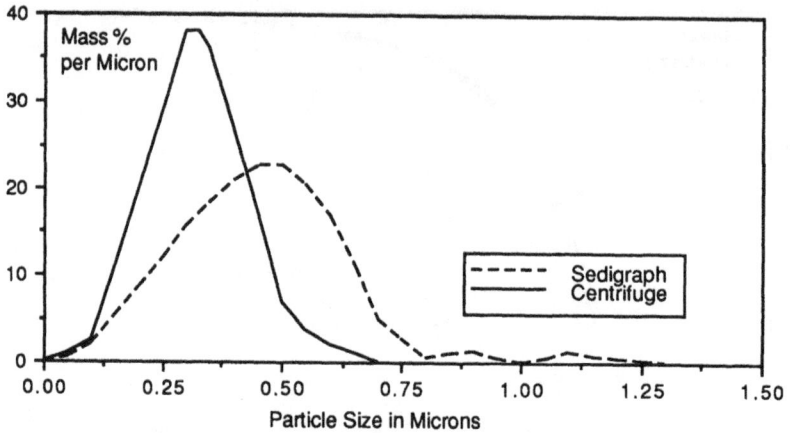

Figure 12 Comparison Between the BI-XDC and the
Sedigraph for a Ceramic Powder

A comparison between frequency plots generated by the
BI-XDC and the Sedigraph (Figure 12) shows that the
BI-XDC gives a finer and narrower distribution than the
Sedigraph which is in accord with visual examination.

6   CONCLUSIONS

An x-ray scanning sedimentometer has been described,
which can operate in either the gravitational or
centrifugal mode. The instrument has the capability
of determining particle size distributions from plus
100μm to less than 0.05μm with a total analysis time,
covering a 15:1 size range, in less than 10 minutes.

The overall reproducibility of 0.87% for 6 analyses of
BCR 66 is comparable with the value of 0.85% found in
previous studies using the Sedigraph 5100.

The mean deviation from Standard BCR 66 data, obtained
with a pipet gravity sedimentometer, is -5.48%.
Statistically identical analyses are obtained with and
without scanning thus verifying the underlying theory;
changing the centrifuge operating conditions does not
affect the generated size distribution.

Powders which span the border between gravitational
and centrifugal regimes can be analyzed under gravity
to give high discrimination at the coarse end and by
centrifuge to give high discrimination at the fine end
and the two distributions blended. The blending
capability of the instrument was tested by carrying
out a gravitational analysis of BCR 70 down to 2 μm
and determining the sub-2 μm distribution by

centrifugal analysis. The data agrees well with standard data, particularly at the ends, with a mean deviation of 7.31%.

The instrument generates a highly reproducible (mean standard deviation for 6 runs = 0.79%), but finer and narrower size distribution than determined by gravitational sedimentation which is in agreement with electron microscope studies

7   ACKNOWLEDGEMENTS

The author wishes to acknowledge the contribution of Arun Ranade of Particle Technology Inc. who assembled and tested the prototype instrument and Monash Kochar of Brookhaven who built the production model under the supervision of Walter Tscharnuter. Experimental work was carried out by Tom Ledwandowski of Du Pont under the supervision of Dave Fields. The author also wishes to thank Reg Davies for his support for this project and to the Du Pont Company for permission to publish.

REFERENCES

1   H.J. Kamack, Anal. Chem., 1951, 23,6,844-50
2   H.J. Kamack, Br. J. Appl. Phys.,1972, 5, 1962-68
3   T. Allen, "Particle Size Measurement", Publ Chapman and Hall, 4th Ed. 1990
4   T. Allen and L. Svarovsky, Proc. Soc. Analyt.Chem., 1972, 9, 2, 38-40
5   T. Allen and R. Davies, "Evaluation of Instruments for Particle Size Analysis", Presented at PARTEC 1989, Nurenberg.17-46, Preprints 1 3
6   T. Allen, R. and Davies, "Evaluation of Instruments for Particle Size Analysis", available from Manager Outside Sales, E. I. du Pont de Nemours & Company, PO Box 6094, Newark, De 19714-6094

Continuous Sieve Cascadography, Very Narrow Band Sizing

T. P. Meloy and M. C. Williams

PAC, 223 WHITE HALL, COMER WVU, MORGANTOWN, WV 26505, USA

ABSTRACT

A continuous method for making size cut of spheres with as little as 1% variation in diameter is described. Using the principles of Cascadography, a continuous two dimensional Sieve Cascadograph is designed and analyzed. Depending on the narrowness of the cut, operating cost are estimated at ten cents to a dollar a pound of material separated. The narrowness of the cut goes up at least as fast as the square of the equipment size. All design data may be obtained from the batch Cascadograph. Powders are described as the fifth state of matter.

I   INTRODUCTION

More accurate size separation methods are needed. Closely sized spheres are in rising demand for the packing of columns and other industrial applications. While the technology for making smaller mono sized spheres is well known, for the larger mono sized spheres there exists no method of making said spheres. To obtain a sample of semi mono sized spheres, sizing methods must be used. Current sizing methods are not able to meet the desired size variation specifications that industry needs.

To size spheres above 100 microns, a number of sizing techniques exist; for example: settling and sieving. Of all the size separation technologies, sieving is the best. The narrowest band size cut that can be made is on screens with a difference of 2^k in sieve opening. Thus, from such a separation, the product particle diameter ratio will range from 1 to 1.2 (1.189). Desired is a sizing method that will decrease said ratio by an order of magnitude. Such a narrow size separation can be made on the continuous Cascadograph described in this paper.

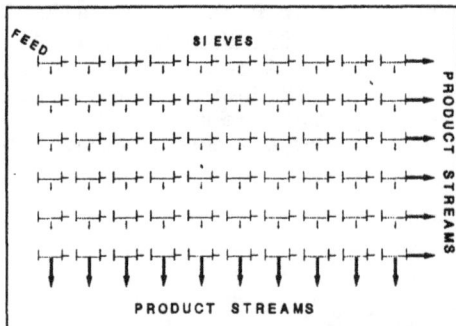

Figure 1   Two dimensional Cascadograph where particles move to the right or down. Note the multiple product streams at the bottom and edge.

Sieve Cascadography is a form of particle chromatography. In gas chromatography, the column is modeled as a series of plates. Each gas molecule is adsorbed on each plate. Each type of molecule has a different residence time on a plate. In a gas chromatograph column, all plates are identical. All molecules are fed into the column at the same time. Molecules with shorter residence times on a plate move through the column more rapidly than those molecules with a longer residence time. In particle chromatography, Sieve Cascadograph, the particles are the molecules and the sieves are the plates. The process, model and mathematics are the same in gas chromatography and Sieve Cascadograph.

Sizing Cascadography is well described in the literature by Durney and Meloy (1985), Meloy et al (1985). A Cascadograph is a stack of mono-size sieves in which the feed is introduced into the top sieve at the start of the separation. Each particle has its own residence time on each sieve. Smaller particles with shorter residence times make their way through the stack rapidly while larger particles with longer residence times move through the stack more slowly. (All feed particles have initially passed through a sieve of the same size.) Eventually all feed particles will pass through the one dimension sieve stack. This type of Cascadography is a batch process where in one may plot particle size versus time of sieving.

A continuous Sieve Cascadograph may be created by a two dimensional array of sieves. Such an 2D array of sieves are shown in Figures 1 & 2. All feed enters the array in the upper left hand sieve. In such a 2D Sieve Cascadograph, a particle may leave its present sieve by going to the sieve below or going to the sieve to the right. Each particle makes a random walk, pedary, through the 2D Sieve Cascadograph. Eventually each particle will exit the 2D Sieve

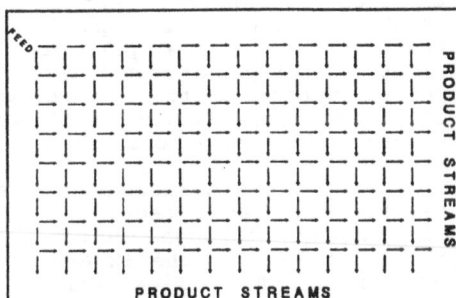

Figure 2   Model of a 2D Sieve Cascadograph. T is the Transfer Tensor, the probability of a given particle moving to the right.

Cascadograph either at the right side or at the bottom. The largest particles exit the 2D Sieve Cascadograph in the top right product stream while the smallest particles exit in the bottom left stream.

For both the sieving and the movement of particles the entire 2D array of sieves is vibrated with an eccentric motion. During the eccentric motion, particles are thrown up in the air and to the right. When the particles alight, they have been moved to the right. Some particles when alighting may pass through the sieve to the sieve below. Thus the particles take many small steps to the right and some large steps down. Depending on the design of the 2D Sieve Cascadograph, there may be between 500 and 5,000 steps to the right and 20 to 200 layers of sieves the particle must pass through on the way to the bottom. (All sieve layers are made from the same size sieve cloth.) A particles pedary is a complex walk usually involving both steps to the right and steps down through the layers of sieves. Each step in its pedary, whether to the right or down, is a random step. During each step, a given particle has a constant chance of moving right or passing through the sieve.

In principle the design of a 2D Sieve Cascadograph is simple as may be seen in Figure 3. There are n layers of sieves. Each layer of sieve has identical size sieve cloth. Within a layer, the sieve cloth is continuous. Particles move along a given layer until they either exit that layer at the right or the fall to the next layer. Particles that fall through a layer either exit that new layer at the right or the fall to the next layer. Particles that fall through all layers of sieve, exit the 2D Sieve Cascadograph at the bottom rather than the right side.

II   MODELING THE 2D SIEVE CASCADOGRAPH

The simplest model of the 2D Sieve Cascadograph assumes that at each shake of the Cascadography, a given particle has the same chance of moving to the right as that particle had during any other shake of the Cascadograph. Moreover, for a given particle, its Transfer Tensor value is the same irrespective of its position in the Cascadograph. This probability is governed by the Transfer Tensor. Each particle has its own Transfer Tensor which, in the case of a sphere, is determined by the diameter of the particle. Larger particles are more likely to move to the right and not pass through the sieve than are smaller particles. This difference in the value of the Transfer Tensor as a function of particle size enables the Cascadograph to achieve the particle separation based on particle diameter.

A 2D Sieve Cascadograph may be viewed as a rectangular array through which particles move in the +x and -y directions. From this image it is but a short leap of logic to viewing a 2D Sieve Cascadograph as a rectangular, two-dimensional network of cells. Excepting the feed cell, each cell in the array has two inputs (feeds) and two outputs (products). Moreover, each cell is connected to its four adjacent neighbors thus making the 2D Sieve Cascadographs a two-dimensional separation network. See Figure 2.

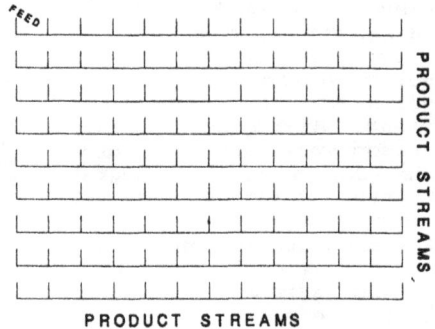

Figure 3   Physical layout of a 2D Sieve Cascadograph. Note the layers of sieves. In each sieve layer the sieve cloth is continuous.

In the model used, each cell receives material from cells left and above and sends material to cells below and to the right. All material in a cell during one cycle is sent on to other adjacent cells in the next cycle - plug flow. There is no feedback or recycling. Because the Transfer Tensors in each cell are assumed to be identical, all cells are assumed to behave identically.

In this Cascadograph model, two different mono size feeds will be considered: the ith and jth components. (Mono size feeds

are feeds containing particles with identical diameters.) Con-
comitantly, each component has its own Transfer Tensor: T_i and
T_j.   A Transfer Tensor is the probability of a particle of the
ith component in a given cell being transferred to the cell to
its right.   One minus the Transfer Tensor, $(1-T_i)$, is the
probability that a particle of the ith component will move to the
lower cell.   There are Transfer Tensors, $(1-T_i)$ and $(1-T_j)$,
respectively,   for   the   ith   and   jth   components.   In   this
Cascadograph model, as mentioned above, the T_i and T_j Transfer
Tensors are the same for each cell in the Cascadograph.

There are mxn cells in the network array: m rows and n
columns.   Each cell is referred to by two numbers, the first
designating the row and the second the column.   The upper left
hand cell is the 1,1 cell because it is in the 1st row and 1st
column. All feed enters the Cascadograph through the 1,1 cell.
From the 1,1 cell the particles first move to either the 1,2 or
2,1 cells and then are dispersed to all other cells in the
Cascadograph model.   For a forty by forty network of cells
particles exit the Cascadograph on the right edge through cells
1,40 to 40,40, and on the bottom through cells 40,1 to 40,40.

To exit the Cascadograph from the 1,40 cell, an ith
component particle must have passed serially through forty cells
always moving to the right.   The probability of this happening
is T_i raised to the 40th power.   In a like manner, for an ith
component particle to exit the Cascadograph from the 40,1 cell,
it must have passed serially through forty cells always moving
down. Thus the probability of this happening is $(1-T_i)$ raised to
the 40th power.

If an ith component particle is to exit the right edge of
the Cascadograph eleven cells down, then it must have made forty
steps to the right and eleven steps down.   In moving across the
Cascadograph surface, this particle made the steps down randomly
interspersed with the steps across the Cascadograph.   This means
there were a total of 31 steps randomly taken, down eleven and
across forty. If one lets m represent the number of steps across
and n the number of steps down, then the probability (a number
between 0 and 1) of a particle exiting at any point on the two
edges of the Cascadograph is:

$$P(m,n) = \frac{(m+n)!}{m!\,n!}\, T_i^m \,(1-T_i)^n \qquad\qquad (1)$$

Equation 1 also expresses the probability that a particle
of the ith component, having a Transfer Tensor of T_i, will appear
in a cell located in the mth column and nth row.   Thus, one may
use Equation 1 to determine the quantitative particle mix in any
cell on the Cascadograph.

By computing the value of Equation 1 for each cell on the
edge of the Cascadograph, the output profile of the Cascadograph
may be modeled. One particle size exits predominantly at the right
edge of the Cascadograph, the other predominantly on the bottom.
The bottom and right edges are conjoined and may be considered

a straight line for the sake of the edge profile. Although the edge (perimeter) of the Cascadograph has right angles in it, the edge may be considered a continuous line and can be "pulled out" into a straight line. For convenience in seeing the output profile of particles on the Cascadograph's edge, later Figures plot the bottom edge and right hand edge as a continuous straight line. For a forty by forty network of cells these two edges are joined at cell forty. In essence, the right hand edge has been rotated down ninety degrees at the hinge point (40th cell) to form the continuous straight line seen in Figure 1.

III   RESULTS

Narrow size separation of spheres is what the 2D Cascadograph is designed to accomplish. Three separation graphs are presented in three different plots. These plots may be seen in Figures 4, 5 and 6. Plotted is the frequency distribution of batches of mono-sized particles versus exit point from the continuous Sieve Cascadograph. In an operating continuous Cascadograph the ratio of the size range on the x-axis is from 1 to 1.18. All figures are taken from relatively small continuous Cascadograph, the largest being a 40 by 40 where as in practice one is more likely to have 100 by 1000 cell continuous Cascadograph.

A 40 x 40 continuous Cascadograph is fed equal quantities of two separate batches of mono-sized spheres. See Figure 4. One set has a Transfer Tensor equal to 0.5, while the second has a Transfer Tensor equal to 0.8. Even in this relatively small continuous Cascadograph a relatively sharp separation is achieved between the two batches of particles.

For comparison, a 20 x 20 continuous Cascadograph is fed equal quantities of the same two separate batches of mono-sized spheres as in Figure 4. See Figure 5. While even in this small continuous Cascadograph a relatively sharp separation is made between the two batches of particles. Large continuous Cascadographs will make far sharper separations.

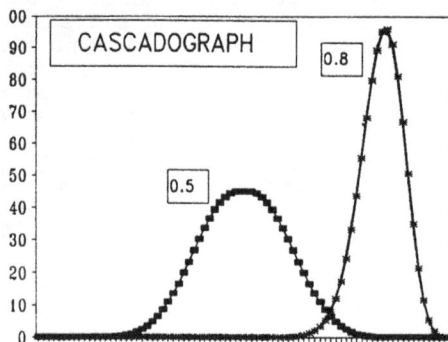

Figure 4   Separation   of   two
feeds   of mono-sized   particles   on a 40x40
2D Cascadograph.

While not shown, in the 2D Cascadograph the sharpness of separation appears to be at least proportional to the number of cells in the Cascadograph. This trend may be seen by examining Equation 1. The continuous Cascadograph, like the gas chromatograph, can be built to make very narrow size distributional cuts.

Finally, Figure 6 shows three different mono-sized feeds to

the 40 x 40 continuous Cascadograph. Once again relatively sharp separations have been accomplished. In a large 2D Cascadograph one can have as many as 100 different well separated bands of very closely sized particles from a feed where the ratio between the largest and smallest particle is 1 to 1.18.

IV   DISCUSSION

As the theoretical results indicate, a continuous 2D Sieve Cascadograph will effectively separate closely sized spheres based on their size. Because the classical batch Sieve Cascadographs works on the separation of particles as a function of time, so then the 2D Sieve Cascadographs will make the separation on a continuous basis. With the proper design of the 2D Sieve Cascadographs, the narrowness of the separation can be increased at least one order of magnitude and probably two orders of magnitude from any size separation methods now extant.

To design a continuous Cascadograph one needs values of the Transfer Tensor as a function of sphere diameter. From the batch Cascadograph one may obtain the needed Transfer Tensor data. From the batch Cascadograph one obtains the time a given size particle requires to traverse the n sieves in the Cascadograph. From this residence time information, one can compute the average residence time on a single sieve. This residence time data versus particle size is all that is needed to design the continuous Cascadograph.

In the continuous Cascadograph, particles which exit the sieves at the right all have the same residence time in said Cascadograph because all particles move to the right with the same velocity irrespective of which sieve layer they may be on. Thus particles having a residence time on a single sieve greater than the residence time in the continuous Cascadograph, will exit on the first level sieve, upper right. Particles having a residence time on a single sieve more than half but less than the full residence time in the continuous Cascadograph will exit right on the second level sieve. In a like manner, particles with residence times between one third and one half, will exit on the third level sieve. Those with residence times between one forth and one third, will exit on the forth level sieve. The above design data defines the diameter of particles exiting at a given level <u>on the right</u>.

Particle exiting the continuous Cascadograph on the bottom, pass through the n levels of sieves in less time than the residence time of the particles exiting on the right. Where a particle exits the bottom is a direct measure of the time require to pass through the n sieve levels in either a batch or a continuous Cascadograph. Thus the position of a particle exiting the bottom is directly correlated with the batch Cascadograph data.

Sufficient data for designing the continuous Cascadograph is available from batch Cascadograph data. Particles exiting at the right are in a different regimen than particle exiting the bottom. Design and operating condition may be varied to control the performance of the continuous Cascadograph. An effect operating 2D Cascadograph appears to be easy to build and effective in operation.

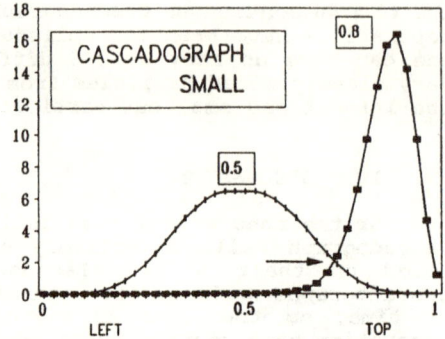

Figure 5 A small 2D Cascadograph makes a less sharp separation. Compare the overlap marked by arrow with that in the previous Figure. Same feeds.

Cost
Operating costs of a 2D Sieve Cascadographs are estimated to be below $0.90/lbs. This value is based on the following factor estimates. Current cost of sieving is $0.10/ton. One hundred layers of sieves increases the cost 100 fold. Being a new machine increases the cost 10 fold. Materials, sieve cloth, increases the cost 3 fold. Low through put increases the cost 2 fold. Design costs increases the cost 3 fold. Multiplying these factors together and dividing by 2,000 lbs/ton, one obtains the cost of the narrow band separation on the 2D Sieve Cascadographs as 90 cents a lbs or less. Actual cost may be significantly lower.

With such an optimistic view of the performance of a 2D Sieve Cascadographs the obvious need is to build and test said device. In spite of its promise, funds for any type of powder science are virtually impossible to obtain. Governments don't have programs for powder science and thus they will not fund such a program. The powder industry, while of growing importance tends to be fragmented and thus will not fund said program. Clearly, with the ever growing recognition of the role powders play in national economies, a powder program focusing attention and bringing resources to the field is needed.

5th State of Matter
In 1928, Irving Langmuir defined plasma as the forth state of matter because these ionized gases obeyed both the laws of hydrodynamics and electromagnetism. Powders are made of solids, the particles become ionized in fluids and thus obey the laws of electromagnetism, and, in fluids, can behave as a liquid. Because of their strange behavior, powders are neither solids, nor gases, nor liquids, nor are they plasma. Powders are the fifth state of matter.

Assumptions
To derive the behavior of the continuous Cascadograph, two explicit and one implicit assumptions were made. The explicit assumptions were that a) Transfer Tensor was constant at all locations in the Cascadograph and b) that the particles moved across a sieve layer as plug flow. The implicit assumption was that particles would behave similarly in the continuous Cascadograph as in the batch Cascadograph.

Assuming that the Transfer Tensor is constant in each cell of the continuous Cascadograph circuit is the design objective and the same assumption made in the analysis of the batch Cascadograph. The major cause for this assumption not to be true would be variations in the openings of the sieve cloth. This variation could be caused by poor quality sieve cloth, abuse of the sieve cloth during construction of the Cascadograph, abuse of the sieve cloth during operation of the Cascadograph, or wear of the sieve cloth during normal operations. Assuming a constant Transfer Tensor is a valid assumption.

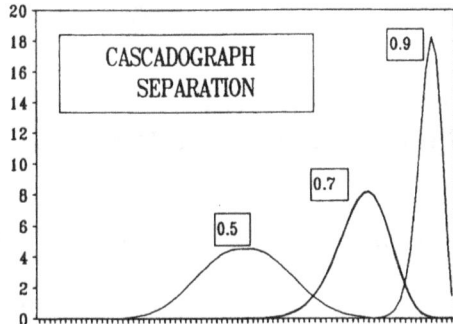

Figure 6 Three mono-size feeds are separated by the 2D Cascadograph. The T_i are respectively: 0.5, 0.7 and 0.9.

Plug flow is a simplifying assumption that speeds calculation. When non plug flow models were run no difference in the output were detected as long as the <u>ratio</u> of the probability to move right to the probability to move down was constant. Since in the 2D Cascadograph as conceived, the residence time in a cell does not effect the output, plug flow is a valid assumption. However, increasing the residence time in a cell does increase the amount of material in the Cascadograph during steady state operation.

V   CONCLUSIONS

A continuous Sieve Cascadograph for separating spheres into narrow size bands has been designed and modeled. With a continuous size distribution of sphere sizes in the feed, a large Cascadograph should be able to produce size fractions with negligible variation in size diameter. Few development problems are expected. Specifically:

1.   A large continuous Sieve Cascadograph should be able to produce particles with 1% size variation.
2.   Operating costs are estimated to be in the range of ten cents to a dollar a pound. For less sharp separations costs would be lower.
3.   All necessary design data for the continuous Cascadograph can be obtained from a batch Cascadograph.
4.   Because they have the characteristic of solids, fluids and plasmas, powders are defined as <u>the fifth state of matter.</u>

VI   BIBLIOGRAPHY

MELOY, T.P., Clark, N.N. Durney, T.E. and Pitchumani, B. "Measuring the Particle Shape Mix in a Powder with the Cascadograph" Chemical Engineering Science, 40 No. 7, 1985, pp. 1077-1084.

Durney, T.E., MELOY,T.P., (1985), "Experimental Proof: Residence Time Distribution in Cascadography", Int. J. Mineral Process, 14, pp 313-17.

Survey on Porosity Determining Methods

E. Robens, K. K. Unger and D. Kumar

INSTITUT FÜR ANORGANISCHE UND ANALYTISCHE CHEMIE DER JOHANNES-
GUTENBERG-UNIVERSITÄT, POSTFACH 3980, D-6500 MAINZ, GERMANY

Results of a pore size analysis depend on the resolution of the measuring method. Reducing the size of the comparison volume, area or length shifts the detection limit to lower values and increases the resulting values of volume and surface area in the pore. One cannot define "true" integral values like porosity, pore volume, mean pore diameter or specific surface area. More reliable information contains distributions of volume and the surface with respect to the pore width or as a summarising value the fractal dimension.

There are two possibilities to investigate the porosity of dried samples either to inspect the porous structure itself or to fill the matrix with a fluid and to inspect the fluid.

An impression of the inner pore structure may be obtained by mechanical or tomographical cuttings whilst direct observation of the surface can use mechanical tracing, the scanning (force) tunnel microscope and optical methods.

Intrusion porometry occurs with a non-wetting liquid whilst in displacement porometry the porous matrix is filled with a wetting liquid.

High performance liquid chromatography (HPLC) is rapidly increasing in its application to gel structures and organic pore systems together with the increasing use of size exclusion chromatography (SEC).

The Use of Catastrophic Tumbling to Specify the 3-D Structure of Large Fineparticles

B. H. Kaye, G. G. Clark, J. Gratton and Y. Liu

INSTITUTE OF FINEPARTICLE RESEARCH, LAURENTIAN UNIVERSITY, SUDBURY, ONTARIO, CANADA

Abstract

In this presentation the concept that a strange attractor from a time series generated by observing the tumbling behaviour of a rock fragment in a slowly rotating cylinder is introduced. It is shown that for a specified set of conditions the tumbling behaviour can be related to the fractal dimension of an agglomerate of rocks tumbling in the cylinder. It was demonstrated that this could form a method characterizing the fractal dimension of a tumbling assembly of rocks. It was further shown that as the speed of tumbling increases that a tumbling rate is reached at which the morphology of the agglomerate no longer influences the tumbling behaviour. This fact could be important when studying the large scale handling of gravel, crushed rock and coal fragments. Work was also presented on a study of the avalanching behaviour of powder on an inclined slope. It was demonstrated that the distribution of the sizes of the powder avalanches on the slope generated a set of data which could be described by a fractal dimension. This could in turn be linked to the shape and size of the powder being poured onto the inclined slope. The implications of these two new techniques for studying the dynamics of fineparticles were explored.

Extending Photon Correlation Spectroscopy to Concentrated Systems

F. K. McNeil-Watson

MALVERN INSTRUMENTS LTD., MALVERN, UK

A number of workers have applied Photon Correlation spectroscopy to the spectrum of light scattered by concentrated dispersions of particles in liquid suspension using fibre optic components to launch the illuminating source and receive the scattered intensity. Auweter and Horn (1) used multimode fibre and a discrete coupler arrangement to separate transmitted and received light, Brown (2) described an arrangement using separate monomode fibres for transmit and receive suitable for dilute systems, Horne (3) has described a method using bundles of multimode fibres and shown that real size information can be extracted from concentrated systems by applying the diffusing wave formulation of Pine et al. (4). Dycott (5) and Wu (6) and his coworkers have applied an integrated monomode fibre system to concentrated systems but used analogue processing techniques.

We now describe an instrument using monomode fibre and an integrated fibre coupler, that therefore has no breaks in the transmission path due to the insertion of bulk optical components. Back reflection of a portion of the incident beam from the output port means that we detect a mixture of scattered light and a "reference beam", the so-called heterodyne detection scheme in which the first order correlation coefficient $G^{(1)}(T)$ is measured directly. The signal is detected by a photomultiplier and processed by a multibit correlator. The fibre end is available to use directly as a probe into a dispersion or may be .mounted in a thermostatted flow cell. Results will be presented showing the performance on model systems as a function of concentration. The use of such an instrument for remote sensing in a variety of applications will be discussed.

References.

1. Auweter H and Horn D 1985 J. Coll. Interface Sci. 105
2. Brown R 1987 Applied Optics 26
3. Horne DS 1989 J Phys D 22
4. Pine DJ et al. 1988 Phys Rev Lett 60
5. Dyott R 1978 Mic. Opt. Acoust 2
6. Wu P et al.1985 J.Coll Interface Sci. 110

Automation of Laser Diffraction Technology

S. Rothele, M. Heuer, W. Witt and U. Kesten

SYMPATEC GMBH., CLAUSTHAL-ZELLERFELD, GERMANY

Having undergone rapid evolution during the past two decades laser diffraction technology has become one of the most important industrially applied methods for particle size analysis worldwide.

At present it is about to establish its dominant position as the standard analytical technique for off-line applications in laboratories of all fields of industry.

The huge potential offered by the principle, and its technical application, also enable realistic solutions for the automation of laboratory analyses e.g. in cooperation with robots.

These automatic instruments (auto-line systems), working in an operational network, allow for complete automation of the analytical cycle but do not include sampling and sample preparation.

The auto-line dispersing systems available for dry and wet products are the central part of the instrument. They react on samples that are filled into the sample container and process the sample without any further operator action, thus assuring highly repeatable results without "operator finger-print".

High sample throughput, with sampling times less than one minute for dry dispersed powders, and under five minutes for suspension analysis, is another important characteristic of the auto-line systems.

As a consequence of the features mentioned, the ultimate aim of the auto-line systems is to release the operator from the more tedious aspects of the analysis, thus allowing him to concentrate his expertise, and time, on the careful preparation of samples and the subsequent evaluation of the data produced.

In-line Particle Size Analyzer for Versatile Use in the Processing Industry

Hans Hoffman, John Hokanson and Ekhard Preikschat

LASENTEC (LASER SENSOR TECHNOLOGY, INC.), BELLEVUE, WA, USA

Recent advances in laser-optics have brought about a new generation of particle size analyzers, which are capable of measuring particle size distributions directly inside a process flow (e.g. inside a reactor vessel or inside a slurry flow) without having to take or dilute a sample. The device measures the size distribution over a range from 1 to 1,000 microns, and provides graphic displays, printer output, as well as 2 programmable analog current (4-20 mA) output signals for use in process control.

This technology provides the critical link between the traditional laboratory measurements and direct "in-process" measurements. It provides real time information for direct process control. The technology is now being used in many different industries, e.g. mining, chemical, pharmaceutical, petro-chemical, pulp & paper, food processing, bio-tech.

Distributional Coincidence Correction for Enhanced Electrozone PSA Precision

Robert H. Berg

PARTICLE DATA INC., ELMHURST, IL, USA

Performance of this elegant, high precision measurement method is enhanced by several advances in hardware and software design. Sensor design features allow avoidance of non-axial transit and removal of the effects of variances in conductivity and transit velocity. Substantial expansion of ranges of particle size and sample concentration for a sensor and the theoretical bases for same have been developed. Correction for secondary coincidence allows sample concentrations without data distortion to be increased twenty-fold or more with corresponding reduction in data acquisition time. Fast, "hands-off", analysis capabilities are provided by auto-detection and clearance of sensor blockages and auto-execution of customized sequences of analytical steps. Operational ease is further aided by standardized size scales, default calibrations, and a full complement of data presentation techniques.

The Design of Control Systems for Industrial Crystallisers

P. G. J. van de Wel, S. M. Lemkowitz, C. J. M. van Wingerden and B. Scarlett

DELFT UNIVERSITY OF TECHNOLOGY, FACULTY OF CHEMICAL ENGINEERING & MATERIALS SCIENCE, PARTICLE TECHNOLOGY GROUP, P.O. BOX 5054, 2600 GA DELFT, THE NETHERLANDS

The design and implementation of control systems for both batch and continuously operated industrial crystallisers can be achieved by mathematical and physical structured models for the process dynamic behaviour and from on-line measurements of the crystal distribution (CSD).

CONTROLLER DESIGN

Batch Crystallisers.

Physical models, based on balance equations & empirical relations of the crystallisation kinetics are used to design feedback controllers that steer the unseeded batch crystalliser from an initial state to a final optimal state.

Continuous crystallisers.

Physical and black box models are used to design controllers that suppress oscillations in the CSD and optimise the produce CSD of continuous crystalliser.

ON-LINE MEASUREMENT OF CSD.

Crystal size distributions (CSD) are measured both by forward and backward light scattering. In the new set-up for forward light scattering the original semi-circular detector is replaced by a CCD-array.
Pixels are grouped from the collected scattered signal and averaged to obtain an optimally scaled inverse problem from which the size distribution is retrieved.

A Monodisperse Aerosol Generator, Using the Taylor Cone, for the Production of 1μm Droplets

G. M. H. Meesters, P. H. W. Vercoulen, J. C. M. Marijnissen and B. Scarlett

DELFT UNIVERSITY OF TECHNOLOGY, FACULTY OF CHEMICAL ENGINEERING & MATERIALS SCIENCE, PARTICLE TECHNOLOGY GROUP, P.O. BOX 5054, 2600 GA DELFT, THE NETHERLANDS

In a strong non-uniform electric field (E=10⁶V/m), a pendant drop will deform into a conically shaped volume of liquid. In a well known paper published in 1964 by G.I. Taylor the conditions were studied under which this conically shaped surface under the influence of an electric field could exist. As published by Meesters et al. (1990), Taylor's derived theory is too simple. The process can be described using the following pressure balance.

$$p = \sigma/(r.\tan\alpha) - 1/2 \in E \qquad (1)$$

with
p = pressure difference across liquid surface	(N/m)
σ/(r.tanα) = pressure due to surface tension	(N/m)
r.tanα = radius of curvature of cone	(m)
1/2εE = electric pressure	(N/m)
ε= dielectric constant of surrounding medium	(F/m)
E = electric field strength	(V/m)

From the tip of this Taylor cone very small droplets (=1μm) are ejected, producing a highly charged aerosol. The highly charged drops can be discharged using a corona discharge of opposite charge.

Propagation Mechanism of Dust Explosions

P. G. J. Van der Wel[1], S. M. Lemkowitz[1] C. J. M. van Wingerden[2] and B. Scarlett[1]

[1]DELFT UNIVERSITY OF TECHNOLOGY, FACULTY OF CHEMICAL ENGINEERING & MATERIALS SCIENCE, DEPARTMENT OF CHEMICAL ENGINEERING. P.O. BOX 5054, 2600 GA DELFT, THE NETHERLANDS
[2]PRINS MAURITS LABORATORY, SECTION GAS AND DUST EXPLOSIONS, TNO, RIJSWIJK, THE NETHERLANDS

Dust explosion research has been largely practical. Because of this approach, and also due to its complex nature, little is known concerning fundamental aspects of dust explosions, such as the propagation mechanism. Therefore reliable predictions can be made only within areas in which thorough practical research has been carried out.

A more fundamental approach, from which extrapolation may well be possible, is to study the macroscopic phenomenon of dust explosions from the viewpoint of the microscopic phenomena occurring during the combustion process of powders dispersed in air. This is the approach of Delft University.

Improvements on a Particle Beam Generator

O. Klievit, J. C. M. Marijnissen, P. J. T. Verheijen and B. Scarlett

DELFT UNIVERSITY OF TECHNOLOGY, FACULTY OF CHEMICAL ENGINEERING & MATERIALS SCIENCE, PARTICLE TECHNOLOGY GROUP, P.O. BOX 5054, 2600 GA DELFT, THE NETHERLANDS

The aim of the project was to develop an apparatus in which both size and chemical composition of individual aerosol particles can be determined on-line in real-time. The first part of the instrument is the particle beam generator, which generates a particle beam by expanding an aerosol through a nozzle into a vacuum. This poster discusses the design and testing of such an apparatus. The aim of future work is to pass the particle beam through a low energy laser beam. The light scattered by the individual particles will provide information on the size of the particle and trigger a high energy Nd:YAG laser. The pulse from this laser vaporizes and partially ionizes the fragments of the incident particle, while a third, UV laser should complete the ionization. A time of flight mass spectrometer will be used to analyze the resulting ions.

List of Exhibitors

Belstock Controls
10 Moss Hall Crescent
Finchley
London N12 8NY
UK

Biotage Europe
Harforde Court
Foxholes Business Park
John Tate Road
Hertford
Herts SG13 7NW
UK

BIRAL
P O Box 2
Portishead
Bristol BS20 9JB
UK

Bromley Instruments Ltd
18 Melrose Road
Biggin Hill
Kent TN16 3DA
UK

Brookhaven Instruments
  Corporation
750 Blue Point Road
Holtsville
NY 11742
USA

Chemlab Scientific Products Ltd
Construction House
Grenfell Avenue
Hornchurch
Essex RM12 4EH
UK

Christison Scientific
  Equipment Ltd
Albany Road
Gateshead
Tyne & Wear NE8 3AT
UK

Coulter Electronics Ltd
Northwell Drive
Luton
Beds LU3 3RH
UK

Dantec Electronics Ltd
Techno House
Redcliffe Way
Bristol BS1 6NU
UK

Horiba Instruments Ltd
1 Harrowden Road
Blackmills
Northampton NN4 0EB
UK

Laser Lines Ltd
Beaumont Close
Banbury
Oxon OX16 7TQ
UK

Leeds & Northrup Ltd
Wharfdale Road
Tyseley
Birmingham B11 2DJ
UK

Malvern Instruments Ltd
Spring Lane South
Malvern
Worcs WR14 1AQ
UK

Micromeritics Ltd
4 The Ringway Centre
Edison Road
Houndmills
Basingstoke
Hampshire RG21 2YH
UK

Pacific Scientific Ltd
11 Manor Courtyard
Hughenden Avenue
High Wycombe
Bucks HP13 5RE
UK

Palas GmbH
Haid-und-Neu-Str 7-9
D-7500 Karlsruhe 1
Germany

Powder Products Ltd
Unit 29
Trent Lane Industrial Estate
Castle Donington
Derby DE7 2NP
UK

PSTIS
Department of Chemical
   Engineering
Loughborough
Leics LE11 3TU
UK

Roth Scientific Co Ltd
Alpha House
Alexandra Road
Farnborough
Hants GU14 6BU
UK

Steptech Instrument Services
   Ltd & Quantachrome Corp
Business and Technology Centre
Bessemer Drive
Stevenage
Herts SG1 2DX
UK

SYMPATEC GmbH
System-Partikel-Technik
Burgstatter Strasse 6
D-3392 Clausthal-Zellerfeld
Germany

SYMPATEC UK/Unit 38
c/o Bury Business Centre
Kay Street
Bury
Lancs BL9 6BU
UK

Whitehouse Scientific
The Whitehouse
Whitchurch Road
Waverton
Chester CH3 7PB
UK

Author Index

Subject Index